Proceedings in Computational Statistics

13th Symposium held
in Bristol,
Great Britain, 1998

Edited by
Roger Payne and Peter Green

With 118 Figures
and 61 Tables

Physica-Verlag
A Springer-Verlag Company

Prof. Roger Payne
Statistics Department
IACR-Rothamsted
Harpenden
Herts AL5 2JQ, United Kingdom

Prof. Peter Green
Department of Mathematics
University of Bristol
Bristol BS8 1TW, United Kingdom

ISBN 3-7908-1131-9 Physica-Verlag Heidelberg New York

Cataloging-in-Publication Data applied for
Die Deutsche Bibliothek – CIP-Einheitsaufnahme
Proceedings in computational statistics: 13th symposium held in Bristol, Great Britain, 1998; with 61 tables / COMPSTAT. Ed. by Roger Payne and Peter Green. – Heidelberg: Physica-Verl., 1998
ISBN 3-7908-1131-9

Printed in Germany

Softcover Design: Erich Kirchner, Heidelberg

SPIN 10684733 88/2202-5 4 3 2 1 0 – Printed on acid-free paper

Preface

This Volume contains the Keynote, Invited and Full Contributed papers presented at COMPSTAT'98. A companion volume (Payne & Lane, 1998) contains papers describing the Short Communications and Posters. COMPSTAT is a one-week conference held every two years under the auspices of the International Association of Statistical Computing, a section of the International Statistical Institute.

COMPSTAT'98 is organised by IACR-Rothamsted, IACR-Long Ashton, the University of Bristol Department of Mathematics and the University of Bath Department of Mathematical Sciences. It is taking place from 24-28 August 1998 at University of Bristol. Previous COMPSTATs (from 1974-1996) were in Vienna, Berlin, Leiden, Edinburgh, Toulouse, Prague, Rome, Copenhagen, Dubrovnik, Neuchatel, Vienna and Barcelona. The conference is the main European forum for developments at the interface between statistics and computing. This was encapsulated as follows in the COMPSTAT'98 Call for Papers.

Statistical computing provides the link between statistical theory and applied statistics. The scientific programme of COMPSTAT ranges over all aspects of this link, from the development and implementation of new computer-based statistical methodology through to innovative applications and software evaluation. The programme should appeal to anyone working in statistics and using computers, whether in universities, industrial companies, research institutes or as software developers.

Selection of the Full Contributions from over 180 submitted abstracts presented a challenging but very interesting task for the Scientific Programme Committee (the *SPC*): Roger Payne (United Kingdom) Chair, Jaromír Antoch (Czech Republic), Gianfranco Galmacci (Italy), Peter Green (United Kingdom), Wolfgang Härdle (Germany), Peter van der Heijden (Netherlands), Albert Prat (Spain), Jean-Pierre Raoult (France), Mats Rudemo (Sweden).

The papers were all refereed by the SPC and their colleagues, with additional help from Antoine de Falguerolles who represented the European Regional Section of the IASC on the Committee. Other Consultative Members of the Committee were Andrew Westlake (ASC: United Kingdom), Chooichiro Asano (EARS IASC: Japan), Christian Guittet (EUROSTAT: Luxembourg), Malcolm Hudson (IASC: Australia) and Carey Priebe (INTERFACE: USA). We would like thank our colleagues on the SPC for all their help. The resulting Proceedings provide a broad overview of the currently active research areas in statistical computing.

We are also especially grateful to the other members of the Local Organising Committee: Peter Lane of IACR-Rothamsted; Gillian Arnold and Phil Brain of IACR-Long Ashton; Steve Brooks of University of Bristol; Andrew Wood with Merrilee Hurn initially, and then Stuart Barber, of University of Bath.

Roger Payne
Peter Green
April 1998

References

(ed.) Payne, R.W. & Lane, P.W. (1998). *COMPSTAT 1996, Proceedings in Computational Statistics, Short Communications and Posters.* Harpenden: IACR-Rothamsted.

Contents

Keynote Papers

Invited Papers

Contributed Papers

Keynote Papers

Analysis of Clustered Multivariate Data from Developmental Toxicity Studies

Geert Molenberghs, Helena Geys, Lieven Declerck, Gerda Claeskens, and Marc Aerts

Biostatistics, Center for Statistics, Limburgs Universitair Centrum, Universitaire Campus , B-3590 Diepenbeek, Belgium

Keywords. Clustered data, dose-response models, exponential family, generalized estimating equations, marginal model, maximum likelihood, pseudo-likelihood

1 Introduction

Society is becoming increasingly concerned about problems related to fertility and pregnancy, birth defects, and developmental abnormalities. Consequently, regulatory agencies such as the U.S. Environmental Protection Agency (EPA) and the Food and Drug Administration (FDA) have placed an increased priority on protecting the public from drugs, chemicals and other environmental exposures that may contribute to these risks. Because human data are generally limited, data from controlled animal experiments are generally used as the basis for regulation. This work is motivated by data collected from studies with a segment II design that involve exposing pregnant animals (rats, mice or rabbits) during the period of major organogenesis and structural development. Dose levels for the Segment II design consist of a control group and 3 or 4 dose groups, each with 20 to 30 pregnant dams. The dams are sacrificed just prior to normal delivery, at which time the uterus is removed and examined for resorptions and fetal deaths. The viable fetuses are examined carefully for many different types of malformations, which are commonly classified into three broad categories: external malformations are those visible by naked eye, for instance missing limbs; skeletal malformations might include missing or malformed bones; visceral malformations affect internal organs such as the heart, the brain, the lungs etc. Each specific malformation is typically recorded as a dichotomous variable (present/absent) and fetuses may have several types. The data, presented in this work, investigate the effects of di (2-ethyhexyl)-phtalate (DEHP) in mice and of ethylene-glycol (EG) in rats. The DEHP study is concerned about the possible toxic effects of phtalic acid esters. These are used extensively as plasticizers for numerous plastic devices. Due to their presence in human and animal tissues, considerable concern has been developed as to their possible toxic effects. EG is a high-volume industrial chemical with diverse applications. It may represent little hazard to human health in normal industrial handling, except ehnt used at elevated temperatures. However, accidental or intentional ingestion is toxic and may result in death.

The analysis of such data raises a number of challenges. Models that try to approximate the complex data generating mechanism of a developmental toxicity study, should take into account the litter effect and the number of viable fetuses, malformation indicators, weight and clustering, as a function of exposure. Furthermore, the size of the litter may be related to outcomes among live fetuses. Finally, one may have to deal with outcomes of a mixed continuous-discrete nature. Scientific interest may be in inference about the dose effect, on implications of model misspecification, on assessment of model fit, etc. In most of these studies it is reasonable to assume that exposure covariates are constant within litters and the primary interest is in evaluating dose-response effects.

2 Accounting for litter effects

2.1 Conditional models

Molenberghs & Ryan (1998) constructed a joint distribution for clustered multivariate binary outcomes, based on a multivariate exponential family model (Cox, 1972). Defining z_{ij} as the number of individuals from cluster i, positive on outcome j and $z_{ijj'}$ as the number, positive on both outcomes j and j', the model is expressed as:

$$f_{\boldsymbol{Y}_i}(\boldsymbol{y}_i;\boldsymbol{\Theta}_i) = \exp\left\{\sum_{j=1}^{M}\theta_{ij}z_{ij}^{(1)} + \sum_{j=1}^{M}\delta_{ij}z_{ij}^{(2)} + \sum_{j<j'}\omega_{ijj'}z_{ijj'}^{(3)} \right. \tag{1}$$
$$\left. + \sum_{j<j'}\gamma_{ijj'}z_{ijj'}^{(4)} - A(\boldsymbol{\Theta}_i)\right\},$$

where $z_{ij}^{(1)} = z_{ij}$, $z_{ij}^{(2)} = -z_{ij}(n_i - z_{ij})$, $z_{ijj'}^{(3)} = 2z_{ijj'} - z_{ij} - z_{ij'}$, $z_{ijj'}^{(4)} = -z_{ij}(n_i - z_{ij'}) - z_{ij'}(n_i - z_{ij}) - z_{ijj'}^{(3)}$. $A(\boldsymbol{\Theta}_i)$ is the normalizing constant, resulting from summing the previous model over all 2^{Mn_i} possible outcomes. The advantage of this model is the flexibility with which both main effects and associations can be modelled, and the absence of constraints on the parameter space which eases interpretability. Further, the fact that the probability model depends explicitly and implicitly on the cluster size is seen as an advantage since it is in line with the observation that litter size itself depends on the level of exposure.

The classical advantages of exponential families can be lost, especially in multivariate settings, where the normalizing constant poses formidable computational requirements. This is especially true for clusters of variable length, because the normalizing constant depends on the cluster size. Arnold & Strauss (1991) propose the use of *pseudo-likelihood* (PL). The principal idea is to replace the joint density by a product of conditional densities *that do not necessarily multiply to the joint distribution*. The advantage of this procedure is that the normalizing constant cancels. Let $(\boldsymbol{y}_1, \ldots, \boldsymbol{y}_N)$ be a set of M-dimensional observation vectors. Define S as the set of all $2^M - 1$ vectors of length M, consisting solely of zeros and ones, with each vector having at least one nonzero entry. Denote by $\boldsymbol{y}_k^{(s)}$ the subvector of $\boldsymbol{y}_k$ corresponding to the components of $s \in S$ that are nonzero. The associated joint density is written as $f_s(\boldsymbol{y}_k^{(s)};\boldsymbol{\Theta}_k)$. Specify a set $\boldsymbol{\delta} = \{\delta_s | s \in S\}$ of real numbers, with at least one nonzero component and define the log pseudo-likelihood as:

$$p\ell = \sum_{i=1}^{N}\sum_{s\in S}\delta_s \ln f_s(\boldsymbol{y}_i^{(s)};\boldsymbol{\Theta}_i), \tag{2}$$

Table 1. Data from DEHP Study in Mice

Dose	Dams	Live	External		Visceral		Skeletal	
			#	%	#	%	#	%
0	30	330	0	0	5	1.5	4	1.2
0.025	26	288	3	1	1	0	1	0
0.050	26	277	15	5.4	20	7.2	12	4.3
0.1	17	137	24	17.5	21	15.3	25	18.2
0.15	9	50	27	54	25	50	24	48

where some (though not all) of the δ_s's may be negative. This must correspond to a product of marginal and conditional densities. Arnold & Strauss (1991) established consistency and asymptotic normality. For clustered, multivariate binary data a convenient PL function is found by replacing the joint density (1), by the product of Mn_i univariate conditional densities describing outcome j for the kth individual in a cluster, given all other outcomes in the cluster:

$$PL(1) = \prod_{i=1}^{N} \prod_{j=1}^{M} \prod_{k=1}^{n_i} f(y_{ijk}|y_{ij'k'}, j' \neq j \text{ or } k' \neq k; \Theta_i). \quad (3)$$

Equation (3) is but one definition of the PL function. Geys, Molenberghs & Ryan (1997b) consider an alternative specification.

Table 1 summarizes the external, visceral and skeletal malformation data of the DEHP study. Table 2 shows the maximum likelihood and pseudo-likelihood estimates of the DEHP study for the univariate conditional model.

Geys, Molenberghs & Ryan (1997a) showed that only a very small efficiency loss was paid for enormous computational gains of pseudo-likelihood over maximum likelihood.

Geys, Molenberghs & Ryan (1997b) also proposed score and likelihood ratio tests for the pseudo-likelihood framework. They are easy to calculate, exhibit a very satisfactory behaviour and provide the necessary tools for model selection. Aerts & Claeskens (1997bc) showed how bootstrap approximations can be used as interesting alternatives to the classical asymptotic chi-squared distributions of these test statistics.

2.2 Marginal models

In this section, we study an exhangeable version of the Bahadur model (Bahadur, 1961), with constant pairwise correlation ρ and zero higher order correlations:

$$f(z_i|\pi_i, \rho_i) = \binom{n_i}{z_i} \pi_i^{z_i} (1-\pi_i)^{n_i - z_i}$$
$$\times \left[1 + \rho_i \left\{ \binom{z_i}{2} \frac{1-\pi_i}{\pi_i} + \binom{n_i - z_i}{2} \frac{\pi_i}{1-\pi_i} - z_i(n_i - z_i) \right\}\right]. \quad (4)$$

Bahadur (1961) discusses restrictions on the correlation parameters. Indeed, depending on the values of π_i and ρ_i, (4) may fail to be nonnegative for some

Table 2. Maximum Likelihood (Model Based Standard Errors; Empirically Corrected Standard Errors) and Pseudo-likelihood (Standard Errors) Estimates for the Conditional Model

Method	Par.	External	Visceral	Skeletal	Collapsed
ML	β_0	-2.81(0.58;0.52)	-2.39(0.50;0.52)	-2.79(0.58;0.77)	-2.04(0.35;0 .42)
	β_d	3.07(0.65;0.62)	2.45(0.55;0.60)	2.91(0.63;0.82)	2.98(0.51;0.66)
	β_a	0.18(0.04;0.04)	0.18(0.04;0.04)	0.17(0.04;0.05)	0.16(0.03;0.03)
PL	β_0	-2.85(0.53)	-2.30(0.50)	-2.41(0.73)	-1.80(0.35)
	β_d	3.24(0.60)	2.55(0.53)	2.52(0.81)	2.95(0.56)
	β_a	0.18(0.04)	0.20(0.04)	0.21(0.05)	0.20(0.03)

Table 3. Maximum Likelihood Estimates (Standard Errors) for the Bahadur Model

Par.	External	Visceral	Skeletal	Collapsed
β_0	-4.93(0.39)	-4.42(0.33)	-4.67(0.39)	-3.83(0.27)
β_d	5.15(0.56)	4.38(0.49)	4.68(0.56)	5.38(0.47)
β_a	0.11(0.03)	0.11(0.02)	0.13(0.03)	0.12(0.03)

outcomes. From the restrictions, it can be deduced that the lower bound approaches zero as the cluster size increases. However, it is important to note that also the upper bound for ρ_i is constrained. Taking a (realistic) cluster of size 12, the upper bound is in the range $(0.09; 0.18)$. A recent study of the constraints has been conducted by Declerck, Aerts & Molenberghs (1997).

Table 3 shows the results of fitting the Bahadur model to the DEHP data. If interest is restricted to the marginal mean function and the pairwise association parameter, one can replace a full likelihood approach by generalized estimating equations (GEE2) where only the first two moments are modelled and working assumptions are adopted about third and fourth order moments (Liang, Zeger & Qaqish, 1992). Consistent point estimates are supplemented with *robust* standard errors. Often, point estimates differ only slightly from their likelihood counterparts, while test statistics may change considerably. The results of applying the GEE2 method to the DEHP data are given in Table 4. Also here, pseudo-likelihood ideas can be useful. Le Cessie & Van Houwelingen (1994) replace the likelihood contribution $f(y_{i1}, \ldots, y_{in_i})$ by the product of all pairwise contributions $f(y_{ij}, y_{ik})$ $(1 \leq j < k \leq n_i)$. Grouping the outcomes for subject i into a vector $\boldsymbol{Y}_i$, the contribution of the ith cluster to the log pseudo-likelihood is $pl_i = \sum_{j<k} \ln f(y_{ij}, y_{ik})$ if it contains more than one observation and an ordinary logistic regression contribution otherwise.

For binary data and taking the exchangeability assumption into account, the log pseudo-likelihood contribution pl_i can be formulated as:

$$pl_i = \binom{z_i}{2} \ln \pi_{i11} + \binom{n_i - z_i}{2} \ln(1 - 2\pi_{i10} + \pi_{i11}) + z_i(n_i - z_i) \ln(\pi_{i10} - \pi_{i11}), \quad (5)$$

Table 4. GEE2 Estimates (Standard Errors) for the Bahadur Model

Par.	External	Visceral	Skeletal	Collapsed
β_0	-4.98(0.37)	-4.49(0.36)	-5.23(0.40)	-5.23(0.40)
β_d	5.29(0.55)	4.52(0.59)	5.35(0.60)	5.35(0.60)
β_a	0.15(0.05)	0.15(0.06)	0.18(0.02)	0.18(0.02)

Table 5. Pseudo-likelihood, GEE2 and GEE1 Estimates (Standard Errors) for the Marginal Odds Ratio Model (Collapsed Outcome)

Method	β_0	β_d	β_a
PL	-3.98(0.30)	5.57(0.61)	1.11(0.27)
GEE2	-3.69(0.25)	5.06(0.51)	0.97(0.23)
GEE1	-4.02(0.31)	5.79(0.62)	0.41(0.34)

where π_{i11} denotes the bivariate probability of observing two successes and π_{i10} is the marginal probability of observing one success. A non-equivalent specification of the pseudo-likelihood contribution for the ith cluster is $p\ell_i^* = p\ell_i/(n_i - 1)$. The factor $1/(n_i - 1)$ corrects for the fact that each response Y_{ij} occurs $n_i - 1$ times in the ith contribution to the PL and it ensures that the PL reduces to full likelihood under independence. Geys, Molenberghs & Lipsitz (1997) explore the connection between these pseudo-likelihoods and generalized estimating equations for marginally specified odds ratio models. They show under which conditions both PL approaches coincide and study the general differences. The relative merits of both methods in terms of computational ease and relative efficiency are assessed. Table 5 shows the parameter estimates obtained by fitting a marginal odds ratio model to the DEHP data (collapsed outcome only), using pseudo-likelihood as well as GEE methods.

2.3 Random-effects models

In random-effects models, the intracluster correlation is assumed to arise from natural heterogeneity in the parameters across litters. Stiratelli, Laird & Ware (1984) assume the parameter vector $\boldsymbol{\beta}$ to be normally distributed. Skellam (1948) assumes the random success probability P_i in cluster i to follow a beta distribution with mean π_i and, given P_i, the outcomes within the ith cluster follow a binomial distribution. This results in the beta-binomial model with marginal distribution

$$f(z_i \mid \pi_i, \rho) = \binom{n_i}{z_i} \frac{B(\pi_i(\rho^{-1} - 1) + z_i, (1 - \pi_i)(\rho^{-1} - 1) + (n_i - z_i))}{B(\pi_i(\rho^{-1} - 1), (1 - \pi_i)(\rho^{-1} - 1))} \quad (6)$$

where $B(.,.)$ denotes the beta function. Note that the beta-binomial model and the Bahadur model have the same first and second order moments and hence they both feature the intraclass correlation coefficient ρ as a measure of association.

Table 6 gives parameter estimates for the beta-binomial model, applied to the DEHP study. Aerts, Declerck & Molenberghs (1997) and Molen-

Table 6. Maximum Likelihood Estimates (Standard Errors) for the Beta-Binomial Model

Par.	External	Visceral	Skeletal	Collapsed
β_0	-4.91(0.42)	-4.38(0.36)	-4.88(0.44)	-3.83(0.31)
β_d	5.20(0.59)	4.42(0.54)	4.92(0.63)	5.59(0.56)
β_a	0.21(0.09)	0.22(0.09)	0.27(0.11)	0.32(0.10)

berghs, Declerck & Aerts (1997) compared the conditional model, the Bahadur model, and the beta-binomial model for parameter estimation, hypothesis testing, and safe dose determination. They concluded that the conditional model is computationally faster and more stable while the beta-binomial model has readily interpretable parameters. In both cases, the likelihood ratio test for no dose effect has satisfactory behaviour. The Bahadur model is hard to use, both from the computational view-point and due to parameter space restrictions (Declerck, Aerts & Molenberghs, 1998).

3 Goodness-of-fit for likelihood based models with clustered binary data

Lipsitz *et al.* (1996) note that for the special case of a binary response, several methods for assessing the goodness-of-fit of binary logistic regression models have been proposed. To construct a crude goodness-of-fit measure for clustered binary data, we adapted the methods proposed by Hosmer & Lemeshow (1989) and Tsiatis (1980). Groups are constructed according to deciles of the predicted malformation probabilities in each covariate region. Given this partition, the goodness-of-fit statistic is formulated by defining $G-1$ group indicators $I_{ik}^{g} = 1$ if $\hat{\pi}_{ik}$ is in region g ($g = 1, \ldots, G-1$) and 0 otherwise. Here, $\hat{\pi}_{ik}$ is the estimated malformation probability of the kth individual within the ith cluster, calculated from the model that takes the clustering between individuals into account. Next, the following model is considered:

$$\ln\left(\frac{\pi_{ik}}{1-\pi_{ik}}\right) = \mathbf{x}_{ik}\boldsymbol{\beta} + \sum_{g=1}^{G-1} I_{ik}^{g}\gamma_g,$$

where $\mathbf{x}_{ik}$ is the covariate vector for the kth individual in the ith cluster and $\boldsymbol{\beta}$ is a vector of regression parameters. The association is modelled similarly as in the model for which the goodness-of-fit is assessed. If the mean structure in the original model is correctly specified, then $\gamma_1 = \ldots = \gamma_{G-1} = 0$. To test the goodness-of-fit of the model, we suggest the use of the likelihood ratio statistic, since it is simple to calculate and has good power properties. For large samples, all estimated expected frequencies should typically be greater than 1 and at least 80 percent should be greater than 5. Otherwise, one can collapse some frequencies, hence reducing the number of groups G (Lipsitz *et al.*, 1996). Hosmer & Lemeshow (1989) noted that $G = 6$ should be a minimum, since a test statistic calculated from fewer than 6 groups will usually have low power and thus indicate that the model fits well. The size of the test is approximately correct. Note that in the goodness-of-fit assessment described above, correlation is essentially treated as a nuisance parameter and interest is focused on the relationship between the covariates and the probability of response. Recent work has shown there may be disadvantages in the use of goodness-of-fit tests based on the ones proposed by Hosmer & Lemeshow (Hosmer *et al.*, 1997). Local likelihood estimators as studied in Aerts & Claeskens (1997a) can also play an important role as a diagnostic tool for parametric modeling and can be used to construct nonparametric test procedures. Developing improved goodness-of-fit test statistics is the topic of current research.

4 Joint modelling of continuous and discrete outcomes

Measurements of both continuous and discrete outcomes are encountered in many statistical problems. In the particular context of teratology studies, quantitative risk assessment is concerned with determining the effect of dose on the probability that an individual is malformed or of low birth weight, both being important measures of teratogenicity. This probability is not additive due to the fact that both outcomes are correlated. Plausible statistical methods should take this into account. Conditional models, such as the ones described by Catalano & Ryan (1992) and Fitzmaurice & Laird (1994) take into account the dependence between weight and malformation, but the correlation is not directly estimated. Joint models allow the outcomes to have different relationships with dose, and incorporate the correlation between outcomes. Regan & Catalano (1998) introduced a correlated probit approach. For each binary outcome it is assumed that there exists an underlying continuous variable and hence the joint distribution of the vector of weight and latent malformation outcomes can be assumed to follow a multivariate normal distribution. A full likelihood approach was developed based on the methodology of Ochi & Prentice (1984). However, since in quantitative risk assessment clustering is a nuisance, one can avoid specifying the full distribution within a litter using GEE ideas to account for litter effects. Another approach is the Plackett-Dale approach, described in Buyse, Molenberghs & Geys (1998). The latent malformation outcomes are assumed to follow a Plackett distribution. First, suppose that all littermates are independent. Assume that the cumulative distributions of the weight outcome X_i and malformation outcome Y_i of a certain individual are given by F_{X_i} and F_{Y_i}. Their dependence can be defined using a global cross-ratio at cutpoint (x, y):

$$\psi_i = \frac{F_{X_i,Y_i}(x,y)(1 - F_{X_i}(x) - F_{Y_i}(y) + F_{X_i,Y_i}(x,y))}{(F_{X_i}(x) - F_{X_i,Y_i}(x,y))(F_{Y_i}(y) - F_{X_i,Y_i}(x,y))}.$$

Now this may be solved for the joint cumulative distribution F_{X_i,Y_i} which has been studied by Plackett (1965):

$$F_{X_i,Y_i}(x,y) = \begin{cases} \frac{1+(F_{X_i}(x)+F_{Y_i}(y))(\psi_i-1)-S(F_{X_i}(x),F_{Y_i}(y),\psi_i)}{2(\psi_i-1)} & \text{if } \psi_i \neq 1, \\ F_{X_i}(x)F_{Y_i}(y) & \text{if } \psi_i = 1, \end{cases}$$

where

$$S(F_{X_i}, F_{Y_i}, \psi_i) = \sqrt{(1 + (\psi_i - 1)(F_{X_i}(x) + F_{Y_i}(y)))^2 + 4\psi_i(1 - \psi_i)F_{X_i}(x)F_{Y_i}(y)}.$$

Based upon this distribution function, we can derive a bivariate Plackett density function $g_i(x, y)$ for mixed continuous-binary outcomes. Suppose the

Table 7. Data from Ethylene Glycol Study in Rats

Dose	Dams	Live	Litter Size		Malf.		Weight	
			Mean	SD	#	%	Mean	SD
0	28	379	13.5	1.50	5	1.3	3.40	0.38
0.125	28	357	12.75	2.03	21	5.8	3.30	0.37
0.250	29	345	11.89	3.26	86	24.9	2.90	0.36
0.500	26	287	11.04	4.01	197	68.6	2.48	0.46

success probability for $\boldsymbol{Y}_i$ is denoted by π_i, then we can define $g_i(x,y)$ by specifying $g_i(x,0)$ and $g_i(x,1)$ such that they sum to $f_{\boldsymbol{X}_i}(x)$. If we define $g_i(x,0) = \partial F_{\boldsymbol{X}_i,\boldsymbol{Y}_i}(x,0)/\partial x$, then this leads to specifying g_i by:

$$g_i(x,0) = \begin{cases} \frac{f_{X_i}(x)}{2}\left(1 - \frac{1+F_{X_i}(x)(\psi_i-1)-F_{Y_i}(y)(\psi_i+1)}{S\left(F_{X_i},1-\pi_i,\psi_i\right)}\right) & \text{if } \psi_i \neq 1, \\ f_{\boldsymbol{X}_i}(x)(1-\pi_i) & \text{if } \psi_i = 1, \end{cases}$$

and

$$g_i(x,1) = f_{X_i}(x) - G_i(x,0).$$

In this formulation we assume $\boldsymbol{X}_i \sim N(\mu_i, \sigma^2)$, with $\mu_i = \beta_0 + \beta_d d_i$ and $\text{logit}(\pi_i) = \alpha_0 + \alpha_d d_i$, where d_i is the exposure level for the ith individual. The global odds ratio is assumed to be constant.

In the case of clustering, we reduce the amount of computation by considering a pseudo-likelihood: $\prod_{i=1}^{N}\prod_{k=1}^{n_i} g(x_{ik}, y_{ik})$. The working assumptions are that littermates within a cluster are independent and we then correct for clustering by using a sandwich estimator for the variance. Both approaches are illustrated on data from a developmental toxicity study of ethylene glycol (EG) in rats, conducted through the NTP (Price *et al.*, 1985). Table 7 summarizes the malformation and fetal weight data from this experiment. The data show clear dose-related trends for both outcomes: the rate of malformation increases with dose and the average fetal weight decreases monotonically with dose. The results of fitting the models considered above to these data with quadratic models for the mean fetal weight and malformation location-scale parameter and linear models for the associations are given in Table 8. Both methods yield similar results, consistent with the summary data. The rate of malformation increases with dose, whereas the mean fetal weight decreases with increasing dose. The fetal weight variances however do not change monotonically with dose, therefore a model that fits them separately for each dose might be more appropriate. The associations between weight and malformation tend to strengthen with increasing dose.

5 Risk assessment

One of the goals of quantitative risk assessment is to determine a safe level of exposure, based on an appropriate dose-response model. The standard

Table 8. Model Fit for Ethylene Glycol Study in Rats

Correlated Probit (GEE)				
dose	mean weight	σ^2	Pr (Malf)	ρ
0.00	3.43	0.13	0.01	-0.06
1.25	3.22	0.15	0.07	-0.16
2.50	2.98	0.17	0.24	-0.27
5.00	2.39	0.21	0.74	-0.48
Dale-Plackett				
dose	mean weight	σ^2	Pr(Malf)	ψ
0.00	3.43	0.13	0.01	0.92
1.25	3.21	0.15	0.07	0.62
2.50	2.98	0.16	0.23	0.41
5.00	2.46	0.19	0.70	0.18

approach requires the specification of an adverse event, along with $p(d)$ representing the probability that this event occurs at dose level d. In developmental toxicity studies $p(d)$ typically represents the probability that a fetus is malformed or of low birth weight. Based on this probability, a common measure for the excess risk over background is defined as

$$r(d) = \frac{p(d) - p(0)}{1 - p(0)}. \tag{7}$$

A benchmark dose is defined as the statistical lower confidence limit on a dose corresponding to a risk in the range of 1% to 10% (Crump, 1984). The virtually safe dose (VSD) can be defined and estimated in several ways (Crump & Howe, 1983). For instance, it can be defined as the lower confidence limit on a dose corresponding to an excess risk of 10^{-4}. The *effective dose* (ED) is defined as the value d that solves $r(d;\boldsymbol{\beta}) = 10^{-4}$. For setting confidence limits in low dose extrapolation, i.e. to determine the VSD, several approaches can be considered. Using the delta method, a Wald based method can be obtained. Several authors have indicated that this method suffers from drawbacks, especially with low dose extrapolation (Crump, 1984; Crump & Howe, 1983; Krewski & Van Ryzin, 1981) whence the method may yield negative lower limits. Alternatively, an upper limit for the risk function can be computed, and thus the dose that corresponds to a q% increased response above background is determined from this upper limit curve. This dose level is referred to as the lower effective dose (LED_q) (Kimmel & Gaylor, 1988). Crump & Howe (1983) recommend using the asymptotic distribution of the likelihood ratio. According to this method, an approximate $100(1-\alpha)$% lower limit for the VSD, denoted by VSD(1), corresponding to an excess risk of 10^{-4} is defined as

$$\min\{d(\boldsymbol{\beta}) : r(d;\boldsymbol{\beta}) = 10^{-4} \text{ over all } \boldsymbol{\beta} \text{ such that } 2(\ell(\hat{\boldsymbol{\beta}}) - \ell(\boldsymbol{\beta})) \leq \chi^2_p(1-\alpha)\},$$

with p the number of model parameters. A second approach, denoted VSD(2), is based on the profile likelihood method (Morgan, 1992). First, construct a profile likelihood based confidence interval for the dose effect parameter β_d. Secondly, transform this interval into an interval for d and check that the

Table 9. Effective Doses and Lower Confidence Limits for DEHP Study; Entirely Model Based Computation; All Quantities Shown Should Be Divided by 10^4

Model	Statistic	External	Visceral	Skeletal	Collapsed
Bahadur	ED	27	19	23	9
	VSD(1)	15	13	14	6
	VSD(2)	18	15	16	7
BB	ED	26	18	27	8
	VSD(1)	14	11	14	6
	VSD(2)	17	14	18	6
Cond.	ED	30	23	28	12
	VSD(1)	19	15	18	9

transformation is monotonic. For the conditional model of Section 2.1, there are situations where the transformation is not monotonic, and hence the VSD(2) will not be applied to this model. In Table 9 VSD(1) and VSD(2) are applied to the DEHP data. In general, VSD(1) yields lower results than VSD(2), and the values obtained with the conditional model are somewhat higher than for both other models.

6 Discussion

Developmental toxicity data, such as those presented in Table 1, yield a wide class of challenging statistical problems. In this paper we have discussed some of these: (1) the choice between parametric modelling family for clustered binary data, (2) joint modelling of binary and continuous responses, (3) goodness-of-fit tools and (4) quantitative risk assessment. Some emphasis has been put on pseudo-likelihood methods.

Several problems have not been discussed and/or need further research. These include satisfactory models for a clustered multivariate response vector, with mixed continuous and discrete components, inclusion of dead and reabsorbed fetuses, whereas the current methods are restricted to viable fetuses. For risk assessment, it is not clear whether safe doses need to be based on the effects seen in individual fetuses, as is customarily done, or in a dam. In the latter case, it is clear that one will often find lower safe doses. Whereas generalized estimating equations and pseudo-likelihood have been explored as alternatives to full likelihood for relatively "large" problems, more work is needed on exact inferential procedures for small data sets or situations with very low risks. Finally, there is a need for more satisfactory goodness-of-fit tools.

Acknowledgements
The authors would like to thank Paul Catalano, Stuart Lipsitz, Meredith Regan, Louise Ryan and Paige Williams for several helpful contributions. This work was supported by the NATO Collaborative Research Grant 950648.

References

Aerts, M. & Claeskens, G. (1997a). Local Polynomial Estimation in Multiparameter Likelihood Models. *Journal of the American Statistical Association*, **92**, 1536–1545.

Aerts, M. & Claeskens, G. (1997b). Bootstrapping Pseudolikelihood Models for Clustered Binary Data, Limburgs Universitair Centrum, Technical Report.

Aerts, M. & Claeskens, G. (1997c). Bootstrap Tests for Misspecified Clustered Binary Data Models, Limburgs Universitair Centrum, Technical Report.

Aerts, M., Declerck, L. & Molenberghs, G. (1997). Likelihood Misspecification and Safe Dose Determination for Clustered Binary Data. *Environmetrics*, **8**, 613–627.

Arnold, B.C. & Strauss, D. (1991). Pseudolikelihood estimation : some examples. *Sankhya B*, **53**, 233-243.

Bahadur, R.R. (1961). A representation of the joint distribution of responses to n dichotomous items. In: *Studies in Item Analysis and Prediction*, (ed. H. Solomon), 158–168. Stanford, California : Stanford University Press.

Buyse, M., Molenberghs, G. & Geys, H. (1998). Mixed Discrete and Continuous Outcomes for the Validation of Surrogate Endpoints in Randomized Experiments, Limburgs Universitair Centrum, Technical Report.

Catalano, P.J. & Ryan, L.M. (1992). Bivariate latent variable models for clustered discrete and continuous outcomes. *Journal of the American Statistical Association*, **87**, 651–658.

Cox, D. R. (1972). The analysis of multivariate binary data. *Applied Statistics*, **21**, 113–120.

Crump, K.S. & Howe, R.B. (1983). A review of methods for calculating statistical confidence limits in low dose extrapolation. In: *Toxicological Risk Assessment, Volume I: Biological and Statistical Criteria*, (ed. D.B. Clayson, D. Krewski & I. Munro), 187–203. Boca Raton: CRC Press.

Crump, K.S. (1984). A New Method for Determining Allowable Daily Intakes. *Fundamental and Applied Toxicology*, **4**, 854–871.

Declerck, L., Aerts, M. & Molenberghs, G. (1998). Behaviour of the Likelihood Ratio Test Statistic Under A Bahadur Model For Exchangeable Binary Data. *Journal of Statistical Computation and Simulation*, in press.

Fitzmaurice, G.M. & Laird, N.M. (1995). Regression Models for a Bivariate Discrete and Continuous Outcome With Clustering. *Journal of the American Statistical Association*, **90**, 845–852.

Geys, H., Molenberghs, G. & Lipsitz, S.R. (1997). A note on the Comparison of Pseudo-likelihood and Generalized Estimating Equations for Marginal Odds Ratio Models, Limburgs Universitair Centrum, Technical Report.

Geys, H., Molenberghs, G. & Ryan, L. (1997). Pseudo-likelihood Inference for Clustered Binary Data. *Communications in Statistics: Theory and Methods*, **26**, 2743–2767.

Geys, H., Molenberghs, G. & Ryan, L. (1997). Pseudo-likelihood Inference for Exponential Family Models, Applied to Toxicology Data, Limburgs Universitair Centrum, Technical Report.

Hosmer, D.W. & Lemeshow, S. (1989). *Applied Logistic Regression*. New York: Wiley.

Hosmer, D.W., Hosmer, T., Lemeshow, S. & Le Cessie, S. (1997). A Comparison of Goodness-of-Fit Tests for the Logistic Regression Model. *Statistics in Medicine*, **16**, 965–980.

Kimmel, C.A. & Gaylor D.W. (1988). Issues in Qualitative and Quantitative Risk Analysis for Developmental Toxicology. *Risk Analysis*, **8**, 15–20.

Krewski, D. & Van Ryzin, J. (1981). Dose-response models for quantal response toxicity data. In: *Statistics and Related Topics*, (ed. M. Csorgo,

D. Dawson, J.N.K. Rao & E. Saleh), 201–231. North-Holland, New York, 201–231.

Le Cessie, S. & Van Houwelingen J.C. (1994). Logistic Regression for Correlated Binary Data. *Applied Statistics*, **43**, 95–108.

Liang, K.Y., Zeger, S.L. & Qaqish, B. (1992). Multivariate Regression Analyses for Categorical Data. *Journal of the Royal Statistical Society B*, **54**, 3-40.

Lipsitz, SR., Fitzmaurice, G.M. & Molenberghs, G. (1996). Goodness-of-fit Tests for Ordinal Response Regression Models. *Applied Statistics*, **45**, 175-190.

Molenberghs, G., Declerck, L. & Aerts, M. (1997) Misspecifying the likelihood for clustered binary data. *Computational Statistics and Data Analysis*, **26**, 327–349.

Molenberghs, G. & Ryan, L.M. (1998). Likelihood inference for clustered multivariate binary data, Limburgs Universitair Centrum, Technical Report.

Morgan, B.J.T. (1992) *Analysis of Quantal Response Data*, Chapman and Hall, London.

Ochi, Y. & Prentice, R.L. (1984). Likelihood Inference in a Correlated Probit Regression Model. *Biometrika*, **71**, 531–543.

Plackett, R.L. (1965). A Class of Bivariate Distributions. *Journal of the American Statistical Association*, **60**, 516–522.

Price, C. J., Kimmel, C. A., Tyl, R. W. & Marr, M. C. (1985). The developmental toxicity of ethylene glycol in rats and mice. *Toxico. Appl. Pharmacol.*, **81**, 113-127.

Regan, M.M. & Catalano P.J. (1998). Likelihood Models for Clustered Binary and Continuous Outcomes: Application to Developmental Toxicology, Harvard School of Public Health, Technical Report.

Skellam, J.G. (1948). A probability distribution derived from the binomial distribution by regarding the probability of a success as variable between the sets of trials. *Journal of the Royal Statistical Society, Series B*, **10**, 257–261.

Stiratelli R., Laird, N. & Ware, J.H. (1984). Random effects models for serial observations with binary response. *Biometrics*, **40**, 961–971.

Tsiatis, A.A. (1980). A note on a goodness-of-fit test for the logistic regression model. *Biometrika*, **67**, 250-251.

Wavelets in Statistics: Some Recent Developments

Bernard W. Silverman

Department of Mathematics, University of Bristol, Bristol BS8 1TW, UK

Abstract. Wavelet methods in statistics are in their infancy as far as the range of problems to which they have been applied is concerned. Some elementary aspects of wavelets are reviewed, concentrating on the discrete wavelet transform, because of its relevance to practical and computational statisticians. Several recent areas of research are discussed, concentrating on extensions of the standard paradigm. A Bayesian approach is natural, because of the notion that the wavelet expansion is sparse. Wavelets are easily applied to regression where the errors are correlated. The combination of these ideas is demonstrated on data generated in the study of ion channel gating, with excellent results. The arrangement and distribution of the data can be quite general, because of a fast algorithm that finds details of the variance structure of the discrete wavelet transform. Furthermore, this algorithm naturally leads to simple methods of dealing with outliers and with heteroscedastic data. Finally, an application to deformable templates demonstrates the wide potential of wavelet methods; this arises from a palaeopathological data set in an arthritis study.

Keywords. correlated data, deformable templates, discrete wavelet transform, generalized linear models, image analysis, ion channel data, nonparametric regression, robustness, smoothing, wavelets

1 Introduction

In recent years, many areas of science and engineering have seen an explosion of activity in wavelet methods, both in theoretical research and in applications. In particular, wavelets are of much current interest in statistics. It is too early to see clearly what will be the real methodological impact of wavelets for statistical practice, but we are at the exciting stage of trying to understand for which problems they will be the method of choice. Wavelets are a particularly appropriate subject for a lecture at COMPSTAT. They combine attractive computational properties with the hope of representing and modelling the nonstationary and inhomogeneous phenomena that more classical statistical approaches have difficulty with.

Of course, wavelets are not, and will never be, a universal statistical panacea, but, for us to gain a picture of their true potential, it may be useful to consider ways in which they can be applied more broadly than has perhaps been the case up to now. Furthermore, people interested in computational statistics should have some knowledge of wavelets, in order to be able to have them as part of their toolbox if an appropriate new area of possible application comes along!

The aim of this paper is to provide an eclectic personal view, rather than an exhaustive survey, of wavelet methods. For those who would like to read more, a good introduction is provided by Strang (1993). More detailed treatments, in increasing order of mathematical sophistication, are given by Chui (1992) and Daubechies (1992), for example. A key computational aspect of wavelet methods is the discrete wavelet transform, which is based on filtering ideas discussed extensively in the engineering literature. Vaidyanathan (1990) and Vetterli & Herley (1992) provide detailed surveys and numerous references. The specific papers referred to below will also give leads to much further literature.

It is my hope that readers will be tempted to experiment with wavelet methods for themselves. Both Splus and Matlab market wavelet modules. For Splus, there is also available the package WaveThresh (Nason, 1993 and subsequently). The package itself is available, free of charge, from the web page `http://www.stats.bris.ac.uk/pub/Software.html`; for a paper describing some of the capabilities of the package and of wavelet methods in the statistical context, see Nason & Silverman (1994).

2 What are wavelets anyway?

In statistical practice, the key concept in the use of wavelets is the discrete wavelet transform (DWT), which plays a role analogous to the fast Fourier transform in conventional frequency-domain analysis. For simplicity, we consider the one-dimensional case; there are simple generalizations to higher dimensions, though there is a 'curse of dimensionality' effect beyond two or three dimensions.

Suppose x_i is a sequence of $N = 2^J$ values. We rename x_i the sequence c^J 'at level J'. For simplicity of exposition we apply periodic boundary conditions to the sequence where necessary, but there are versions of the procedure that allow for other treatments at the boundaries. The transform works by applying two linear filters to c^J, each of which produces a sequence of length $N/2$. One filter produces a **smooth** c^{J-1}, sampled at half the rate of the original sequence, while the other captures the high frequency **detail** d^{J-1}. We then apply the same process to c^{J-1} to produce vectors c^{J-2} and d^{J-2} of length 2^{J-2}, and so on. The discrete wavelet transform of x, denoted by $\mathcal{W}x$, is the array of vectors $(c^0, d^0, d^1, \ldots, d^{J-1})$; the total of the lengths of these vectors is N. On a time scale where observation x_i is taken at time i/N, the coefficient d^j_k gives information about x on *scale* 2^{-j} (i.e., near frequency 2^j) near *position* $k2^{-j}$.

A key issue is the choice of the filters. By judicious construction, we can ensure that the following properties hold:

1. The wavelet transform is orthonormal, so that $\mathcal{W}x$ is a rotation of the original vector x.
2. The filters are of short support, as a result of which the transform can be carried out in a small multiple of N operations.
3. The transform can be inverted, again in $O(N)$ operations, by reconstructing c^j from (c^{j-1}, d^{j-1}) successively for $j = 1, 2, \ldots, J$.
4. The filters annihilate low degree polynomials.

It is a mathematical consequence of these properties that a very useful class of sequences x have *economical* wavelet expansions, in that only a small number of nonzero coefficients is necessary to approximate x closely. This will be the case not only for smooth sequences, but also for wider classes of sequences or images that have varying smoothness or regularity properties, for example having smooth parts between discontinuities or being textured in some places and smooth in others. Such sequences cause trouble for standard time and frequency domain methods, and typically require many Fourier coefficients to approximate them well. But as long as you know which ones to use, you may not need many nonzero wavelet coefficients to get a good approximation. This economy of representation means that wavelets give very good *compression* of images and functions, and they are widely used for this purpose.

The major use of wavelets in a statistical setting has been in function estimation. Given noisy observations of a function f on a regular grid

$$Y_i = f(i/N) + \text{error},$$

suppose that $f(i/N)$ has an economical wavelet expansion. The DWT of Y will be a noisy version of the DWT of f, and, if we knew which coefficients of f to estimate, we could use the corresponding coefficients in the DWT of Y to estimate them. There have been many proposals for doing this, but the basic notion is that large coefficients in $\mathcal{W}Y$ include signal, while smaller ones are just noise, so if we only retain coefficients that achieve some *threshold* then we should get a good estimate. Simple choices of threshold give excellent asymptotic properties, including good adaptivity to different behaviour in f.

For finite sample sizes, various issues are important, for example the exact choice of the threshold and the choice of a *primary resolution level* below which we let the coefficients through regardless. Considerable improvements can be obtained by using a *translation-equivariant* or *stationary* wavelet transform; this is an overdetermined transform in which there is the same number of coefficients at each level. It includes the DWT coefficients for every choice of starting point of the sequence, and can be calculated in $O(N \log N)$ operations. See Coifman & Donoho (1995) or Nason & Silverman (1995) for further discussion.

3 Bayesian approaches

The notion that an unknown function is likely to have an economical wavelet expansion is naturally formalized into a prior distribution on the wavelet coefficients. Several authors have considered a mixture prior, for example where each coefficient d^j_k independently has prior probability $(1 - \pi)$ of being zero and π of being drawn from a $N(0, \tau^2)$ distribution. Typically, the parameters π and τ^2 are chosen to depend on the level j of the coefficient d^j_k but not on the position k. To reduce the number of hyperparameters further, one can choose τ^2 and π proportional to $2^{-\alpha j}$ and $2^{-\beta j}$ respectively. The parameters α and β then correspond to smoothness properties of the unknown function; see Abramovich, Sapatinas & Silverman (1998) for a detailed discussion. Their *BayesThresh* approach chooses the constants of proportionality from the data in a natural way.

The BayesThresh paradigm uses the posterior *median* to summarize the posterior distribution. This approach gives a true thresholding rule, in that all wavelet coefficients in the data below a certain threshold are set exactly to zero. (It also has a natural justification in terms of the loss function that is appropriate for inhomogeneous functions.) Figures shown in Abramovich *et al.* (1998) show that BayesThresh performs well on a range of simulated and real data examples. It filters out noise without oversmoothing important detail such as sharp spikes in the signal. The values $\alpha = 0.5$ and $\beta = 1$ generally give good results. For fixed β, increasing α increases the order of the derivatives that are considered in evaluating the smoothness of a function, and so one would expect to use a larger value of α for phenomena where irregularities are more likely to show up in derivatives than in the function itself, though in practice the differences in the estimates are not dramatic.

Another possible approach to the choice of π and τ^2 is by a marginal maximum likelihood (MML) or empirical Bayes approach. The mixture model for the true wavelet coefficients translates into a normal mixture for the wavelet coefficients of the data at each level. Johnstone & Silverman (1998) give an algorithm, related to the EM algorithm, that allows this likelihood to be maximized easily. The performance of this approach is at least as good as BayesThresh, and in addition there is evidence that the MML approach adapts to the unknown function's smoothness, by performing as well as BayesThresh for the best hyperparameter α.

4 Estimators for data with correlated noise

4.1 Level-dependent thresholding

Johnstone & Silverman (1997) consider the extension of wavelet thresholding to data with correlated noise. One of the appeals of wavelets in this context is that the wavelet transform often decorrelates the noise process. For example, Figure 1 of Johnstone & Silverman (1997) shows the wavelet transform of a process with long-range correlation. Because of the correlation structure, the variance of the wavelet coefficients is no longer the same at each level, but decreases as the resolution level increases. For this particular noise structure, the correlations between coefficients, both within and across levels, are small; see Figure 2 of the paper. So the wavelet transform maps a highly dependent series to an array of approximately independent coefficients whose variances depend only on their level.

How should one deal with a wavelet array of coefficients where the noise is of this kind? A natural approach is to use *level-dependent* thresholding, using for each coefficient a threshold that is proportional to its standard deviation. Since, for a stationary noise process, the variances will be constant within levels, these standard deviations can be estimated from data considering the wavelet coefficients level by level rather than all together. The idea of using level-dependent thresholding can equally be applied within the translation-equivariant wavelet transform.

Note that the Bayesian approaches described above will also process each level differently, but because of properties of the prior model for the unknown function rather than those of the noise. Johnstone & Silverman (1998) investigate a synthesis of the Bayesian approach with the type of level-dependent noise induced by correlated noise in the original signal. Furthermore they also work in the context of the translation-equivariant transform. The resulting procedure is not a strict Bayes procedure, because it treats dependent wavelet coefficients as if they were independent (both in the prior and in the data) and for more subtle reasons of consistency between models at different positions of the origin in the DWT. Nevertheless, as we shall see below, it performs excellently in practice.

4.2 An example from neurophysiology

Johnstone & Silverman (1997, 1998) present an example from neurophysiology, using data supplied by R. Eisenberg and R. Levis. The context of the data is the important problem in molecular physiology of the detection and measurement of the picoamp currents that flow through the single membrane channels that control movement in and out of cells. The generated data, the first 2048 values of which are shown in Figure 1, is intended to represent most of the relevant challenges in processing such single channel data. The data consist of a step function randomly switching between values 0 ("off") and 1 ("on"), measured in the presence of additive, correlated noise. The 'true' step signal is also shown in Figure 1.

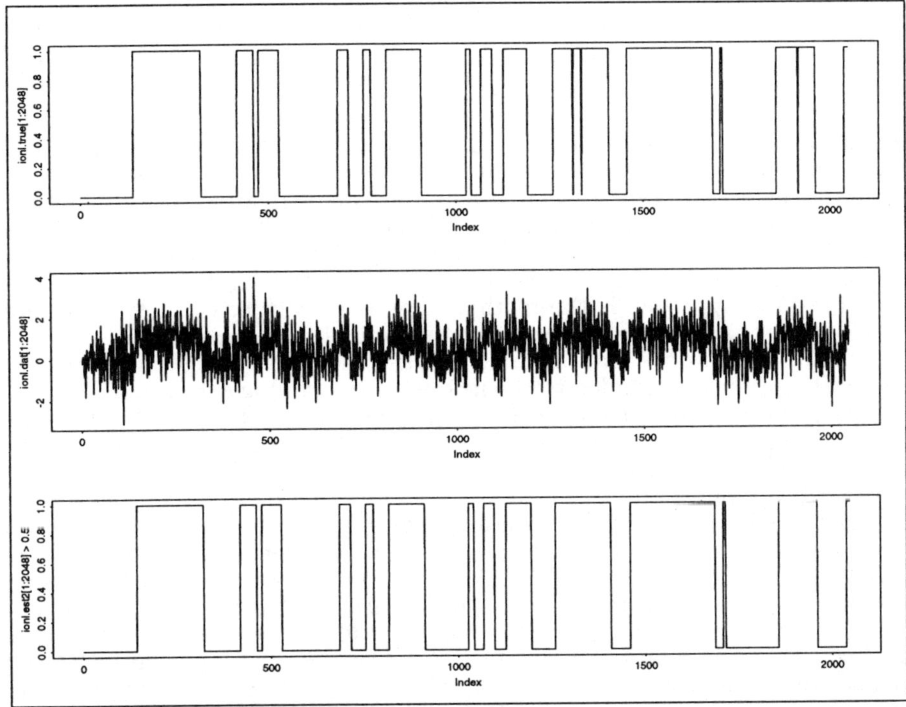

Fig. 1. Top: the 'true' ion channel signal for time points 1 to 2048; middle: the corresponding section of generated data (on a smaller vertical scale); bottom: the estimate obtained by the TI-MML method. Reproduced from Johnstone & Silverman (1998) with the permission of the authors

This generated example differs in kind from the simulated data often considered in statistical studies, in that its underlying model is carefully selected by practitioners directly involved in routine collection and analysis of real data. The obvious advantage of using a generated data set rather than an actual data set obtained in practice is that the 'truth' is known, and so it is possible to quantify the performance of the various methods.

Various approaches were applied to the estimation of the original (0-1) signal from these data. The first approach was to classify the original data pointwise, regardless of the time structure. Four wavelet-based approaches were used, all based on the level-dependent thresholding methods for

Table 1. Comparison of various methods for the neurophysiology data. The performances shown are the averages over 10 successive segments each of length 4096. Method (a) estimates the signal to be zero if the observed value is less than 0.5 and one otherwise

Method	**Errors**	**Percent**
a) Pointwise classification of raw data	1180.3	28.8%
b) Wavelet thresholding, thresholds $\hat{\sigma}_j\sqrt{2\log n}$ at levels $j \geq 6$	643.1	15.7%
c) As (b), but using translation-equivariant wavelet transform	350.7	8.6%
d) As (c), but using SURE thresholds at levels $j = 6, 7, 8$	95.9	2.3%
e) Using the TI-MML approach at the three highest levels	83.8	2.05%
f) Special-purpose algorithm provided by Eisenberg and Levis	82.2	2.01%

correlated data. In each case the estimate obtained by the wavelet approach was rounded to the nearest of 0 and 1 to yield the estimate itself. Median absolute deviation estimates were used to give an estimate $\hat{\sigma}_j$ of the standard deviation of the wavelet coefficients at each level j. The methods used are given in Table 1. The SURE (Stein unbiased risk) method is based on Stein (1981) and is described in detail by Johnstone & Silverman (1997). The TI-MML approach is the method proposed by Johnstone & Silverman (1998), which uses an as-if-independent empirical Bayes approach within the translation-equivariant transform. The first 2048 values as estimated by the TI-MML approach (rounded to the nearest of 0 and 1) are shown in Figure 1. A particular challenge in this context is to identify the pattern of 'openings' and 'closings' and it can be seen that the wavelet method misses only three extremely short 'closings', each of length only 2 time points.

Table 1 compares the error rates of various methods on ten successive segments of length 4096 from the original data record. Also shown is the error rate of the current special-purpose method developed specifically by Eisenberg and Levis for this problem. It can be seen that the TI-MML approach is virtually as good as the special-purpose filtering method, and that a method using SURE thresholds at moderate levels and universal thresholds at high levels is nearly as good. It is very encouraging that relatively straightforward adaptation of general wavelet thresholding approaches achieves good results in this special context. The special-purpose method has no promise of working on any other kind of data, whereas the method we propose is part of a very generally applicable toolkit.

5 Coefficient-dependent thresholding

Suppose that we apply the DWT to a vector c^J that has general covariance matrix Σ^J. How much can we easily find about the variance properties of the transform? Kovac & Silverman (1998) develop a fast algorithm for the case where Σ^J is a band matrix, certainly not necessarily stationary. The algorithm essentially works by applying the filters used in the DWT successively to the rows and columns of the variance matrix at each level of the transform. It yields the variance matrix of all the vectors c^j and d^j, in an amount of time linear in the length 2^J of c^J. In particular, the algorithm gives the variances of the individual wavelet coefficients, and thereby allows level-dependent thresholding to be extended to *coefficient-dependent* thresholding.

The natural approach is to use thresholds proportional to individual standard deviations, a method given theoretical support by results of Johnstone & Silverman (1997).

The algorithm potentially has very wide applicability. For example, consider the nonparametric regression problem where we have observations

$$Y_i = f(t_i) + \text{error}$$

where the t_i are irregularly spaced. A simple approach is to interpolate the observations to a fine regular grid of size 2^J to give a vector c^J, and then to continue with the wavelet thresholding paradigm. If a linear interpolation approach is used, the variance structure of c^J can be inferred from that of the original data; if the original data are assumed to be independent and identically distributed, then c^J will have a band-limited variance matrix with bandwidth depending on the size of the largest gap in the t_i sequence. Such an approach can also be used if the t_i are on a regular grid that is not of size 2^J, as an alternative to extending to a grid of size 2^J by some boundary extension approach.

The method naturally allows wavelet regression techniques to be applied to data that have known non-stationary covariance structure, whether or not the points t_i are regularly spaced. It also allows for observations to be deleted or given varying weights, for example within robust procedures that downweight outlying observations. For details and examples, see Kovac & Silverman (1998).

An exciting possibility, yet to be developed in any detail, is the use of the method within the framework of generalized linear model (GLIM) dependence. As explained in Green & Silverman (1994, Chapters 5 and 6), for example, nonparametric smoothing ideas can be incorporated into the GLIM framework by assuming that one or more of the dependences on the covariates is a curve rather than simply linear. Such nonparametric GLIMs can be fitted by solving a sequence of weighted regression problems, each having the same structure as a standard nonparametric regression with unequal variances. Using a wavelet estimator of the kind set out in this paper allows for the fitting of dependences whose behaviour has inhomogeneous smoothness properties. Because of the nonlinear nature of the wavelet smoothing, there may be problems with the convergence of this iteration, and the detailed investigation of this convergence is, at the time of writing, a subject for future research.

6 Wavelet methods for deformable templates

6.1 Images collected in palaeopathology

There are many problems nowadays where an observed image can be modelled as a deformed version of a standard image, or template. The assumption is that the image is a realization of the template, perhaps with additional variability that is also of interest. My own interest in this issue stems from a palaeopathological study of skeletons temporarily excavated from a cemetery in Humberside. Of particular interest to the palaeopathology group in the Rheumatology Department at Bristol University is the information that can be gained about patterns of osteoarthritis of the knee. As part of his PhD research project, Lee Shepstone has collected a considerable number of images of the kind shown in Figure 2, using the experimental setup shown in Figure 3. Further details of the work described in this section are given by Downie, Shepstone & Silverman (1998).

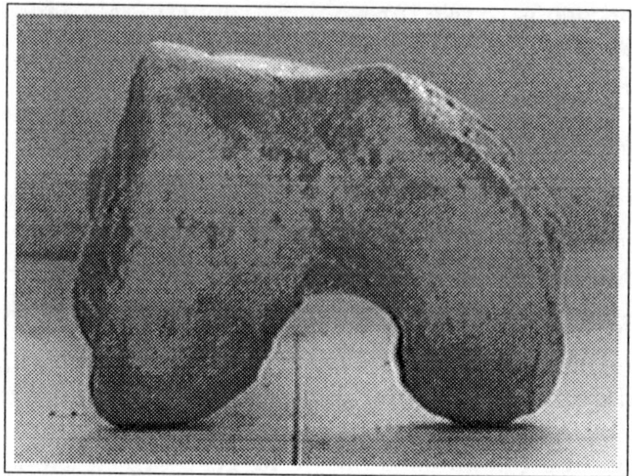

Fig. 2. A typical image of the femoral condyles from the distal end of the femur. Reproduced from Downie *et al.* (1998) with the permission of the authors

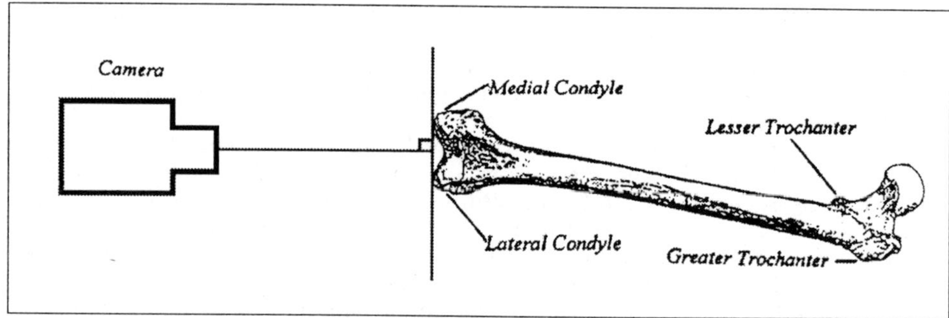

Fig. 3. Lee Shepstone's experimental setup for collecting the femur image data. Reproduced from Downie *et al.* (1998) with the permission of the authors

The important features of these bones as far as the osteoarthritis study is concerned are the shape of the bone and the occurrence and position of various changes, notably eburnation (polished bone, caused by loss of cartilage) and osteophytes (bony outgrowths). The images are digitized as pixel images and are marked up by comparison with the original bone to label the pixels corresponding to the areas of these changes. The aim of any study of the deformations is twofold: firstly, to give a standard mapping to relate positions on various bones, and secondly, to gain information about the shape of individual bones. For the first purpose, we are interested only in the *effect* of the deformation, but for the second the details of the deformation itself are important.

6.2 Models for deformations

Amit, Grenander & Piccioni (1991) model deformations as follows. Let I and T be functions on the unit square $\mathcal{U}$, representing the image and the template respectively. In our particular application they will be zero-one functions. The deformation is defined by a two-dimensional *deformation function* f such that, for u in $\mathcal{U}$, $u + f(u)$ is also in $\mathcal{U}$. The aim is then to get a good fit of the image $I(u)$ to the deformed template $T(u + f(u))$ measuring discrepancy by summed squared difference over the pixels in the image. Amit *et al.* (1991) model f by expanding it as a bivariate Fourier series. A Bayesian approach is then applied, under the assumption that the errors in individual pixels are independently normally distributed. A prior distribution is placed on the Fourier coefficients by assuming them to have independent normal distributions, with variances that reflect the smoothness assumptions about the deformation.

The deformation f is a vector of two functions (f_x, f_y), giving the coordinates of the deformation in the x and y directions. Both $f_x(u)$ and $f_y(u)$ are defined for each u in the unit square. In our work we expand each of them as a two-dimensional wavelet series, which may be more appropriate than a Fourier expansion because it is reasonable to assume that deformations will have localized features. In two dimensions, the wavelet multiresolution analysis of an array of values yields coefficients w_κ, where the index $\kappa = (j, k_1, k_2, \ell)$. This coefficient gives information about the array near position (k_1, k_2) on scale j. Three orthogonal aspects of local behaviour are modelled, indexed by ℓ in $\{1, 2, 3\}$, corresponding to horizontal, vertical and diagonal orientation.

To model the notion that the deformation has an economical wavelet expansion, a mixture prior of the kind described in Section 3 was used. Because the assumption of normal identically distributed errors is not realistic, we prefer to consider our method as being a penalized least squares method with a Bayesian motivation, rather than a formal Bayesian approach. A particular bone unaffected by any pathology was arbitrarily designated as the template. Experimentation with several different approaches indicates that the best computational method for estimating the deformation within this framework is an iterated coefficient-wise maximization. Figure 4 shows the template, the image, and the deformed template, for a particular test image.

6.3 Gaining information from the wavelet model

Figure 5 demonstrates the information available in the wavelet expansion of the deformation. Only 27 of the possible 2048 coefficients are nonzero, indicating the extreme economy of representation of the deformation. For each of these coefficients, a number equal to the level j of the coefficient is plotted at the position (k_1, k_2). The size at which the number is plotted gives the absolute size of the coefficient; the orientation ℓ is indicated by colours invisible on this copy, but available on the world-wide web version of Downie *et al.* (1998).

The figure shows that most of the nonzero coefficients are near the outline of the image, because of the localization properties of the wavelets. At the top of the image in the y component, coefficients at all resolution levels are present, indicating the presence of both broad scale and detailed warping effects. The deformation is dominated by two coefficients, showing that the main effects are a fairly fine-scale effect at the middle of the top of the image, and a larger scale deformation centred in the interior of the image. The full implications of this type of plot remain a subject for future research; in some contexts the coefficients and their values will be candidates for subsequent statistical analysis, while elsewhere they will be valuable for the insight they give into the position and scale of important aspects of the deformation.

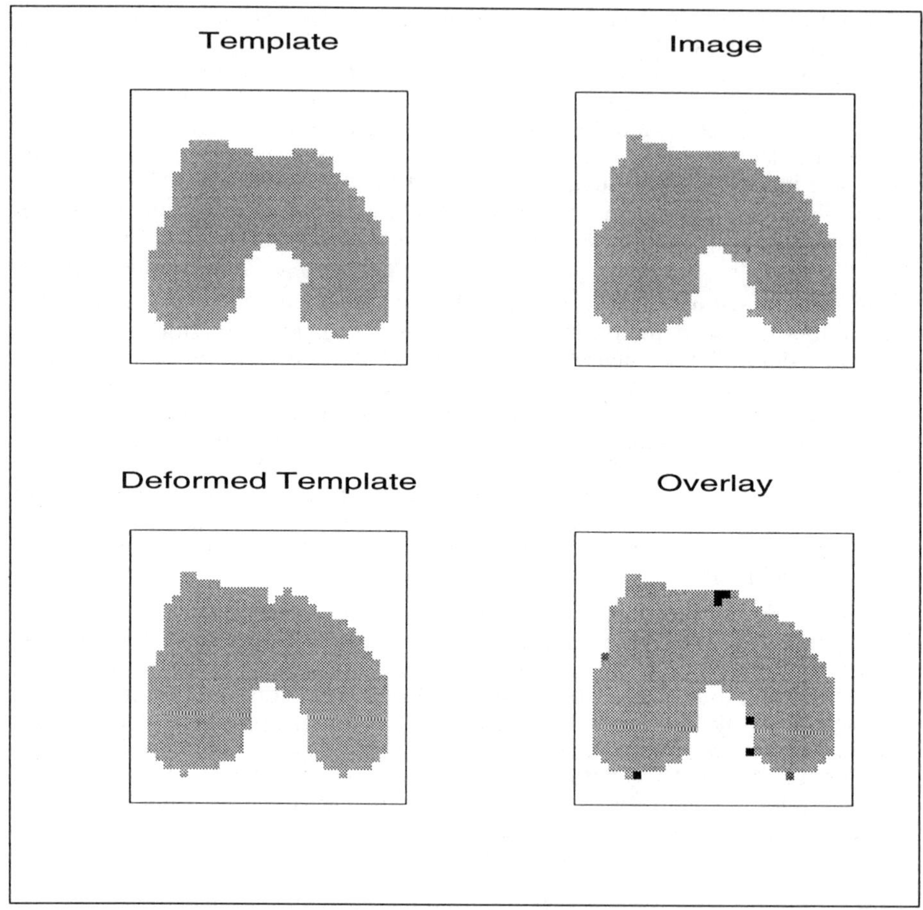

Fig. 4. For a particular femoral image, the template, the image itself, the template after applying the estimated deformation, and an overlay of the deformed template and the image, showing the pixels where the two do not agree. Reproduced from Downie *et al.* (1998) with the permission of the authors

7 Concluding remarks

In this paper, I have presented only a small selection of ideas for using wavelets beyond the classical case of nonparametric regression with 2^n equally-spaced data points with i.i.d. normal errors. Much of this work is provisional, and a great deal of further research is needed to fine-tune existing methods, to develop methodology for a broader range of statistical problems, and to make a careful assessment, both practical and theoretical, of the advantages and disadvantages of wavelet methods. The wavelet approach, and other related methods, will be of growing importance in computational statistics in the next decade, and I eagerly await many future developments.

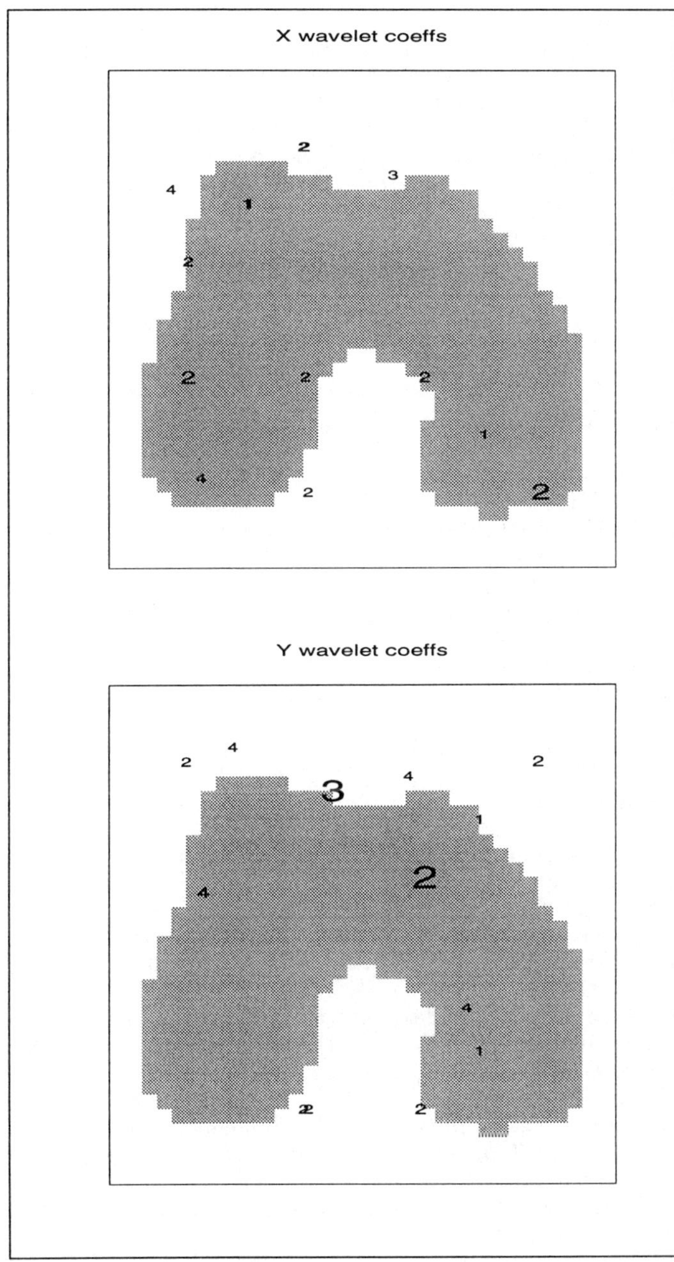

Fig. 5. The wavelet coefficient positions and sizes for the deformation shown in Figure 4. The top figure shows the x coordinate of the deformation, the bottom the y coordinate. The numbers denote the scale of the particular coefficient, and are plotted at the centre of the support of the corresponding wavelet. The printed size of each number indicates the absolute size of the wavelet coefficient. Reproduced from Downie *et al.* (1998) with the permission of the authors

Acknowledgements
This paper was written while the author was a Fellow at the Center for Advanced Study in the Behavioral Sciences, Stanford, California, supported by National Science Foundation grant number SBR-9601236. For their contributions to the material discussed, and for their continuing intellectual stimulation and support, he is most grateful to Felix Abramovich, Tim Downie, Peter Green, Iain Johnstone, Arne Kovac, Guy Nason, Theofanis Sapatinas, and Lee Shepstone. He is also very grateful to Kathleen Much for her most helpful comments on the manuscript.

References

The starred reports are as yet unpublished and are available from `http://www.stats.bris.ac.uk/~bernard`

Abramovich, F., Sapatinas, T. & Silverman, B.W. (1998). Wavelet thresholding via a Bayesian approach. *J. Roy. Statist. Soc.* B, **60**, in press.

Amit, Y., Grenander, U. & Piccioni, M. (1991). Structural image restoration through deformable templates. *J. Amer. Statist. Assoc.*, **86**, 376–387.

Chui, C.K. (1992). *An Introduction to Wavelets.* London: Academic Press.

Coifman, R.R. & Donoho, D.L. (1995). Translation-invariant denoising. In: *Wavelets and Statistics, Lecture Notes in Statistics 103*, (ed. A. Antoniadis & G. Oppenheim), 125–150. Heidelberg: Springer Verlag.

Daubechies, I. (1992). *Ten Lectures on Wavelets.* Philadelphia: SIAM.

Downie, T.R., Shepstone, L. & Silverman, B.W. (1998). Economical representation of image deformation functions using a wavelet mixture model*.

Green, P.J. & Silverman, B.W. (1994). *Nonparametric Regression and Generalized Linear Models: A Roughness Penalty Approach.* London: Chapman and Hall.

Johnstone, I.M. & Silverman, B.W. (1997). Wavelet threshold estimators for data with correlated noise. *J. Roy. Statist. Soc.* B, **59**, 319–351.

Johnstone, I.M. & Silverman, B.W. (1998). Empirical Bayes approaches to wavelet regression*.

Kovac, A. & Silverman, B.W. (1998). Extending the scope of wavelet regression methods by coefficient-dependent thresholding*.

Nason, G.P. (1993 and subsequently). The WaveThresh software package for Splus. Department of Mathematics, University of Bristol.

Nason, G.P. & Silverman, B.W. (1994). The discrete wavelet transform in S. *J. Comp. Graph. Stat.*, **3**, 163–191.

Nason, G.P. & Silverman, B.W. (1995). The stationary wavelet transform and some statistical applications. In: *Wavelets and Statistics, Lecture Notes in Statistics 103*, (ed. A. Antoniadis & G. Oppenheim), 281–300. Heidelberg: Springer Verlag.

Stein, C. (1981). Estimation of the mean of a multivariate normal distribution. *Ann. Statist.*, **9**, 1135–1151.

Strang, G. (1993). Wavelet transforms versus Fourier transforms. *Bull. Amer. Math. Soc.*, new series **28**, 288–305.

Vaidyanathan, P.P. (1990). Multirate digital filters, filter banks, polyphase networks, and applications: a tutorial. *Proc. IEEE*, **78**, 56–93.

Vetterli, M. & Herley, C. (1992). Wavelets and filter banks: theory and design. *IEEE Trans. Sig. Proc.*, **40**, 2207–2232.

Invited Papers

Mortality Pattern Prediction in Worker Cohorts

Kryštof Eben[1,2], Jiří Vondráček[1] & Keith Binks[2]

[1] Institute of Computer Science, Czech Academy of Sciences
Pod vodárenskou věží 2, 182 07 Prague, Czech Republic

[2] Westlakes Scientific Consulting, Moor Row, Cumbria CA24 3LN, England

1 Introduction

As an integral part of their occupational health care system, companies may be interested in the future disease-specific mortality patterns in their workforce. The mortality pattern is characterised by deaths from different competing disease groupings e.g . cancer of the respiratory system, cardiovascular disease, etc. In addition to monitoring the health of the workforce, mortality pattern predictions are central to the estimation of future pension, insurance or compensation liability.

In this paper, the main emphasis is upon the assessment of smoking habits of the workers. This can help to construct more precise mortality predictions over the long term.

We have analysed the mortality experience of the cohort of employees of British Nuclear Fuels plc (BNFL), who have worked at the Sellafield site since 1948.

Models for the age-time specific mortality patterns have been developed using information available for the workers, including smoking status and crude measure of socio-economic status.

Information on smoking habits is available only for a subgroup of the workers. For this subgroup smoking data are not necessarily complete or even comprehensive during the period of employment. For the remaining workers there is no information on their smoking habits. We are, therefore, faced with several types of missing information. For example, there can be several decades between a worker leaving employment and his subsequent death; during which time there is no information on his smoking habits. We have treated the process of smoking as a stochastic covariate; we have modelled the process of smoking cessation, and the influence of the possibly stopped smoking process, on the mortality pattern.

1.1 Information on smoking habits

Smoking habit information for workers has been recorded since 1948, but more frequent measurements began in the sixties. A period of relatively low coverage then followed, and since then, the majority of the information has been recorded throughout the eighties and nineties. Since 1948 smoking information has been available for 13,000 workers; however, over 50% of them contribute only one or two measurements.

Information on smoking habits is collected from regular medical examinations. It has the form of approximate number of cigarettes smoked per day, both at the time of the examination and in the past.

We assume that the data gathered in this way since 1948 are "missing at random" (Rubin, 1976). In this case, the proportion of smokers, never-smokers and ex-smokers should be the same among the workers with known smoking status, and those with smoking status unknown. Also, the intensity of smoking in smokers could be estimated on the basis of the information available. We will adopt this assumption hereafter.

2 Estimation of parameters and construction of the prediction function

Consider a smoking-related disease, e.g. lung cancer. We have to construct a partial likelihood for the parameters in a model of mortality of the disease and to predict the future numbers of deaths. We shall suppose that smoking affects mortality through the cumulative number of cigarettes smoked up to the present time.†

The cumulative number of cigarettes is a stochastic covariate, only measurable during the lifetime of the worker and correlated with lifetime.

If the cumulative number of cigarettes is missing, it is not possible to replace it by a satisfactory surrogate covariate for several reasons. In particular, there would be no adjustment for smoking cessation. For the Sellafield cohort we have found that over 40 percent of smokers stop smoking during their lifetime. When constructing a surrogate covariate, one has to extrapolate the last known intensity of smoking sometimes 20 or more years to the future. Such a surrogate causes an overestimate of risk in workers who stop smoking after they leave employment or retire and whose true risk is considerably lower. Consequently, the presence of survivors with high estimated risk can cause underestimation of the corresponding parameter in the model, i.e. of the effect of smoking. We shall see later that we can estimate the intensity of stopping smoking from the database and adjust for that effect.

In what follows, we shall construct the prediction functions for mortality patterns along the lines of Jewell & Nielsen (1993).

Our principal time scale t is age. Calendar time will enter into the models as a covariate. Time $t = 0$ means in practice an age of 20, since there is a very low number of deaths under this age and the age of 20 is the mean reported age of starting smoking in our cohort.

For a worker, let t^* be the date of his last dated smoking information, if such a date exists. We have three categories of workers in the database. For each category we have to construct a different prediction function:

a) a current smoker at t^*. Here we have to adjust for possible smoking cessation.
b) an ex-smoker at t^*. We do not allow anyone to resume smoking in our models. The prediction function is constructed in a standard manner. In our model we identify a never-smoker by having zero intensity of smoking. Therefore, never-smokers arise as "ex-smokers" with $t^* = 0$.
c) There is no smoking information at all. Here we have to base the prediction function on the percentage of never-smokers and age- and time-specific intensities of smoking in the cohort.

---

† For some diseases, lung cancer especially, a latency period follows after exposure to cigarette smoke. Therefore, for practical purposes the cumulative number of cigarettes need to be lagged.

2.1 Prediction functions for current smokers at t^*

We start with case a); the computations are generalizations of the material presented in Jewell & Kalbfleisch (1992). For a current smoker at t^*, let τ denote the random instant of possible smoking cessation, and $X(.)$ the cumulative number of cigarettes. We suppose that X is a counting process (Andersen *et al.*,1992, Section II.4). Its intensity $\mu(.)$ is itself a stochastic quantity varying between workers. We suppose the process X is known from the beginning until t^*, thus ignoring the error and missing observations before t^*. X is unobserved after t^*. Many workers tend to keep their average number of cigarettes smoked per day fixed. For the remaining workers the changes in the average number of cigarettes per day are hardly predictable. Therefore we assume $\mu(.)$ to be known after t^* and for practice we freeze $\mu(.)$ at its last known value.

Let $X^\tau(.)$ be the stopped process, with stopping time τ; i.e. $X^\tau(u) = X(\min(\tau, u))$. Further, let $\bar{X}(t)$ be the trajectory of X from 0 to t:

$$\bar{X}(t) = \{X(u), 0 \leq u \leq t\}.$$

We shall consider I different causes of death (in practice they will be groups of causes of death). For an individual j and death cause i, let $N_{ij}(t)$ be the usual counting process which jumps if the individual j dies of cause i at time t. Let $\lambda_{ij}(t)$ be the mortality rate corresponding to $N_{ij}(t)$. We are interested in forecasting the aggregated processes $N_{i.}(t) = \sum_{j=1}^{J} N_{ij}(t)$ for causes $i = 1, ..., I$ in the cohort of J individuals. In what follows we shall suppress the index j whenever possible.

Definition. The mortality pattern of an individual is the vector $\boldsymbol{\lambda}(t) = (\lambda_1(t), ..., \lambda_I(t))'$ of mortality rates corresponding to the vector of counting processes $\boldsymbol{N}(t) = (N_1(t), ..., N_I(t))'$. ∎

We adopt the following additive relative risk model for the effect of smoking:

$$\lambda_i(t) \equiv \lambda_i(t|X) = \lambda_{0i}(t)(1 + \beta_i X^\tau(t) e^{-\gamma_i (t-\tau)^+}), \quad i = 1, ..., I, \tag{1}$$

where $\lambda_{0i}(t)$ is a suitable baseline hazard. In other words, the excess relative risk is proportional to the cumulative number of cigarettes in current smokers and is damped exponentially in ex-smokers. The constant γ_i can be chosen so as to cancel the excess relative risk after a sufficiently long period without smoking. We do not explicitly write other covariates and use only one time scale for the present.

We shall now construct the prediction function for the mortality patterns of the members of the cohort. The horizon of prediction will be $s \geq 0$. As a special case $s = 0$ we get the partial likelihood for estimation of the parameters.

Let $T_1, .., T_I$ be potential survival times, conditionally independent when the trajectory of the smoking process is fixed. Denote $T = \min(T_1, ..., T_I)$ the actual overall survival time.

Let F_{t+s} be the sigma-algebra generated by the sample paths of X up to $t+s$ and the indicator variable $I(T > t+s)$ of overall survival T up to $t+s$. For the conditional expectation $\lambda_{iF_{t+s}}(t+s) \equiv \mathrm{E}[\lambda_i(t+s)|F_{t+s}]$ we have

$$\lambda_{iF_{t+s}}(t+s) = \lambda_i(t+s) I(T > t+s).$$

Let $\tilde{F}_t$ be a sigma-algebra generated by $\bar{X}(t^*)$ and $I(T > t)$, i.e. by the observed sample paths of X up to t^* and the indicator variable of overall survival T up to t.

A forecast of the mortality pattern is a vector of prediction functions in the sense of Jewell & Nielsen (1993) , Theorem 1:

$$\mathbf{f}^*(s,\tilde{F}_t) = \mathrm{E}[\boldsymbol{\lambda}_{F_{t+s}}(t+s)|\tilde{F}_t], \tag{2}$$

where $\boldsymbol{\lambda}_{F_{t+s}}(t+s) = (\lambda_{1F_{t+s}}(t+s), ..., \lambda_{IF_{t+s}}(t+s))'$. The prediction functions are densities of residual lifetime after t, given the information $\bar{X}^T(t^*, t+s)$ on smoking.

In practice we measure time in years and want to estimate the vector $\mathbf{N}.(t+s) = (N_1.(t+s), ..., N_I.(t+s))'$, i.e. the number of deaths from causes $1, ..., I$, for future single years, for several years ahead. In our predictions for the Sellafield cohort we need a horizon of 15 years. The estimate is simply the sum of the individual forecast vectors over all workers in the cohort, who are alive at time t:

$$\hat{\mathbf{N}}.(t+s) = \sum_{j=1}^{J} \mathbf{f}_j^*(s,\tilde{F}_t).$$

Let $\bar{X}(t^*, t+s)$ be a trajectory of X, starting at t^* and ending at $t+s$. Let $\bar{X}^T(t^*, t+s)$ be a segment of that trajectory, starting at t^* and ending at $\min(T, t+s)$. We shall write $P(\bar{X}^T(t^*, t+s)|\bar{X}(t^*), T > t)$ for the probability measure on the segments. Obviously $\mathbf{f}^*(s,\tilde{F}_t) = 0$ when $T \leq t$ and from now on we shall deal with the distribution of segments, conditional on $\bar{X}(t^*)$ and $T > t$.

(2) can be written explicitly as

$$\mathbf{f}^*(s,\tilde{F}_t) = I(T > t) \int \boldsymbol{\lambda}(t+s) I(T > t+s) dP(\bar{X}^T(t^*, t+s)|\bar{X}(t^*), T > t),$$

where $\boldsymbol{\lambda}(t+s) = (\lambda_1(t+s), ..., \lambda_I(t+s))'$.

As usual we write $\Lambda_i(t) = \int_0^t \lambda_i(u)du$ for the cumulative hazard of the i–th cause and $\Lambda_.(t) = \sum_{i=1}^{I} \Lambda_i(t)$ for the cumulative hazard of overall survival T, similarly for $\Lambda_{0.}$ etc.. We shall now factorize the distribution on the segments in order to compute the prediction function.

Each segment of a trajectory of X has the following “building elements”:

- the overall survival T which gives the endpoint $\min(T, t+s)$
- the stopping time τ of the smoking process and the derived variable $\xi = \min(\tau, t+s)$, which has the same distribution as τ except for an atom in $t+s$
- the number m of jumps of the counting process X between t^* and ξ and the vector of their instants $\boldsymbol{w} = (w_1, ..., w_m)$.

Thus we write

$$\begin{aligned}
&dP(\bar{X}^T(t^*, t+s)|\bar{X}(t^*), T > t) \equiv dP(T, \xi, m, \boldsymbol{w}|\bar{X}(t^*), T > t)\\
&= dP(T|\xi, m, \boldsymbol{w}, \bar{X}(t^*), T > t) dP(\xi, m, \boldsymbol{w}|\bar{X}(t^*), T > t)\\
&= dP(T|\xi, m, \boldsymbol{w}, \bar{X}(t^*), T > t) \frac{1}{P(T > t|\bar{X}(t^*))}\\
&\quad \times P(T > t|\xi, m, \boldsymbol{w}, \bar{X}(t^*)) dP(m, \boldsymbol{w}|\xi, \bar{X}(t^*)) dP(\xi|\bar{X}(t^*))\\
&= \frac{1}{P(T > t|\bar{X}(t^*))} dP(T|\xi, m, \boldsymbol{w}, \bar{X}(t^*), T > t)\\
&\quad \times e^{-\Lambda_.(t)} dP(m, \boldsymbol{w}|\xi, \bar{X}(t^*)) dP(\xi|\bar{X}(t^*)).
\end{aligned} \tag{3}$$

Returning back to the prediction function, we have

$$\begin{aligned}
\mathbf{f}^*(s, \tilde{F}_t) &= \frac{1}{P(T > t|\bar{X}(t^*))} \\
&\quad \times \int_{\xi,m,\boldsymbol{w}} \boldsymbol{\lambda}(t+s) \int_{T=t}^{\infty} I(T > t+s) dP(T|\xi, m, \boldsymbol{w}, \bar{X}(t^*), T > t) \\
&\quad \times e^{-\Lambda.(t)} dP(m, \boldsymbol{w}|\xi, \bar{X}(t^*)) dP(\xi|\bar{X}(t^*)) \\
&= \frac{1}{P(T > t|\bar{X}(t^*))} \int_{\xi,m,\boldsymbol{w}} \boldsymbol{\lambda}(t+s) e^{-(\Lambda.(t+s) - \Lambda.(t))} \qquad (4) \\
&\quad \times e^{-\Lambda.(t)} dP(m, \boldsymbol{w}|\xi, \bar{X}(t^*)) dP(\xi|\bar{X}(t^*)) \\
&= \frac{1}{P(T > t|\bar{X}(t^*))} \int_{\xi} \int_{m,\boldsymbol{w}} \boldsymbol{\lambda}(t+s) e^{-\Lambda.(t+s)} dP(m, \boldsymbol{w}|\xi, \bar{X}(t^*)) dP(\xi|\bar{X}(t^*))
\end{aligned}$$

The distribution $P(m, \boldsymbol{w}|\xi, \bar{X}(t^*))$ is the ordinary likelihood associated with the counting process X with intensity $\mu(.)$, observed in the interval $(t^*, \xi]$ (see (1) and (2) in Lindsey,1995 or Andersen *et al.*,1992, p.223) :

$$dP(m, \boldsymbol{w}|\xi, \bar{X}(t^*)) = (\prod_{j=1}^{m} \mu(w_j)) e^{-\int_{t^*}^{\xi} \mu(v) dv}$$

Therefore the inner integral in (4) equals

$$\begin{aligned}
&\mathrm{E}_{m,\boldsymbol{w}}[\boldsymbol{\lambda}(t+s) e^{-\Lambda.(t+s)}] \qquad (5) \\
&\quad = \sum_{m=0}^{\infty} \int_{t^* \le w_1 < \ldots < w_m \le \xi} \boldsymbol{\lambda}(t+s) e^{-\Lambda.(t+s)} (\prod_{j=1}^{m} \mu(w_j)) e^{-\int_{t^*}^{\xi} \mu(v) dv} d\boldsymbol{w}.
\end{aligned}$$

The computation of (5) is a slight generalization of Jewell & Kalbfleisch (1992) and we defer it to the appendix. The result is stated in the following proposition.

Proposition. Let the intensities λ_i be given by (1). Then

$$\begin{aligned}
&\mathrm{E}_{m,\boldsymbol{w}}[\lambda_i(t+s) e^{-\Lambda.(t+s)}] \\
&= e^{-\int_{t^*}^{\xi} \mu(v) dv} e^{-\Lambda.(t^*)} e^{-(\Lambda_0.(t+s) - \Lambda_0.(t^*) + X(t^*)(R_0(t+s) - R_0(t^*)))} \\
&\quad \times \lambda_{0i}(t+s) e^{Q(t^*, t+s, \xi, \boldsymbol{\beta})} [1 + \beta_i e^{-\gamma_i(t+s-\xi)} (X(t^*) + Q(t^*, t+s, \xi, \boldsymbol{\beta}))],
\end{aligned}$$

where $\quad \boldsymbol{\beta} = (\beta_1, ..., \beta_I)', \qquad R_0(t) = \int_0^t \sum_{i=1}^{I} \beta_i \lambda_{0i}(u) e^{-\gamma_i (u-\xi)^+} du$

and

$$Q \equiv Q(t^*, t+s, \xi, \boldsymbol{\beta}) = \exp\{-(R_0(t+s) - R_0(\xi))\} \int_{v=t^*}^{\xi} \mu(v) e^{-(R_0(\xi) - R_0(v))} dv.$$

■

It remains to integrate $\mathrm{E}_{m,\boldsymbol{w}}[\lambda_i(t+s)e^{-\Lambda_{\cdot}(t+s)}]$ over ξ and to compute the probability $P(T > t|\bar{X}(t^*))$. To this point, consider prediction functions with zero prediction horizon, $s = 0$. We can repeat the steps starting from (3) for this special case, i.e. for segments of trajectories starting at t^* and ending at t. To avoid any confusion, we put

$$\zeta = \min(t, \tau),$$

ζ now plays the role of ξ. It is

$$\begin{aligned} P(T > t|\bar{X}(t^*)) &= \int_{\zeta,m,\boldsymbol{w}} dP(T > t, \zeta, m, \boldsymbol{w}|\bar{X}(t^*)) \\ &= \int_{\zeta}\Big(\int_{m,\boldsymbol{w}} e^{-\Lambda_{\cdot}(t)} dP(m, \boldsymbol{w}|\zeta, \bar{X}(t^*))\Big) dP(\zeta|\bar{X}(t^*)) \\ &= \int_{\zeta} \mathrm{E}_{m,\boldsymbol{w}}[e^{-\Lambda_{\cdot}(t)}] dP(\zeta|\bar{X}(t^*)). \end{aligned} \tag{6}$$

Identical steps which lead from (A3) to (A4) (see Appendix) give

$$\begin{aligned} &\mathrm{E}_{m,\boldsymbol{w}}[e^{-\Lambda_{\cdot}(t)}] = e^{-\int_{t^*}^{\zeta}\mu(v)dv} e^{-\Lambda_{\cdot}(t^*)} e^{-\sum_{i=1}^{I}(1+\beta_i X(t^*))(\Lambda_{0i}(\zeta)-\Lambda_{0i}(t^*))} \\ &\quad \times \exp\{-[\Lambda_{0\cdot}(t) - \Lambda_{0\cdot}(\zeta) + X(t^*)(R_0(t) - R_0(\zeta))]\} e^{Q(t^*,t,\zeta,\boldsymbol{\beta})} \\ &= e^{-\int_{t^*}^{\zeta}\mu(v)dv} e^{-\Lambda_{\cdot}(t^*)} e^{-(\Lambda_{0\cdot}(t)-\Lambda_{0\cdot}(t^*)+X(t^*)(R_0(t)-R_0(t^*))} e^{Q(t^*,t,\zeta,\boldsymbol{\beta})} \end{aligned} \tag{7}$$

Summarizing our results, we obtained

$$\mathbf{f}^*(s, \tilde{F}_t) = \frac{I(T > t)}{P(T > t|\bar{X}(t^*))} \int_{\xi} \mathrm{E}_{m,\boldsymbol{w}}[\boldsymbol{\lambda}(t+s)e^{-\Lambda_{\cdot}(t+s)}] dP(\xi|\bar{X}(t^*)), \tag{8}$$

where $\mathrm{E}_{m,\boldsymbol{w}}[\boldsymbol{\lambda}(t+s)e^{-\Lambda_{\cdot}(t+s)}]$ is given by the Proposition and $P(T > t|\bar{X}(t^*))$ is given by (6) and (7).

2.2 Prediction functions for ex-smokers and when no smoking information is available

Computing prediction functions of ex-smokers, i.e. workers who have stopped smoking at time τ before t^*, presents no difficulty. It will be based simply on the hazard

$$\lambda_i(t|X) = \lambda_{0i}(t)(1 + \beta_i X(\tau)e^{-\gamma(t-\tau)^+}),$$

where now $\tau < t^*$ is known. The σ-algebra $\tilde{F}_t$ equals F_t and we suppose we know the whole trajectory of $X^{\tau}(t)$. Thus

$$\mathbf{f}^*(s, \tilde{F}_t) = I(T > t)\frac{\boldsymbol{\lambda}(t+s)}{e^{-\Lambda_{\cdot}(t)}} e^{-\Lambda_{\cdot}(t+s)},$$

i.e. the prediction function equals the density of residual lifetime under the competing risks $1, ..., I$ (see Jewell & Nielsen, 1993, Example 1).

Finally, we consider the case of workers who do not have any smoking related information available. For them $t^* = 0$, $X(t^*) = 0$ and the σ-algebra F_{t+s} has different generators:

1) the indicator variable $I(T > t+s)$ as before

2) the indicator variable of being a never-smoker at time 0

3) the trajectories $\mu(v), 0 \le v \le \min(\tau, T, t+s)$ of the smoking intensity.

In Section 2 we supposed that $\mu(.)$ is fixed and known even in the future (due to the latency period, this is in some cases true at least for a few years in advance), but we admitted the possibility of smoking cessation. For practical purposes, we have now to shrink the set of possible trajectories of smoking intensity. We shall suppose that $\mu(v) = \mu(1) = \mu$ is constant up to the possible instant τ of smoking cessation. Thus F_{t+s} is generated by the random variables $I(T > t+s), I(\text{never-smoker}), \mu, \tau$. In our model being a never-smoker is equivalent to $\mu = 0$ (although including a further indicator covariate can be considered). The generator under 2) thus contributes to the distribution of μ an atom in $\mu = 0$ of probability $P(\text{never-smoker})$.

Finally we have to suppose that the smoking cessation instant τ and μ are independent, so that also ξ and μ are independent.

The factorization (3) is now

$$
\begin{aligned}
& dP(\bar{X}^T(0,t+s)|T>t) \equiv dP(T,\xi,m,\boldsymbol{w},\mu|T>t) \\
&= dP(T|\xi,m,\boldsymbol{w},\mu,T>t)dP(\xi,m,\boldsymbol{w},\mu|T>t) \\
&= dP(T|\xi,m,\boldsymbol{w},\mu,T>t)\frac{1}{P(T>t)}P(T>t|\xi,m,\boldsymbol{w},\mu)dP(m,\boldsymbol{w},\xi,\mu) \\
&= \frac{1}{P(T>t)}dP(T|\xi,m,\boldsymbol{w},\mu,T>t)e^{-\Lambda_{.}(t)}dP(m,\boldsymbol{w}|\xi,\mu)dP(\xi)dP(\mu).
\end{aligned}
$$

The expectation $\mathrm{E}_{m,\boldsymbol{w}}[\boldsymbol{\lambda}(t+s)e^{-\Lambda_{.}(t+s)}]$ is taken conditionally on (μ,ξ) and it has elements

$$
\begin{aligned}
& \mathrm{E}_{m,\boldsymbol{w}}[\lambda_i(t+s)e^{-\Lambda_{.}(t+s)}] \\
&= e^{-\int_0^{\xi}\mu(v)dv}e^{-\sum_{k=1}^{I}\Lambda_{0k}(\xi)}\exp\{-[\Lambda_{0.}(t+s)-\Lambda_{0.}(\xi)]\} \\
&\quad \times \lambda_{0i}(t+s)e^{Q(0,t+s,\xi,\boldsymbol{\beta})}[1+\beta_i e^{-\gamma_i(t+s-\xi)}Q(0,t+s,\xi,\boldsymbol{\beta})] \\
&= e^{-\int_0^{\xi}\mu(v)dv}e^{-\Lambda_{0.}(t+s)} \\
&\quad \times \lambda_{0i}(t+s)e^{Q(0,t+s,\xi,\boldsymbol{\beta})}[1+\beta_i e^{-\gamma_i(t+s-\xi)}Q(0,t+s,\xi,\boldsymbol{\beta})].
\end{aligned}
$$

Similarly as in (6) and (7) we have for the marginal survival function $P(T > t)$ when no smoking information is available

$$\int_{\zeta,\mu}\int_{m,\boldsymbol{w},} e^{-\Lambda_{\cdot}(t)} dP(m,\boldsymbol{w}|\zeta,\mu)dP(\zeta)dP(\mu) \\ = \int_{\zeta,\mu} e^{-\int_0^\zeta \mu(v)dv} e^{-\Lambda_{0\cdot}(t)} e^{Q(0,t,\zeta,\beta)} dP(\zeta)dP(\mu). \tag{9}$$

Altogether

$$\mathbf{f}^*(s,\tilde{F}_t) = I(T>t)\frac{1}{P(T>t)}\int_{\xi,\mu} \mathrm{E}_{m,\boldsymbol{w}}[\boldsymbol{\lambda}(t+s)e^{-\Lambda_{\cdot}(t+s)}]dP(\xi)dP(\mu),$$

where $\mathrm{E}_{m,\boldsymbol{w}}[\boldsymbol{\lambda}(t+s)e^{-\Lambda_{\cdot}(t+s)}]$ and $P(T>t)$ have been redefined above.

3 Computational aspects

As is usual in the analysis of mortality rates, we work with yearly data. We use the unit of one year for both age and calendar time.

To fit the above model, we utilize the approach suggested by Lindsey (1996). The problem is treated as a Poisson regression with hazard functions outside the class of generalized linear models (c.f. Section 6 in Lindsey's paper). Such a regression can be accomplished e.g. by SAS procedure MODEL. The integrals appear as sums.

For the workers with no smoking information at all, the distribution of the intensity μ as well as the probability of being a never-smoker, both at the age of 20, has to be estimated. These estimates have been provided from the part of the worker cohort where some information was available. To reduce the amount of computations, we assume that μ is gamma-distributed. We can then change the order of integration in (9) and integrate first over μ, obtaining explicit expressions depending on the parameters of the gamma distribution. Thus μ plays the role of frailty with a known distribution.

The probability of being a never-smoker at the age of 20, the intensity of stopping smoking as well as the parameters of the gamma distribution depend on calendar time. Moreover, there are issues related to calendar time which are independent of smoking, e.g. the decline of number of deaths from cardiovascular diseases after intensive care units started to function in the seventies. These effects have to be represented by suitable covariates and included into the model.

3.1 An illustrative example

We illustrate the method on a simplified analysis of mortality experience of the cohort of employees of British Nuclear Fuels plc (BNFL), who have worked at the Sellafield site since 1948. For the purpose of this example we grouped the death causes into the following rough categories:

- cancer of respiratory system
- diseases of circulatory system
- other diseases.

The death rates in the first two categories are going to be influenced by smoking represented as above by the cumulative number of cigarettes, lagged 5 years, whereas "other" diseases were treated as smoking-independent. The effect of smoking was found to be statistically significant.

We formed the following covariates to model time trends:

- a power function of birth cohort (year of birth) entering into the model as a proportional factor was used to explain the trend in mortality from "other" diseases. This corresponds to a steady decline in overall mortality.
- a sigmoid function (arctan) was used to model the intervention which caused the decline in mortality from diseases of circulatory system in the eighties.

No calendar time trend was used to model mortality from cancers of the respiratory system; time trends have been modelled exclusively as trends in smoking habits.

Further, all causes were influenced by a fixed covariate, available in all workers, namely a crude measure of socioeconomic status distinguishing the "industrial" and "non-industrial" workers.

The baseline was chosen as Gompertz in accordance with the common use in demography. The fit was done for the period 1968-1995.

The figures below show the observed and fitted age-specific mortality rates and observed, fitted and predicted absolute numbers of deaths from the three causes above.

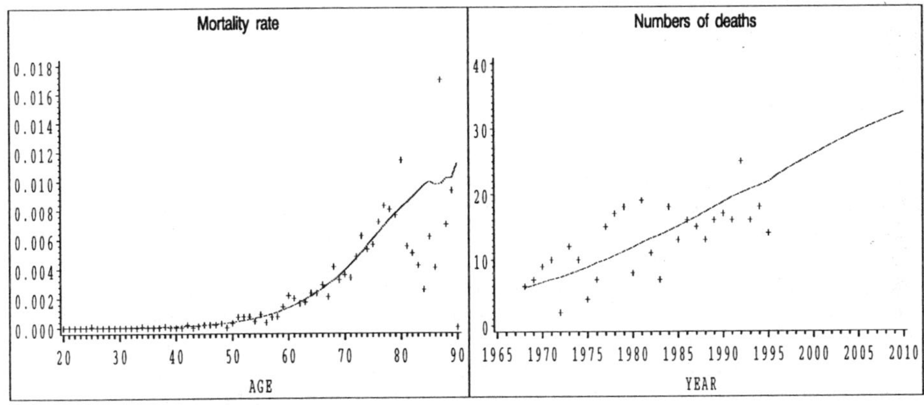

Fig 1. Cancer of respiratory system

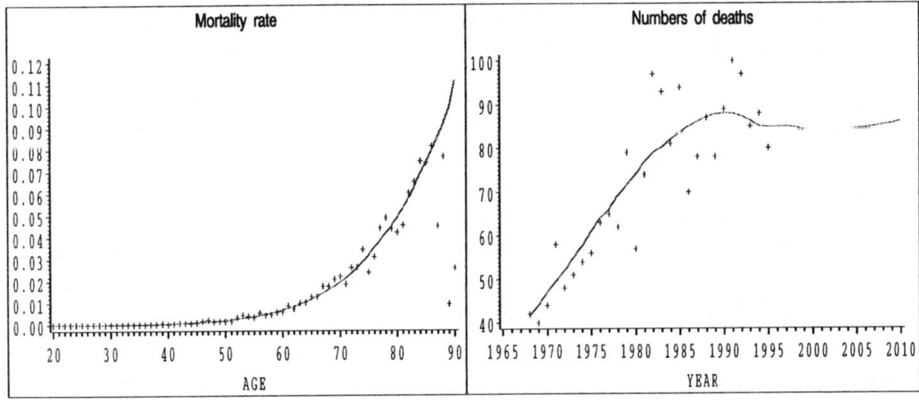

Fig 2. Diseases of circulatory system

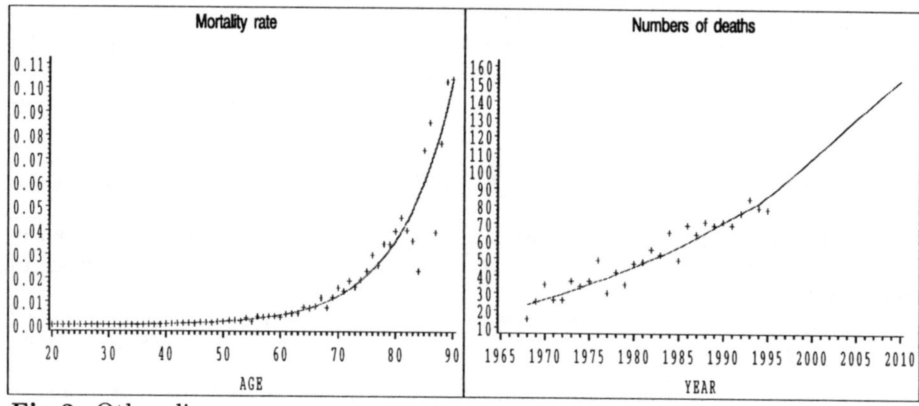

Fig 3. Other diseases

The forecasts are derived from individual risks of each worker in the cohort. They reflect the age structure of the cohort and on other conditions specific for the company. This is useful for the purposes of management, e.g. for company planning of healthcare. This type of analysis, however, is not intended to provide precise quantification of adverse effect of smoking on health.

4 Conclusion

For practical purposes other variables have to be taken into account. In particular occupational radiation doses of the nuclear facility workers, the time since entry (or other representation of the "healthy worker effect") and other variables may be included in the models. Also, a much finer grouping of death causes is necessary.

We used the same method to analyse the mortality experience of the population of England and Wales. This provides an indispensable reference data, since there are low numbers of cases in the cohort in the oldest-old age category. Moreover, calendar time trends are better identifiable. Combining results for the population and for the cohort is a matter of further investigation.

The model appears to be a reasonable tool for obtaining prediction of the mortality patterns and modelling the effect of smoking. Its control parameters have a natural interpretation and may be predicted with the help of expert information. Its nonlinearity, however, requires a careful analysis and tuning of all the control parameters in order to obtain reliable results.

References

Andersen, P. K., Borgan, Oernulf, Gill, R. D. & Keiding, N. (1993). *Statistical models based on counting processes. Springer Series in Statistics.* New York: Springer Verlag.

Jewell, N. P. & Kalbfleisch, J.D. (1992). Marker models in survival analysis and applications to issues associated with AIDS. In: *AIDS Epidemiology: Methodological Issues.* (ed. N.P. Jewell, K. Dietz & V. Farewell), 231-255. Boston: Birkhäuser.

Jewell, N. P. & Nielsen, J.P. (1993). A framework for consistent prediction rule based on markers.*Biometrika,* **80**, 153-164.

Lindsey, J.K. (1995). Fitting Parametric Counting Processes by using Log-linear Models. *Applied Statistics,* **44**, 201-212.

Rubin, D.B. (1976): Inference and missing data. *Biometrika,* **63**, 581-92.

Appendix. The computations leading to the Proposition of Section 2.1. Recall that we have to integrate over the number m of jumps of X in instants $w_1, ..., w_m$ between t^* and ξ, and over their instants $w_1, ..., w_m$. For our model with intensities given by (1) we have

$$\Lambda_.(t+s) = \Lambda_.(t^*)$$
$$+ \int_{t^*}^{t+s} \sum_{i=1}^{I} [\lambda_{0i}(u)\left(1 + \beta_i e^{-\gamma_i (u-\xi)^+} (X(t^*) + \sum_{j=1}^{m} I(w_j \leq u))\right)]du.$$

Splitting the region of integration in ξ and having in mind that $e^{-\gamma_i(u-\xi)^+} = 1$ for $u \leq \xi$, we come to

$$\Lambda_.(t+s) = \Lambda_.(t^*) + \sum_{i=1}^{I} (1 + \beta_i X(t^*))(\Lambda_{0i}(\xi) - \Lambda_{0i}(t^*))$$
$$+ \sum_{i=1}^{I} \int_{\xi}^{t+s} \lambda_{0i}(u)(1 + \beta_i e^{-\gamma_i(u-\xi)} X(t^*))du \qquad \text{(A1)}$$
$$+ \sum_{i=1}^{I} \beta_i \int_{t^*}^{t+s} \lambda_{0i}(u) e^{-\gamma_i(u-\xi)^+} \sum_{j=1}^{m} I(w_j \leq u)du.$$

The last summand in (A1) can be further rewritten as

$$\sum_{i=1}^{I} \beta_i \sum_{j=1}^{m} \int_{t^*}^{t+s} \lambda_{0i}(u) e^{-\gamma_i(u-\xi)^+} I(w_j \leq u)du$$
$$= \sum_{i=1}^{I} \beta_i \sum_{j=1}^{m} \int_{w_j}^{t+s} \lambda_{0i}(u) e^{-\gamma_i(u-\xi)^+} du$$
$$= \sum_{i=1}^{I} \beta_i \sum_{j=1}^{m} (\int_{w_j}^{\xi} \lambda_{0i}(u)du + \int_{\xi}^{t+s} \lambda_{0i}(u) e^{-\gamma_i(u-\xi)} du) \qquad \text{(A2)}$$
$$= \sum_{i=1}^{I} \beta_i (m \int_{\xi}^{t+s} \lambda_{0i}(u) e^{-\gamma_i(u-\xi)} du + \sum_{j=1}^{m} \int_{w_j}^{\xi} \lambda_{0i}(u)du).$$

Inserting from (A1) and (A2) into (5) in Section 2.1, we get

$$\mathrm{E}_{m,w}[\lambda(t+s) e^{-\Lambda_.(t+s)}] = e^{-\Lambda_.(t^*)} e^{-\sum_{i=1}^{I} (1+\beta_i X(t^*))(\Lambda_{0i}(\xi) - \Lambda_{0i}(t^*))}$$
$$\times \exp\{-\sum_{i=1}^{I} \int_{\xi}^{t+s} \lambda_{0i}(u)(1 + \beta_i e^{-\gamma_i(u-\xi)} X(t^*))du\} \qquad \text{(A3)}$$
$$\times \sum_{m=0}^{\infty} \lambda(t+s) \exp\{-m \int_{\xi}^{t+s} \sum_{i=1}^{I} \beta_i \lambda_{0i}(u) e^{-\gamma_i(u-\xi)} du\}$$
$$\times \int_{t^* \leq w_1 < ... < w_m \leq \xi} \exp\{-\sum_{j=1}^{m} \int_{w_j}^{\xi} \sum_{i=1}^{I} \beta_i \lambda_{0i}(u)du\} (\prod_{j=1}^{m} \mu(w_j)) e^{-\int_{t^*}^{\xi} \mu(v)dv} dw.$$

Put

$$R_0(t) = \int_0^t \sum_{i=1}^{I} \beta_i \lambda_{0i}(u) e^{-\gamma_i (u-\xi)^+} du.$$

Similarly as in Jewell & Kalbfleisch (1992) we can use the symmetry of the integrand with respect to permutations of $w_1, .., w_m$ and we obtain for the integral over $\boldsymbol{w}$ in (A3)

$$\int_{t^* \le w_1 < \ldots < w_m \le \xi} \left(\prod_{j=1}^{m} \mu(w_j) \exp\{-(R_0(\xi) - R_0(w_j))\} \right) e^{-\int_{t^*}^{\xi} \mu(v)dv} d\boldsymbol{w}$$
$$= e^{-\int_{t^*}^{\xi} \mu(v)dv} \frac{1}{m!} \left[\int_{v=t^*}^{\xi} \mu(v) e^{-(R_0(\xi) - R_0(v))} \right]^m .$$

Put

$$Q \equiv Q(t^*, t+s, \xi, \boldsymbol{\beta}) = \exp\{-(R_0(t+s) - R_0(\xi))\} \int_{v=t^*}^{\xi} \mu(v) e^{-(R_0(\xi) - R_0(v))} dv.$$

Then the sum over m in (A3) equals

$$e^{-\int_{t^*}^{\xi} \mu(v)dv} \sum_{m=0}^{\infty} \boldsymbol{\lambda}(t+s) \frac{Q^m}{m!}.$$

The vector $\sum_{m=0}^{\infty} \boldsymbol{\lambda}(t+s) \frac{Q^m}{m!}$ has elements

$$\sum_{m=0}^{\infty} \lambda_{0i}(t+s)[1 + \beta_i e^{-\gamma_i(t+s-\xi)}(X(t^*) + m)] \frac{Q^m}{m!}$$
$$= \lambda_{0i}(t+s)[(1 + \beta_i e^{-\gamma_i(t+s-\xi)} X(t^*)) e^Q + \beta_i e^{-\gamma_i(t+s-\xi)} Q e^Q]$$
$$= \lambda_{0i}(t+s) e^Q [1 + \beta_i e^{-\gamma_i(t+s-\xi)}(X(t^*) + Q)].$$

Altogether we have for the elements of the vector $\mathrm{E}_{m,\boldsymbol{w}}[\boldsymbol{\lambda}(t+s) e^{-\Lambda.(t+s)}]$:

$$\begin{aligned}
&\mathrm{E}_{m,\boldsymbol{w}}[\lambda_i(t+s) e^{-\Lambda.(t+s)}] \\
&= e^{-\int_{t^*}^{\xi} \mu(v)dv} e^{-\Lambda.(t^*)} e^{-\sum_{k=1}^{I} (1+\beta_k X(t^*))(\Lambda_{0k}(\xi) - \Lambda_{0k}(t^*))} \\
&\quad \times \exp\{-[\Lambda_{0.}(t+s) - \Lambda_{0.}(\xi) + X(t^*)(R_0(t+s) - R_0(\xi))]\} \\
&\quad \times \lambda_{0i}(t+s) e^{Q(t^*, t+s, \xi, \boldsymbol{\beta})} [1 + \beta_i e^{-\gamma_i(t+s-\xi)}(X(t^*) + Q(t^*, t+s, \xi, \boldsymbol{\beta}))] \\
&= e^{-\int_{t^*}^{\xi} \mu(v)dv} e^{-\Lambda.(t^*)} e^{-(\Lambda_{0.}(t+s) - \Lambda_{0.}(t^*) + X(t^*)(R_0(t+s) - R_0(t^*)))} \\
&\quad \times \lambda_{0i}(t+s) e^{Q(t^*, t+s, \xi, \boldsymbol{\beta})} [1 + \beta_i e^{-\gamma_i(t+s-\xi)}(X(t^*) + Q(t^*, t+s, \xi, \boldsymbol{\beta}))].
\end{aligned} \tag{A4}$$

Design Algorithms for Correlated Data

J. Eccleston and B. Chan

Department of Mathematics,
The University of Queensland,
Brisbane, QLD, 4072,
Australia

Abstract. The design of experiments when the data are correlated is an area of increasing research interest, particularly since methods for the analysis of experiments which incorporate a correlation structure for errors or over the plots are becoming more generally applied. In this paper the design of experiments of a two-dimensional layout (rows and columns) with a spatial process is investigated. Algorithms based on simulated annealing and Tabu search are developed for constructing optimal designs. The robustness of designs with respect to correlation structure is examined. A number of examples are considered including a practical example of an early generation variety trial example is given.

Keywords. Optimal design, correlated data, spatial processes, simulated annealing, Tabu

1 Introduction

The design of experiments when the data are correlated has been an area of increasing research interest over the last decade. Theoretical results have proved difficult to obtain, and there are few results available for many typical practical situations, for example, field trials for crop variety experiments and early generation trials which often have a large number of treatments and a two-dimensional plot layout.

An algorithmic procedure of constructing designs when the best design is not known or in non-standard situations, can have many benefits. An efficient design can be found for the particular situation at hand rather than trying to adapt the situation to a standard design. Also unequal replication and block sizes can be addressed and non-standard optimality criteria can be considered. Algorithmic methods have been widely utilised in the case of uncorrelated errors; for an extensive list of references see Martin & Eccleston (1997)

Algorithms for the construction of designs when the data are correlated according to a prescribed correlation structure have been proposed by Russell & Eccleston (1987a, b), Zergaw (1989), and Martin & Eccleston (1997). The last paper developed an algorithm based on simulated annealing (written in

Matlab). The other two are quite restrictive in the designs considered, and generally discuss only within block permutations in block designs. Here we consider a very general design situation, namely a two-dimensional layout and correlation structures that include spatial processes. The analysis, that is commonly referred to as spatial analysis, is via generalised least squares, see Cullis & Gleeson (1991), and more recently Gilmour *et al.* (1997).

In the next section some preliminary issues are dealt with including the model, the correlation structure and optimality criterion. Search routines and design algorithms are outlined in Section 3. Section 4 examines some examples and considers the robustness of the results. Here robustness refers to the optimality of a design when the assumed correlation structure is not the true structure - roughly speaking how removed from the true correlation structure can the assumed one be without the constructed design becoming very poor and inefficient. A practical application to an early generation trial is given in Section 5. A discussion of the two search methods and the designs obtained is given in the final section.

2 Preliminaries

There are many possible models that can be considered, however, here we concentrate on a generally accepted model for spatial analysis of field type experiments (Gilmour *et al.*, 1997). The model for a two-dimensional layout, a p by q row-column array, with row and column effects as well as the spatial process is assumed and is represented as follows:

$$E(Y) = X\tau$$

where Y is a $(n\times 1)$ vector of observations, X is the $(n\times v)$ design matrix (the treatment allocation to experimental plots) and τ a vector of v treatment effects, also $n = pq$. The correlation structure of the data is defined through the variance-covariance matrix of Y, namely Var(Y), and a general form is:

$$Var(Y) = \sigma^2_e I + \sigma^2_r R + \sigma^2_c C + \sigma^2_s V$$

where σ^2_e, σ^2_r, σ^2_c and σ^2_s are the respective variance components for random error, row effects, column effects and the spatial process. The matrices R and C describe the row and column layouts respectively, and I is an identity matrix of appropriate dimension. The observations are ordered by rows (arbitrarily) thus $R = I_p \otimes J_q$ and $C = J_p \otimes I_q$ where J_p is a $p\times p$ matrix of ones.

The correlation structure between plots is defined through the correlation matrix, V. The most commonly assumed form of V is a separable process, that is the kronecker product of the processes for rows and for columns. Here V is defined as $V_c \otimes V_r$ where V_r and V_c are the within row and column correlation structures respectively. Gilmour *et al.* (1997) recommend the use of auto-regressive processes of lag 1 for both components of V, that is the spatial process is described by an AR(1) by AR(1) separable process.

A strength of our approach is its flexibility with respect to the correlation process, since many can be included. The algorithms discussed can consider processes such as auto-regressive, moving average, linear variance and even independence for either or both components, also it is not necessary for rows and columns to have the same correlation structure. Block designs can easily be accommodated by a one-dimensional process, sometimes referred to as a linear process, simply by deleting R or C and assigning one of the components of V as an identity matrix. If one wishes to concentrate only on the spatial process then Var(Y) becomes σ^2_sV alone.

The generalised least squares analysis gives as an estimate of τ

$$\hat{\tau} = (X'\Lambda^{-1}X)^{-1}X'\Lambda^{-1}Y$$

with variance

$$\mathrm{Var}(\hat{\tau}) = \sigma^2(X'\Lambda^{-1}X)^{-1}$$

where Λ is Var(Y). The analysis is usually performed using the REML procedure, see Gilmour *et al.* (1997). Further to cater for an overall mean the variance-covariance matrix is adjusted by the usual sweep operation to $\Lambda^* = \Lambda^{-1}-\Lambda^{-1}\mathbf{1}\mathbf{1}'\Lambda^{-1}/(\mathbf{1}'\Lambda^{-1}\mathbf{1})$ where **1** is a vector of ones of appropriate dimension.

Obviously the relative values of the variance components σ^2_e, σ^2_r, σ^2_c and σ^2_s and the correlations, ρ_r and ρ_c, influence a design layout. Generally the variance components may not be known but the relativities between them may be able to be expressed as ratios with respect to σ^2_e.

The criterion we use to determine the best design is the well-known A-optimality, which minimises the average of the variances of all elementary contrasts between the treatments. This can be shown to be equivalent to minimising the trace of $(X'\Lambda^*X)^{-1}$. The A-value of a design is thus its trace$(X'\Lambda^*X)^{-1}$. If the major interest is in a particular set of contrasts $L'\tau$, as would be the case in an early generation trial where test treatments are compared to one or more controls, then A-optimality is easily adjusted to minimise the average variance of the contrasts $L'\tau$. This can be shown to be equivalent to minimising the trace of $L'(X'\Lambda^*X)^{-1}L$. If there is more than one control treatment then a linear combination of the several optimality values may be taken as an overall optimality measure.

Since for almost all cases we consider the optimal design is unknown, our goal is to find the design which has the best A-value for a specified correlation structure via a search algorithm. This requires a method to move from one design to another, which is done by interchanging the treatments between a pair of plots to obtain a new design. Such a procedure is computationally intensive and in many practical situations designs are much larger than those often considered in theoretical research, therefore it is necessary to employ search routines that are efficient and quick. We discuss two methods, one based on simulated annealing and the other Tabu, in the next section. Both of these procedures are quite simple and it is this that makes them so widely applicable. Simulated annealing or modified versions of it, have been used for finding optimal or near optimal block designs and row-column designs, see references above. However, the authors do not know of any applications of Tabu search to the design of experiments.

3 Simulated annealing and Tabu algorithms

The number of possible designs (for even small designs) is such that an exhaustive search is usually unrealistic and unmanageable. Clearly methods which efficiently search the design space (the set of all possible designs) are necessary. The algorithms presented here are in the context of a spatial process over a row-column experimental layout.

Once an initial design is constructed it is not difficult to devise a method to move to new designs by interchanging (swapping) treatments between plots. The problem of becoming stuck at a local optimal design which was encountered in early design algorithms for uncorrelated data (see Jones & Eccleston, 1980) can be avoided by using a procedure which allows designs to move away from local optima. We shall discuss briefly two such search methods; simulated annealing (Aarts & van Laarhoven, 1989) and Tabu (Glover, 1989, 1990), which although quite different in their approach are both very effective.

Simulated annealing has the ability to avoid being trapped in a local optimal design by accepting inferior designs on the basis of a well-defined probability rule. While Tabu is somewhat like a type of steepest descent routine but with a procedure for moving to a new design at each iteration and a rule to prohibit returning to a previously obtained design. The application of these methods to the construction of designs for correlated data is developed below.

3.1 Simulated annealing

Simulated annealing is a search method which has been widely used in recent years due to its efficiency in finding good solutions in a relatively short period of time. The algorithm described below is one of many possible variations on the theme of simulated annealing.

The objective function (denoted by f) to be minimised in the annealing process is the A-value of the design, the solution returned at the conclusion of the search will be the one with the best A-optimality found by the algorithm. Initially a starting design, D, is generated by randomly allocating treatments to the n plots in the design. This design is then perturbed to a new feasible solution, D′, by randomly interchanging two treatments. If D′ has a smaller A-value, then the operating design D is updated to D′. Otherwise the change in A-value is calculated and D′ is accepted with probability $exp(-\Delta f/t)$, where t is the annealing temperature associated with the objective function and Δf the change in the objective function (A-value). The temperature t is initially set high and is slowly decreased during the process according to some specified rule. The process of generating random perturbations continues while the best overall design is noted. When the temperature reaches some small number ε the annealing process stops and returns the best solution found.

The steepest descent follows at the conclusion of the annealing process. The best design found by the annealing process is used as the starting design in a steepest descent routine. Every possible pairwise swap of treatments of this design is considered, and for each the A-value is calculated, if there is an improvement, we update the best design and start the process again from that design. The process is repeated until eventually no improvements have been recorded in a complete run.

Prior to the start of the annealing algorithm the initial design is perturbed a large number of times to find the maximum consecutive difference in the A-value, denoted by *maxdiff*. The value *maxdiff* is used to determine the initial and final temperatures, which are given by $-maxdiff/ln(initprob)$ and $-maxdiff/ln(\varepsilon)$ respectively, for some specified values *initprob* and ε (both between 0 and 1).

The algorithm starts with the initial design, and a specified number of perturbations, M, are generated at each temperature, with designs accepted according to the above specified rule, and the temperature is decreased according to the cooling schedule. Two commonly used methods are *geometric* cooling, where $t_{i+1}=\alpha_g t_i$, i being the iteration number of the process and α_g is the cooling constant, and *Lundy* cooling, where $t_{i+1}=t_0/(1+i\alpha_L t_0/maxdiff)$. By varying the values of α_g and α_L the search may be hastened or slowed.

SIMULATED ANNEALING ROUTINE

1. Select an initial random design D and let $D^* = D$ where D^* denotes the best design currently found. Begin with t large.
2. Initialise the iteration number $i = 1$.
3. Generate a perturbation, D', of the operating design D by a random interchange of two treatments. Calculate Δf.
4. Let $D = D'$ with probability $exp(-\Delta f/t)$.
5. If D is the better than D^* then let $D^* = D$.
6. Increment i by 1 and return to step 3 until the specified number of iterations, M, have been executed.
7. Reduce the temperature. If the temperature reaches some small specified number ε then stop. Otherwise set $D = D^*$ and return to step 2.
8. Try all possible pairwise interchanges of treatments in D^*. If there is an improvement, select the best and update D^*. Repeat until no further improvement is obtained in a complete run.

3.2 Tabu search

Another search method which we have found to be effective in finding optimal designs is Tabu. Once an initial random design D is obtained, every possible pairwise swap of treatments in the design is considered and the corresponding

A-values are calculated. From these new designs, the best, say D′ that has not been previously accepted is selected. The operating design is updated to this new design D′ regardless of whether or not it is better, hence allowing for the possibility of escaping out of a local minimum. During the process, the best overall design, D^*, is noted. This process eventually stops when the iteration number reaches some specified number.

To be able to distinguish whether a design has been previously accepted by the algorithm, it is necessary to store all accepted designs under the Tabu set, T. A feasible solution is accepted only if it is not in the Tabu set.

The specified number of iterations executed before the process is stopped can vary greatly from design to design. Too small a value may cause optimal solutions not be obtained, too high a value may cause optimal solutions be found well before the process is stopped and hence a large amount of time is wasted. Unfortunately there appears to be no definitive method as to the best choice of this number, however, a reasonable guess may be found after some experimentation.

TABU SEARCH

1. Select an initial random design D and let $D^* = D$ where D^* denotes the best design currently found. Set the iteration counter k=1 and begin with T empty.
2. Investigate every possible pairwise swap of treatments in the operating design D and select $D' \notin T$ such that D′ is the best of these designs in terms of A-optimality.
3. Let D = D′. If D′ is better than D^* then let D^*=D′.
4. If the specified number of iterations has elapsed then stop. Otherwise set $T = T \cup D'$, k = k+1 and return to step 2.

4 Results

Nearest neighbours between treatments are believed to play an important role when data are spatially correlated, however, different correlation structures cause other properties to come into play. This is best illustrated with an example. The examples below have in general been obtained from either search algorithms.

4.1 Spatial process only

Consider an experiment with 5 treatments in a 5×5 array. In order to understand the effect of a spatial process on a design let us suppose the underlying correlation structure follows a separable AR(1)*AR(1) process only. (Examples including row and column variance components are considered later in this section.) Let ρ_r denote the correlation between rows and ρ_c the correlation between columns. We shall consider the cases when $\rho_r = \rho_c$ and $\rho_r \neq \rho_c$, and without loss of generality set $\sigma^2_s = 1$.

For case 1, $\rho_r = \rho_c = 0.5$ (say), design A is the optimal design found and has an A-value 0.2255 and for case 2 where $\rho_r \neq \rho_c$, let $\rho_r = 0.5$ and $\rho_c = 0.9$, the best A-value found is 0.0404 with B as the corresponding design:

(A)	(B)
5 3 5 3 2	3 1 5 1 4
2 5 3 2 3	2 3 1 4 3
4 2 1 3 4	3 2 4 3 5
2 1 5 4 1	2 1 2 5 4
1 5 4 1 4	1 2 5 4 5

Both designs, A and B, have 17 self-diagonal neighbours. A large number of self-diagonals has been observed for all designs we have considered under an AR(1)*AR(1) process. From the correlation matrix, it can be seen that elements corresponding to diagonal neighbours in the design have a positive influence on the information matrix $C = X'\Lambda^*X$. It follows that a large number of self-diagonals, and a certain degree of nearest neighbour balance, would lead to a good design. As well as these two properties, good designs tend to have treatments appearing as equally frequent as possible on corner plots and edge plots.

Suppose now that the correlation structure follows a separable MA(1)*MA(1) process. Then for the design of 5 treatments in a 5×5 array the following designs are found. For the first case where $\rho_r = \rho_c = 0.3$ (say) and the second case where $\rho_r \neq \rho_c$, let $\rho_r = 0.3$ and $\rho_c = 0.5$, the optimal A-values found are 0.4094 and 0.1567 respectively and are as follows:

(C)	(D)
3 1 3 4 3	5 4 5 4 5
1 3 2 3 5	3 1 3 1 3
4 2 1 5 2	2 5 2 5 2
1 4 2 1 5	4 3 4 3 4
4 5 4 5 2	1 2 1 2 1

The designs C and D are very different in structure compared to A and B. In general, optimal designs for MA(1)*MA(1) tend to have a large number of self-second neighbours (either in rows or in columns). In the case where $\rho_r = \rho_c$ self-diagonals and nearest neighbours continue to play, but in the second case where the correlations are different, the self-diagonals appear to be dominated by self second neighbours in the direction of the larger correlation. This can be seen from design C having 14 self-diagonals, 5 self-second neighbours in rows, 6 self-second neighbours in columns and design D having no self-diagonals, 15 self-second neighbours in rows and no self-second neighbours in columns.

Consider now the linear variance (LV) model with ψ_r and ψ_c as the parameter values between rows and columns respectively, as defined in Williams (1986), again for the design of 5 treatments in a 5×5 array. For the case where $\psi_r = \psi_c = 1$ (say), the best design found (E) has an A-value 2.2618. When $\psi_r \neq \psi_c$, let $\psi_r = 1$ and $\psi_c = 5$, then design F below is the best found and has an A-value of 4.9243:

3 2 5 1 4	4 5 2 1 3
2 4 3 5 1	5 4 1 3 2
1 3 4 2 5	4 1 5 2 3
4 5 1 3 2	2 4 3 5 1
5 4 2 1 3	1 2 4 3 5
(E)	(F)

Design E has 11 self-diagonals, no self-second neighbours in either direction, and design F has 14 self-diagonals, and no self-second neighbours. The LV model behaves similar to an AR(1)*AR(1) process, except that the number of self-diagonals is slightly less. (A similar comment was made by Martin, Eccleston & Jones (1998) regarding factorial designs.)

From the examples above, we can derive the following general results. Under an AR(1)*AR(1) structure, self-diagonals, nearest neighbour balance, corner balance, edge balance are all important factors in an optimal design. For a MA(1)*MA(1) process, self-second neighbours appear to be important, however in the case where row and column correlations are equal, self-diagonals and nearest neighbours play a role as well. For a LV model, again self-diagonals and nearest neighbour balance determine a good design, however, these properties are not as strong as that for an AR(1)*AR(1) process.

Often in practice, the correlation structure is either unknown or estimated with uncertainty. It would therefore be useful to have some sort of indication of how good or bad a design is when the assumed correlation structure is in fact quite different to the true structure. Two things can go wrong, either the assumed structure is correct but the correlations are incorrect or the assumed structure is totally different, allowing for either correct or incorrect correlation values. Efficiency tables can therefore be set up for optimal designs under the assumed structure against the true underlying structure, where efficiency of a design D is defined as the ratio of the A-value of the optimal design under the true structure over the A-value of design D under the same structure, where D is the design used under the assumed structure.

Efficiency tables are drawn for the 5×5 example with various combinations of row and column correlations and a summary of findings is presented in the table below.

As demonstrated in the above example, A-optimal designs are quite robust to various types of structures and correlations except in situations where a MA(1)*MA(1) process is involved with large correlations. This could be due to the nature of the MA(1) structure, because when the correlation gets close to a critical value, the correlation matrix becomes nearly negative definite.

For the case when the row and column numbers are quite different, say the number of columns is larger than the number of rows, the following findings were obtained for a 4×10 design with 4 treatments. Rows were either completely or near completely neighbour balanced under an AR(1)*AR(1) process in general. Optimal designs for an AR(1)*AR(1) process have extremely good efficiencies (very close to 100%) under the same structure but different correlations. Efficiency tables for all three structures follow similar trends as for the 5×5 example, except whenever a MA(1)*MA(1) process with high correlations is involved, the efficiencies can drop as low as 15%.

True Underlying Structure	Assumed Underlying Structure
AR(1)*AR(1) ($0.1 \le \rho_r, \rho_c \le 0.9$)	AR(1)*AR(1) - Very efficient with values between 90-100%. MA(1)*MA(1) - Efficiencies range from 50-100%, low efficiencies occur at high correlations and high efficiencies occur at low correlations. LV Model - Very efficient with values between 85-100%.
MA(1)*MA(1) ($0.1 \le \rho_r, \rho_c \le 0.5$)	AR(1)*AR(1) - Efficiencies range from 55-100%, low efficiencies only occur at high correlations. MA(1)*MA(1) - Efficiencies can be very high or very low, ranging from 35-100%, low efficiencies occur when the true values are such that $\rho_r > \rho_c$ but the assumed values are actually $\rho_r < \rho_c$ and vice versa. LV Model - Efficiencies range from 50-100%, low efficiencies occur at high correlations.
LV Model ($0.5 \le \psi_r, \psi_c \le 5$)	AR(1)*AR(1) - Very efficient, values range from 80-100%. MA(1)*MA(1) - Efficiencies range from 40-100% with lowest efficiencies occurring at high correlations. LV Model - Very efficient, values range from 90-100%.

4.2 Spatial process with row and column effects

Let us now consider examples where row and/or column effects are involved as well as the spatial process, for convenience only the AR(1)*AR(1) structure is considered. Assume without loss of generality $\sigma^2_e = 1$ and so the variance components σ^2_r, σ^2_c and σ^2_s are ratios with respect to σ^2_e. By varying the values of these ratios we are effectively changing the weights of the row and column random effects and the spatial process respectively. Obviously these weights will influence the layout of the optimal design. Let us now illustrate this with an example on the 5×5 design. Suppose the values for σ^2_r, σ^2_c and σ^2_s are 1, 1, and 1 respectively with $\rho_r = \rho_c = 0.5$, then the best design found with an A-value of 1.1172 is given below:

2 4 3 1 5
3 1 5 2 4
5 2 4 3 1
4 3 1 5 2
1 5 2 4 3

(G)

Comparing this design to design A, we notice that some treatments occur more than once in the rows and columns of A (non-binary), while G is a Latin square and most importantly has no self-diagonals. By introducing a random row and column effect, each treatment is restricted to appear at most once in each row and each column and as a result, may destroy some of the properties induced by the spatial process. This is not desired in practice, since there are reasons to believe the spatial process plays a more important role in the spatial analysis than random row and column effects, see Gilmour *et al.* (1997). A solution would be to put heavier weight on the spatial process, that is, increasing the value of σ^2_s so that the variance is dominated by the spatial component but yet row and column random effects still play a role. If the value of σ^2_s increases to 3 some self-diagonals are observed even though the design is still a Latin square (H). The number of self-diagonals in this design is 6 and has an A-value of 1.6969. As the value of σ^2_s further increases, the spatial process becomes dominant and the number of self-diagonals gradually increases. When $\sigma^2_s = 5$ the best design found (J) has 14 self-diagonals and non-binary columns, and the Latin square feature starts to vanish:

```
5 1 3 2 4        1 3 4 5 2
1 2 4 3 5        3 1 5 2 4
4 3 1 5 2        1 5 3 4 2
3 5 2 4 1        4 1 2 3 5
2 4 5 1 3        5 4 1 2 3
   (H)              (J)
```

5 Application - early generation trial

Many field trials involve hundreds of treatments and are usually run in stages where treatments with low yield are excluded from further stages and only those with high yield are retained for further testing. A common approach is to test new treatments which are replicated only once against some known control treatments which may be replicated several times. Information provided by the Queensland Department of Primary Industries (QDPI) where a wheat breeding program is carried out suggests that early generation trials usually consist of 200-500 genotypes with embedded controls, plots are usually thin and long, and laid out in the row direction. Due to the size of the plots, the values of ρ_r and ρ_c range from 0.2-0.7 and 0.0-0.3 respectively. An example which is much smaller than what might be used in practice, but sufficiently large to illustrate some basic ideas follows.

Consider 40 test treatments and 2 controls (1 and 2) in a 10×5 array, so that each control is replicated 5 times. In the past, experimenters would embed the controls diagonally across the plots so that they are reasonably spread within columns. The transpose of the following design would be reasonable:

```
1  7 11 15 19  2 27 31 35 39
3  1 12 16 20 23  2 32 36 40
4  8  1 17 21 24 28  2 37 41
5  9 13  1 22 25 29 33  2 42
6 10 14 18  1 26 30 34 38  2
             (K)
```

To avoid complications the variance is assumed to compose a spatial AR(1)*AR(1) process only and σ^2_s is 1 without loss of generality. The cases where random row and column effects are present are briefly referred to later in the section. Now let the values of ρ_r and ρ_c be 0.6 and 0.2 respectively. Design K has an A-value of 73.8441; however, one would not expect this to be the optimal design as the neighbouring structure is quite poor. All of the test treatments that are neighbours with control treatments, except for 19 and 26, appear twice as nearest neighbours with a control (once in rows and once in columns), however, the majority of other test treatments are not nearest neighbours of control treatments at all, for example treatments 15, 36, etc. If the two controls are alternated then the design has a much better neighbour structure and A-value is 69.7145. Assuming spatial correlation, our algorithms yield a better design (L) with A-value 67.9790, as well as a better neighbouring structure. Notice that the control treatments tend to appear on diagonal plots and that each column still contains each of the controls exactly once. The transpose of the design is as follows:

3	8	2	15	18	23	1	31	34	38
4	9	12	1	19	24	28	2	35	39
5	10	2	16	20	25	29	32	1	40
6	1	13	17	21	26	2	33	36	41
7	11	14	2	22	27	30	1	37	42

(L)

When row and/or column random effects are included, designs tend to have control treatments distributed evenly in rows and/or columns. However, there appears to be fewer diagonal neighbours between the controls. These are only preliminary impressions from the examples we have considered and more thorough research is required.

6 Conclusion

The algorithms described and used here have wide applicability and produce designs which are reasonable. The resulting designs exhibit characteristics involving the neighbour structure, corner and edge plots and diagonal neighbour structure all of which are important but dependent on the particular values for the various parameters in the variance-covariance structure.

Both the search methods, simulated annealing and Tabu, appear to be very useful in obtaining designs for correlated data. However, there is little guidance in the literature as to the implementation of an optimal cooling scheme for simulated annealing and how to determine the number of iterations for a Tabu routine. A recommendation as to which of the two methods is better is not possible at this time. Suffice it to say that both appear to work well and require some experimentation.

References

Aarts, E.H.L. & van Laarhoven, P.J.M. (1989). Simulated annealing: an introduction. *Statistica Neerlandica*, **43**, 31-52.

Cullis, B.R. & Gleeson, A.C. (1991). Spatial analysis of field experiments - an extension to two dimensions. *Biometrics*, **47**, 1449-1460.

Gilmour, A.R., Cullis, B.R. & Verbyla, A.P. (1997). Accounting for natural and extraneous variation in the analysis of field experiments. *Journal of Agricultural, Biological, and Environmental Statistics*, **2**, 269-293.

Glover, F. (1989). Tabu search - Part I. *ORSA Journal on Computing,* **1**, 190 -206.

Glover, F. (1990). Tabu search - Part II. *ORSA Journal on Computing,* **2**, 4-32.

Jones, B. & Eccleston, J. (1980). Exchange and interchange procedures to search for optimal designs. *Journal of the Royal Statistical Society Series B*, **42**, 238-243.

Martin, R.J. & Eccleston, J.A. (1997). Construction of optimal and near optimal designs for dependent observations using simulated annealing, *The Australian Journal of Statistics,* accepted (under revision), *Research Report*, University of Sheffield, Department of probability and statistics.

Martin, R.J., Eccleston, J.A. & Jones, G. (1998). Some results on multi-level factorial designs with dependent observations. *Journal of Statistical Planning and Inference*, in press.q

Russell, K.G. & Eccleston, J.A. (1987a). The construction of optimal balanced incomplete block designs when adjacent observations are correlated. *The Australian Journal of Statistics*, **29**, 84-90.

Russell, K.G. & Eccleston, J.A. (1987b). The construction of optimal incomplete block designs when observations within a block are correlated. *The Australian Journal of Statistics*, **29**, 293-302.

Williams, E.R. (1986). A neighbour model for field experiments, *Biometrika*, **73**, 279-287.

Zergaw, G. (1989). A sequential method of constructing optimal block designs. *The Australian Journal of Statistics*, **31**, 333-342.

(Co)Variance Structures for Linear Models in the Analysis of Plant Improvement Data

Arthur R Gilmour,[1] Brian R Cullis,[2] Alison B Frensham,[2] and Robin Thompson[3]

[1] Orange Agricultural Institute, NSW Agriculture, Forest Road, Orange, NSW, 2800, Australia
[2] Wagga Wagga Agricultural Institute, NSW Agriculture, Wagga Wagga, NSW, 2650, Australia
[3] Statistics Department, IACR-Rothamsted, Harpenden AL5 2JQ, UK

Abstract. Plant improvement programs involve the evaluation of a large number of genotypes (varieties) in a series of designed experiments known as multi-environment trials (MET). The combined analysis of MET data is a complex statistical problem which requires extensions to the standard linear mixed model. The analysis must accommodate spatial correlation structures for the plot errors from each trial and appropriate genetic covariance structures. *ASReml* (Gilmour, Cullis, Welham & Thompson, 1998) provides a broad range of variance structures for both the errors and the random effects in a linear mixed model. The gains in statistical efficiency resulting from the use of more complex but more realistic variance structures are large. With *ASReml* they can be achieved at very little extra cost since the algorithm and use of sparse matrix methods ensures timely analyses. In this paper the computational strategy of *ASReml* will be described and some of the scope of the program will be demonstrated in the analysis of a MET data set.

Keywords. Average information, multi-environment trials, sparsity, spatial analysis

1 Introduction

Cullis, Gogel, Verbyla & Thompson (1998) present a spatial mixed model analysis for MET data. This is an extension of the spatial analysis of a single field trial (see Gilmour, Cullis & Verbyla, 1997) in which plot error variation is partitioned into three major sources and modelled accordingly. The sources are non-stationary large scale variation across the field, extraneous variation (often induced by experimental procedure) and stationary local trend. The first two sources can be accommodated by including appropriate terms in the model such as design factors and polynomial functions of the spatial coordinates of the field plots. Local trend is accommodated using a covariance structure. Experience has shown that a first-order separable autoregressive process (denoted AR1×AR1) is often appropriate. The decomposition of error variation provides a simpler, more plausible approach than the original methodology of Gleeson & Cullis (1987) and Cullis & Gleeson (1991) in which error variation as a whole was modelled using a covariance structure. A wide range of ARIMA processes was needed for this purpose. There are strong similarities with the Gilmour *et al.* (1997) approach and that advocated by

Cressie (1991) for the analysis of geostatistical spatial data. The notable difference is that in the former, extraneous variation is identified and removed as a separate source.

We now extend the mixed model for MET data of Cullis *et al.* (1997) to accommodate more general covariance structures for the genetic effects. The yields from different trials can be regarded as different traits for each genotype. Thus there is an underlying covariance matrix which links the genetic effects in different trials. That is, $\boldsymbol{G} = \{\gamma_{ij}\}$ where γ_{ii} is the genetic variance in trial i and γ_{ij} is the genetic covariance between trials i and j. The core routines in *ASReml* which are also being implemented within GENSTAT (Payne *et al.* 1998), enable the simultaneous estimation of the genetic (co)variances and the spatial parameters associated with the error variance structures for individual trials.

In this paper we present the extended linear mixed model which incorporates the analysis described above as a special case. We briefly outline the computational strategy involved in fitting this model using the Average Information algorithm (Gilmour, Thompson & Cullis, 1995). The analysis of a MET data set using *ASReml* is presented in Section 4. We conclude with a discussion of future work.

2 Description of models and AI algorithm

We consider the model

$$\boldsymbol{y} = \boldsymbol{X}\boldsymbol{\tau} + \boldsymbol{Z}\boldsymbol{u} + \boldsymbol{e} \tag{1}$$

where $\boldsymbol{y}$ is an n vector of data, $\boldsymbol{X}$ and $\boldsymbol{Z}$ are design matrices, $\boldsymbol{\tau}$ and $\boldsymbol{u}$ are t and b vectors of fixed and random effects and $\boldsymbol{e}$ is the error term. We assume, for convenience of presentation, that $\boldsymbol{X}$ has full column rank and $\boldsymbol{Z} = [\boldsymbol{Z}_1, \ldots, \boldsymbol{Z}_b]$ where each $\boldsymbol{Z}_i$ represents the design matrix for the ith random factor and $\boldsymbol{u}' = [\boldsymbol{u}_1', \ldots, \boldsymbol{u}_b']$ where $\boldsymbol{u}_i$ is a q_i vector. Note $q = \sum q_i$. We assume further that

$$\begin{bmatrix} \boldsymbol{u} \\ \boldsymbol{e} \end{bmatrix} \sim \mathbf{N}\left(0, \begin{bmatrix} \boldsymbol{G}(\boldsymbol{\gamma}) & 0 \\ 0 & \boldsymbol{R}(\boldsymbol{\phi}) \end{bmatrix}\right)$$

where $\boldsymbol{\gamma}$ is a vector of variance parameters relating to $\boldsymbol{u}$ and $\boldsymbol{\phi}$ is the vector of variance parameters relating to $\boldsymbol{e}$.

2.1 Variance structures for the errors

The vector $\boldsymbol{e}$ may comprise a series of subvectors indexed by a factor, generically labelled sections. These subvectors, $\boldsymbol{e}_j, j = 1, \ldots, p$, represent the errors for each of several sections. We assume that the elements of these subvectors are uncorrelated between sections, but are correlated within sections. Thus, the matrix $\boldsymbol{R}$ is compactly written as $\boldsymbol{R} = \oplus_{j=1}^{p} \boldsymbol{R}_j$, where $\boldsymbol{R}_j$ is the variance matrix of the errors for section j and is a function of the variance and covariance parameters for section j.

Cullis *et al.* (1997) consider the spatial analysis of MET data in which the jth trial consists of n_j plots arranged in a rectangular grid of n_{r_j} rows by n_{c_j} columns ($n_j = n_{r_j} \times n_{c_j}$). In this application sections correspond to trials. The error variance matrix for the jth trial is given by

$$\boldsymbol{R}_j = \boldsymbol{R}_j(\boldsymbol{\phi}_j) = \sigma_j^2 \Sigma_j(\boldsymbol{\rho}_j) + \psi_j \boldsymbol{I}_{n_j}$$

where Σ_j is the spatial correlation matrix for local trend and is a function of parameters $\boldsymbol{\rho}_j$, σ_j^2 is the variance of this trend process and ψ_j is the variance of a white noise or "measurement error" process (which may be excluded). This model allows for between trial error variance heterogeneity (through σ_j^2) and a different spatial correlation structure for each trial.

Each component matrix, Σ_j (or $\boldsymbol{R}_j$ if there is no measurement error) is assumed to be the kronecker product of one, two (or more) component matrices. This assumption is also known as separability (see Martin, 1979) and results in significant computational efficiencies. These matrices are indexed for each section by factors labelled for example, layers, columns and rows in the data. These factors must uniquely define the experimental units and the field ordering, in a spatial or temporal sense. In the MET analysis the data are assumed ordered as rows within columns within trials and $\Sigma_j = \Sigma_{c_j} \otimes \Sigma_{r_j}$, where the component matrices Σ_{c_j} and Σ_{r_j} are the correlation matrices corresponding to trend along the columns and rows of the field.

In *ASReml* possible model choices for the component matrices include identity, autoregressive (and generalisations of this for unequally spaced data) and moving average. Welham *et al.* (1998) discuss a range of applications using these structures.

2.2 Variance structures for random factors

We generally assume that the random effects vector $\boldsymbol{u}$ is comprised of b subvectors; each of these subvectors (of length q_i) represents specific random effects in equation (1). Separate components of $\boldsymbol{u}$ are assumed uncorrelated and follow Gaussian distributions, with zero mean and variance matrices denoted by $\boldsymbol{G}_i$. Thus $\boldsymbol{G} = \oplus_{i=1}^{b} \boldsymbol{G}_i$.

Each submatrix, $\boldsymbol{G}_i$ is assumed to be the kronecker product of one, two (or more) component matrices. These matrices are indexed for each of the factors constituting the term in the linear model. For example, the term *site.genotype* has two factors and so the matrix $\boldsymbol{G}_i$ is comprised of two component matrices defining the variance structure for each factor in the term.

Models for the component matrices $\boldsymbol{G}_i$ include models for standard random factors, for which $\boldsymbol{G}_i = \gamma_i \boldsymbol{I}_{q_i}$, where γ_i is the variance component for the factor, or correlated random factors, for which $\boldsymbol{G}_i = \boldsymbol{G}_{i1} \otimes \boldsymbol{G}_{i2} \cdots \otimes \boldsymbol{G}_{it_i}$, where t_i is the number of factors in the *ith* random term. The vector $\boldsymbol{u}_i$ is therefore assumed to be the vector representation of a t_i array. For example for *site.genotype*, the vector $\boldsymbol{u}_i$ is simply the vec of a matrix with columns defined by *site* and rows defined by *genotype*.

A range of models for **G** component structures is currently available in *ASReml*. These include identity, uniform, unstructured, factor analytic, antedependence and banded and heterogeneous forms for uniform and factor analytic.

2.3 Estimation: AI algorithm

Following Gilmour *et al.* (1995), the mixed model equations are

$$\begin{bmatrix} \boldsymbol{X}'\boldsymbol{R}^{-1}\boldsymbol{X} & \boldsymbol{X}'\boldsymbol{R}^{-1}\boldsymbol{Z} \\ \boldsymbol{Z}'\boldsymbol{R}^{-1}\boldsymbol{X} & \boldsymbol{Z}'\boldsymbol{R}^{-1}\boldsymbol{Z} + \boldsymbol{G}^{-1} \end{bmatrix} \begin{bmatrix} \hat{\boldsymbol{\tau}} \\ \tilde{\boldsymbol{u}} \end{bmatrix} = \begin{bmatrix} \boldsymbol{X}'\boldsymbol{R}^{-1}\boldsymbol{y} \\ \boldsymbol{Z}'\boldsymbol{R}^{-1}\boldsymbol{y} \end{bmatrix} \quad (2)$$

The solution of (2) requires values for $\boldsymbol{\gamma}$ and $\boldsymbol{\phi}$. In practice we replace $\boldsymbol{\gamma}$ and $\boldsymbol{\phi}$ by their REML estimates $\tilde{\boldsymbol{\gamma}}$ and $\tilde{\boldsymbol{\phi}}$ which maximise the likelihood of error contrasts.

The residual log likelihood can be written as

$$\begin{aligned}\ell &= -\tfrac{1}{2}(\log\det \boldsymbol{X}'\boldsymbol{H}^{-1}\boldsymbol{X} + \log\det \boldsymbol{H} + \boldsymbol{y}'\boldsymbol{P}\boldsymbol{y}) \\ &= -\tfrac{1}{2}(\log\det \boldsymbol{C} + \log\det \boldsymbol{R} + \log\det \boldsymbol{G} + \boldsymbol{y}'\boldsymbol{P}\boldsymbol{y}) \qquad (3)\end{aligned}$$

where $\boldsymbol{H} = \boldsymbol{R} + \boldsymbol{Z}\boldsymbol{G}\boldsymbol{Z}'$, $\boldsymbol{C}$ is the coefficient matrix in (2) and

$$\begin{aligned}\boldsymbol{P} &= \boldsymbol{H}^{-1} - \boldsymbol{H}^{-1}\boldsymbol{X}(\boldsymbol{X}'\boldsymbol{H}^{-1}\boldsymbol{X})^{-1}\boldsymbol{X}'\boldsymbol{H}^{-1} \\ &= \boldsymbol{R}^{-1} - \boldsymbol{R}^{-1}\boldsymbol{W}\boldsymbol{C}^{-1}\boldsymbol{W}'\boldsymbol{R}^{-1}\end{aligned}$$

where $\boldsymbol{W} = [\boldsymbol{X} \ \ \boldsymbol{Z}]$. Letting $\boldsymbol{\kappa} = (\boldsymbol{\gamma}, \boldsymbol{\phi})$, the REML estimate of κ_i satisfies

$$\partial\ell/\partial\kappa_i = -\tfrac{1}{2}\left(\operatorname{tr}\left[\boldsymbol{P}\dot{\boldsymbol{H}}_i\right] - \boldsymbol{y}'\boldsymbol{P}\dot{\boldsymbol{H}}_i\boldsymbol{P}\boldsymbol{y}\right) = 0 \qquad (4)$$

where $\dot{\boldsymbol{H}}_i = \partial\boldsymbol{H}/\partial\kappa_i$.

In general the solution to (4) requires an iterative scheme. Given an initial estimate, an update of $\boldsymbol{\kappa}$, using the FS algorithm, is

$$\boldsymbol{\kappa}^{(0)} + \boldsymbol{B}^{(0)}\partial\ell/\partial\boldsymbol{\kappa}(\boldsymbol{\kappa} = \boldsymbol{\kappa}^{(0)})$$

where $\boldsymbol{B}$ is the inverse of the expected information matrix of $\boldsymbol{\kappa}$.

The elements of the observed information matrix are

$$\begin{aligned}-\partial^2\ell/\partial\kappa_i\partial\kappa_j = \tfrac{1}{2}\operatorname{tr}\left[\boldsymbol{P}\ddot{\boldsymbol{H}}_{ij}\right] - \tfrac{1}{2}\operatorname{tr}\left[\boldsymbol{P}\dot{\boldsymbol{H}}_i\boldsymbol{P}\dot{\boldsymbol{H}}_j\right] \\ + \boldsymbol{y}'\boldsymbol{P}\dot{\boldsymbol{H}}_i\boldsymbol{P}\dot{\boldsymbol{H}}_j\boldsymbol{P}\boldsymbol{y} - \tfrac{1}{2}\boldsymbol{y}'\boldsymbol{P}\ddot{\boldsymbol{H}}_{ij}\boldsymbol{P}\boldsymbol{y} \qquad (5)\end{aligned}$$

where $\ddot{\boldsymbol{H}}_{ij} = \partial^2\boldsymbol{H}/\partial\kappa_i\partial\kappa_j$

The elements of the expected information matrix are

$$\mathcal{E}\left(-\partial^2\ell/\partial\kappa_i\partial\kappa_j\right) = \tfrac{1}{2}\operatorname{tr}\left[\boldsymbol{P}\dot{\boldsymbol{H}}_i\boldsymbol{P}\dot{\boldsymbol{H}}_j\right] \qquad (6)$$

The evaluation of some traces in either (5) or (6) can be either not feasible or very computer intensive. We therefore consider the matrix denoted by $\mathcal{I}_A$ which is a simplified average of the terms in (5) and (6). The elements of $\mathcal{I}_A$ are

$$\mathcal{I}_A(\kappa_i, \kappa_j) = \tfrac{1}{2}\boldsymbol{y}'\boldsymbol{P}\dot{\boldsymbol{H}}_i\boldsymbol{P}\dot{\boldsymbol{H}}_j\boldsymbol{P}\boldsymbol{y}$$

This is obtained by averaging (5) and (6) and approximating $\boldsymbol{y}'\boldsymbol{P}\ddot{\boldsymbol{H}}_{ij}\boldsymbol{P}\boldsymbol{y}$ by its expectation, $\operatorname{tr}\left[\boldsymbol{P}\ddot{\boldsymbol{H}}_{ij}\right]$ in those cases when $\ddot{\boldsymbol{H}}_{ij} \neq 0$. For variance components models (i.e. those linear in $\boldsymbol{H}$), the terms in $\mathcal{I}_A$ are exact averages of those in (5) and (6). We call this matrix the average information (AI) matrix and use it in place of the expected information matrix to update $\boldsymbol{\kappa}$.

If we let k be the number of parameters in $\boldsymbol{\kappa}$ and define $\boldsymbol{Y} = [\boldsymbol{y}_0, \boldsymbol{y}_1, \ldots, \boldsymbol{y}_k]$, where $\boldsymbol{y}_i, i > 0$ is the 'working' variate for κ_i and is given by

$$\boldsymbol{y}_i = \dot{\boldsymbol{H}}_i\boldsymbol{P}\boldsymbol{y} = \dot{\boldsymbol{H}}_i\boldsymbol{R}^{-1}\tilde{\boldsymbol{e}} \qquad (7)$$

where $\tilde{\boldsymbol{e}} = \boldsymbol{y} - \boldsymbol{X}\hat{\boldsymbol{\tau}} - \boldsymbol{Z}\tilde{\boldsymbol{u}}$, $\hat{\boldsymbol{\tau}}$ and $\tilde{\boldsymbol{u}}$ are solutions to (2) and $\boldsymbol{y}_0 = \boldsymbol{y}$, the data vector, then the $\mathcal{I}_A$ matrix is the partition of the (scaled) residual sums of squares and products matrix $\tfrac{1}{2}\boldsymbol{Y}'\boldsymbol{P}\boldsymbol{Y}$ corresponding to $[\boldsymbol{y}_1, \ldots, \boldsymbol{y}_k]$.

3 *ASReml* computing strategies

3.1 Sparsity

A serious problem in some REML algorithms has been the inability to handle moderately large problems. Animal breeders developed programs (e.g. DFREML, Meyer, 1990) which utilized sparsity to handle large models but these were not widely applicable to other situations. One feature of the Average Information algorithm is that it is amenable to sparse matrix computations. The iterative solution for $\boldsymbol{\kappa}$ implies that the matrix $\mathbf{C}$ must be reformed each iteration. The calculation of $\mathbf{C}$ involves the design matrices $\mathbf{X}$ and $\mathbf{Z}$ which *ASReml* forms once only and holds in sparse form. Some variance structures for $\mathbf{R}$ and $\mathbf{G}$ have sparse inverses which helps to maintain sparsity in $\mathbf{C}$. Common structures with sparse inverses include identity, diagonal, autoregressive and low order antedependence structures.

3.2 Algorithm

The central subroutine in *ASReml*

- forms the mixed model sums of squares and cross products matrices in (2) and stores in sparse form
- adds $\mathbf{G}^{-1}$ to the part of $\mathbf{C}$ which corresponds to $\boldsymbol{Z}$
- determines an order for solving the equations which will retain a high level of sparsity
- performs absorption keeping intermediate results needed for the calculation of the scores and log likelihood
- backsolves for $\hat{\tau}$ and $\tilde{\boldsymbol{u}}$ and calculates the residuals
- forms working variables and their cross-products
- does absorption of working variables to obtain the matrix $[\boldsymbol{Y}'\boldsymbol{P}\boldsymbol{Y}]$, from which the AI matrix and parts of the scores are obtained
- completes the sparse inverse of the coefficient matrix; cells not needed in forming the scores are not computed
- calculates the scores and updates the variance parameters

3.3 The score and average information matrices

Equations (4) and (7) give general forms for the score and working variable for κ_i. For a random effects variance parameter γ_{ij} associated with the ith random factor $\mathbf{u}_i$ the working variable is given by $\boldsymbol{y}_{ij} = \mathbf{Z}_i \dot{\mathbf{G}}_{ij} {\mathbf{G}_i}^{-1} \tilde{\mathbf{u}}_i$, where $\dot{\mathbf{G}}_{ij} = \partial \mathbf{G}_i / \partial \gamma_{ij}$. The score can be written as

$$-\tfrac{1}{2}\left(\mathrm{tr}[{\mathbf{G}_i}^{-1}\dot{\mathbf{G}}_{ij}] - \mathrm{tr}[{\mathbf{G}_i}^{-1}\dot{\mathbf{G}}_{ij}{\mathbf{G}_i}^{-1}\mathbf{C}^{Z_i Z_i}] - \boldsymbol{y}_0'\boldsymbol{P}\boldsymbol{y}_{ij}\right)$$

where $\mathbf{C}^{Z_i Z_i}$ is the block of the inverse of the coefficient matrix in the mixed model equations corresponding to $\mathbf{Z}_i$.

For a variance parameter ϕ_{ij} associated with the ith section the working variable is given by $\boldsymbol{y}_{ij} = \dot{\mathbf{R}}_{ij}{\mathbf{R}_i}^{-1}\tilde{\mathbf{e}}_i$, where $\dot{\mathbf{R}}_{ij} = \partial \mathbf{R}_i / \partial \phi_{ij}$. The score is

$$-\tfrac{1}{2}\left(\mathrm{tr}[{\mathbf{R}_i}^{-1}\dot{\mathbf{R}}_{ij}] - \mathrm{tr}[\boldsymbol{W}'\mathbf{R}^{-1}\dot{\mathbf{R}}^*_{ij}\mathbf{R}^{-1}\boldsymbol{W}\mathbf{C}^{-1}] - \boldsymbol{y}_0'\mathbf{P}\boldsymbol{y}_{ij}\right)$$

where $\dot{\mathbf{R}}^*_{ij} = \mathrm{diag}\left(0 \ldots \dot{\mathbf{R}}_{ij} \ldots 0\right)$.

3.4 Ordering for solution

A feature of the average information algorithm is that it does not require the whole inverse coefficient matrix ($\mathbf{C}^{-1}$) to be formed. In forming the score for γ_{ij} we need the trace of a product of two matrices $\boldsymbol{A} = \mathbf{G}_i{}^{-1}\dot{\mathbf{G}}_{ij}\mathbf{G}_i{}^{-1}$ and $\boldsymbol{B} = \mathbf{C}^{Z_iZ_i}$. Since $\mathbf{G}_i{}^{-1}\dot{\mathbf{G}}_{ij}\mathbf{G}_i{}^{-1}$ is at least as sparse as the partition of $\mathbf{C}$ corresponding to $\mathbf{Z}_i$ and since the trace of the product $\boldsymbol{AB}$ is the sum of each element in $\boldsymbol{A}$ multiplied by the corresponding element in $\boldsymbol{B}$, we only need to form elements in $\mathbf{C}^{Z_iZ_i}$ which correspond to non-zero elements in the partition of $\mathbf{C}$ corresponding to $\mathbf{Z}_i$. Similar logic applies to the score for ϕ_{ij}.

The mixed model equations are therefore reordered so that $\mathbf{C}^{-1}$ is kept sparse. In a large system of equations, choosing the order can itself be computer intensive. *ASReml* seeks to maintain sparsity by absorbing one of the sparsest equations at each step. While greater sparsity might be achieved by determining the number of new cells filled by absorbing each row, this would greatly increase the time to determine the order. The order is determined on the first iteration only.

4 Example: Analysis of MET data from a South Australian oat breeding program

We present an analysis of a set of yield data from the 1996 series of Stage 3 variety evaluation trials from the SARDI oat breeding program (data kindly provided by Dr. Pamela Zwer). The material being tested consisted of 96 genotypes (predominantly new material from the breeding program but including some existing commercial varieties). Genotypes were grown in replicated trials at 7 sites in South Australia and a single site in Victoria (Lake Bolac). Stage 3 is the penultimate stage of testing in the program. A small number (less than 10) of the top yielding new genotypes in these trials will be selected to proceed to wide-spread testing (Stage 4) at a larger number of sites and over two consecutive seasons. New genotypes which yield well in Stages 3 and 4 may be recommended for commercial use.

All 1996 Stage 3 trials were designed as alpha lattices with 3 replicates (Wanilla had 2 replicates), the total layout comprising 24 rows by 12 columns (Wanilla laid out as 32 rows by 6 columns). This MET data set is balanced, with all 96 genotypes being grown at all sites. We stress that lack of balance presents no difficulties – the same approach to analysis is adopted.

In the linear mixed model for MET data we fit a (fixed) *site* main effect and random *site.genotype* effects. A *genotype* main effect is not fitted so that the latter represent the genotype effects at each site rather than site by genotype interactions. We choose to do this since in the presence of large scale differences between sites (as is the case in the example – see trial mean yields in Table 1) the genotype main effect may be a mis-leading measure of overall genotype yield. We return to this issue later. We use a variance structure for *site.genotype* of the form $\mathbf{G} = \mathbf{G}_s \otimes \mathbf{G}_g$, where $\mathbf{G}_s{}^{(8\times8)}$ and $\mathbf{G}_g{}^{(96\times96)}$ are the component matrices for sites and genotypes respectively. We choose $\mathbf{G}_g = \mathbf{I}$ but other forms are possible.

The first step in the analysis is to determine appropriate spatial models for each site. For this purpose we regard the genetic effects at different sites as independent. This is analogous to conducting 8 separate analyses and is achieved using $\mathbf{G}_s$ of the form diag(γ_i) where γ_i is the genetic variance for the ith site. Our first choice for the spatial models is an AR1×AR1 for each site. The *ASReml* input code for this analysis is given overleaf:

```
SA Oats 1996 : stage 3          # Job Title

# Names of data fields with sizes of factors (row and column
# factors have levels given by maximum value across sites);
# fc2 is a 2 level factor for extraneous variation
  genotype 96 row 32 column 12  yield   site 8   fc2 2

# Name of data file which has one header line to skip
# Data file was sorted as rows within columns within sites
oats3.asdat !skip 1

# Linear model with fixed site and random site.genotype effects
# missing values (mv) to be estimated
yield ~ site mv !r site.genotype

# Variance model specification. R-structures:
# 8 sections (sites) with 2 dimensions (rows, columns);
# one G-structure (site.genotype)
8 2 1

# R-structure for each site: AR1 for columns, AR1 for rows,
# starting values for AR parameters and variance of
# process (!S2) are given
12 column AR .193 !S2=.139     #Error Model for Birdwood
24 row AR .098
12 column AR .72 !S2=1.164     #Error Model for Kybybolite
24 row AR .93
12 column AR .202 !S2=.083     #Error Model for Mallala
24 row AR .052
12 column AR .097 !S2=.037     #Error Model for Palmer
24 row AR -.014
12 column AR -.082 !S2=.074    #Error Model for Pinery
24 row AR -.040
12 column AR .434 !S2=.052     #Error Model for Turretfield
24 row AR .696
6 column AR -.065 !S2=.189     #Error Model for Wanilla
32 row AR .433
12 column AR .077 !S2=.128     #Error Model for Lake Bolac
24 row AR .013

# G-structure for site.genotype (2 factors in term)
site.genotype 2
site 0 DIAG .2 .2 .1 .1 .01 .1 .1 .1  # G matrix for sites,
                                      # starting values given
genotype 0 I                          # G matrix for genotypes
```

We then use diagnostics on the residuals for each site from this model to assess the adequacy of the spatial model. Plots of residuals against row/column number for each site can be used to check for potential outliers. Gogel (1997) has more formal approaches for the detection of outliers. A key diagnostic for the spatial model is the sample variogram (see Gilmour *et al.*, 1997 for details). Figure 1 displays the sample variogram for each site after the fit of the AR1×AR1 models.

The theoretical variogram for a stationary AR1×AR1 process is monoton-

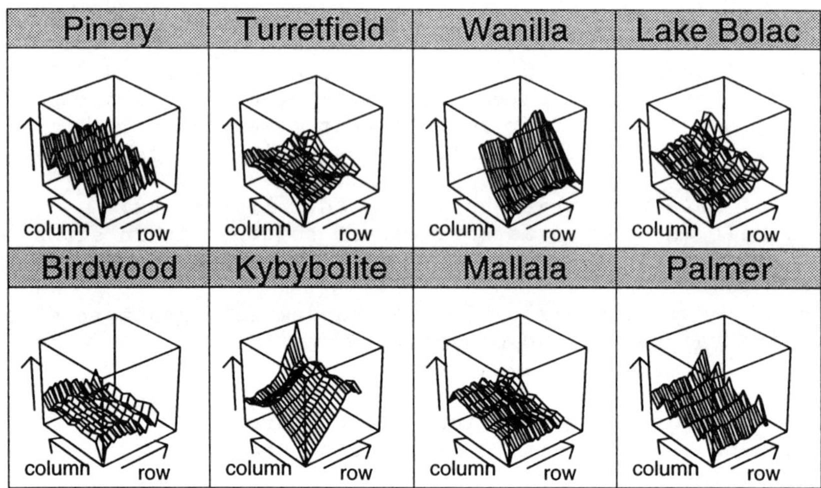

Fig. 1. MET analysis of oat data: variograms for AR1×AR1 models for each site

ically increasing from the origin and reaches a plateau which is the variance of the process. Many of the sample variograms in Figure 1 are not consistent with this. There is evidence of non-stationarity, e.g. across rows at site 2 (Kybybolite) and across columns at site 7 (Wanilla) with the variograms failing to asymptote in these directions. Non-stationarity can be removed by fitting polynomial functions or cubic smoothing splines (see Verbyla, Cullis, Kenward & Welham, 1998) to the row/column co-ordinates at these sites. To fit a linear regression to the rows at site 2 we include the term *site|2/lin(row)* in the model.

The step-like appearance across columns in the variograms for sites 4 (Palmer) and 5 (Pinery) suggests systematic extraneous variation. In discussions with the breeder it was found that there was variation in the lengths of plots, with plots in even numbered columns being shorter than plots in odd numbered columns. This was accommodated by fitting a 2 level factor, *fc2* (values of 1 for even columns; 2 for odd) for these sites. So the term *site|4/c(fc2)* was included, for example. Note that the "c" here imposes sum to zero constraints on the factor.

There was also evidence of extraneous variation aligned with rows and columns at other sites but it was not systematic so was removed by fitting random row and column effects. Measurement error was included for sites with strong spatial correlation, namely sites 2 (Kybybolite) and 6 (Turretfield). This was achieved by including the term *site|2/row.col* in the model, for example.

After several runs of the analysis in which such terms were added and the resultant variograms re-checked, plausible spatial models were achieved. Figure 2 shows the resultant sample variograms. The fitting of row and column regressions and factors in the model has removed most of the non-stationarity and extraneous variation. There is now much closer agreement with the theoretical AR1×AR1 form.

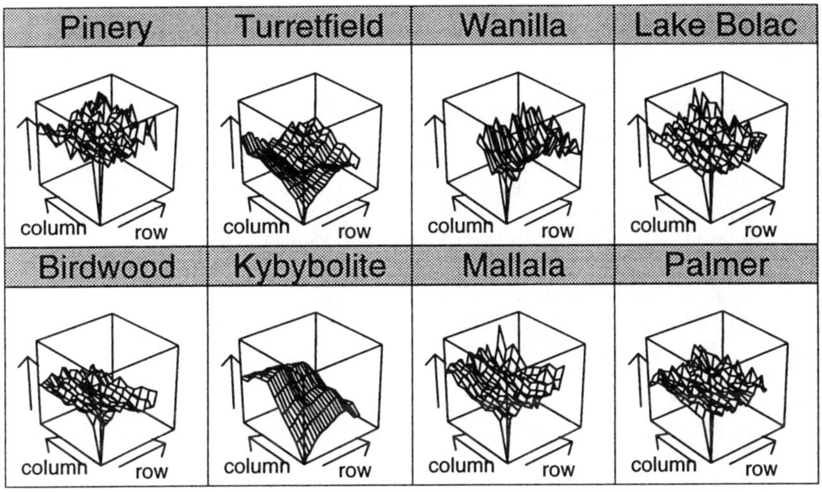

Fig. 2. MET analysis of oat data: variograms for final spatial models

The key in the current approach to spatial analysis is the identification of an appropriate variance structure for the plot errors. There is no longer a dichotomy between spatial analysis and traditional methods such as randomised complete block (RCB) and incomplete block (IB) analyses. The latter provide legitimate variance models which would be adopted in the spatial approach if found to be consistent with the data. Comparisons of spatial with RCB or IB analysis in terms of an efficiency measure are therefore inappropriate. Previous studies have often focussed on such comparisons however, so for purely illustrative purposes we present the estimated heritability (ratio of genetic to total variance) for both the spatial and IB analysis of each trial:

	Birdwood	K'bolite	Mallala	Palmer	Pinery	T'field	Wanilla	L. Bolac
IB:	0.76	0.55	0.63	0.72	0	0.77	0.22	0.66
spatial:	0.80	0.65	0.67	0.75	0.11	0.79	0.33	0.67

The heritability provides an efficiency measure for analyses in which genotype effects are assumed random. In the example the value is higher for the spatial compared to the IB analysis in each case. More important however is the fact that the variance modelling process pointed to the spatial structures rather than IB as being most appropriate for these data. A crucial issue is the connection between design and analysis. Experience has shown that although RCB and IB models may correspond to the experimental design they rarely provide a good fit to the underlying variance structure. A structure which is frequently appropriate corresponds to a separable AR1×AR1 process with random row and column effects. The construction of designs consistent with this model is the subject of current research.

The spatial analysis using an unstructured form for $\mathbf{G}_s$ was then performed. Thus the full genetic covariance matrix across sites was estimated. The *AS-Reml* code for this model is given below:

```
SA Oats 1996 : stage 3
  genotype 96 row 32 column 12  yield   site 8   fc2 2
oats3.asdat !skip 1
yield ~ site|2/lin(row) site|8/lin(row) site|2/lin(col) ,
        site|3/lin(col) site|7/lin(col) site|8/lin(col),
        site|4/c(fc2)   site|5/c(fc2)   site mv,
     !r site.genotype,
        site|1/row .028 site|1/col .043,
        site|3/col .038 site|4/col .013 site|5/col .022,
        site|7/col .024 site|8/col .042,
        site|2/row.col .172  site|6/row.col .034
8 2 1
12 column AR .193 !S2=.139     #Error Model for Birdwood
24 row AR .098
12 column AR .72 !S2=1.164     #Error Model for Kybybolite
24 row AR .93
12 column AR .202 !S2=.083     #Error Model for Mallala
24 row AR .052
12 column AR .097 !S2=.037     #Error Model for Palmer
24 row AR -.014
12 column AR -.082 !S2=.074    #Error Model for Pinery
24 row AR -.040
12 column AR .434 !S2=.052     #Error Model for Turretfield
24 row AR .696
6 column AR -.065 !S2=.189     #Error Model for Wanilla
32 row AR .433
12 column AR .077 !S2=.128     #Error Model for Lake Bolac
24 row AR .013

# unstructured G matrix; starting values as lower triangle
site.genotype 2
site 0 US  0.2023 !+35
   0.00773 0.17460
   0.05138 0.00915 0.06101
   0.03754 0.00669 0.04448 0.04431
   0.01218 0.00217 0.01443 0.01054 0.004338
   0.02757 0.00491 0.03267 0.02387 0.007742 0.07832
   0.01759 0.00313 0.02084 0.01523 0.004939 0.01118 0.03708
   0.01632 .00291 .01934 .01413 .004582 .01037 .006618 .09189
genotype 0 I
```

This model involves 24 spatial parameters (16 autoregressive correlation parameters and 8 variances), 9 variance parameters associated with random extraneous variation and measurement error, 8 genetic variances and 28 genetic covariances. It defines 1672 equations to solve and took 40 minutes to converge in six iterations from the starting values given on a Sparcstation 10 with 160Mb RAM. Table 1 contains the REML estimates of these parameters (except those associated with extraneous variation and measurement error).

Table 1. Variance parameter estimates from MET analysis of oat data. Error variance parameters: spatial variance and autoregressive correlations (ρ). Genetic variance matrix: variances on diagonal; correlations above, covariances below

Site	Birdwood	K'bolite	Mallala	Palmer	Pinery	T'field	Wanilla	L. Bolac
yield (t/ha)	6.0	4.5	2.6	1.5	3.6	3.6	5.1	2.6
Spatial variance parameters								
variance	0.1362	1.1766	0.0810	0.0374	0.0741	0.0568	0.1900	0.1278
column ρ	.17	.73	.18	.09	-.09	.43	0	.07
row ρ	.08	.94	.09	.05	-.04	.63	.43	.02
Genetic Variance Matrix								
Birdwood	**0.2013**	0.509	0.442	0.228	0.685	0.519	0.460	0.479
Kybybolite	0.0954	**0.1746**	-0.001	0.010	0.177	0.110	0.039	0.367
Mallala	0.0483	-0.0001	**0.0595**	0.950	0.797	0.387	0.479	0.222
Palmer	0.0215	0.0008	0.0488	**0.0444**	0.695	0.317	0.176	0.052
Pinery	0.0216	0.0052	0.0137	0.0103	**0.0049**	0.873	0.240	-.369
Turretfield	0.0651	0.0128	0.0264	0.0187	0.0172	**0.0782**	0.604	0.490
Wanilla	0.0390	0.0031	0.0221	0.0070	0.0032	0.0320	**0.0358**	0.416
Lake Bolac	0.0653	0.0466	0.0164	0.0033	-0.0079	0.0416	0.0239	**0.0922**

5 Current work and future developments

The use of the unstructured form for the genetic variance matrix in an analysis of MET data has several drawbacks. First, estimation may be inefficient due to the large number of variance parameters involved. A more parsimonious representation may be preferred. Second, the analysis provides no overall measure of yield performance for each genotype. As discussed in the example, such a measure is vital for the selection of the best genotypes. Current work in which a particular model for the genetic variance matrix is proposed, resolves both of these issues. The approach provides a parsimonious interpretation of genotype by trial interaction and a sensible estimate of overall performance for each genotype.

Recent developments under consideration for inclusion in *ASReml* are

- adjustments to the standard errors for model effects and tests of fixed effects as a result of the fact that they are based on estimated variance parameters (see Kenward & Roger, 1997)
- modelling of variances as a function of explanatory variables (see Frensham, Cullis & Verbyla, 1997). While *ASReml* has some facility for defining relationships, more general modelling of variances is possible
- procedures for the identification of outliers in spatial mixed models (see Gogel, 1997)
- random effects in non-Gaussian settings

Issues requiring further investigation are the stability of alternative variance parameterisations (for example, unstructured forms compared with Cholesky or lower rank representations) and choices for starting values for variance parameters in complicated forms such as unstructured.

6 Conclusion

ASReml broadens the scope for using more realistic variance models when fitting linear mixed models. In the current paper this has been demonstrated in relation to data from a series of plant improvement trials. There are numerous other examples which require non-standard variance structures, including the analysis of repeated measures data, random coefficient models and the fitting of cubic smoothing splines. *ASReml* has been used for all such applications.

References

Cressie, N.A.C. (1991) Statistics for spatial data. New York: John Wiley & Sons.

Cullis, B.R. & Gleeson, A. C. (1989) Efficiency of neighbour analysis for replicated variety trials in Australia. *J. Agric. Sci. Camb.*, **113**, 233–239.

Cullis, B.R. & Gleeson, A.C. (1991) Spatial analysis of field experiments—an extension to two dimensions. *Biometrics*, **47**,1449–1460.

Cullis, B.R., Gogel, B.J., Verbyla, A.P. & Thompson, R. (1998). Spatial analysis of multi-environment early generation trials. *Biometrics*, **54**, in press.

Frensham, A.B., Cullis, B.R. & Verbyla, A.P. (1997). Genotype by environment variance heterogeneity in a two stage analysis. *Biometrics*, **53**, 1373-1383.

Gilmour, A.R., Thompson, R. & Cullis, B.R. (1995). Average Information REML, an efficient algorithm for variance parameter estimation in linear mixed models. *Biometrics*, **51**, 1440-1450.

Gilmour, A.R., Cullis, B.R. & Verbyla, A.P. (1997) Accounting for natural and extraneous variation in the analysis of field experiments. *Journal of Agricultural, Biological, and Environmental Statistics*, **2**, 269-293

Gilmour, A.R., Cullis, B.R., Welham, S.J. & Thompson, R, (1998). *AS-REML*. Program user manual printed by NSW Agriculture, Orange Agricultural Institute, Forest Road, Orange, NSW, 2800, Australia.

Gleeson, A. C. & Cullis, B. R. (1987) Residual maximum likelihood (REML) estimation of a neighbour model for field experiments. *Biometrics*, **43**, 277–288.

Gogel, B.J. Spatial analysis of multi-environment trials. (1997) Ph.D. thesis. Department of Statistics, University of Adelaide.

Kenward, M.G. & Roger, J.H. (1997) The precision of fixed effects estimates from restricted maximum likelihood. *Biometrics*, **53**, 983-997.

Martin, R.J. (1979) A subclass of lattice processes applied to a problem in planar sampling. *Biometrika*, **66**, 209–217.

Meyer, K. (1990) DFREML Users Manual.

Payne, R.W., Lane, P.W., Baird, D.B., Gilmour, A.R., Harding, S.A., Morgan, G.W., Murray, D.A., Thompson, R., Todd, A.D., Tunnicliffe-Wilson, G., Webster, R., Welham, S.J. & White, R.P. (1998) *Genstat 5 Release 4.1 Reference Manual Supplement.* Numerical Algorithms Group, Oxford.

Verbyla, A.P., Cullis, B.R., Kenward, M.G. & Welham, S.J. (1998) The analysis of designed experiments and longitudinal data using smoothing splines. *Journal of the Royal Statisitical Society, Series C*, in press.

Welham, S.J., Thompson, R. & Gilmour, A.R., (1998). A general form for specification of correlated error models, with allowance for heterogeneity. In: *COMPSTAT98 Proceedings in Computational Statistics.* Heidelberg: Physica-Verlag, in press.

Optimal Scaling Methods for Graphical Display of Multivariate Data

Jacqueline J. Meulman

Department of Data Theory, Leiden University, P.O. Box 9555, 2300 RB Leiden, The Netherlands

Abstract. Various methods are discussed that emphasize the graphical representation of results from the analysis of multivariate data. Specifically, qualitative (or categorical) multivariate data will be considered, where observed variables are only partially known, and dealt with by using an optimal scaling framework that replaces nominal and ordinal variables by optimally quantified variables. The approach to graphical display that is advocated is closely related to techniques of multidimensional scaling, and involves the choice of particular standardizations and metrics to be used.

> *"Many measurement models in the behavioral sciences are based on geometric representations of the observed behavior. Frequently this geometric representation is a one-dimensional scale but it need not be, and multidimensional representations are becoming more common. The points on these scales or in these spaces may represent individuals or stimuli or both, and the relations among the points reflect the observations according to some rule." (Coombs, Dawes & Tversky, 1970, p. 32).*

Keywords. Categorical data, qualitative data, data analysis, principal components analysis, correspondence analysis, optimal scaling, lower-rank approximation, graphical display, biplot, triplot, ordinal scaling level, nominal scaling level

1 Introduction

This paper will discuss various possibilities for graphical display of multivariate ordinal and nominal data. The structure of this overview was inspired by the introductory paragraph of Gabriel & Odoroff (1986) which reads: "Biplots seem to be the only graphic display which simultaneously show both the scatter of the units (rows) and the configuration of variables (columns), and do so in a way that allows recovery of the observations (matrix elements) themselves (Gabriel, 1971, ...). By contrast, most other techniques display properties of EITHER the rows OR the columns separately, but not of both together (...). For example, multidimensional scaling plots show inter-row differences as distances but do not display the columns. Similarly, factor analytic plots show the correlations between columns but ignore the rows. Correspondence analysis (Greenacre, 1984) does display both row and column markers, but does so in a manner that makes recovery of the elements quite difficult". The following sections will address biplots in factor analysis and components analysis, biplots in multidimensional scaling, biplots in correspondence analysis, and alternative metrics and biplots in the analysis of correspondence tables.

To conclude the introduction, the data that will be used for empirical illustration will briefly be described. The data are from a paper by Rietveld, Boon & Meulman (1997), and this particular study was undertaken to document a seasonal variation in genital infections and in precursor lesions of cervical carcinoma, as detected in cervical smears. The data were collected in Leiden, the Netherlands, where summer and winter are separated by spring and fall, each lasting about the same amount of time. A series of 504,093 cervical smears were obtained from a routine cytology laboratory over a 9-year observation span (January 1983-January 1992). The cervical smears were examined for infections - *Monilia, Trichomonas, Actinomyces, Human Papilloma Virus* (HPV), and *Chlamydia* - as well as for mild, moderate, and severe dysplasias, carcinoma in situ and squamous carcinoma. Mild, moderate, and severe dysplasias were grouped into "dysplasia" (DYS), and the carcinomas in situ and invasive carcinomas into "carcinoma" (CAR). The counts were corrected for factors influencing the number of screenings per month by expressing them as rates observed per 1,000 smears.

2 Some fundamentals about joint graphical display

The prefix "bi" in the term biplot refers to two sets of different entities, and not to two dimensions, as is sometimes erroneously assumed. A recent book on biplots is Gower & Hand (1996). In the analysis of multivariate data, the m variables are usually represented as vectors (arrows) and the n individual units of observation (in the sequel denoted by the neutral term *objects*) as points in the same low-dimensional space. The orthogonal projection of the object points onto the variable vectors gives an approximation of the columns of the data matrix. Algebraically, the approximation is given by the inner product of the object scores and the variable scores. The classic reference to the basic notion of lower-rank approximation is usually Eckart & Young (1936), but this reference is challenged by Stewart (1993) who remarks that Schmidt (1907) was much earlier. The idea that biplots would be the only graphical joint display of rows and columns is rather gratuitous, and at best a tautology, since this is how a biplot would be defined, at least, if we add the restriction of inner product approximation. The joint representation of rows and columns as points and vectors in a common space originates with Tucker (1960), and has found extremely interesting applications in the analysis of preference data (Carroll, 1972), before the display became well-known as the biplot (through Gabriel, 1971). In the psychometric literature, the particular representation is known as the vector model. When analyzing preference data, it is crucial to realize that the subjects (respondents) are not necessarily the objects in an analysis where the variables are the ordering mechanism. Because judges order a number of options according to their liking, the individual judges should be given the role of variables in the analysis, and the options the role of objects.

3 Biplots in principal components analysis and factor analysis

There is a large amount of literature on obtaining factor scores in the case of "proper" factor analysis, which is the analysis based on the correlation or the covariance matrix, and where due to the estimation of the so-called communalities of the variables, the factor scores can not be determined uniquely. Here we shall concentrate on the close relative of factor analysis, principal components analysis (PCA), since it makes much more sense to use PCA if one wishes to obtain a joint representation since otherwise one would have to deal with the non-uniqueness of the object scores from a factor analysis.

If PCA would be needlessly restricted to the analysis of the correlation or covariance matrix, we would indeed lose the objects in the analysis. It is much more interesting, however, to view PCA as a bilinear model (as is done in Kruskal, 1978), where the prefix "bi" refers again to two sets of entities, the objects and the variables.

Assuming the variables are standardized, we write the singular value decomposition $\mathbf{Q} = \mathbf{K}\Lambda\mathbf{L}'$ to minimize

$$\sigma(\mathbf{X}; \mathbf{A}) = ||\mathbf{Q} - \mathbf{X}\mathbf{A}'||^2, \tag{1}$$

over $\mathbf{X}$ and $\mathbf{A}$: the observed scores in the m-dimensional space $\mathbf{Q}$ are approximated by the inner product of the p-dimensional object scores $\mathbf{X}$ and the p-dimensional variable scores $\mathbf{A}$, with p much smaller than m. The jth row in $\mathbf{A}$ (denoted by $\mathbf{a}'_j$) gives the coordinates to display the variable $\mathbf{q}_j$ in the space $\mathbf{X}$. Because the fit in a joint representation is defined on inner products ($\mathbf{X}\mathbf{A}' \approx \mathbf{Q}$), a coherent choice of normalization has to be made. Usually, the object scores are normalized to have means of zero and variance of one ($\mathbf{X} = \mathbf{K}_p$); the coherent normalization implies that the column scores ($\mathbf{A} = \mathbf{L}_p\Lambda_p$) are component loadings (correlations) between the variables and the p dimensions of the space fitted to the objects. If two variables have a decent fit, the angle between the vectors approximates their correlation. When the object scores are normalized, one refrains from the classical scaling distance interpretation with respect to the objects (as in Gower's 1966 principal coordinates analysis). To attain the latter, one should rescale the row scores dimensionwise by using the singular values ($\mathbf{X} = \mathbf{K}_p\Lambda_p$), and normalize the column scores ($\mathbf{A} = \mathbf{L}_p$), keeping the inner product fixed. Many intermediate solutions exist, as well as more extreme ones, and they are all equally valid as long as the singular values are partitioned properly over the row and column scores, i.e., $\mathbf{X} = \mathbf{K}_p\Lambda_p^q$, and $\mathbf{A} = \mathbf{L}_p\Lambda_p^{(1-q)}$.

We do not have to assume that the matrix $\mathbf{Q}$ contains fixed columns; it may contain any ordinal or nominal transformation of variables in a given matrix $\mathbf{Z}$. Columns in $\mathbf{Q}$ have to be as close as possible to a particular linear combination $\mathbf{XA}'$. Incorporating optimal scaling amounts to the minimization of $||\mathbf{XA}' - \mathbf{Q}||^2$ over $\mathbf{X}$, over $\mathbf{A}$, and over nonlinear functions $\mathbf{q}_j = \phi_j(\mathbf{z}_j), j = 1, \ldots, m$. If $\mathbf{Z}$ contains categorical variables, each column $\mathbf{z}_j$ defines a binary indicator matrix $\mathbf{G}_j$ with n rows and l_j columns, where l_j denotes the number of categories. Elements z_{ij} then define elements $g_{ir(j)}$ as follows: $z_{ij} = r \longrightarrow g_{ir(j)} = 1; z_{ij} \neq r \longrightarrow g_{ir(j)} = 0$, where $r = 1, \ldots, l_j$ is the running index indicating a category number in variable j. If category quantifications are denoted by $\mathbf{y}_j$, then a variable $\mathbf{q}_j$ can be written as $\mathbf{G}_j\mathbf{y}_j$. The functions $\phi_j(\mathbf{z}_j)$ are called transformations for ordinal variables, and scalings, scorings or quantifications for nominal variables.

In the optimal scaling process, we make a distinction between rank=1 and rank=p optimal scaling, where p denotes the chosen dimensionality in the solution. Rank=p optimal scaling has famous predecessors in the techniques proposed in Fisher (1938, 1940), Guttman (1941), Burt (1950), and Hayashi (1952). Benzécri's "analyse des correspondances (multiple)" became known as (multiple) correspondence analysis (Greenacre, 1984); other names for the same technique are dual scaling (Nishisato, 1980), and homogeneity analysis (Gifi, 1990). Rank=p optimal scaling implies a nominal scaling level, taking only categorical information into account, and is associated with a centroid model. A categorical variable is represented by a set of category points; rank=p optimal scaling locates a category point in the centre of gravity (centroid) of the associated objects. When rank-orders (among the categories) are to be taken into account as well, we choose an ordinal scaling level. The latter uses least squares monotonic regression or monotonic regression splines, and is usually associated with rank=1 optimal scaling, or a vector model. The ordinal scaling level originates with nonmetric multidimensional scaling (Kruskal, 1964), and was subsequently applied to regression analysis (Kruskal, 1965) and factor analysis (Kruskal & Shepard, 1974). Although rank=p is the most commonly used scaling level for nominal data, rank=1 scaling through least squares regression (Gifi, 1990) or nonmonotonic regression splines can be applied as well. Rank=1 optimal scaling (nominal and ordinal) fits category points as markers on a vector through the origin. The coordinates in the space of $\mathbf{X}$ are given by $\mathbf{y}_j\mathbf{a}_j'/\mathbf{a}_j'\mathbf{a}_j$.

4 Biplots in multidimensional scaling

Meulman (1986, 1992) described an approach that analyzes multivariate data through the derived distances between the n objects. As in the previous section, the data matrix $\mathbf{Q}$ may contain quantifications of categorical or transformations of ordinal variables. In the distance approach to multivariate analysis, the variables are used to define an observation or measurement space in which the objects are located according to their scores, and the distances in this observation space are proximities to be approximated by distances between object points in a low-dimensional representation space. When $D(\bullet)$ represents an $n \times n$ matrix with Euclidean distances, the objective function $||D(\mathbf{Q}) - D(\mathbf{X})||^2$ is minimized over the coordinates matrix $\mathbf{X}$ and the optimally scaled matrix $\mathbf{Q} = \{\mathbf{q}_j\} = \{\phi_j(\mathbf{z}_j), j = 1, \ldots, m\}$.

Although the primary aim is to represent distances between objects optimally, one usually wishes to know how the location of the object points is related to the variables generating the inter-object proximities. A straightforward way to project a set of variables in a given configuration is through the use of multiple regression, Kruskal & Wish (1978), called "property fitting" in Carroll (1972). In the regression, the columns in $\mathbf{X}$ are the independent variables, and the weight vector $\mathbf{a}_j$ obtained from the regression represents the (dependent) variable $\mathbf{q}_j$. To be precise, the coordinates in $\mathbf{a}_j$ give the endpoint of the vector that represents the variable $\mathbf{q}_j$ in the p-space $\mathbf{X}$. A comment on obtaining the biplot is this manner, is that different rationales are used for fitting objects on the one hand and variables on the other: object points are obtained using least squares distance fitting and vector coordinates through multiple regression, which makes the method incoherent. A possible alternative and coherent method was discussed in Meulman (1998). Because distances $D(\bullet)$ are invariant under rotation, $D(\mathbf{X}) = D(\mathbf{XA}')$ if $\mathbf{A}'\mathbf{A} = \mathbf{I}$. Thus,

$$\mathrm{STRESS}(\mathbf{X}) = ||D(\mathbf{Q}) - D(\mathbf{X})||^2 = ||D(\mathbf{Q}) - D(\mathbf{XA}')||^2, \tag{2}$$

where $\mathbf{A}$ is a rotation matrix of order $m \times p$. Since rotation matrices are usually of order $p \times p$, we will call $\mathbf{A}$ a *rotation-expansion* matrix, because the transformation preserves the distances in an expanded space. Obtaining the vector endpoints, amounts to an orthogonal Procrustes problem of order $p \times p$: define the eigenvalue decomposition of the $p \times p$ matrix $\mathbf{X}'\mathbf{QQ}'\mathbf{X}$ as $\mathbf{X}'\mathbf{QQ}'\mathbf{X} = \mathbf{L}\Lambda^2\mathbf{L}'$, then $\mathbf{A}$ is found by $\mathbf{A} = \mathbf{Q}'\mathbf{XL}\Lambda^{-1}\mathbf{L}'$. The coordinates to display the variable $\mathbf{q}_j$ in the space $\mathbf{X}$ are again given by the row vector $\mathbf{a}_j'$; the data elements $\{q_{ij}\}$ are represented as points in $\mathbf{X}$ by plotting $\mathbf{q}_j\mathbf{a}_j'/\mathbf{a}_j'\mathbf{a}_j$. The m-dimensional coordinate system $\mathbf{XA}'$ can be used to evaluate the MDS solution directly in terms of the (transformed) variables, with the squared Pearson correlation coefficient (variance-accounted-for) as a natural measure of goodness-of-fit.

5 Biplots in correspondence analysis

The starting point in this section is a two-way table $\mathbf{F}$ containing positive entries. The most common example of such a table is undoubtedly a two-way contingency table expressing the relationships between the categories of two categorical variables A and B, where the entry f_{ij} denotes the count of individual units of observation (objects) falling in category i of variable

A and in category j of variable B. An exemplary approach to the analysis of a contingency table is the study of the interaction. Independence is the key concept here, since lack of independence implies interaction between the categories of the two variables. The objective of modelling is then to replace the observed frequencies by estimates that satisfy certain regularity properties by using specific assumptions to link the observed counts to a particular structure. Instead of loglinear modelling, for example, this paper will use other strategies to smooth the empirical frequencies, and this is by fitting particular lower-rank approximations that allow for graphical display.

Correspondence analysis displays the residuals of the independence model, defined on the margins of the table. The matrix $\mathbf{E}$ with expected frequencies $\{e_{ij}\}$ is given by $\frac{1}{N}\mathbf{M}_r\mathbf{u}_r\mathbf{u}_c'\mathbf{M}_c$, where $\mathbf{M}_r$ and $\mathbf{M}_c$ denote diagonal matrices with the row respectively the column marginals on the main diagonal, and $\mathbf{u}_r$ and $\mathbf{u}_c$ are vectors of 1's of size R (number of rows) and C (number of columns), respectively. The grand total is denoted by N. In terms of lower-rank approximation, the inner product $\mathbf{XY}'$ of row scores $\mathbf{X}$ and column scores $\mathbf{Y}$ approximates $\mathbf{M}_r^{-1}(\mathbf{F} - \frac{1}{N}\mathbf{M}_r\mathbf{u}_r\mathbf{u}_c'\mathbf{M}_c)\mathbf{M}_c^{-1}$.

In contrast with the vector model used in principal components analysis, the results of a correspondence analysis are usually displayed in an alternative way: both row and column objects are represented as points. This way of representation is associated with the so-called unfolding/ideal point model (originally developed for preference data), where proximity relations between row and column entries are represented as distances (the closer a row object to a column object, the larger the preference; for example, see Heiser, 1987). In correspondence analysis, the unfolding/ideal point interpretation induces complications and confusion (for example, see the discussion between Carroll, Green & Schaffer, 1986, and Greenacre, 1989). A major factor is the indeterminacy in choosing a coherent normalization of row and column scores. This indeterminacy is directly associated to the infinite number of choices in the inner product approximation discussed for PCA in Section 3, and therefore it would be much more appropriate to interpret the correspondence analysis results in terms of the vector model as well. Different, yet *coherent* normalizations will give *different* sets of interpoint distances, but always the *same* inner products (projections). Whether the rows should be displayed as vectors or the columns depends on the particular correspondence table analyzed.

6 Alternative metrics and graphical display in the analysis of correspondence tables

We have seen in the previous section that correspondence analysis uses both the row means and the column means for standardization (in $\frac{1}{N}\mathbf{M}_r\mathbf{u}_r\mathbf{u}_c'\mathbf{M}_c$), and the χ^2 metric $\mathbf{M}_r^{-1}$ and $\mathbf{M}_c^{-1}$ for the rows and columns, respectively. This particular use of the marginal frequencies of the table results from the use of correspondence analysis in the study of (in)dependence between two categorical variables. Classical "Analyse des Correspondances" à la Benzécri

(1992), however, stands for a much more general technique to analyze any kind of positive measure of correspondence. From this point of view, we need not be confined to the particular use of the margins. Write a general inner product approximation of a correspondence table $\mathbf{P}$ as:

$$\mathbf{XY}' \approx \mathbf{W}_r^{-1}(\mathbf{P} - \frac{1}{N}\mathbf{D}_r\mathbf{u}_r\mathbf{u}_c'\mathbf{D}_c)\mathbf{W}_c^{-1}, \tag{3}$$

with $\mathbf{W}_r$ and $\mathbf{D}_r$ diagonal matrices for the rows, and $\mathbf{W}_c$ and $\mathbf{D}_c$ diagonal matrices for the columns. Then a number of simple alternatives for standard correspondence analysis exist. First, set $\mathbf{W}_r = \frac{N}{R}\mathbf{I}_r, \mathbf{W}_c = \frac{N}{C}\mathbf{I}_c, \mathbf{D}_r = \mathbf{W}_r$ and $\mathbf{D}_c = \mathbf{M}_c$. Then the inner product $\mathbf{XY}'$ approximates (a properly scaled version of) the matrix $\mathbf{P}$ corrected for its column means. Similarly, when $\mathbf{W}_r = \frac{N}{R}\mathbf{I}_r, \mathbf{W}_c = \frac{N}{C}\mathbf{I}_c, \mathbf{D}_r = \mathbf{M}_r$ and $\mathbf{D}_c = \mathbf{W}_c$, $\mathbf{XY}'$ approximates (a properly scaled version of) $\mathbf{P}$ corrected for its row means. If $\mathbf{P}$ contains proportions that sum to 1 in its rows (or its columns), the weight matrices simplify to $\mathbf{W}_r = \mathbf{I}_r, \mathbf{D}_r = \mathbf{I}_r, \mathbf{W}_c = \frac{R}{C}\mathbf{I}_c$, and $\mathbf{D}_c = \mathbf{W}_c$ (or $\mathbf{W}_r = \frac{C}{R}\mathbf{I}_r, \mathbf{D}_r = \mathbf{W}_r, \mathbf{W}_c = \mathbf{I}_c$, and $\mathbf{D}_c = \mathbf{I}_c$). In these four alternatives, the row and column metrics $\mathbf{W}_r^{-1}$ and $\mathbf{W}_c^{-1}$ are functions of the identity matrices $\mathbf{I}_r$ and $\mathbf{I}_c$, and thus we are dealing with ordinary Euclidean or Pythagorean distances, in contrast to the χ^2 distances in correspondence analysis. Of course, the latter fits in (3) as well by setting $\mathbf{W}_r = \mathbf{M}_r, \mathbf{W}_c = \mathbf{M}_c, \mathbf{D}_r = \mathbf{W}_r$, and $\mathbf{D}_c = \mathbf{W}_c$. (The five options are included in the 8.0 version of CORRESPONDENCE in SPSS Categories.)

There are various ways to display the approximation (in $\mathbf{XY}'$) of the relation between rows and columns in a correspondence table. First, when a column is represented in p-space by a vector, the end point of the vector is given by $\mathbf{y}_j$, where $\mathbf{y}_j$ contains the elements of the jth row in $\mathbf{Y}$. The coordinates of the projections of the row points in $\mathbf{X}$ onto the vector $\mathbf{y}_j$ in the joint space are then given in the $R \times p$ matrix $\mathbf{X}\mathbf{y}_j\mathbf{y}_j'/\mathbf{y}_j'\mathbf{y}_j$. Second, the approximation $\mathbf{X}\mathbf{y}_j$ can be viewed as a smooth version of a column in the correspondence table, under the assumption that the relation between rows and columns can be displayed in p-dimensional space with a decent fit. These estimated values can subsequently be plotted on separate uni-dimensional scales, one for each column, with markers given by the row labels, and ordered on the scale. Third, the same fitted values can be displayed against the row labels in a fixed predetermined order to display the columns as a function of the rows. These three possibilities will be shown in the next section.

7 Graphical representation of seasonal fluctuation

For the seasonal fluctuation data, we have rates available on the monthly incidence of seven diseases over nine years, so the data form a 12 x 7 x 9 correspondence table that contains various sources of variation. Rates vary per month, per year, and per type of disease. Although the differences per year and per type of disease are interesting in their own right, these are not the sources of variation we are interested in here. The major source of variation under study being the differences between seasons, the data were first scaled per disease over each 12 month period to have a column sum of one. Next, proportions were averaged over nine years, giving a 12 x 7 matrix that sums to one. Finally four seasons were created: winter was defined from January - March, spring from April - June, summer from July - September, and fall from October - December. The matrix **E** with expected frequencies and given by $\frac{1}{N}\mathbf{D}_r\mathbf{u}_r\mathbf{u}_c'\mathbf{D}_c$, expresses the null-hypothesis of nonexistent seasonal effects (proportions are equal over seasons). The (Observed - Expected) data can be perfectly displayed in three dimensions; a two-dimensional solution accounts for 99% of the total variance. The graphical display is given in Figure 1. It was

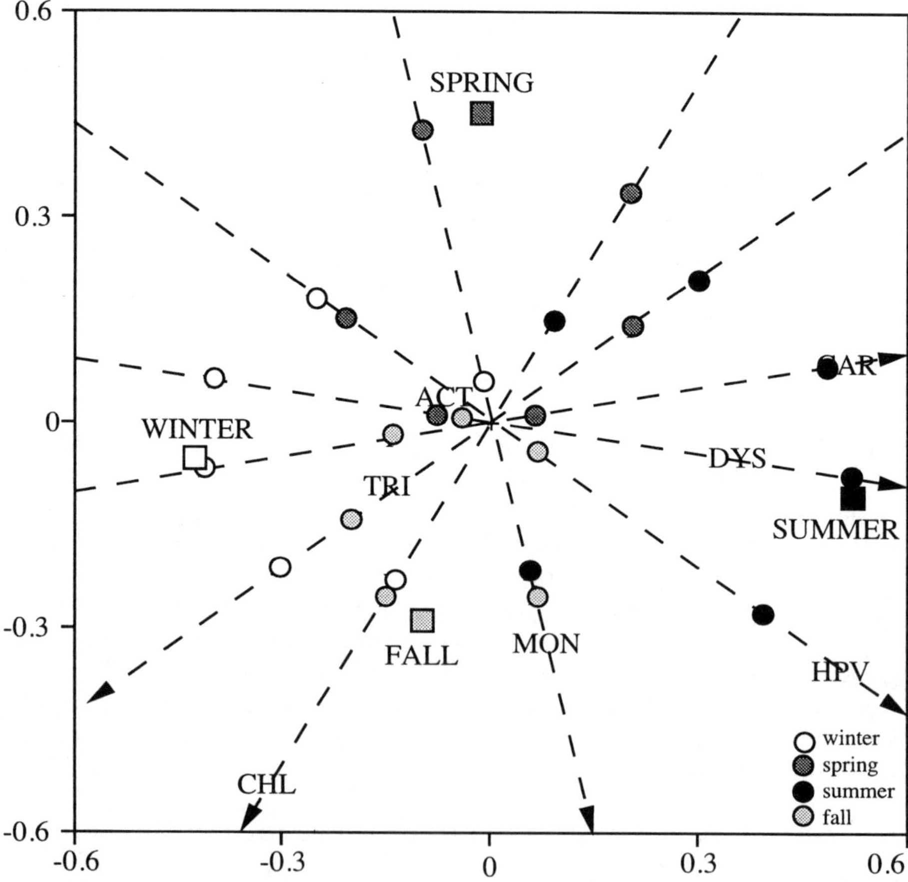

Fig. 1. Graphical display of the results of the relational data analysis of the seasonal disease data: seasons are represented as points, diseases as vectors, and the different markers symbolize the relation between each disease and the four seasons

deemed appropriate to display the seasons as points, indicated by an open square, and the diseases as vectors. Because the Euclidean metric was used, distances from the origin indicate the size of the seasonal variation. ACT and TRI show little variation; CHL, HPV and CAR a lot. The orthogonal projections (indicated by dots) of the seasons on a particular vector represents the prevalences of the disease during the year. The solution reveals a winter versus summer dimension (the horizontal axis) and a spring versus fall dimension (on the vertical axis). Starting with CAR, and going clockwise, the markers representing the seasons show that CAR and DYS occur especially in summer (and not in winter), HPV especially in summer (and not in spring and winter), MON in fall and summer (and not in spring), CHL most in fall and winter (and not in spring). TRI shows little variation, but occurs most in winter (and not in summer).

In Figure 2, this cyclic information has been used to order the diseases through the year, and plot the variation of each disease with its low and high point as the two endpoints of a uni-dimensional scale. The information

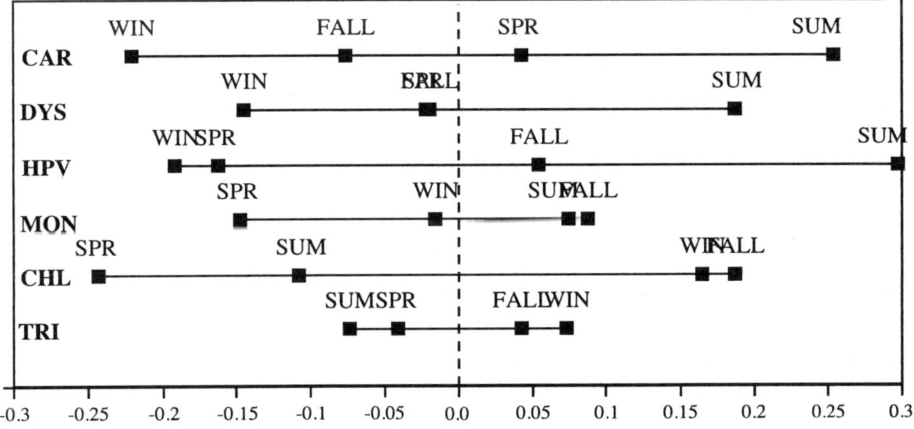

Fig. 2. An alternative way to represent the relation between seasons and diseases: separate ordering and spacing of fitted rates on uni-dimensional scales

that is displayed here is exactly the same as in Figure 1, but the seasonal differences can much more easily be compared. A final representation is given in Figure 3, where again the same information is displayed, but now the order of the seasons in the year has been kept fixed on the horizontal axes, while the variation is represented as a curve. This representation very nicely shows that through the year, diseases come and go, with CAR, DYS, and HPV in the first row having their peak in summer, and MON, CHL, and TRI in the second row having their peak in fall and winter.

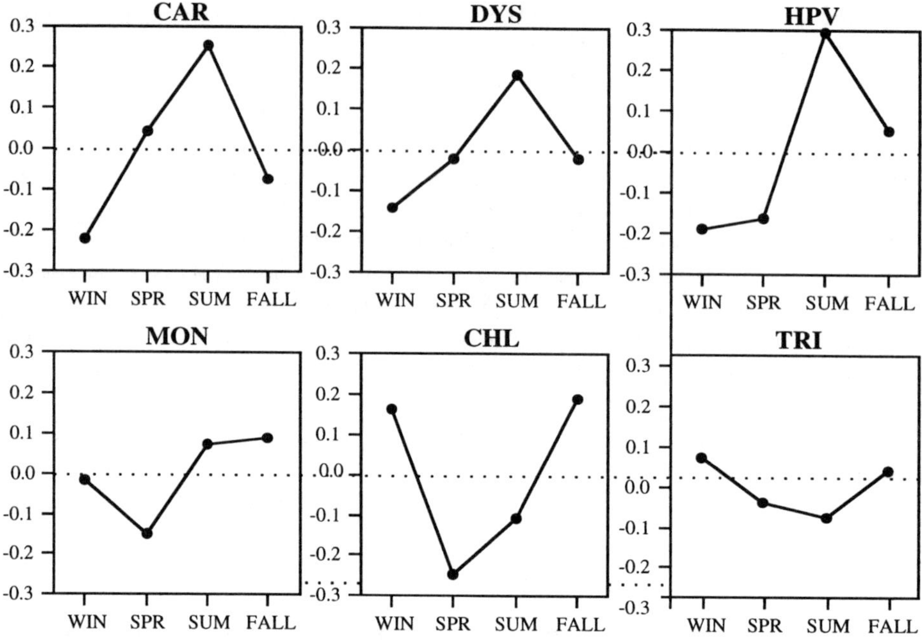

Fig. 3. A third way to represent the relation between seasons and diseases: fitted rates as functions of the seasons

8 Conclusion

Multivariate data analysis offers a lot of possibilities of graphical display. We have focused on the representation of rows as points, and columns as vectors in a common p-dimensional space. We have shown that this so-called vector model originates from the psychometric literature, and have argued that the results of a correspondence analysis, or the analysis of a correspondence table in general, can best be represented by a vector model as well, instead of a display that uses points for both the rows and the columns. The class of data analysis techniques discussed, includes categorical and ordinal variables. The latter can be represented as a set of points in the joint space of objects and variables. Those points are either located in the centroid of the appropriate objects (this is called rank=p optimal scaling), or on a straight line (vector) through the origin (rank=1 optimal scaling). The different approaches can be applied for different variables in the same analysis. In this manner, secondary biplots can also be derived. When a categorical group variable (like a response variable in discriminant analysis) is displayed as a set of centroids, these points can subsequently be projected on any vector representing an ordinal variable in the analysis. In this way, the graph not only displays the relationship between objects and variables and objects and groups, but between groups and variables as well, and we could call this display a *triplot*.

The alternative graphical displays proposed in Section 6, and shown in Section 7 in the analysis of a correspondence table, can also be used to display the latter relationship in multivariate data (for an example, see Van der Ham, Meulman, Van Strien & Van Engeland, 1997).

References

Benzécri, J.-P. (1992). *Correspondence analysis handbook.* New York: Marcel Dekkker.

Burt, C. (1950). The factorial analysis of qualitative data. *British Journal of Psychology,* **3**, 166-185.

Carroll, J.D. (1972). Individual differences and multidimensional scaling, In: *Multidimensional scaling: Theory and applications in the behavioral sciences* (ed. R.N. Shepard, A.K. Romney & S.B. Nerlove), Vol. 1, 105-155. New York and London: Seminar Press.

Carroll, J.D., Green, P.E., & Schaffer, C.M. (1986). Interpoint distances comparisons in correspondence analysis. *Journal of Marketing Research,* **23**, 271-280.

Coombs, C.H., Dawes, R.M., & Tversky, A. (1970). *Mathematical psychology: An elementary introduction.* Englewood Cliffs, NJ: Prentice-Hall.

Eckart, C. & Young, G. (1936). The approximation of one matrix by another of lower rank. *Psychometrika,* **1**, 211-218.

Fisher, R.A. (1938). *Statistical methods for research workers.* Edinburgh: Oliver & Boyd.

Fisher, R.A. (1940). The precision of discriminant functions. *Annals of Eugenics,* **10,** 422-429.

Gabriel, K.R. (1971). The biplot graphic display of matrices with application to principal components analysis. *Biometrika,* **58,** 453-467.

Gabriel, K.R. & Odoroff G. (1986). Some diagnoses of models by 3-D biplots. In: *Multidimensional data analysis* (ed. J. de Leeuw, W.J. Heiser, J.J. Meulman & F. Critchley), 91-111. Leiden: DSWO Press.

Gifi, A. (1990). *Nonlinear multivariate analysis.* Chichester: John Wiley & Sons.

Gower, J.C. (1966). Some distance properties of latent roots and vector methods used in multivariate analysis. *Biometrika,* **53,** 325-338.

Gower, J.C. & Hand, D.J. (1996). *Biplots.* London: Chapman & Hall.

Greenacre, M.J. (1984). *Theory and applications of correspondence analysis.* London: Academic Press.

Greenacre, M.J. (1989). The Caroll-Green-Schaffer scaling in correspondence analysis: A theoretical and empirical appraisal. *Journal of Marketing Research,* **26**, 358-365.

Guttman, L. (1941). The quantification of a class of attributes: a theory and method of scale construction. In: *The prediction of personal adjustment* (ed. P. Horst *et al.*), 319-348. New York: Social Science Research Council.

Hayashi, C. (1952). On the prediction of phenomena from qualitative data and the quantification of qualitative data from the mathematico-statistical point of view. *Annals of the Institute of Statistical Mathematics*, **2**, 93-96.

Heiser, W.J. (1987). Joint ordination of species and sites: the unfolding technique. In: *Developments in numerical ecology* (ed. P. Legendre & L. Legendre), 189-221. New York: Springer.

Kruskal, J.B. (1964). Multidimensional scaling by optimizing goodness of fit to a nonmetric hypothesis. *Psychometrika*, **29**, 1-28.

Kruskal, J.B. (1965). Analysis of factorial experiments by estimating monotone transformations of the data. *Journal of the Royal Statistical Society Series B*, **27**, 251-263.

Kruskal, J.B. (1978): Factor analysis and principal components analysis: bilinear methods. In: *International encyclopedia of statistics* (ed. W.H. Kruskal & J.M. Tanur), 307-330. New York: The Free Press.

Kruskal, J.B. & Shepard, R.N. (1974). A nonmetric variety of linear factor analysis. *Psychometrika*, **39**, 123-157.

Kruskal, J.B. & Wish, M. (1978). *Multidimensional scaling.* Newbury Park, CA: Sage.

Meulman, J.J. (1986). *A distance approach to nonlinear multivariate analysis.* Leiden: DSWO Press.

Meulman, J.J. (1992). The integration of multidimensional scaling and multivariate analysis with optimal transformations of the variables. *Psychometrika*, **57**, 539-565.

Meulman, J.J., (1998). A distance-based biplot for multidimensional scaling of multivariate data. In: *Data science, classification, and related methods* (ed. C. Hayashi, N. Ohsumi & Y. Baba), 506-517. Tokyo: Springer Verlag.

Nishisato, S. (1980). *Analysis of categorical data: Dual scaling and its applications.* Toronto: University of Toronto Press.

Rietveld, W.J., Boon, M.E. & Meulman, J.J. (1997). Seasonal fluctuations in the cervical smear detection rates for (pre)malignant changes and for infections. *Diagnostic Cytopathology*, **17**, 452-455.

Stewart, G.W. (1993). On the early history of the singular value decomposition. *SIAM Review*, **35**, 551-566.

Schmidt (1907). Zur Theorie der linearen und nichtlinearen Integralgleichungen. I. Teil. Entwicklung willkürlichen Funktionen nach System vorgeschriebener. *Mathematische Annalen*, **63**, 433-476.

Tucker, L. R (1960). Intra-individual and inter-individual multidimensionality. In: *Psychological scaling: theory and applications* (ed. H. Gulliksen & S. Messick), 155-167. New York: Wiley.

Van der Ham, Th., Meulman, J.J., Van Strien, D.C. & Van Engeland, H. (1997). Empirically based subgrouping of eating disorders in adolescents: a longitudinal perspective. *British Journal of Psychiatry*, **170**, 363-368.

Computer–Assisted Statistics Teaching in Network Environments

Marlene Müller

Institute for Statistics and Econometrics, Humboldt University Berlin, Spandauer Str. 1, D–10178 Berlin, Germany

Abstract. The paper presents the use of interactive tools and interactive graphical displays in introductory and advanced statistics courses. All the examples presented can be used with the statistical computing environment XploRe, either from a Java applet over WWW or in an generic standalone version on the users local computer.

Keywords. Computer–assisted learning of statistics, interactive displays, Java interface, XploRe

1 Introduction

In this article, a number of examples[1] are presented which give an idea how the environment XploRe is used to support and complement courses in introductory and advanced statistics.

For about two years now, the Institute for Statistics and Econometrics has been providing our teaching material in HTML, PostScript or PDF format. (See `http://wotan.wiwi.hu-berlin.de`.) All these materials can be downloaded by students. In parallel, we tried to provide a collection of accompanying XploRe macros for some of the courses.

In the following, some profiles for computer–assisted teaching are sketched. Two types of statistics courses are relevant here:

- Introductory courses:
  Students learn the basic elements and methods.
- Advanced courses:
  Students deal with particular statistical problems.

For both sorts of courses, different aspects of computer–support play a role for us:

Introductory statistics
Only a few computer-based examples are used in these courses. Computer–assisted teaching is meant to complement the course and is not (yet) an integral part of the course. A main reason for this is the fact, that introductory statistics at the Economics Department of Humboldt University is taught for an audience of about 300 students per year.

In consequence, supporting computer programs are primarily presented by the teacher. Their main object is to study properties of statistical objects (e.g. variables, distributions) and methods (e.g. linear regression).

To encourage the students, to try the programs themselves, the material should be easily accessible (WWW), mostly hardware independent, and easy to use.

[1] See `http://wotan.wiwi.hu-berlin.de/~marlene/x4interactive/`

Advanced statistics courses
Those courses, in which statistical software is directly used by students, cover multivariate statistical methods, non- and semiparametric modelling, option pricing and interactive statistics. Course scripts in electronic form are available for the courses "Applied Multivariate statistical Analysis" and "Non- and Semiparametric Modelling".

During these courses, students do not just use computer programs to study properties of statistical objects and methods. Additionally, they should learn to apply "serious" statistical methods to real world examples. This goes up to an introduction into programming the methods themselves.

An ideal framework for computer–assisted teaching would hence be a statistical software system that allows a smooth transition from introductory teachware examples to real statistical applications. This means, students start from software that they can later use for serious statistical analysis. During this transition process, they learn by modifying given programs, how to apply a statistical oriented programming language for the implementation of their own procedures.

2 The software: XploRe

The profiles for the use of a statistical software to assist teaching imply a number of requirements:

- for introductory statistics, routines should be mostly self–explaining,
- for advanced courses, several levels of complexity should be possible: from simple and easy–to–modify macros to full–featured applications,
- easy to access software,
- network capabilities, in particular WWW integration,
- high level programming language with interactive and graphic tools.

XploRe is an interactive computational environment for statistics which meets the above requirements. For an introduction to the software see Schmelzer, Kötter, Klinke & Härdle (1996). A central aspect of XploRe is that it can be used either as standalone version as well as within a local network or the Internet. More exactly, XploRe comes in several flavours:

(1) Generic (standalone) versions are available for Unix/X11 (Solaris/Sparc, Linux/PC, other Unices) and for MS Windows (95/NT for PC).
(2) A Java client version is available, to be used with a XploRe server running on a workstation. The server might run on a remote machine. The XploRe Java client runs under Java 1.1. Virtual Java machines are available for a wide number of operating systems.
(3) A Java applet version can be used from the XploRe Web site which provides access to XploRe from any Web browser supporting Java applets. (`http://www.xplore-stat.de/WWWJava/x4java.html`)
(4) A CGI interface version can be used from the XploRe Web site, giving access to XploRe to any browser that supports forms. (`http://www.xplore-stat.de/x4www.html`)

To use the generic version (1) or the Java client version (2), a local copy of the software needs to be present on the user's computer. To use the Java applet version (3) or the CGI version (4) only a Web browser (providing Java applets/forms) and an Internet connection are necessary.

The latter makes XploRe of particular interest for students: standard Web browsers offer both Java applets and forms, and students have easy access to

XploRe from the university PC pool, from their home PC or even an Internet cafe. In this case, there is no need to own or download the software itself. Additionally, the XploRe help system is provided on-line in HTML[2].

3 Teaching material on the Web

All XploRe routines for teaching can be downloaded from the Web. (See `http://wotan.wiwi.hu-berlin.de` and follow the links to Statistics and Lehrmaterial, Figure 1.) The macro collection for all courses consists of routines that can be used independently from each other.

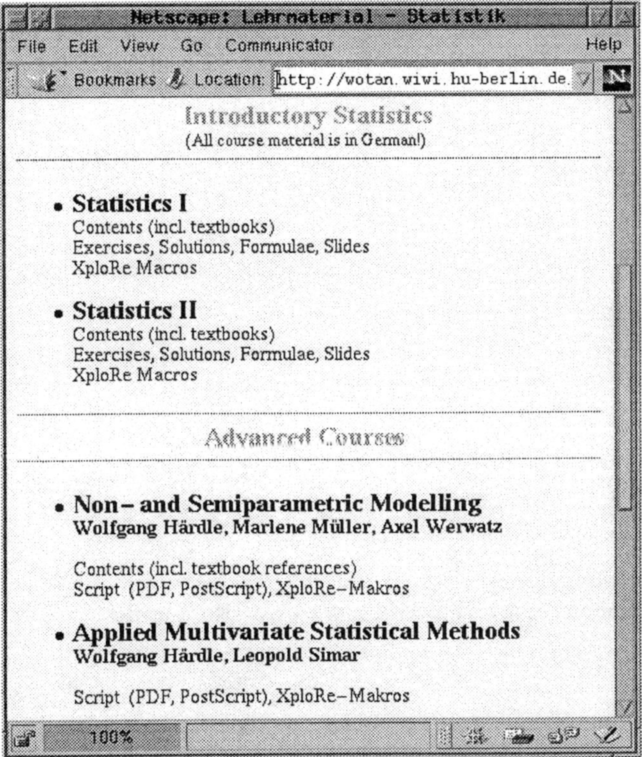

Fig. 1. Web pages with course material

Electronic course script versions are available for "Applied Multivariate statistical Analysis" and "Non- and Semiparametric Modelling". When using the HTML or PDF versions of these courses, it is possible to directly access the XploRe routines used for the examples, since hyperlinks point to them in the text (Figure 2).

[2] `http://www.xplore-stat.de/help/_Xpl_Start.html`

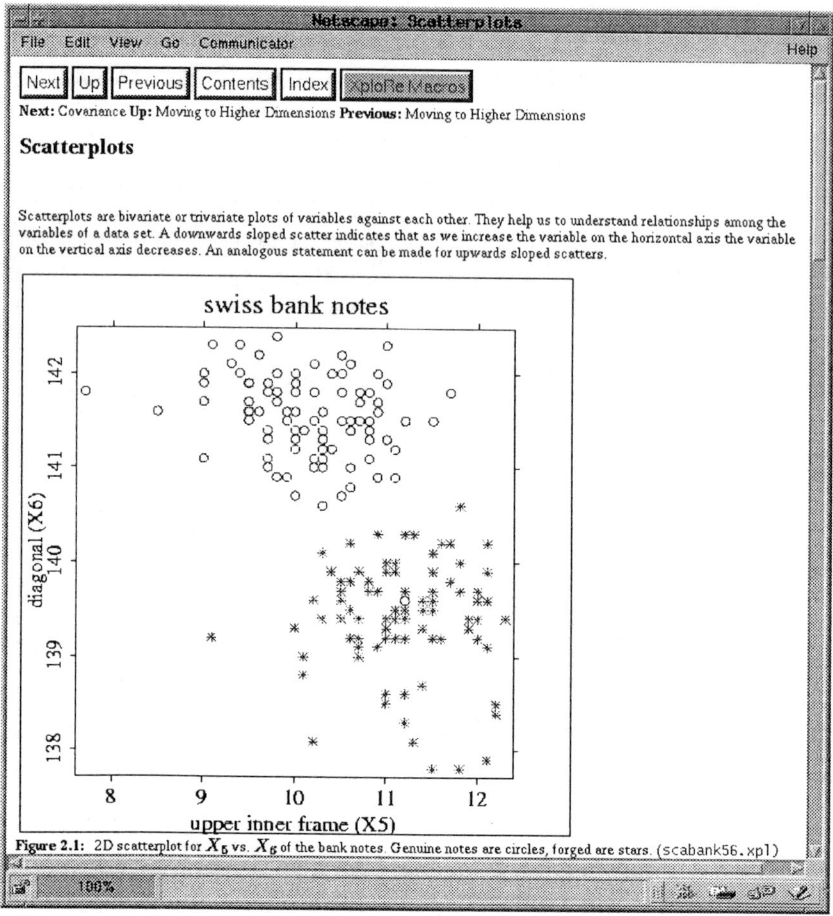

Fig. 2. Electronic course script with link to XploRe routines

When clicking the button "XploRe macros" from the electronic script, a new browser window appears presenting a list of all XploRe routines belonging to this course (Figure 3). It is the same page that could have chosen directly from the teaching material Web page. Each macro is given as a HTML file (in the style of XploRe help files) as well as in XploRe source form. The HTML contains links to related XploRe commands, libraries and functions. The source code can be directly downloaded, this is particularly useful when the students are expected to edit the source themselves in order to understand the programming or to modify certain parameters. The Java applet version of XploRe can be directly used from these "XploRe Macros" pages.

For some examples in the electronic script version, a direct hyperlink to a XploRe routine is given (as to `scabank56.xpl` in the figure caption in

Figure 2). In these cases, the XploRe routine consists of the code that has been used to create the graphic in the script.

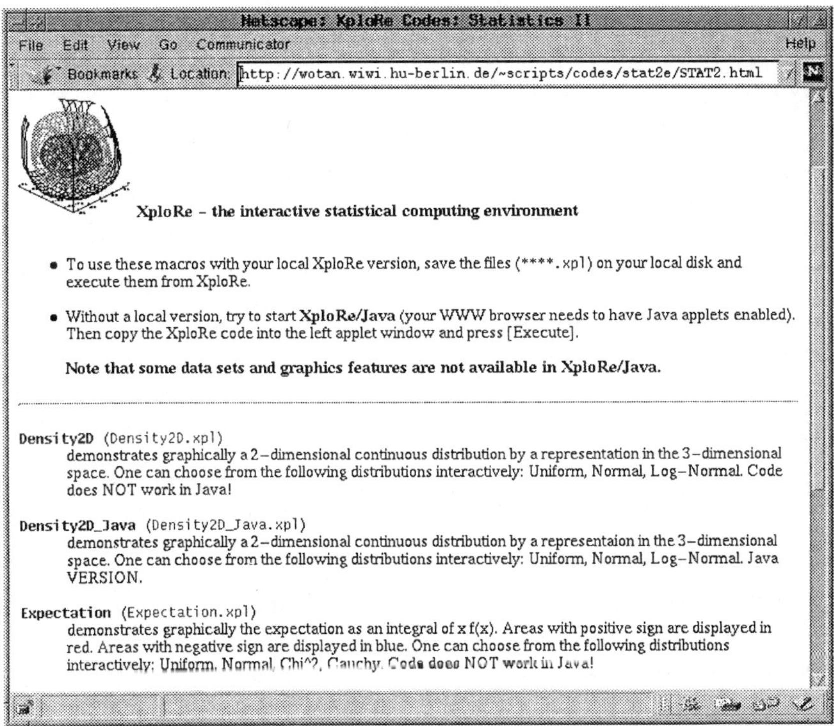

Fig. 3. XploRe routines on the Web

Below, source code for this example is printed. The data used here are the Swiss bank note data from Flury & Riedwyl (1988), which are used as a running example for the multivariate analysis course. The sample consists of 200 observations and two subgroups: 100 bank notes are genuine and 100 are forged. The problem connected with these data is to find a discrimination rule to separate the two groups. It is clear from Figure 2 that the two variables which are plotted (variables 5 and 6, as can be seen from the XploRe source), already give a good separation of the data. This is not so for other combinations of variables, and the students can easily verify this by modifying the code:

```
; -----------------------------------------------------
; Library        AMSA
; -----------------------------------------------------
;  See_also        createdisplay setmaskp show setgopt
; -----------------------------------------------------
;   Macro          scabank56
```

```
; ----------------------------------------------------------
;   Description  scabank56 computes a two dimensional
;                scatterplot of X5 vs. X6 (upper inner
;                frame vs. diagonal) of the Swiss bank
;                notes ("bank2.dat")
; ----------------------------------------------------------
  x=read("bank2")                      ; reads the bank data
  Scatterplot=createdisplay(1, 1)
  layout=3*matrix(100)|12*matrix(100)
  color=1*matrix(100)|4*matrix(100)
  xx=x[,5:6]                           ; variables 5 and 6
  setmaskp(x56, color, layout, 8)  ; mask vector
  show(Scatterplot, 1, 1, xx)          ; 2D plot of variables
  setgopt(Scatterplot, 1, 1, "title", "swiss bank notes")
```

The header of the above lines of code is used to create the corresponding HTML page which then also hyperlinks to the referred XploRe commands (`createdisplay`, `setmaskp`, `show`, `setgopt`).

The above example code is a very simple one, and the interaction of the students with the software is restricted to the direct modification of the code. This requires at least some basic knowledge of programming XploRe and therefore serves as one of the starting examples in the process of learning XploRe.

More interaction is required, when introducing statistical concepts and objects to students beginning with statistics. The next two sections give more insight about the tools that are available in this case.

4 Interactive routines in the Java applet version

Interactive tools for XploRe are partially still under development. However, two basic features are available in virtually all XploRe versions (except the CGI version):

- `readvalue`:
  A input box to enter and modify parameters.
- `selectitem`:
  A selection box to choose from a number of options.

These two features already allow a lot of interaction for the user. Their use in teachware routines will be presented in the following. All the examples can be downloaded and executed either in the Java applet version or in a local XploRe version. The look of the examples will be a little different depending on the GUI used, however the basic handling is identical over all versions and platforms.

A very typical object to start with is the shape of a normal density depending on the location μ and scale parameter σ. This example is realized in the macro `NormalDensity.xpl`[3]. Figure 4 presents a screenshot of its use in the Java applet version. After loading the Java applet, a windows appears, presenting two subwindows: for Input (left) and Output (right). Although an additional text editor is available, it is also possible to edit the code in the Input frame. To run the macro `NormalDensity.xpl`, it needs to be copied into the Input frame and executed (by pressing the Execute button). For more information on the Java applet version see Kötter (1997).

[3] See `http://wotan.wiwi.hu-berlin.de/~marlene/x4interactive/`

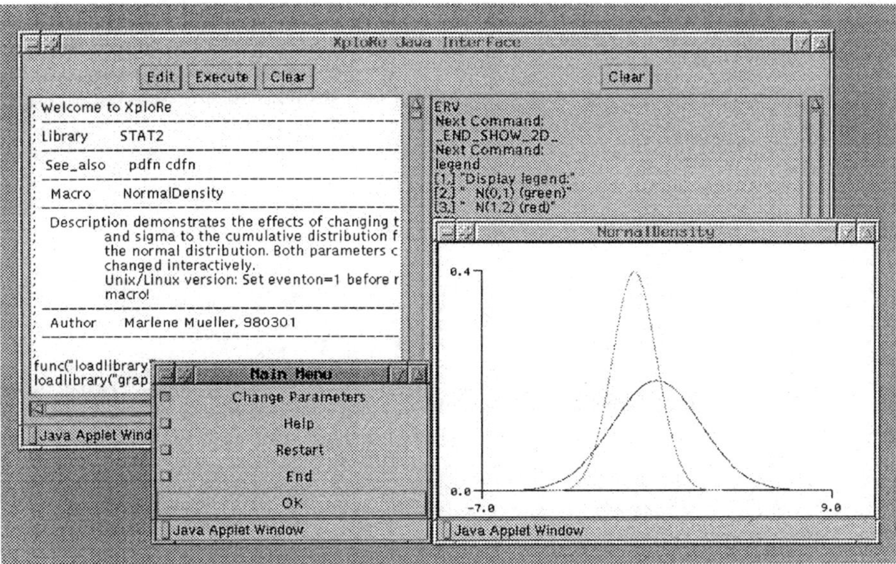

Fig. 4. Normal densities in Java applet

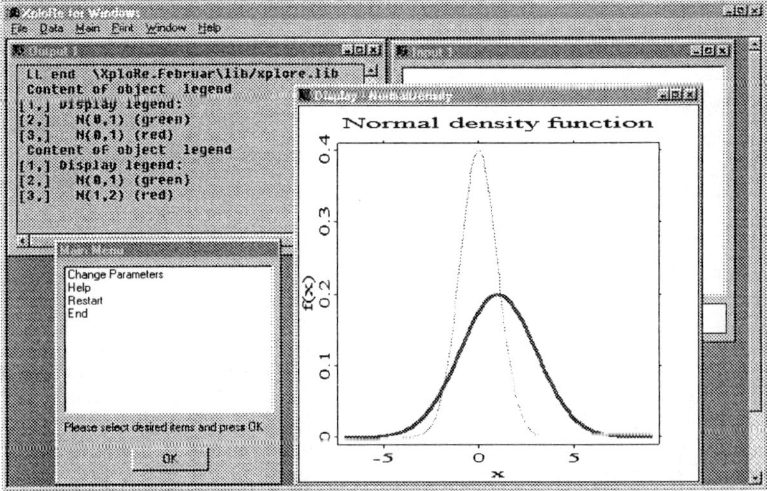

Fig. 5. Normal densities in Windows version

The result is a graphical display showing the standard normal density function (in red) and separately a **selectitem** box (headed "Main menu"). The selection box offers 4 items: "Change parameters" to modify μ and σ, a

"Help" item to printout some explanation, a "Restart" button to reset the parameters, and "End" to stop the macro execution. When choosing "Change parameters", a `readvalue` box appears that allows us to enter new parameters. The routine `NormalDensity.xpl` is so designed, that the starting density (in green) always remains in the plot and serves as a reference curve. In this way, the effect of deviation from the initial parameters μ and σ can easily be checked.

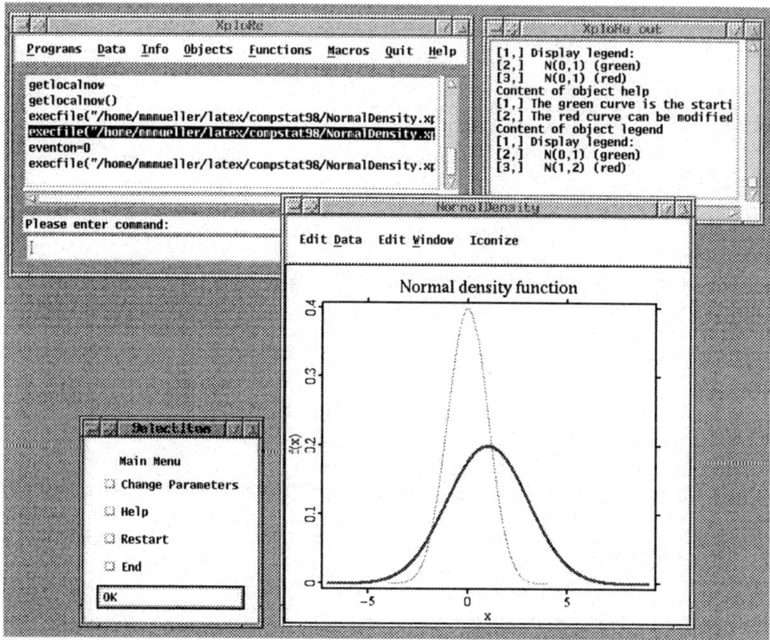

Fig. 6. Normal densities in Unix version

The basic visual difference in the generic Windows and Unix versions of XploRe (compared to the Java versions) consists of the fact that they have separate Input and Output windows. Still, `NormalDensity.xpl` can be used in the same way, Figure 5 and 6 show a screenshot of running this example under the Windows 95 and Linux operating systems, respectively. A collection with more macros, covering also linear and kernel regression, histograms and nonparametric density estimation as well as descriptive statistics, can be found at `http://wotan.wiwi.hu-berlin.de/~marlene/x4interactive/`.

5 Interactive displays in the Unix environment

The Unix versions of XploRe offer some extended interaction possibilities. A number of interesting examples, in particular on the combination of text and graphics, can be found in Schmelzer (1997). We will focus in this section, on how to use some of this features to improve the teachware routines from the previous section.

The key command used in the following, is the XploRe command

- `readevent`:
  Reads mouse clicks or keyboard events.

This allows us interactively to change certain features in a graphical display by simply clicking on them. An example is the following routine, that can be used to explore different types of data. The macro `Credits.xpl` allows us to choose between one of five variables: personal id, credit worthiness, purpose of credit, monthly payments (from 1=low to 4=high) and amount of credit. The data is a subsample of 25 individuals from the credit scoring data of Fahrmeir & Tutz (1994). This small number of observations is chosen intentionally, to allow the students to verify the result easily "by hand". The variables represent the different types: binary, discrete, ordinal and continuous.

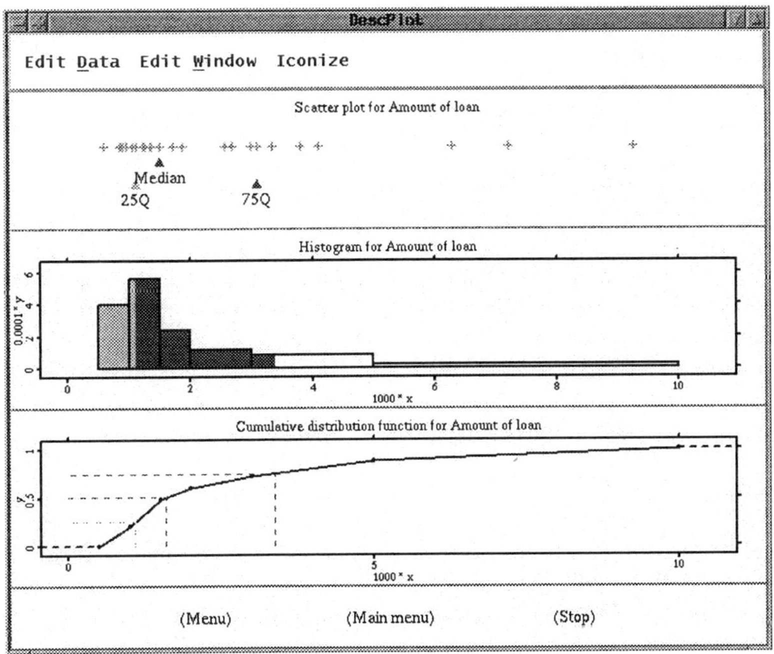

Fig. 7. Descriptive statistics of Credit data

The purpose of the macro is to show the different techniques for explorative analysis. Different tools can be used, for example a barchart can be computed for each variable, although it may not be reasonable in all cases. Figure 7 shows an explorative analysis of the continuous variable `amount`, the amount of the credit taken by the individual.

Up to this point the usage of `Credits.xpl` is quite similar to the previous example. When calling the macro, a menu appears that offers all the variables to analyse. If a variable has been chosen, a second menu offers the graphical tools: barchart, scatterplot histogram, distribution function. Additionally median and quantiles can be chosen, which appear in the latter three graphics (see Figures 7, 8: 25% quantile appears in orange, median in red, 75% quantile in brown). Moreover, a frequency table and mean, variance and other statistical characteristics can be printed in the Output window.

As an advanced interactive feature, the histogram borders can be changed by clicking directly on the graphic. Figure 8 shows the histogram and distribution function after two additional class borders have been. Also the menu has now moved into the graphics: the lowest panel in Figure 7 allows us to return back to the previous selection boxes. The brackets around Menu, Main Menu and End are used to indicate that clicking on these words will result in an action.

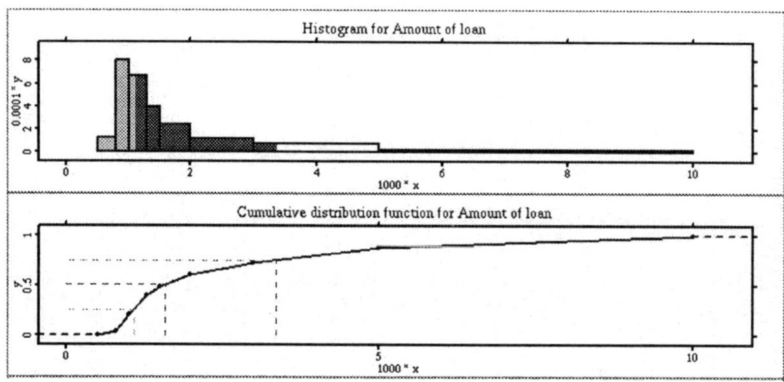

Fig. 8. Descriptive statistics of Credit data with modified histogram and distribution function

A similar technique of an interactive display could be used for the normal density example from the previous section. Figure 9 displays this routine in modified form now. Again words in brackets refer to active areas of the window. Moreover arrows allow us to slightly move the parameters up and down.

In fact, Figure 9 is based on the same source code as Figures 4, 6 and 5. An interface function entirely written in the XploRe programming language was used to create both types of appearance, in dependence of the setting of an global variable (`eventon`). For this interface function, only a few subprocedures have to be provided that produce the display (or two displays), the legend and additional menu items as well as the starting parameters and information in which way parameters may be modified. More details from the source[4] of `NormalDensity.xpl`.

[4] See `http://wotan.wiwi.hu-berlin.de/~marlene/x4interactive/`

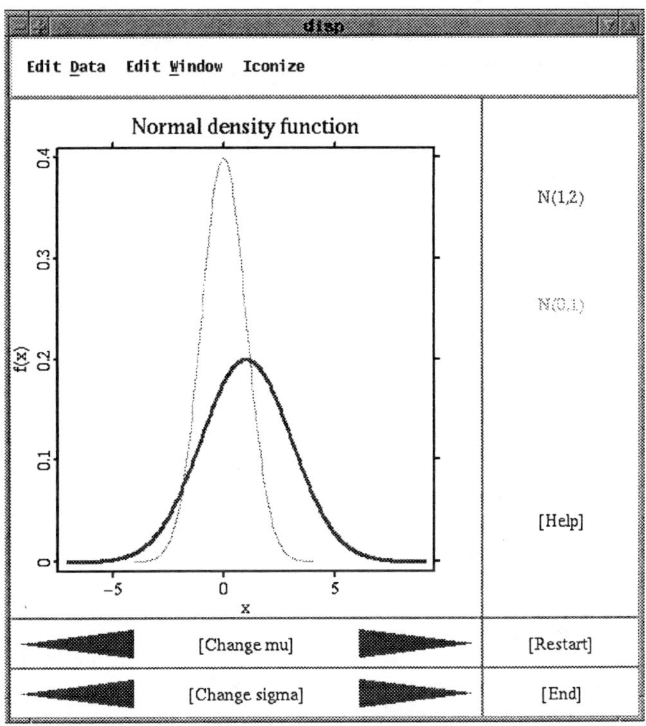

Fig. 9. Normal densities in an interactive display

Another example is the macro `NormalExpectation.xpl` that shows the use of two graphics displays to be modified by the parameters. The upper display corresponds to the density $f(x)$ in `NormalDensity.xpl`, the lower plots the function $xf(x)$ in order to study the modification of the expectation, when the parameters change (Figure 10). The expectation (the integral) can here be read from the area between the curve and the horizontal zero axis.

Acknowledgements

I wish to thank my colleagues at the Institute of Statistics and Econometrics, Humboldt University, Berlin, which have greatly influenced and supported this work on computer–assisted teaching. In particular, I would like to thank the XploRe programming team (Sigbert Klinke, Thomas Kötter and Swetlana Schmelzer) who made all these useful functions available to XploRe.

References

Fahrmeir, L. & Tutz, G. (1994). *Multivariate Statistical Modelling Based on Generalized Linear Models.* New York: Springer–Verlag.

Flury, B. & Riedwyl, H. (1988). *Multivariate Statistics. A Practical Approach.* London: Chapman & Hall.

Härdle, W. & Simar, L. (1998). *Applied Multivariate Statistical Methods.*

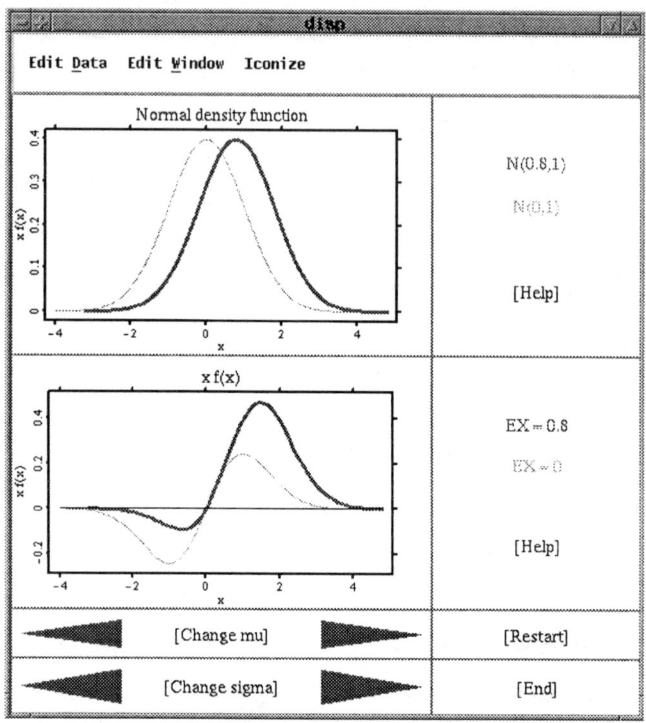

Fig. 10. Normal expectations in an interactive display

Course Script, Institute for Statistics and Econometrics, Humboldt University Berlin.

Härdle, W., Müller, M. & A. Werwatz (1998). *Non- and Semiparametric Modelling.* Course Script, Institute for Statistics and Econometrics, Humboldt University Berlin.

Kötter, T. (1997). Interactive Interfaces of Statistical Software for the Internet. In: *SoftStat '97 – Advances in Statistical Software* (ed. F. Faulbaum & W. Bandilla), 153–158. Stuttgart: Lucius & Lucius.

Müller, M. (1998). Teaching Statistics with XploRe. *Maths&Stats Newsletter*, 21–24. Glasgow: CTI Statistics.
(`http://www.stats.gla.ac.uk/cti/activities/articles.html`)

Schmelzer S. (1997). Processing text information in the highly interactive XploRe environment. *Working paper.*
(`http://wotan.wiwi.hu-berlin.de/~swetlana/papers/art1.ps`)

Schmelzer, S., Kötter, T., Klinke, S. & Härdle W. (1996). A New Generation of a Statistical Computing Environment on the Net. In: *COMPSTAT 1996 Proceedings in Computational Statistics* (ed. A. Prat), 135–148. Heidelberg: Physica–Verlag.

Modelling Bacterial Genomes Using Hidden Markov Models

Florence Muri

Laboratoire de Statistique Médicale, UA CNRS 1323, Université Paris V, 45 rue des Saints-Pères, 75006, Paris, France

Abstract. Long DNA sequences are often heterogeneous in composition. Hidden Markov models are then good statistical tools to identify homogeneous regions of the sequences. We compare different identification algorithms for hidden Markov chains and present some applications to bacterial genomes to illustrate the method.

Keywords. Hidden Markov models, EM algorithm, MCMC methods, DNA sequences, heterogeneity of DNA sequences

1 Introduction

With the great number of sequencing projects, biologists now have large sets of DNA sequences and need statistical tools to analyse all this information. A DNA sequence is a long succession of four nucleotides or bases, Adenine, Cytosine, Guanine and Thymine and can be represented by a finite series $y_1, \cdots, y_n$, each base y_t taken from the alphabet $\mathcal{Y} = \{A, C, G, T\}$. It turns out that an important heterogeneity exists along the genome. Statistical models based on the homogeneity assumption all along the sequence are thus not very realistic. Our purpose is to use a model taking into account the observed heterogeneity, to identify homogeneous regions in the DNA sequence. The break points which delimit these regions may thus separate parts of the genome with different functional or structural properties.

In the hidden Markov chain approach that we propose to use, one assumes that the DNA sequence has a mosaic structure composed of homogeneous regions and that there is a finite number q of models providing a good description of each region. The regions succession is described by an unobservable q-state Markov chain (the hidden states chain). Hence, the bases of the sequence appear with a law which depends on the hidden state. The aim is to reconstruct these regions from the DNA sequence and to estimate the parameters of the q models to characterize the identified regions.

Churchill (1989) has used the EM algorithm (introduced by Dempster *et al.*, 1977, for incomplete data) to compute the maximum likelihood estimate of such a model and to identify homogeneous regions in DNA sequences. To avoid some drawbacks of this procedure, we consider Markov Chain Monte Carlo methods (see for instance Geman & Geman, 1984; Robert, 1996) by firstly considering two stochastic versions of EM, the SEM algorithm, introduced by Celeux & Diebolt (1985) in the mixture setup, and the EM à la Gibbs algorithm, introduced by Robert *et al.* (1993) for hidden Markov chains. We also propose a Bayesian estimation using Gibbs sampling. The performances of these algorithms are discussed and compared by simulations. We then apply these methods to identify homogeneous regions in DNA sequences of two bacteriophages, *lambda* and *bIL67* and of the *B. subtilis* bacterium.

2 Hidden Markov Model

Hidden Markov models are characterized by two processes (see for instance Rabiner, 1989): the hidden states process $s = (s_1, \cdots, s_n)$ such that $s_t \in \mathcal{S} = \{1, \cdots, q\}$ (which in our setup governs the arrangement of the q possible regions along the sequence) and the observed process $y = (y_1, \cdots, y_n)$, $y_t \in \mathcal{Y} = \{A, C, G, T\}$, corresponding to the observed DNA sequence. The states are generated according to an homogeneous first order Markov chain whose transition matrix is denoted by

$$A = \Big(a(u, v) = P(s_t = v \mid s_{t-1} = u), \quad 1 \leq u, v \leq q, \quad \forall t = 1, 2, \cdots \Big)$$

and with an initial distribution equal to $a = \big(a(u) = P(s_1 = u),\ 1 \leq u \leq q \big)$. The bases appear in the sequence with a law which depends on the hidden states (that is which of the q possible regions they belong to).

The *M1-M0* model assumes that, conditionnally on the state s_t, the bases are drawn independently with probability

$$B = \Big(b(u, i) = P(y_t = i \mid s_t = u), 1 \leq u \leq q,\ i \in \mathcal{Y}, \quad \forall t = 1, 2, \cdots \Big)$$

Hence, this model takes into account the bases' composition in the sequence and corresponds to the classical hidden Markov model described in the literature. More generally, the *M1-Mk* model assumes an order k Markovian dependence between the observations conditionnally on the hidden states, with transitions

$$B = \Big(b(u, i_1, \cdots, i_k, i_{k+1}) = P(y_t = i_{k+1} \mid y_{t-1} = i_k, \cdots, y_{t-k} = i_1, s_t = u), \\ 1 \leq u \leq q,\ (i_1, \cdots, i_k, i_{k+1}) \in \mathcal{Y}^{k+1}, \quad \forall t = 1, 2, \cdots, \Big)$$

This model, introduced by Churchill (1989) allows us to take account of the local structure in k-nucleotides of the DNA sequence.
The parameters of the model are denoted by $\theta = (A, B)$ and belong to a space Θ (for the *M1-M0* model, Θ is the $q \times q + q \times |\mathcal{Y}|$-dimensional space of the stochastic matrices A and B).

Hidden Markov chains are thus missing data models and mixture models with dependent data. Let $f(y \mid \theta)$ be the likelihood of the incomplete data y and $g(y, s \mid \theta)$ the likelihood of the complete data (y, s) related to f by

$$f(y \mid \theta) = \sum_{s \in \mathcal{S}^n} g(y, s \mid \theta) \ .$$

In the *M1-M0* model, the incomplete data likelihood is

$$f(y \mid \theta) = \sum_{s \in \mathcal{S}^n} a(s_1) b(s_1, y_1) \prod_{t=2}^{n} a(s_{t-1}, s_t) b(s_t, y_t)$$

and is thus untractable for large values of n. The aim is to reconstruct the hidden states to identify homogeneous regions in the sequence and to estimate θ to characterize the identified regions.

3 Identification algorithms of hidden chains

We consider two approaches: maximum likelihood estimation[1] and Bayesian estimation. Because of the missing data s, the estimation from the incomplete likelihood $f(y \mid \theta)$ is difficult to perform. The solution is to augment the data, by assigning a value to the hidden states s_t and to work with the complete likelihood $g(y, s \mid \theta)$. We present different iterative procedures of the hidden chains (see for instance Rabiner, 1989; Qian & Titterington, 1990; Robert *et al.*, 1993; Archer & Titterington, 1995, for a review). Given a starting point $\theta^{(0)}$ and the common value $\theta^{(m)}$ of the parameter, these algorithms alternate two steps :

1. assign a value to the hidden states $s^{(m+1)} = \left(s_1^{(m+1)}, \cdots, s_n^{(m+1)}\right)$ from $\theta^{(m)}$ and the DNA sequence y;
2. given $s^{(m+1)}$, update $\theta^{(m)}$ by $\theta^{(m+1)}$ on the basis of the complete likelihood $g(y, s^{(m+1)} \mid \theta^{(m)})$.

In the following, the notation $y_{t_1}^{t_2}$ will refer to the $t_2 - t_1 + 1$ consecutive bases $y_{t_1}, y_{t_1+1}, \cdots, y_{t_2-1}, y_{t_2}$. To simplify, we present the algorithms in the *M1-M0* model.

3.1 EM algorithm

The EM algorithm for hidden Markov model is known as the Baum-Welch algorithm (Baum *et al.*, 1970) and consists of

E step: calculate $P(s_{t-1} = u, s_t = v \mid y, \theta^{(m)})$ for all positions t;
M step: choose $\theta^{(m+1)} = \arg\max_{\theta \in \Theta} E\left(\log g(y, s \mid \theta) \mid y, \theta^{(m)}\right)$.

The probabilities in the *E step* are calculated by a "forward-backward" recurrence on the sequence positions (see for instance Rabiner, 1989; Churchill, 1989) derived from the filtering probabilities

$$P(s_t = v \mid y_1^t, \theta) \propto b(v, y_t) \sum_{u=1}^{q} a(u, v) P(s_{t-1} = u \mid y_1^{t-1}, \theta) \tag{1}$$

[1] The consistency and normality results that justify the maximum likelihood approach, have been proved in the *M1-M0* model by Baum & Petrie (1966) and extended to the *M1-M1* model by Muri (1997).

We obtain

$$\begin{aligned} P(s_n = u \mid y, \theta) &= P(s_n = v \mid y_1^n, \theta) \\ P(s_{t-1} = u \mid y, \theta) &= \sum_{v=1}^{q} P(s_{t-1} = u, s_t = v \mid y, \theta) \quad \text{for } t = n, \cdots, 2 \\ &\propto \sum_{v=1}^{q} a(u, v) P(s_{t-1} = u \mid y_1^{t-1}, \theta) P(s_t = v \mid y, \theta) \end{aligned}$$

The maximization in the *M-step* is straightforward and leads to the classical estimates

$$a^{(m+1)}(u, v) = \frac{\sum_{t=2}^{n} P(s_{t-1} = u, s_t = v \mid y, \theta^{(m)})}{\sum_{t=2}^{n} P(s_{t-1} = u \mid y, \theta^{(m)})}$$

$$b^{(m+1)}(i, v) = \frac{\sum_{t=1}^{n} P(s_t = v \mid y, \theta^{(m)}) 1_{\{y_t = i\}}}{\sum_{t=1}^{n} P(s_t = v \mid y, \theta^{(m)})}$$

In the mixture setup, Redner & Walker (1984) proved that every limit point of the sequence $(\theta^{(m)})_{m \geq 0}$ generated by EM satisfies the incomplete log-likelihood equations and that $(\theta^{(m)})_{m \geq 0}$ converges towards the maximum likelihood estimate if the starting point $\theta^{(0)}$ is not too far from the true value. Muri (1997) extends this result to hidden Markov models, under the condition that $(\theta^{(m)})_{m \geq 0}$ is contained in a compact neighbourhood of the true value. From a practical point of view, EM can thus converge to a local maximum.

The E and M *steps* are alternated until an iteration M for which we state convergence[2]: θ is then estimated by $\theta^{(M)}$ and for all positions t in the sequence, the probability of the state s_t to be $v = 1, \cdots, q$, by $\hat{P}(s_t = v) = P(s_t = v \mid y, \theta^{(M)})$.

3.2 MCMC algorithms

To avoid the drawbacks of EM (such as poor stabilization), we consider MCMC alternatives, using Gibbs sampling, to identify the hidden chains.

3.2.1 Maximum likelihood estimation

We present here two stochastic versions of the EM algorithm, the SEM and the EM à la Gibbs algorithms, that can be considered as maximum-likelihood versions of the Gibbs sampling. These two algorithms consists of

E step: simulate the hidden states $s^{(m+1)} = \left(s_1^{(m+1)}, \cdots, s_n^{(m+1)}\right)$;

M step: choose $\theta^{(m+1)} = \arg\max_{\theta \in \Theta} \log g(y, s^{(m+1)} \mid \theta)$.

SEM and EM à la Gibbs only differ in the way of simulating the states at the *E step*.

[2] The stopping rule is $|\log f(y \mid \theta^{(M+1)}) - \log f(y \mid \theta^{(M)})| \leq \epsilon$, for a given ϵ.

For the SEM algorithm, the states s_t are simulated according to their joint conditional law $\pi(s \mid y, \theta^{(m)})$ derived from the relation

$$P(s_1, \cdots, s_n \mid y, \theta) = P(s_n \mid y, \theta) \ldots P(s_t \mid s_{t+1}^n, y, \theta) \ldots P(s_1 \mid s_2^n, y, \theta) .$$

As EM, the *E step* needs a backward-forward formula (see for instance Qian & Titterington, 1990) to calculate for $t = n, \cdots, 1$

$$\begin{aligned} P(s_n = u \mid y, \theta) &= P(s_n = u \mid y_1^n, \theta) \\ P(s_t = u \mid s_{t+1}^n, y, \theta) &\propto a(u, s_{t+1}) P(s_t = u \mid y_1^t, \theta) \end{aligned} \qquad (2)$$

where the filtering probabilities $P(s_t = u \mid y_1^t, \theta)$ are calculated by the forward recurrence (1).

The EM à la Gibbs algorithm avoids this time-consuming recurrence by simulating the states $s_t^{(m+1)}$ component by component according to the conditional distribution $\pi(s_t \mid s_{t'<t}^{(m+1)}, s_{t''>t}^{(m)}, \theta^{(m)}, y)$. In a *M1-M0* model, those distributions are given by

$$\begin{aligned} &\pi(s_1 \mid s_2, y_1, \theta) \propto a(s_1) b(s_1, y_1) a(s_1, s_2) \\ &\pi(s_t \mid s_{t' \neq t}, y, \theta) = \pi(s_t \mid s_{t-1}, s_{t+1}, y_t, \theta) \propto a(s_{t-1}, s_t) b(s_t, y_t) a(s_t, s_{t+1}) \\ &\pi(s_n \mid s_{n-1}, y_n, \theta) \propto a(s_{n-1}, s_n) b(s_n, y_n) \end{aligned}$$

and are much easier to simulate from than (2).

The maximization in the *M-step*, for both algorithms, leads to the estimates

$$a^{(m+1)}(u, v) = \frac{\sum_{t=2}^n 1_{\{s_{t-1}^{(m+1)} = u, s_t^{(m+1)} = v\}}}{\sum_{t=2}^n 1_{\{s_{t-1}^{(m+1)} = u\}}}$$

$$b^{(m+1)}(i, v) = \frac{\sum_{t=1}^n 1_{\{s_t^{(m+1)} = v\}} 1_{\{y_t = i\}}}{\sum_{t=1}^n 1_{\{s_t^{(m+1)} = v\}}}$$

Some convergence results, in this frame, are given by Robert *et al.* (1993). These two algorithms generate two chains in parallel, the state chain $(s^{(m)})_{m \geq 0}$ which is a uniform ergodic Markov chain with finite state space $\{1, \cdots, q\}^n$, and the parameter chain $(\theta^{(m)})_{m \geq 0}$. A duality principle proved by Robert *et al.* (1993) between these two chains, states that the chain $(\theta^{(m)})_{m \geq 0}$ is also uniformly ergodic and converges to a limit distribution $\pi(\theta \mid y)$. We estimate θ by the average $\hat{\theta} = \frac{1}{M} \sum_{m=1}^M \theta^{(m)}$ and the state probabilities $\hat{P}(s_t = v)$ by the average $\frac{1}{M} \sum_{m=1}^M 1_{\{s_t^{(m)} = v\}}$. The limit distribution $\pi(\theta \mid y)$ has no real statistical meaning, but Muri (1997) verifies by simulations that $\pi(\theta \mid y)$ should be centered around the consistent solution of the incomplete likelihood equations (such a result has been proved in the mixture setup by Celeux & Diebolt, 1985, under strong conditions); it means that $\hat{\theta}$, which estimates the theoretical mean $\int_\Theta \theta \pi(\theta \mid y) d\theta$, is near the true value of the parameter.

3.2.2 Bayesian estimation

Bayesian estimation relies on the posterior distribution $\pi(\theta \mid y)$ deduced from a prior distribution $\pi(\theta)$ and the likelihood $f(y \mid \theta)$. We consider independent Dirichlet priors $\mathcal{D}(\alpha(u,1),\ldots,\alpha(u,q))$ for each row a_u of the state transition matrix A and $\mathcal{D}(\beta(u,A),\ldots,\beta(u,T))$ for the rows b_u of the observation probability matrix B, $1 \leq u \leq q$. With a quadratic cost, the Bayesian estimate of θ is the posterior mean $\int_\Theta \theta\pi(\theta \mid y)d\theta$. Instead of simulating θ from $\pi(\theta \mid y)$, we use the conditional posterior distribution $\pi(\theta \mid y, s) \sim \pi(\theta)g(y, s \mid \theta)$, which is much simpler to calculate. The algorithm alternates between two steps:

step 1. simulate $s^{(m+1)} \sim \pi(s \mid y, \theta^{(m)})$ (like SEM)
or for all t, $s_t^{(m+1)} \sim \pi(s_t \mid s_{t-1}^{(m+1)}, s_{t+1}^{(m)}, y, \theta^{(m)})$ (like *EM à la Gibbs*)

step 2. simulate $\theta^{(m+1)} \sim \pi(\theta \mid y, s^{(m+1)})$.

With Dirichlet priors, the posteriors $\pi(\theta \mid y, s)$ are still Dirichlet distributions and the *step* 2 consists of simulating each row, $1 \leq u \leq q$

$$a_u \sim \mathcal{D}\left(\alpha(u,1) + \sum_{t=2}^{n} 1_{\{s_{t-1}=u, s_t=1\}}, \cdots, \alpha(u,q) + \sum_{t=2}^{n} 1_{\{s_{t-1}=u, s_t=q\}}\right)$$

$$b_u \sim \mathcal{D}\left(\beta(u,A) + \sum_{t=1}^{n} 1_{\{s_t=u, y_t=A\}}, \ldots, \beta(u,T) + \sum_{t=1}^{n} 1_{\{s_t=u, y_t=T\}}\right).$$

Gibbs sampling generates again two chains in parallel, $(s^{(m)})_{m\geq 0}$ and $(\theta^{(m)})_{m\geq 0}$. The duality principle states that $(\theta^{(m)})_{m\geq 0}$ converges to its limit distribution $\pi(\theta \mid y)$, which, in the Bayesian frame, is the posterior distribution. We will then estimate the posterior mean, by $\frac{1}{M}\sum_{m=1}^{M} \theta^{(m)}$ and the states probabilities by $\frac{1}{M}\sum_{m=1}^{M} 1_{\{s_t^{(m)}=v\}}$.

3.3 Algorithm comparison

We present simulations results to compare the performances of these algorithms. The results obtained by Muri (1997) show the good behaviour of the EM algorithm when the q states of the hidden chain are characterized by quite different observation transitions B and when the states are sufficiently represented in the sequence, that means when homogeneous regions of reasonable size (not too short, not too long) alternate regularly all along the sequence. MCMC methods also give good results but become stable more slowly than EM. In the other cases, EM is more sensitive to the starting point and can converge very slowly or on the contrary, can become stuck very quickly at a point far from the true value (these results are similar to those obtained by Celeux & Diebolt, 1985, in the mixture setup). MCMC methods then allow, for a sufficient number of iterations, escaping from such points and avoiding poor stabilization. Note that the methods not based on the backward-forward simulation of the states (EM à la Gibbs and its Bayesian counterpart) are faster for each iteration, but require more iterations to give results as good as the other algorithms; we will use these two algorithms in case of alternance of short regions. When the real hidden states chain is composed of few long regions, we will prefer to use a Bayesian estimation with a backward-forward states simulation. Note that when the iteration number is large, EM à la Gibbs and SEM give similar results (this is also the case with the Bayesian estimation) and, as we work with very large datasets, the results obtained by Bayesian and maximum likelihood estimation are quite similar.

4 Applications

We present three applications of these methods to bacterial genomes. The results are performed with the software *RHOM*[3](Research of HOMogeneous regions of DNA sequences) created to allow biologists to use these methods in an automatic way. *RHOM* produces a graphic display of the estimated state probabilities as a function of the sequence position for all the states considered; this representation allows us to visualise the detected homogeneous regions. The precise localisation of these regions and the corresponding parameter estimates (not given here) are also available.

4.1 *Lambda* bacteriophage

The *lambda* bacteriophage is a parasite of the *Escherichia coli* bacterium of length 48502 *bp*. We successively study the case of a $q = 2$, 3 or 4 hidden states *M1-M0* model.

Case of $q = 2$ hidden states

Figure 1 shows the existence of two homogeneous regions clearly delimited (the estimated probabilities are near 0 or 1 all along the sequence). These two regions are characterized by quite different base composition (corresponding to the probability estimation $b(v,j) = P(y_t = j \mid s_t = v), \forall 1 \leq v \leq 2, j \in \mathcal{Y}$): the first region is rich in T and A whereas the second region has a high content of G and a low content of T.

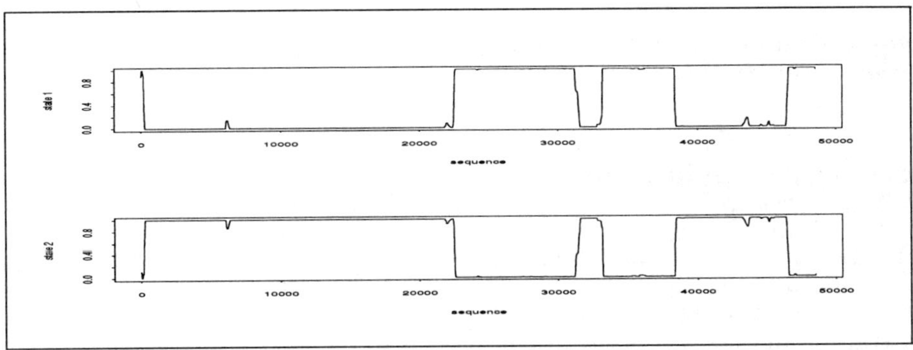

Fig. 1. The estimated probabilities $\hat{P}(s_t = v)$ for $v = 1, 2$, of the states s_t in a 2 hidden state *M1-M0* model, are plotted against the sequence position t: identification of two homogeneous regions of the *lambda* bacteriophage

[3] the software will be soon available to the Web site http://www-bia.inra.fr/J/AB/genome/RHOM/welcome.html

From a biological point of view, these two regions should correspond to different transcription senses of the *lambda* genes (this characterization was proposed by Churchill, 1992): the first region should contain the genes located on one of the DNA strands and the second one the genes located on the other strand.

Case of $q = 3$ hidden states
Figure 2 presents the three homogeneous regions obtained by fitting a 3 hidden state model. When changing from 2 to 3 hidden states, the first region seems to be conserved while the second one seems to be split into two. These results tend to prove that the *lambda* heterogeneity is strongly linked to the existence of this first region which is distinguishable from the rest of the sequence (and thus should have particular properties). Fitting 3 hidden states refines the results obtained on the second region: the second and third regions are both poor in T and the high content in G of the second region identified by a 2 hidden state model, is almost explained by the first long range, which corresponds to the whole second region identified by a 3 hidden state model.

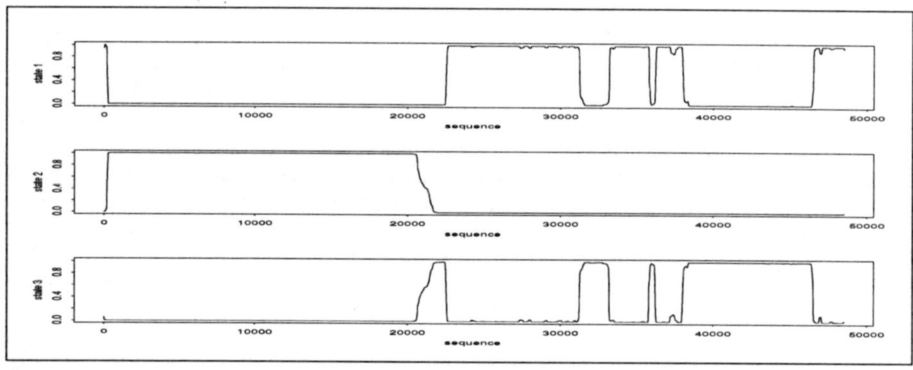

Fig. 2. Identification of three homogeneous regions of the *lambda* phage with a 3 hidden state *M1-M0* model

Case of $q = 4$ hidden states
When we change from 3 to 4 hidden states, Figure 3 shows the conservation of the first and third regions when the second one is split in two. Note that even the short regions are well-delimited, and thus could have a biological meaning. To confirm the conservation and the split of the identified regions, we compute the total variation between the 9 considered states; the distance between a state u and state v is then defined by $d(u,v) = \frac{1}{2}\sum_{i\in\mathcal{Y}} |b(u,i) - b(v,i)|$. Figure 4 shows the plane representation of these distances and clearly illustrate this phenomenon.

Note that the study of the *lambda* bacteriophage in the *M1-M1* and *M1-M2* models, leads to the same conclusions (split and conservation phenomena) as in the *M1-M0* model, and that the identified regions are quite similar. However, increasing the order of the model provides an additional characterization (corresponding to a particular dinucleotides or trinucleotides composition).

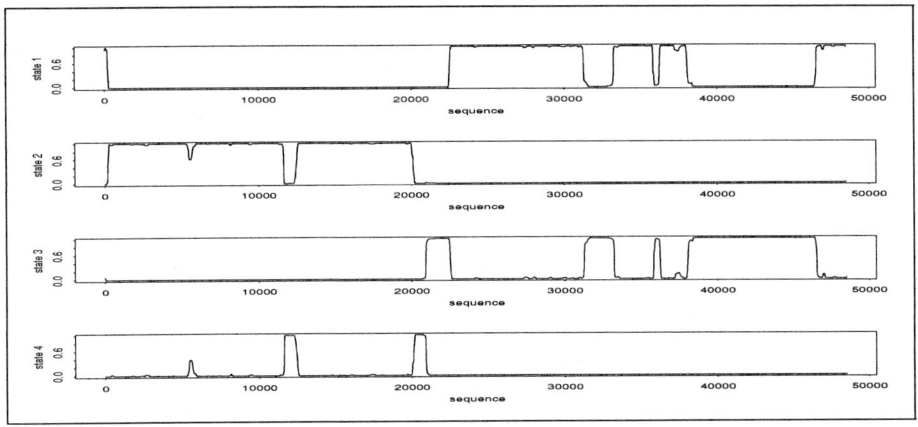

Fig. 3. Identification of four homogeneous regions of the *lambda* phage with a 4 hidden state *M1-M0* model

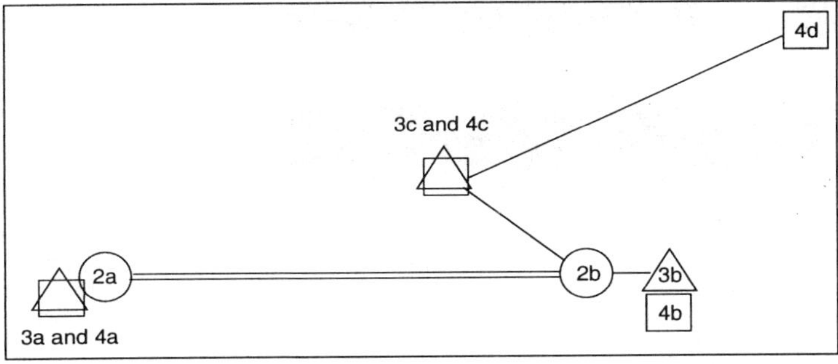

Fig. 4. Plane representation of the total variation distances between the 9 states in the *M1-M0* model; the fit of 2 hidden states ($2a$ and $2b$) is represented by a circle, of 3 hidden states ($3a$, $3b$ and $3c$) by a triangle and of 4 hidden states ($4a$, $4b$, $4c$ and $4d$) by a square

4.2 *bIL67* bacteriophage

The *bIL67* bacteriophage consists of 22195 *bp* and is a parasite of the *Lactococcus lactis* bacterium. We only present the results of the *M1-M0* model (the conclusions are nearly the same when we increase the model order). The analysis of a 2 hidden state model reveals two homogeneous regions clearly identified (represented in the Figure 5): *bIL67* genome is clear-cut in two. The biological meaning of these two regions should be the same as the one advanced for the *lambda* bacteriophage.

As Figure 6 shows, the transition from 2 to 3 hidden states suggests that the *bIL67* bacteriophage is made up almost entirely of two homogeneous regions. In fact, the *bIL67* sequence resists a decomposition further than 2 hidden states.

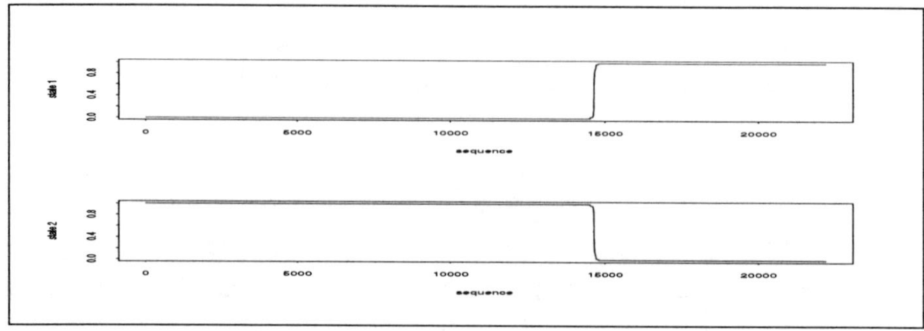

Fig. 5. Identification of two homogeneous regions of the *bIL*67 bacteriophage in a 2 hidden state *M1-M0* model

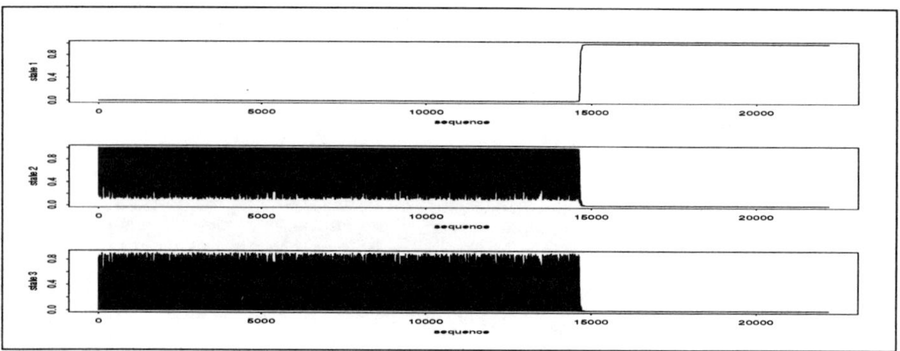

Fig. 6. The fit a of 3 hidden state *M1-M0* model shows that the first region is conserved but that it's impossible to clearly delimit the 2 and 3 regions. The *bIL*67 heterogeneity seems to be essentially explained by the presence of two homogeneous regions

4.3 *B. subtilis* bacterium

We present results obtained for a 118620 *bp* extract of the *Bacillus subtilis* bacterium (*YAC* contig4).

Results in the *M1-M0* model

Figure 7 presents the results with a 2 hidden state model. This analysis shows that no region is clearly identified: 14397 ranges of the first or second regions (of respective length 4 *bp* and 7 *bp*) alternate all along the sequence. The conclusions remain the same if we fit a 3 or 4 hidden state *M1-M0* model.

The study of *B. subtilis* in the *M1-M1* model leads to the same properties as in the *M1-M0* model. These results allow us to assume some homogenity of *B. subtilis* if we only take into account the nucleotides or dinucleotides composition in the modelling.

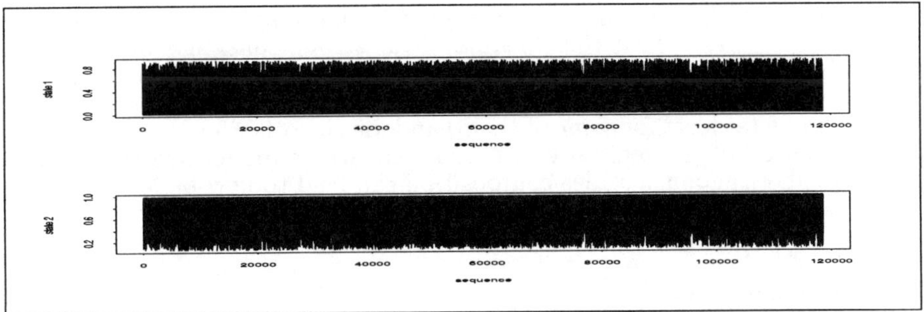

Fig. 7. The fit of a 2 hidden state *M1-M0* model does not allow the identification of well-delimited regions and thus justifies inferring a certain homogeneity of *B. subtilis* when we only take into account the base composition in the modelling

Results in the *M1-M2* model

In this model, the homogeneous regions that we could identify, will be characterized by a particular trinucleotides composition (corresponding to the transitions estimation $b(u,i,j,k) = P(y_t = k \mid y_{t-1} = j, y_{t-2} = i, s_t = u)$ for $u = 1, \cdots, q$ and $i, j, k \in \mathcal{Y}$).

The study of *B. subtilis* in a 2 hidden state model reveals two homogeneous regions, as shown in Figure 8. The second region is less represented in the sequence, as it consists of only 17370 sites, but all the ranges are well delimited. As with the *lambda* phage, when we fit a 3 or 4 hidden state model, we observe conservation of one region (the second one) and splitting of the other one.

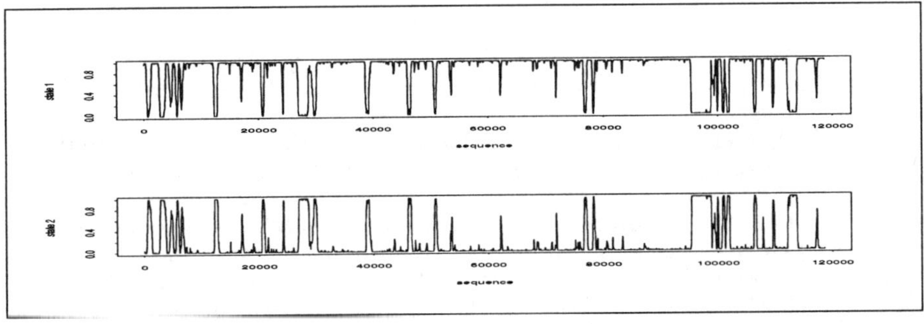

Fig. 8. Identification of two homogeneous regions of the *B. subtilis* bacterium with a 2 hidden state *M1-M2* model

5 Discussion

The comparison by simulation of these identification procedures for hidden Markov chains, and the discussion of their validity, show how strongly we need stochastic or Bayesian alternatives to the EM algorithm. Moreover, the applications to bacterial genomes show the robustness of the detected regions when we change the hidden state number and the model order that models the bases' succession. Note that in spite of the strong coherence between the results obtained from various orders, each model can provide interesting indications for the biologist: on the one hand, the change of the region sizes and their locations (at least for some of them) and, on the other hand, the different characterization of the regions obtained from various models, corresponding to a particular oligonucleotides composition can lead to new biological interpretations. Hidden Markov models are relatively easy to understand and to interpret for the geneticists and allow extraction of real information.

Acknowledgements
I am grateful to Élisabeth de Turckheim and Bernard Prum for their valuable comments and discussions and to the referees for their helpul comments on an earlier version of the paper.

References

Archer, G. & Titterington, D. (1995). Parameter estimation for hidden Markov chains. Technical report, University of Glasgow, Department of Statistics.

Baum, L.E. & Petrie, T. (1966). Statistical Inference for Probabilistic Functions of finite State Markov Chains. *The Annals of Mathematical Statistics*, **37**, 1554-1563.

Baum, L.E., Petrie, T., Soules, G. & Weiss, N. (1970). A Maximization Technique Occuring in the Statistical Analysis of Probabilistic Functions of Markov Chains. *The Annals of Mathematical Statistics*, **41**, 1, 164-171.

Celeux, G. & Diebolt, J. (1985). The SEM algorithm: a probabilistic teacher algorithm derived from the EM algorithm for the mixture problem. *Computational Statistics Quarterly*, **2**, 73-82.

Churchill, G.A. (1989). Stochastic Models for Heterogeneous DNA Sequences. *Bulletin of Mathematical Biology*, **51**, 1, 79-94.

Churchill, G.A. (1992). Hidden Markov chains and the analysis of genome structure. *Computers chem.*, **16**, 2, 107-115.

Dempster, A.P., Laird, N.M. & Rubin, D.B. (1977). Maximum Likelihood from Incomplete Data via the EM Algorithm. *J. Royal Statist. Soc. Ser. B*, **39**, 1-38.

Geman, S. & Geman, D. (1984). Stochastic relaxation, Gibbs distributions and the Bayesian restoration of images. *IEEE Trans. Pattn. Anal. Mach. Intel.* **6**, 721-741.

Muri, F. (1997). *Comparaison d'algorithmes d'identification de chaînes de Markov cachées et application à la détection de régions homogènes dans les séquences d'ADN*. PhD thesis, Université René Descartes, Paris V.

Qian, W. & Titterington, D.M. (1990). Parameter estimation for hidden Gibbs chains. *Statist. Probab. Lett.*, **10**, 49-58.

Rabiner, L.R.A (1989). Tutorial on Hidden Markov Models and Selected Applications in Speech Recognition. *Proceedings of the IEE*, **77**, 257-286.

Redner, R.A. & Walker, H.F. (1984). Mixture Densities, Maximum Likelihood and the EM Algorithm. *SIAM Review*, **26**, 2, 195-239.

Robert, C.P., Celeux, G. & Diebolt, J. (1993). Bayesian estimation of hidden Markov chains: A stochastic implementation. *Statist. Probab. Lett.*, **16**, 77-83.

Robert, C.P. (1996). *Méthodes de Monte Carlo par Chaînes de Markov*. Economica.

MCMC Specifics for Latent Variable Models

Christian P. Robert
CREST, INSEE, Timbre J340, 75675 Paris cedex 14, France

Abstract. We derive from previous analyses of specific latent variable models an overall review, under the theme of their strong connections with simulation-based statistical methods. These connections go both ways: latent variable models were instrumental in designing these new methods, whose convergence properties and convergence diagnostic tools are specific to these models, and hybrid methods like simulated maximum likelihood primarily apply in such settings.

Keywords. asymptotic normality, convergence monitoring, diagnostics, mixture models, simulated likelihood, stochastic volatility

1 Introduction

Latent variable models have long been a bottleneck for statistical inference, in the senses that their involuted structure was prohibiting exact processing and that the approximations available until recently were not necessarily satisfactory. Tailored statistical methods like the EM algorithm and Data Augmentation have been designed mainly to deal with such models and the goal of this paper is to emphasize both the efficiency of simulation methods when dealing with latent variable models and the possibility of exploiting the special structure of latent variable models in convergence assessment of the corresponding MCMC algorithms. Moreover, the connection also goes the other way, in the sense that latent variable models were instrumental in the derivation of these simulation methods and still suggest new approaches to simulation and convergence diagnostics.

A general definition of latent variable models is to use a marginal representation,

$$x \sim f(x) = \int g(x,z)dz, \tag{1}$$

where only x is observed. However, even though examples like the Student's t distribution enjoy such a marginal representation, since

$$x|z \sim \mathcal{N}(0, z^{-2}), \qquad z \sim \mathcal{G}a(\nu/2, \nu/2),$$

they cannot be considered as latent variable models, in the sense that the *latent variable* z must have some meaning within the model. In addition, latent variable models are such that the dimension of z increases with the sample size or the dimension of x. The examples in the following sections illustrate the characteristics of these models, while stressing their links with both likelihood and Bayesian simulation methods. In fact, latent variable models are such that inference requires the simulation of the corresponding latent variable z. As described in Billio *et al.* (1998), the class of latent variable models encompasses a wide variety of models in Econometrics and Finance.

2 Mixtures of distributions

Mixtures are a typical case of latent variable model, whose link with specific statistical algorithms can be traced back to the EM algorithm of Dempster *et al.* (1977) and the stochastic extension of Broniatowski *et al.* (1983), SEM, was directly sparked by an interest in mixtures. These structures are intended to represent heterogeneous structures or to work as approximate nonparametric models, through the representation

$$x \sim f(x) = \sum_{i=1}^{k} p_i f(x|\theta_i). \tag{2}$$

The inferential goals of a mixture analysis may be (i) to get information on the homogeneity classes, that is to determine from which components the x_j's are (most likely) issued from; (ii) to draw inference on the parameters p_i, θ_i of the mixture; (iii) to determine the degree of heterogeneity of the mixture or the level of complexity of an unknown distribution, by getting an estimate of the number of components, k.

The latent variable structure of a mixture is associated with the component indicator, that is in the hierarchical representation of (2),

$$x|z \sim f(x|\theta_z), \qquad z \sim \mathcal{M}(1; p_1, \ldots, p_k). \tag{3}$$

As stressed in the literature (Titterington *et al.*, 1985; Robert, 1996), there are several types of difficulties with the analysis of mixtures:

- mixtures meet "standard" identifiability problems, since they are invariant under permutation of the indices. Besides, they are also only weakly identifiable, in the sense that there is always a non-zero probability $(1 - p_i)^n$ that no observation comes from the i-th component.
- Independent improper priors on the parameters of (2) cannot be used, since if $\pi(\theta, p)$ is an improper prior, then, for every n,

$$\int \pi(\theta, p|x_1, \ldots, x_n) d\theta \, dp = \infty.$$

- The likelihood function is not bounded in most setups. For instance, in the case of a normal mixture (4),

$$\lim_{\sigma_1 \to 0} \mathrm{L}(\mu, \sigma, p|x) = \infty\,.$$

- The geometry of the parameter space is quite involved and nonlinear, and invalidates usual likelihood ratio tests.
- The likelihood has a highly multimodal surface which usually hinders the implementation of standard optimization methods.
- The posterior distributions may have closed form expressions but they are useless in practice, due to the combinatorial explosion of the likelihood.

These issues have been addressed by Mengersen & Robert (1996) who propose a reparameterisation of mixture models to overcome the difficulties with an improper prior modelling. For instance, a *normal mixture*,

$$\sum_{i=1}^{k} p_i \mathcal{N}(\mu_i, \sigma_i^2)\,, \tag{4}$$

can be written as

$$
\begin{aligned}
&q_1\mathcal{N}(\theta_1,\tau_1^2) + (1-q_1)q_2\mathcal{N}(\theta_1+\tau_1\theta_2,\tau_1^2\tau_2^2) \\
&+ (1-q_1)(1-q_2)q_3\mathcal{N}(\theta_1+\tau_1\theta_2+\tau_1\tau_2\theta_3,\tau_1^2\tau_2^2\tau_3^2) + \ldots \\
&+ (1-q_1)\cdots(1-q_{k-1})\mathcal{N}(\theta_1+\ldots+\tau_1\cdots\theta_k,\tau_1^2\cdots\tau_k^2),
\end{aligned}
$$

thus expressing each component as a perturbation of the previous component. This new parameterization is obviously in one-to-one correspondence with the original expression, but a main incentive for change is to allow for independent improper priors on the perturbations, since Robert & Titterington (1998) have shown that the prior distribution ($u = 2, \ldots, k$)

$$
\pi(\theta_1,\tau_1) = 1/\tau\,,\ \tau_u \sim \mathcal{U}_{[0,1]}\,,\ \theta_u \sim \mathcal{N}(0,\zeta^2)\,,\qquad \zeta > 0, \tag{5}
$$

was associated with a proper posterior distribution, for every sample $x_1, \ldots, x_n$, under the identifiability constraint $\tau_2 \leq 1, \ldots, \tau_k \leq 1$. A similar reparameterisation applies for *exponential mixtures*,

$$
\sum_{i=1}^k p_i\,\mathcal{E}xp(\lambda_i) = q_1\,\mathcal{E}xp(\tau_1) + (1-q_1)q_2\,\mathcal{E}xp(\tau_1\tau_2) + \ldots\,,
$$

and allows for improper priors like $\pi(\tau, q) \propto \tau_1^{-1}$, under the identifiability restriction $\tau_2 \leq 1, \ldots, \tau_k \leq 1$, as shown by Gruet *et al.* (1998). This prior also works for *Poisson mixtures*,

$$
q_1\,\mathcal{P}(\tau_1) + (1-q_1)q_2\,\mathcal{P}(\tau_1\tau_2) + \ldots
$$

As mentioned earlier, mixture models are strongly related to MCMC techniques, more exactly with *Data Augmentation* methods in the spirit of Diebolt & Ip (1981) or Tanner & Wong (1987), in that they somehow call for the completion of the model in an exponential family model, for lack of manageable alternatives. (See Robert & Casella, 1998, for a detailed treatment of MCMC methods.) The *completion* step in the MCMC algorithm comes with the generation of the indicator variables $z_1, \ldots, z_n$ in (3), as

$$
z_i^{(t)} \sim P(z_i = j|x_i, \theta^{(t)}, p^{(t)})\,,
$$

at iteration t of the algorithm. For exponential families, the parameters can then be directly simulated. For instance, in the exponential case, the parameter step of the Data Augmentation algorithm is a succession of simple Gibbs steps,

$$
\begin{aligned}
\tau_1^{(t)} &\sim \mathcal{G}a(n, \textstyle\sum_j \tau_2\cdots\tau_j n_j \bar{x}_j)\,, \\
\tau_i^{(t)} &\sim \mathcal{G}a(n_i+\ldots+n_k, \textstyle\sum_{j\geq i} \lambda_j n_j \bar{x}_j/\tau_i)\,\mathbb{I}_{\tau_i<1}\,, && i > 1 \\
q_i^{(t)} &\sim \mathcal{B}e(n_i+1, n_{i+1}+\ldots+n_k)\,. && i \geq 1
\end{aligned}
$$

(For some parameterizations, Metropolis–Hastings steps may also be required, see Robert & Mengersen, 1995.) As discussed in Robert (1996), this simple derivation of an MCMC algorithm for mixtures of distributions does not necessarily lead to good convergence properties. Figure 1 provides a control panel for convergence properties in the exponential case, with plots of the estimated density and cdf, of a comparison between the average of the allocated components $z_j^{(t)}$ and the estimated expected allocated component, $\mathbb{E}[z_j|x_1,\ldots,x_n,\hat{\tau},\hat{q}]$, and of a so-called *allocation map*, where the successive allocations $z_j^{(t)}$ of each observation are represented by grey levels. While the first three graphs are quite conclusive about convergence, the allocation map gives a different picture of strong instability in the allocation of the observations, which can be imputed either to slow mixing in the MCMC algorithm or to weak identifiability structures, as discussed in Gruet *et al.* (1998).

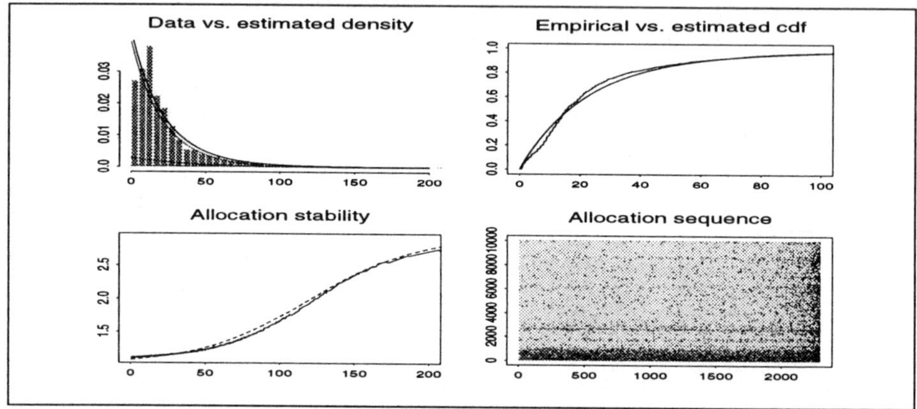

Fig. 1. Control panel for the estimation of an exponential mixture with 3 components and 2228 observations: density *(upper left)*, cdf *(upper right)*, averaged vs. expected allocations *(lower left)*, and allocation map *(lower right)*. [The grey levels represent the successive components allocated to each observation.] (*Source:* Gruet *et al.*, 1998.)

Note also that the reparameterisation is perfectly fitted for reversible jump MCMC techniques, as in Gruet *et al.* (1998), since the other components are not modified by the so-called *split and merge* moves.

3 Hidden Markov models

Our second example adds another degree of complexity through a Markov dependence between the z_j's as, for instance, in the normal case,

$$P(z_t = u|z_j\ ,j < t) = p_{z_{t-1}u}, \qquad x_t|z,x_j\ j \neq t\ \sim \mathcal{N}(\mu_{z_t},\sigma^2_{z_t}),$$

thus preserving conditional independence between the observations. *Hidden Markov models* are commonly used in signal processing and Econometrics.

The analysis of Section 2 applies to this structure, with a reparameterisation of the location-scale parameters, as

$$\mu_j = \theta_1 + \tau_1\theta_2 + \ldots + \tau_1 \ldots \tau_{j-1}\theta_j, \quad \sigma_j = \tau_1 \ldots \tau_j\ .$$

Under the identifiability condition $\tau_2 \leq 1,\ldots,\tau_k \leq 1$, Robert & Titterington (1998) show that the prior (5) is valid in this setting. (See Robert & Titterington, 1998, for the analysis of a Poisson hidden Markov model.) The rows $p_{i\cdot}$ of the transition matrix $\mathbb{P}$ are distributed from a Dirichlet prior.

The Gibbs steps are straightforward, with normal distributions,

$$\mathcal{N}\left(\frac{\zeta^2 n_u \overline{x}_u + \mu_{u+1} + \mu_{u-1}\sigma_u^2/\sigma_{u-1}^2}{\zeta^2 n_u + 1 + \sigma_u^2/\sigma_{u-1}^2}, \frac{\zeta^2 \sigma_u^2}{\zeta^2 n_u + 1 + \sigma_u^2/\sigma_{u-1}^2}\right),$$

on the μ_u's, and truncated gamma distributions,

$$\mathcal{G}a^T\left(\frac{n_u - 1}{2}, \frac{n_u(\overline{x}_u - \mu_u)^2 + s_u^2 + (\mu_{u+1} - \mu_u)^2\zeta^{-2}}{2}\right),$$

on the σ_u^{-2}'s. The latent states can be simulated one-by-one, that is by taking advantage of the Markovian structure $(1 < i < k)$

$$P(z_i = u | \ldots, z_{i-1}, z_{i+1}, \ldots, \mathbb{P}) \propto p_{z_{i-1}u} p_{u z_{i+1}} f(x_i|\mu_u, \sigma_u), \tag{6}$$

but an alternative often advocated in the signal processing literature is to make use of *forward–backward* formulae (see, e.g., Baum *et al.*, 1970),

$$\begin{aligned} P(x_{j+1}, \ldots, x_n \mid z_j = i, \mathbb{P}) &= \sum\nolimits_\ell p_{i\ell} P(x_{j+1}, \ldots, x_n | z_{j+1} = \ell, \mathbb{P}) \qquad (7) \\ &\propto \sum\nolimits_\ell p_{i\ell} f(x_{j+1}|\mu_\ell, \sigma_\ell) P(x_{j+2}, \ldots, x_n | z_{j+1} = \ell, \mathbb{P}) . \end{aligned}$$

The conditional distribution of z_1 can be derived directly as

$$P(z_1 = i | x_1, \ldots, x_n, \mathbb{P}) \propto \pi_i P(x_2, \ldots, x_n | z_1 = i, \mathbb{P}) f(x_1|\mu_i, \sigma_i) .$$

Once z_1 is generated, the distributions of z_2 conditional on z_1, ..., of z_j conditional on $z_1, \ldots, z_{j-1}$, follow from (7). Robert *et al.* (1998) show that both approaches are quite similar in terms of general convergence properties and that only finer convergence diagnostics, such as those presented below, can discriminate in favour of backward updating.

While EM-type methods can be seen as precursors of MCMC methods, it is also possible to derive maximum likelihood estimates by MCMC techniques, bypassing some difficulties like the dependence on initial conditions. Section 5 presents a general approach, but we first recall a technique which computes MLEs as limits of (formal) Bayes estimates. Christened *prior feedback* in Robert (1993), it has been proposed under other names in the literature (see Robert & Casella, 1998). The sequence of priors can be chosen as $\pi_m(\xi) \propto \pi(\xi) f^m(x|\xi)$ with a arbitrary prior $\pi(\xi)$. The resulting (pseudo-) Bayes estimates $\delta_m(x)$ correspond to regular Bayes estimates for m replications of the original sample and the method can be controlled by monitoring the stabilization (in m) of the $\delta_m(x)$'s. (See Robert & Titterington, 1998, for illustrations in hidden Markov models.) While apparently involuted, this method applies to latent variable models, where the stabilization of the prior feedback estimator of the density occurs quite quickly.

4 Switching ARMA models

Another extension is to create an additional dependence in the observed variables, as in an ARMA structure which depends on a latent state Markov chain $s_t \in \{0, 1, \ldots, M\}$,

$$y_t = \gamma_{s_t} + \sum\nolimits_{i=1}^p \varphi_i^{s_t}(y_{t-i} - \gamma_{s_{t-i}}) + \sum\nolimits_{j=0}^q \theta_j^{s_t} \sigma_{s_{t-j}} \epsilon_{t-j} , \qquad t \geq 1, \tag{8}$$

with

$$P(s_t = i | s_{t-1} = j) = \pi_{ij} .$$

This model was proposed by Hamilton (1989) to describe highly heterogeneous datasets in Econometrics. Processing (8) involves several difficulties:

- The likelihood function is not available in closed form.
- The model is not Markovian, although it can be imbedded in an artificial Markov model, at high cost in terms of computation time.
- A full latent variable structure, besides the completion of the latent states, is not available, in the sense that the generation of the ϵ_t's does not lead to an easier resolution.
- Both stationarity and identifiability requirements on the ARMA model imply complicated constraints on the φ_i's and θ_j's, while the stationarity constraint on the *switching ARMA model* (8) is generally unknown.

Consider for instance the following special cases:

$$y_t = \gamma_{s_t} + \sigma_{s_t}\epsilon_t - \theta\sigma_{s_{t-1}}\epsilon_{t-1}, \qquad [\mathrm{MA}_1]$$
$$y_t = \gamma_{s_t} + \varphi(y_{t-1} - \gamma_{s_{t-1}}) + \sigma_{s_t}\epsilon_t - \theta\sigma_{s_{t-1}}\epsilon_{t-1}, \qquad [\mathrm{ARMA}_1]$$
$$y_t = \mu + \varphi_{s_t}(y_{t-1} - \mu) + \sigma\epsilon_t - \theta_{s_t}\sigma\epsilon_{t-1}, \qquad [\mathrm{ARMA}_2]$$

Prior distributions based on the same representation as in Sections 2 and 3 can be used, integrating the identifiability requirement $\sigma_1 < \sigma_0$ and the stationarity constraints $|\theta_j| < 1$, $|\varphi_i| < 1$, as

$$\pi(\xi) \propto \sigma_0^{-3} e^{-(\gamma_0-\gamma_1)^2\zeta/2\sigma_0^2} \mathbb{I}_{\sigma_0>\sigma_1}$$

for $[\mathrm{MA}_1]$ and $[\mathrm{ARMA}_1]$, while $\pi(\xi) \propto 1/\sigma$ works for $[\mathrm{ARMA}_2]$. Higher order (in p and q) models require a reparameterisation, derived in Barnett *et al.* (1996), which naturally integrates these constraints, while allowing for a flat prior on the new parameters.

In this case, the MCMC implementation is not as straightforward as in the previous sections, due to the MA structure. For instance, in the $[\mathrm{MA}_1]$ model, while the formal Gibbs sampler reads like

1. `Generate the missing states` (s_t) `from`

$$f(s_0, .., s_T|\xi, y_0, .., y_T) \propto \prod_{t=1}^{T} \exp\{-(y_t - \gamma_{s_t} + \theta\eta_{t-1})^2/2\sigma_{s_t}^2\}\sigma_{s_t}^{-1}\pi_{s_{t-1}s_t},$$

`with` $\eta_t = \sigma_{s_t}\epsilon_t = y_t - \gamma_{s_t} + \theta\eta_{t-1}$, $t \geq 1$.

2. `Generate the parameters` ξ `from` $f(\xi|s_0, \ldots, s_T, y_0, \ldots, y_T)$.

a direct Gibbs sampling approach cannot be implemented because of the non-Markovian dependence between the s_t's. A solution proposed in Billio *et al.* (1998a) is to devise a Metropolis–Hastings algorithm whose proposal is based on the elimination of the inconvenient terms in the s_t's, by replacing the "true" innovations η_t with pseudo-innovations $\hat{\eta}_t$ based on previous values. The pseudo-innovations $\hat{\eta}_t$ can then be used in a proposal distribution, which is corrected through a Metropolis–Hastings acceptance step.

For instance, for the generation of the *latent variables* of the $[\mathrm{MA}_1]$ model, at iteration $m+1$, define $(t > 0)$

$$\eta_0^{(m+1)} = 0, \quad \eta_t^{(m+1)} = y_t - \gamma^{(m)}_{s_t^{(m+1)}} + \theta^{(m)}\eta_{t-1}^{(m+1)},$$

between the generations of $s_t^{(m+1)}$ and of $s_{t+1}^{(m+1)}$. The latent variable $s_{t+1}^{(m+1)}$ is then generated from

$$p(s_t) \propto \exp\{-(y_t - \gamma_{s_t}^{(m)} + \theta^{(m)}\eta_{t-1}^{(m+1)})^2/2(\sigma_{s_t}^{(m)})^2\}(\sigma_{s_t}^{(m)})^{-1}\pi_{s_{t-1}^{(m+1)}s_t}\pi_{s_t s_{t+1}^{(m)}} \cdot$$

The whole vector $(s_t^{(m+1)})$ is accepted in a Metropolis–Hastings step. An alternative approach is to accept each $s_t^{(m+1)}$ individually, but, while giving similar performance in practice, the global approach is closer to the completion idea, and achieves reasonable acceptance rates in the examples treated by Billio *et al.* (1998a).

Similarly, the *MA coefficient* of the [MA$_1$] model can be processed at iteration $m+1$ by first defining $(t > 0)$

$$\hat{\eta}_t^{(m)} = y_t - \gamma_{s_t^{(m+1)}}^{(m+1)} + \theta^{(m)}\hat{\eta}_{t-1}^{(m)}$$

as pseudo-innovations, then generating $\tilde{\theta}$ from $\mathcal{N}(\beta^{(m)}, \tau^{(m)})$, with

$$\beta^{(m)} = -\frac{\sum_{t=1}^T \left[y_t - \gamma_{s_t^{(m+1)}}^{(m+1)}\right]^2 \hat{\eta}_{t-1}^{(m)} \left(\sigma_{s_t^{(m+1)}}^{(m+1)}\right)^{-2}}{\sum_{t=1}^T \left(\sigma_{s_t^{(m+1)}}^{(m+1)}\right)^{-2} \left(\hat{\eta}_t^{(m)}\right)^2},$$

$$\tau^{(m)} = \tau_\theta^2 \Big/ \left[\sum_{t=1}^T \left(\sigma_{s_t^{(m+1)}}^{(m+1)}\right)^{-2} \left(\hat{\eta}_{t-1}^{(m)}\right)^2\right].$$

and taking

$$\hat{\theta} = \begin{cases} \tilde{\theta} & \text{if } |\tilde{\theta}| < 1, \\ 1/\tilde{\theta} & \text{otherwise,} \end{cases}$$

as the proposed value. (The scale parameter τ_θ is used to achieve an optimal acceptance rate in Metropolis–Hastings algorithm.)

5 Stochastic volatility

The last example presented in this paper is the *stochastic volatility* model, where the latent variables are more complex than in the previous section, since they are also continuous. A particular case is as follows $(t = 1, \ldots, T)$:

$$y_t^* = a + by_{t-1}^* + c\varepsilon_t^*, \quad y_t = \exp(0.5\, y_t^*)\varepsilon_t. \tag{9}$$

Note that, somehow, the information about the parameters is contained in the latent variables y_t^* rather than in the observables y_t. These models are quite common in the modelling of financial data.

The MCMC implementation reflects the greater complexity of the model, in the sense that the latent variables y_t^*, while necessary, cannot be directly simulated from the conditional distributions $f(y_t^*|y_{t-1}^*, y_{t+1}^*, y^t, \theta)$. The parameter θ is straightforward to simulate from the completed AR(1) model.

Metropolis–Hastings schemes can be based on approximations of

$$f(y_t^*|y_{t-1}^*, y_{t+1}^*, y^t, \theta) \propto \exp\left\{-(y_t^* - a - by_{t-1}^*)^2/2c^2 \right.$$
$$\left. -(y_{t+1}^* - a - by_t^*)^2/2c^2 - y_t^*/2 - y_t^2 e^{-y_t^*}/2\right\},$$

either via a linearisation of $(y_t^* + y_t^2 e^{-y_t^*}/2)$ in $(y_t^* - \log(y_t^2))^2/4$, or, as in Jacquier *et al.* (1996), via a gamma approximation. In the examples of Billio *et al.* (1998b), the normal approximation does much better.

Once a satisfactory MCMC algorithm has been tested and calibrated for Bayesian inference, it can provide an approximation of the likelihood function by the *Simulated Likelihood Ratio* method of Billio *et al.* (1998b). The principle behind this technique is to represent, for a (complete) joint distribution

$$(y^{*T}, y^T) = (y_1^*, \ldots, y_T^*, y_1, \ldots, y_T) \sim f(y^T|\theta),$$

with observables $(y_1, \ldots, y_T)$, the (observed) likelihood ratio as

$$\frac{f(y^T|\theta)}{f(y^T|\bar{\theta})} = \mathbb{E}_{\bar{\theta}}\left[\frac{f(y^{*T-k}, y^T|\theta)}{f(y^{*T-k}, y^T|\bar{\theta})}\middle| y^T\right], \tag{10}$$

for any $k \in \{1, ..., T\}$ and an arbitrary $\bar{\theta}$. The practical implementation requires simulation, since, if the S $y^{*T-k}(s)$ are iid from $f(y^{*T-k}|y^T, \bar{\theta})$,

$$\frac{1}{S}\sum_{s=1}^{S} \frac{f(y^{*T-k}(s), y^T|\theta)}{f(y^{*T-k}(s), y^T|\bar{\theta})} \tag{11}$$

converges to (10) by a standard importance sampling argument. (See Geyer, 1996, for an earlier proposal.). For the stochastic volatility model, the approximation (11) is an average of the terms

$$\frac{f(y, y_i^*|\theta)}{f(y, y_i^*|\bar{\theta})} \propto \left(\frac{\bar{c}}{c}\right)^T \prod_{t=1}^{T} \frac{\exp\{-(y_{it}^* - a - by_{i(t-1)}^*)^2/2c^2\}}{\exp\{-(y_{it}^* - \bar{a} - \bar{b}y_{i(t-1)}^*)^2/2\bar{c}^2\}}.$$

The method strongly differs from simulated EM methods (Diebolt & Ip, 1996; Lavielle & Mouline, 1997) in that it only requires one simulation run and one optimization run, for a fixed value of $\bar{\theta}$. It also provides a full and smooth evaluation of the likelihood surface, whatever the simulation size S in (11). Moreover, since (10) involves a likelihood ratio, it partly avoids numerical problems in computation of likelihoods.

As in every importance sampling method, there are some constraints for the variance of (11) to be finite: the method is only efficient for θ's such that

$$\mathbb{E}_{\bar{\theta}}\left[\left(\frac{f(y^{*T-k}, y^T|\theta)}{f(y^{*T-k}, y^T|\bar{\theta})}\right)^2 \middle| y^T\right] < \infty. \tag{12}$$

In the stochastic volatility model, (12) is satisfied if $\bar{c}^{-2} < 2c^{-2}$ and $\bar{b}^2 < b^2$.

In the implementation of the method, the starting point is very influential on its performance. When $\bar{\theta}$ is far from the true value θ_0, the value maximising (11) often is far closer to $\bar{\theta}$ than to the true value θ_0. Geyer (1996) suggested the implementation of recursive versions of the method where the solution of one iteration is used as the next $\bar{\theta}$. However, the resulting fixed points are not always close to θ_0 and, for the stochastic volatility model, the best starting value is the noninformative Bayes estimate (see Billio *et al.*, 1998).

6 Specific convergence properties

We won't recall here the general convergence properties of MCMC algorithms, referring the reader to Tierney (1994), Roberts & Rosenthal (1997), or Robert & Casella (1998) for details. As shown in the previous examples, it often occurs, however, that latent variable models produce two chains $(z^{(t)})$ and $(\theta^{(t)})$, which are in duality, either in the strong sense of Data Augmentation, or in the weaker sense that $\theta^{(t)} \sim \pi(\theta|z^{(t)})$. In such cases, Diebolt & Robert (1994) note that the probabilistic properties of $(z^{(t)})$ transfer to the chain $(\theta^{(t)})$ by a *Duality Principle*, as in the following examples:

- If $(z^{(t)})$ is ergodic with stationary distribution $\tilde{f}$ (respectively geometrically ergodic with rate ϱ), $(\theta^{(t)})$ is ergodic (geometrically ergodic with rate ϱ) for every $\pi(\cdot|z)$ and its stationary distribution is
$$\tilde{\pi}(\theta) = \int \pi(\theta|z)\tilde{f}(z)dz.$$
- If the chain $(z^{(t)})$ is α-mixing (respectively β-mixing), the chain $(\theta^{(t)})$ is also α-mixing (β-mixing).
- When $(z^{(t)})$ is a finite state space Markov chain, as in many latent variable models, $(z^{(t)})$ is geometrically ergodic and the CLT applies.

Although applicable in greater generality, *Rao–Blackwellisation* is also a technique which naturally applies to latent variable models. As an alternative to the empirical average, Gelfand & Smith (1990) proposed

$$\delta_{rb} = \frac{1}{T} \sum_{t=1}^{T} \mathbb{E}\left[h(z_1)|z_2^{(t)}, \cdots, z_p^{(t)}\right].$$

Liu, Wong & Kong (1994) showed that rigorous domination by δ_{rb} occurs for Data Augmentation simulation schemes. Rao–Blackwellisation also provides density estimators for the marginal distributions of z_i $(i = 1, \ldots, p)$ as

$$\frac{1}{T} \sum_{t=1}^{T} g_i(z_i|z_j^{(t)}, j \neq i).$$

Moreover, latent variable structures are usually associated with closed-form conditional expectations, especially in the finite case, even though the corresponding improvement may be negligible (see Robert & Mengersen, 1995).

7 Convergence assessment

Once again, we refer the reader to the relevant literature (Cowles & Carlin 1996; Brooks & Roberts, 1998; Mengersen *et al.*, 1998) for reviews on convergence assessment, including the evaluation of approximations of $\mathbb{E}^f[h(z)]$. For instance, the *Riemann quadrature* method of Philippe (1997) is particularly well-adapted to latent variable models. This method, which only works well in dimension 1, is a mix of numerical analysis and Monte Carlo methods, either in its original form,

$$\hat{\theta}_T = \sum_{t=2}^{T} \left\{z_{(t)} - z_{(t-1)}\right\} h(z_{(t)}) f(z_{(t)}), \qquad z_{(1)} \leq \cdots \leq z_{(T)},$$

or in its Rao-Blackwellised version,

$$\hat{\theta}_T = \frac{1}{T}\sum_{t=2}^{T}\{z_{(t)} - z_{(t-1)}\}\sum_{u=1}^{T} h(z_{(t)}, y_u)g(z_{(t)}|y_u)\,. \tag{13}$$

Moreover, it provides a *control variate device*, since

$$\lim_{T\to\infty}\frac{1}{T}\sum_{t=2}^{T}\{z_{(t)} - z_{(t-1)}\}\sum_{u=1}^{T} g(z_{(t)}|y_u) = 1\,,$$

for every coordinate z of the (Gibbs) Markov chain. The approximation (13) leads to an additional convergence monitor since the different coordinates give different estimates which converge to the same quantity (see Gruet *et al.*, 1998, for illustrations).

While studying convergence for hidden Markov chains, Robert *et al.* (1998) devised a series of nonparametric tests for convergence monitoring which apply in a wide range of settings. Figure 2 presents a related control panel for a three component normal hidden Markov model, including minimum p-values of

- a Spearman test of independence between $\theta^{(0)}$ and $\theta^{(t)}$, implemented on 100 parallel replications;
- a Kolmogorov-Smirnov test of stationarity for $(\theta^{(1)}, \cdots, \theta^{(T)})$ against $(\theta^{(T+1)}, \cdots, \theta^{(2T)})$, implemented through subsampling of both parts to achieve (quasi-)independence;
- a Kolmogorov-Smirnov test of normality of $\sqrt{T}(\bar{\theta}_T - \mathbb{E}^\pi[\theta|x])$, described in Section 8 *(lower right)*.

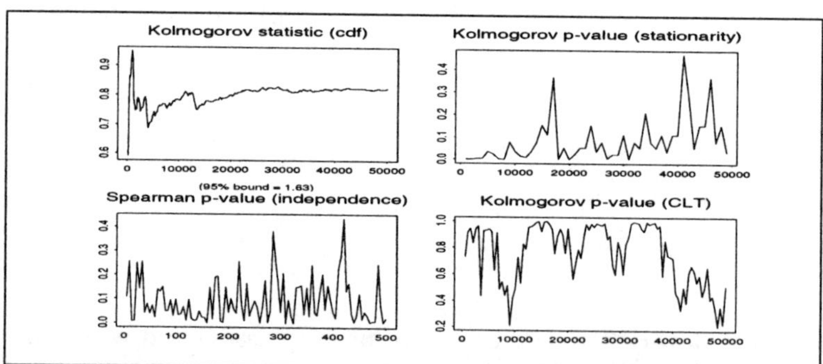

Fig. 2. Control panel for the convergence of the Gibbs Markov chain associated with a 3 latent states normal hidden Markov model and a backward (global) updating of the latent variables. (*Source:* Robert *et al.*, 1998.)

8 CLT assessment

MCMC algorithms associated with latent variable models produce Markov chains of large dimensions, through the completion of the sample. While these completed samples are not necessarily of direct interest, the Duality Principle shows that they govern the convergence properties of the overall chain and may induce stronger convergence properties when the latent variables are finite or discrete. They also give a finer picture of the stability of the chain through allocation maps and are central the general convergence diagnostic of Robert *et al.* (1998).

This method tests for asymptotic normality of standardised averages, while estimating the mean and variance by the empirical averages, $\widehat{\mu}_T$ and $\widehat{V}_T$. Instead of relying on the standard CLT, the diagnostic subsamples the chain $(z^{(t)})$ at increasingly distant epochs to achieve asymptotic independence, that is at at times t_k such that $t_{k+1} - t_k - 1 \sim \mathcal{P}(\nu k^d)$, $\nu \geq 1$ and $d > 0$. When the chain $(z^{(t)})$ is ergodic and geometrically α-mixing, if a Lyapounov-type condition on h applies, then

$$S_T = (N_T \widehat{V}_T)^{-1/2} \sum_{k=1}^{N_T} (h(z^{(t_k)}) - \widehat{\mu}_T) \rightsquigarrow \mathcal{N}(0,1) .$$

This result is valid in general but it is particularly attractive in our models as geometric α-mixing holds when $(z^{(t)})$ is finite, and the whole vector of latent variables can be tested for normality. Indeed, the $(z_j^{(t)})$'s are subsampled independently (in j) and the standardised sums,

$$\zeta_j = N_j^{-1/2} \widehat{V}_j^{-1/2} \sum_{k=1}^{N_j} (z_j^{(t_{jk})} - \widehat{\mu}_j) ,$$

are asymptotically i.i.d. $\mathcal{N}(0,1)$. Nonparametric tests like the Kolmogorov-Smirnov or the Shapiro-Wilks procedures can then test the normality of the sample $(\zeta_1, \ldots, \zeta_n)$, as illustrated in Robert *et al.* (1998).

Acknowledgements
Comments from Professor Titterington and from an anonymous referee are gratefully acknowledged.

References

Barnett, G., Kohn, R. & Sheather, S. (1996). Bayesian estimation of an autoregressive model using Markov chain Monte Carlo. *J. Econometrics*, **74**, 237–254.

Baum, L.E., Petrie, T., Soules, G. & Weiss, N. (1970). A maximization technique occurring in the statistical analysis of probabilistic functions of Markov chains. *Ann. Math. Statist.*, **41**, 164–171.

Billio, M., Monfort, A., & Robert, C.P. (1998a). Bayesian estimation of switching ARMA models. *J. Econometrics*, in press.

Billio, M., Monfort, A., & Robert, C.P. (1998b). The Simulated Likelihood Ratio method. Doc. travail CREST, Insee, Paris.

Broniatowski, M., Celeux, G. & Diebolt, J. (1983). Reconnaissance de mélan ges de densités par un algorithme d'apprentissage probabiliste. In: *Data Analysis and Informatics*, **3** (ed. E. Diday). Amsterdam: North-Holland.

Brooks, S. & Roberts, G. (1998). Diagnosing convergence of Markov chain Monte Carlo algorithms. *Canad. J. Statist.*, in press.

Cowles, M.K. & Carlin, B.P. (1996). Markov Chain Monte-Carlo convergence diagnostics: a comparative study. *J. Amer. Statist. Assoc.* , **91**, 883–904.

Dempster, A.P., Laird, N.M. & Rubin, D.B. (1977). Maximum likelihood from incomplete data via the EM algorithm. *J. Royal Statist. Soc.* (Ser. B), **39**, 1–38.

Diebolt, J. & Ip, E.H.S. (1996). Stochastic EM: method and application. In: *Markov chain Monte-Carlo in Practice* (ed. W.R. Gilks, S.T. Richardson & D.J. Spiegelhalter), 259–274. London: Chapman and Hall.

Diebolt, J. & Robert, C.P. (1994). Estimation of finite mixture distributions by Bayesian sampling. *J. Royal Statist. Soc.* (Ser. B), **56**, 363–375.

Gelfand, A.E. & Smith, A.F.M. (1990). Sampling based approaches to calculating marginal densities. *J. Amer. Statist. Assoc.* , **85**, 398–409.

Geyer, C.J. (1996). Estimation and optimization of functions. In: *Markov chain Monte-Carlo in Practice* (ed. W.R. Gilks, S.T. Richardson & D.J. Spiegelhalter), 241–258. London: Chapman and Hall.

Gruet, M.A., Philippe, A., & Robert, C.P. (1998). MCMC control spreadsheets for exponential mixture estimation. Doc. Travail CREST, INSEE.

Hamilton, J.D (1989). A new approach to the economic analysis of nonstationary time series and the business cycle. *Econometrica*, **57**(2), 357-384.

Jaquier, E., Polson, N.G. & Rossi, P.E. (1994). Bayesian analysis of stochastic volatility models. *J. Business Economic Stat.*, **12**, 371–417.

Lavielle, M. & Moulines, E. (1997). On a stochastic approximation version of the EM algorithm. *Statist. Comput.*, **7**(4), 229–236.

Liu, J.S., Wong, W.H. & Kong, A. (1994). Covariance structure of the Gibbs sampler with applications to the comparisons of estimators and sampling schemes. *Biometrika*, **81**, 27–40.

Mengersen, K.L. & Robert, C.P. (1996). Testing for mixtures: a Bayesian entropic approach. In: *Bayesian Statistics 5* (ed. J.O. Berger, J.M. Bernardo, A.P. Dawid, D.V. Lindley & A.F.M. Smith), 255–276. Oxford: Oxford University Press.

Mengersen, K.L., Robert, C.P., & Guihenneuc-Jouyaux, C. (1998). MCMC Convergence Diagnostics: a Review. In: *Bayesian Statistics 6* (ed. J.O. Berger, J.M. Bernardo, A.P. Dawid, D.V. Lindley & A.F.M. Smith), in press. Oxford: Oxford University Press.

Philippe, A. (1997). Processing simulation output by Riemann sums. *J. Statist. Comput. Simul.*, **59**(4), 295–314.

Robert, C.P. (1993). Prior Feedback: A Bayesian approach to maximum likelihood estimation. *Comput. Statist.*, **8**, 279–294.

Robert, C.P. (1996). Inference in mixture models. In: *Markov Chain Monte Carlo in Practice* (ed. W.R. Gilks, S. Richardson & D.J. Spiegelhalter), 441–464. London: Chapman and Hall.

Robert, C.P. & Casella, G. (1998). *Monte Carlo Statistical Methods.* New York: Springer Verlag.

Robert, C.P. & Mengersen, K.L. (1998). Reparametrization issues in mixture estimation and their bearings on the Gibbs sampler. *Comput. Statist. & Data Anal.*, in press.

Robert, C.P., Rydén, T. & Titterington, D.M. (1998). Convergence controls for MCMC algorithms, with applications to hidden Markov chains. Tech. Report 98–2, Uni. of Glasgow.

Robert, C.P. & Titterington, M. (1998). Resampling schemes for hidden Markov models and their application for maximum likelihood estimation. *Statist. Comput.*, in press.

Roberts, G.O. & Rosenthal, J.S. (1997). Markov chain Monte Carlo: some practical implications of theoretical results. Tech. Report, Stats. Lab., U. of Cambridge.

Rubinstein, R.Y. (1981). *Simulation and the Monte Carlo Method.* New York: J. Wiley.

Tanner, M. & Wong, W. (1987). The calculation of posterior distributions by data augmentation. *J. Amer. Statist. Assoc.* , **82**, 528–550.

Tierney, L. (1994). Markov chains for exploring posterior distributions. *Ann. Statist.*, **22**, 1701–1786.

Titterington, D.M., Smith, A.F.M. & Makov, U.E. (1985). *Statistical Analysis of Finite Mixture Distributions*. New York: J. Wiley.

Exploratory Versus Decision Trees

Roberta Siciliano

Dipartimento di Matematica e Statistica, Università di Napoli Federico II

Abstract. Tree-structured methods using recursive partitioning procedures provide a powerful analysis tool for exploring the structure of data and for predicting the outcomes of new cases. Some attention is given to partitioning algorithms and this paper in particular recalls two-stage segmentation. The step from exploratory to decision trees requires the definition of simplification methods and statistical criteria to define the final rule for classifying/predicting unseen cases. Alternative strategies are considered, which result either in the selection of a method among the others or in the definition of a compromise among them.

Keywords. Classification, regression, two-stage segmentation, pruning

1 Introduction

Segmentation methodology can be viewed as a heuristic data analysis tool to be used for large data sets characterized by high-dimensionality and nonstandard structure. These particular features mark segmentation as a nonparametric approach to data analysis where no hypothesis can be made on the variable distribution nor can the tree-structure resulting from the segmentation procedure be modelled parametrically. The dependence of a response variable on a given set of predictors is the only assumption required for segmentation of objects. This consists of partitioning the objects into a number of latent classes (on the basis of the manifest variables) in recursive way so that a tree-structure is produced. Two main targets can be achieved with the tree: *classification* and *regression* according to the type of the response variable, i.e. either categorical or numerical.

Many nonparametric methods and segmentation procedures have been stimulated by real problems of data analysis. The most appealing aspect for the user of segmentation is that the final tree provides a comprehensive description of the phenomenon in the different contexts of the application such as marketing, credit scoring, finance, medical diagnosis, etc. In fact, users very often accept statistical results only if these confirm theoretical hypotheses on the phenomenon derived from prior knowledge. Thus, several open questions arise when using such heuristic tools; in segmentation the most difficult ones to answer concern which tree to consider for explanation of the dependence data structure, and how to evaluate the accuracy of the final tree classifier/predictor if this should be extended to unseen objects without considering any "inferential dogma". This latter aspect considers segmentation methodology not only as an exploratory tool but also as a confirmatory nonparametric model.

This paper deals with the above questions. A drawback of segmentation methods is that their performance, as evaluated by crude measures such as error rate estimates, greatly depends on the type as well as on the quality of data sets. We discuss how to take account of other statistical criteria to satisfy either an exploratory or a confirmatory purpose in tree-structures methodology. A distinction is made between the two problems involved in investigating the data sets: that is whether to explore dependency, or to predict and decide about future responses on the basis of the selected predictors. *Exploration* can be obtained by performing a segmentation of the objects until a given stopping rule defines the final partition of the objects to interpret. *Confirmation* is a completely different problem that requires definition of decision rules, usually obtained by performing a pruning procedure soon after a segmentation procedure. Pruning consists of simplifying trees in order to remove the most unreliable branches and improve the accuracy of the rule for classifying fresh cases. Unfortunately, a weak point in the construction of decision trees is in the sensitivity of the classification/prediction rules as measured by the size of the tree and its accuracy to the type of data set as well as to the adopted pruning procedure. In other words, the ability of a decision tree to detect cases and take the right decisions cannot be evaluated only by a statistical index; it requires a more sophisticated type of analysis. Furthermore, as in statistical inference where the power of a testing procedure is judged with respect to changes of the alternative hypotheses, similarly, the induction by decision trees is strongly dependent on the hypotheses to verify and their alternatives. For instance, in classification trees the number of response classes and the prior distribution of cases among the classes has influence on the quality of the final rule.

The work in this paper derives from our belief that statistical modelling can fruitfully be used to complement exploratory trees and decision trees induction, in order to define a new paradigm of analysis based on *semiparametric* classification and the regression trees approach.

2 Exploratory trees

2.1 Two-stage segmentation

Let $(Y, \mathbf{X})$ be a multivariate random variable where $\mathbf{X}$ is the vector of M predictors $(X_1, \ldots, X_m, \ldots, X_M)$ taking values in $\mathcal{X} \subset \mathcal{R}^{\mathcal{M}}$ and Y is the criterion variable taking values either in the set of prior classes $\mathcal{C} = \{1, \ldots, j, \ldots, J\}$ (if categorical) or in the real space (if numerical). On the basis of a sample of N objects taken from the distribution of $(Y, \mathbf{X})$ a simple goal of exploratory trees is to uncover the predictive structure of the problem, understanding which variables and which interactions of variables are the most significant to explain the dependent variable.

Our focus is on data sets whose dimensionality requires some sort of variable selection and no linearity among the variables can be assumed. In regression analysis, flexible tools are given by nonparametric approaches such as generalized additive models (Hastie & Tibshirani, 1990), kernel neighbour discrimination, etc. (Hand, 1997). Tree methods consist of a recursive partitioning procedure of objects into K disjoint latent classes such that objects are internally homogeneous within the classes and externally heterogeneous among the classes with respect to the response variable Y. Trees are thus constructed by repeated partitions of subsets of $\mathcal{X}$ into descendant subsets beginning with $\mathcal{X}$ itself. Usually, such a recursive procedure terminates according to a stopping rule based either on a fixed low number of objects in the current node or on a fixed low value of the criterion used for segmentation of the subsamples. Terminal nodes of the tree constitute the final partition of the N objects into some desirable subgroups; to each terminal node is assigned either a value (i.e. the average of the response numerical variable in regression trees) or a class (i.e. the modal response category in classification trees). In the following, we assume that K is fixed for each node of the tree (i.e. for $K = 2$ and $K = 3$ we have respectively binary and ternary trees), although in other approaches related to expert systems K is a parameter to be determined node by node (Quinlan, 1986).

We recall and generalize to some extent the two-stage segmentation introduced by Mola & Siciliano (1992, 1994) which relies on the assumption that a predictor X_m is not merely used as generator of partitions but it also plays a global role in the analysis: we evaluate the *global effect* of X_m on the response variable Y as well as the *partial effect* of any partition p generated by X_m, for $p \in P_m$ where P_m is the set of all partitions of the X_m modalities into K latent classes. In the first stage, a *variable selection* criterion is applied to find one or more predictors that are the most predictive for the response variable. On the basis of the set of partitions generated by the selected predictor(s), in the second stage, a *partitioning* criterion is considered to find the best partition of the objects at the given node. The criteria to be used in the two stages depend on the nature of the variables, the tool of interpretation and the desired description in the final output. The partitioning algorithm takes account of the computational cost induced by the recursive nature of the procedure and the number of possible partitions at each node of the tree.

2.2 Fast segmentation algorithm

A standard approach for segmentation of objects considers the relative reduction in impurity of the response variable Y when passing from the parent node to its descendants. Let $e_Y(t)$ be any impurity measure (heterogeneity for a categorical response variable or variation for a numerical response variable) of Y at node t, and let $e_Y(i|t)$ be the same impurity measure for the conditional distribution of Y given the modality i of X_m for $i \in I_m$. Denote by $\pi(i|t)$ the probability of an object in node t having modality i of any predictor X_m; denote by $\pi(k|t)$ the probability of an object in node t falling into the descendant k. For a general and unified formulation of segmentation criteria we define the *global impurity reduction factor* for the predictor X_m as

$$\omega_{Y|X_m}(t) = \sum_{i \in I_m} \pi(i|t) e_Y(i|t), \tag{1}$$

and the *local impurity reduction factor* for any partition p of X_m into K classes as

$$\omega_{Y|p}(t) = \sum_{k \in K} \pi(k|t) e_Y(k|t), \tag{2}$$

where any k indicates a subset of the I_m predictor's modalities. The common practice is to evaluate only (2) for all partitions of all predictors. The factor (2) includes as special cases several partitioning criteria; for instance, CART (Breiman *et al.*, 1984), ID3 (Quinlan, 1986), CN2 (Clark & Niblett, 1989). Variants of these criteria can also be considered, such as those proposed by Taylor & Silverman (1993) for the problem of small splits as well as by Aluja-Banet & Nafria (1998) for data diagnostics (see also Clark & Pregibon,

1992). It is worth noting that the relative reduction in impurity, as defined by $(e_Y(t) - \omega_{Y|p})/e_Y(t)$, can also be shown to be related to well known statistical measures (Siciliano & Mola, 1997), i.e. Pearson's square correlation coefficient for regression trees, the conditional entropy index and the predictability τ index for classification trees. In the same vein, the global node impurity can be evaluated following a modelling approach (Siciliano & Mola, 1994). An example is provided by logistic classification (Mola, Klaschka & Siciliano, 1996): a pseudo-$\bar{R}^2$ measure is defined by $\bar{R}^2(t) = [LL_0(t) - LL_s(t)]/LL_0(t)$, where $LL_s(t)$ is the likelihood ratio statistic of the logistic submodel s assigned to the node t and the likelihood $LL_0(t)$ under the trivial model with no predictors. The $\bar{R}^2$ measure takes values in $[0,1]$ and measures the percent of the "uncertainty in the data". The local impurity reduction factor when passing to K descendants is evaluated by the fit of the submodel within each subgroup, namely $\sum_{k \in K} \pi(k|t)\bar{R}^2(t_k)$ where all significant predictors contribute to the definition of the best partition.

Discarding the two-stage segmentation criterion, the optimal solution at each node is obtained by finding the minimum of (2) among the best partitions of all predictors

$$min_{m \in M}\{min_{p \in P_m} \omega_{Y|p}(t)\}. \tag{3}$$

This means that the best partition is not necessarily found to be generated by the best predictor that provides the highest global impurity reduction (1). However, several simulation studies show that in two-class problems for example the best split in binary trees is generated by one of the first three best predictors with probability near to 1, and by one of the first two best predictors with probability near to 0.95 (Siciliano & Mola, 1998a). A predictor with high predictability power on the response variable, i.e. with high global impurity reduction, has high probability to generate the best partition. These results might justify the application of the standard two-stage algorithm with general criteria (1) and (2) in stages I and II respectively: the selection of some (but not all) the best predictors saves a lot of computing time! Nevertheless, a suitable property involving both (1) and (2) allows us to guarantee that the optimal solution can be found with savings in computation time and without necessarily trying out all predictors. For an impurity measure satisfying the condition $e_Y(k|t) \geq e_Y(i|t)$ for $i \in k$, the property $\omega_{Y|X_m}(t) \leq \omega_{Y|p}(t)$ holds for any $p \in P_m$ of X_m. This property allows to define a *fast segmentation algorithm* based on the following rules:

(a) iterate the standard two-stage algorithm using (1) and (2) selecting one predictor at a time and each time considering the predictors that have not been selected previously;

(b) stop the iterations when the current best predictor in the order, namely $X_{(v)}$ at iteration v, does not satisfy the condition $\omega_{Y|X_{(v)}}(t) \geq \omega_{Y|p^*_{(v-1)}}(t)$ where $p^*_{(v-1)}$ is the best partition at iteration $(v-1)$.

The fast segmentation algorithm finds the optimal partition according to (3) but with substantial time savings in terms of the reduced number of partitions to be tried out at each node to find the best partition (Siciliano & Mola, 1996). Simulation studies show that the relative reduction in the average number of partitions analyzed by the fast algorithm with respect to the standard approach increases as the number of predictor modalities and the number of objects at a given node increase (Mola & Siciliano, 1997). Further theoretical results about the computational efficiency of fast-like algorithms can be found in Klaschka, Siciliano & Antoch (1998).

2.3 Visualizing tree nodes by factorial segmentation

Factorial segmentation for categorical/categorized variables is an alternative partitioning procedure for growing exploratory trees on the basis of a reduced-rank factorial model such as *nonsymmetric correspondence analysis* (Lauro & D'Ambra, 1984; Siciliano, Mooijaart & van der Heijden, 1993). This model decomposes, through a generalized singular value decomposition, the Goodman and Kruskal predictability τ index of the cross-classification of Y with a predictor X_m into a number of factorial terms.

A two-stage criterion is applied where in the first stage the best predictor is selected by maximizing the τ index. A factorial representation of the dependence of the response categories on the best predictor categories is assigned to each node of the tree. A partitioning criterion can be defined on the basis of the first factorial axis, which retains the highest percentage of inertia of the response variable due to the best predictor. This inertia, which corresponds to the reduction in impurity of Y due to X_m, is related to the square singular values and the square predictor coordinates as follows:

$$e_Y(t) - \omega_{Y|X_m}(t) \cong \lambda_1^2(t) = \sum_i \pi(i|t)(\lambda_1(t) r_{i1}(t))^2, \tag{4}$$

where $\lambda_1(t)$ is the highest singular value, corresponding to the first set of predictor scores $r_{i1}(t)$; the coordinates of the predictor's categories given by $\lambda_1(t)r_{i1}(t)$ for $i \in I_m$ satisfy the centring condition $\sum_i \lambda_1(t) r_{i1}(t)\pi(i|t) = 0$. Dividing each side of equation (4) by ${\lambda_1}^2(t)$ we obtain predictability measures (or contributions of the predictor categories to the reduction of impurity explained by the first factorial axis); namely $\sum_i \pi(i|t){r_{i1}}^2(t) = 1$. We distinguish between *strong categories* with $|r_{i1}(t)| \geq 1$, and *weak categories* with $|r_{i1}(t)| < 1$.

The predictor categories with a negative coordinate will predict response categories different from those predictor categories with a positive coordinate, so that using their sign yields a binary splitting criterion (Mola & Siciliano, 1998). In practice, however, we can have coordinates close to zero but with different signs, in which case using them to make different predictions does not make sense. Therefore, we use these intermediate predictor categories to define an additional subgroup, leading us to a ternary partitioning criterion (Siciliano & Mola, 1998b). Thus, we partition objects into three subgroups according to the partitions of the predictor's categories into three subgroups: the first includes categories such that $r_{i1}(t) \geq 1$ (strong left categories), the second includes categories such that $|r_{i1}(t)| < 1$ (weak categories), the third includes categories such that $r_{i1}(t) \leq -1$ (strong right categories). The middle subgroup includes objects in which the response variable is not strongly characterized by any category of the best predictor. During the partitioning procedure there can be empty subgroups and there can be combinations of more predictors to form compound variables, namely multiple questions for the objects.

2.4 Latent budget trees

Standard procedures for classification trees are not always feasible in the multiclass case with a *multiclass response variable*. Typical problems for example in binary trees arise from the poor strength of the splits in classifying objects over different classes, so that some of these never occur in the final leaves of the tree, and are thus never explained by the exploratory tree. An alternative tree-growing procedure is provided by *latent budget trees*, which are characterized by a sequence of latent budget models assigned to the nodes of the tree (Siciliano, 1998). The latent budget model at node t for a cross-classification of Y with a predictor X_m is a reduced-rank decomposition of the I_m theoretical budgets $\pi(j|i,t)$ as a mixture of latent budgets $\pi(j|k,t)$ with mixing parameters $\pi(k|i,t)$:

$$\pi(j|i,t) = \sum_{k \in K} \pi(k|i,t)\pi(j|k,t), \tag{5}$$

where k is the latent class, usually $K = 2$ for binary trees, $K = 3$ for ternary trees, and the parameters are conditional probabilities summing to one over the first index. The procedure selects at each node the best (compound) predictor on the basis of the AIC criterion or its modified version $AIC^* = G^2(X_m) - 2df$, where $G^2(X_m)$ is the likelihood ratio statistic for testing the latent budget model with predictor X_m against the saturated model, and df are its degrees of freedom. The partitioning criterion considers the mixing parameters representing a synthesis of the I_m predictor's modalities into K subgroups: each predictor's category is assigned to the k-th latent budget which presents the *highest mixing parameter estimate*, i.e. the probability estimate to fall into the k-th descendant given that it has the i-th predictor modality. The partition of the categories induces a partition of the objects into K subgroups. A further aid to the interpretation of the tree is provided by comparing the *latent budget parameter estimates* with the independence hypothesis: for the k-th latent budget the response categories which depart more from independence are those better predicted by the latent budget and thus by the given partition. Notice that through the Bayes rule the latent budget parameter $\pi(j|k,t)$ can be viewed as the posterior probability of falling in class j once the object is assigned to the k-th descendant. Thus, starting from the root node of the tree, at which the prior probability estimates given by the group proportions of the response variable are assigned, the posterior probability estimates are recursively updated and related through a chain of conditional probability estimates to yield the definition of the final posterior classifications at the terminal nodes of the latent budget tree.

3 Decision trees

Exploratory trees can fruitfully be used to investigate the structure of data but they cannot straightforwardly be used for induction purposes. The main reason is that exploratory trees are accurate and effective with respect to the *training data set* used for growing the tree but they might perform poorly when applied for classifying/predicting fresh cases which have not been used in the growing phase.

A further step is required for *decision tree induction* relying on the hypothesis of the *uncertainty* in the data due to *noise* and *residual variation* (Mingers, 1989b). Simplifying the tree is necessary to remove the most unreliable branches and improve understandability. The pioneer approach to simplification was based on arresting the recursive partitioning procedure according to some stopping rule (*prepruning*). A more recent approach consists of forming the totally expanded tree and retrospectively removing some of the branches (*postpruning*). The result can be either *a set of optimally pruned trees* (on which basis the final decision rule is defined) or just *one best pruned tree* (which thus represents the final rule).

Decision tree induction definitely has an important purpose represented by *understandability*: the tree structure for induction needs to be simple and not so large; this is a difficult task especially for binary trees since a predictor may reappear (even though in a restricted form) many times down a branch. At the same time, a further requirement is given by *identifiability*: on one hand, terminal branches of the expanded tree reflect particular features of the training set causing *overfitting*; on the other hand, overpruned trees do not necessarily allow us to identify all the response classes/values (*underfitting*).

The goal of simplification for decision trees is thus inferential, i.e. to define the structural part of the tree model, reducing the size of the tree while retaining its accuracy. Basically, the idea of Mingers (1989a) that the performance of the simplification method in terms of accuracy is independent from the partitioning criterion used in the tree growing procedure has been contradicted by Buntine & Niblett (1992). The choice of the most suitable method for simplifying trees depends not only on the partitioning criterion, and thus on the expanded tree from which to start simplifying, but also on the objective and the kind of data sets. Thus, exploratory trees become an important preliminary step for decision trees induction. In simplification procedures it is worthwhile to distinguish between *optimality criteria for pruning* the tree and *criteria for selecting* the best decision tree. These two processes do not necessarily coincide and often require independent data sets (*training set* and *test set*). In addition, a *validation data set* can be required to assess the quality of the final decision rule (Hand, 1997). In this respect, segmentation with pruning and assessment can be viewed as stages of any computational model building process based on supervised learning algorithms like expert systems and neural networks (Sethi, 1990; Russel, 1993). Particular attention should be given to prior data processing and final validation, which should coherently be defined according to the problem (Mola *et al.*, 1997).

3.1 Criteria for pruning

Postpruning algorithms can work from the bottom of the tree to the top (*down-top postpruning*) or vice versa (*top-down postpruning*). The training set is often used for pruning whereas the test set is used for selection of the final decision rule; this is the case with the error-complexity pruning of CART (Breiman *et al.*) and the critical value pruning (Mingers, 1989b). Nevertheless, some methods require only the training set, such as the pessimistic error pruning and the error based pruning (Quinlan, 1987, 1993) and also the minimum error pruning (Cestnik & Bratko, 1991) and the cross-validation method of CART; other methods require only the test set, such as the reduced error pruning (Quinlan, 1987).

Denote by T_t the branch departing from the node t and having $|\tilde{T}_t|$ terminal nodes. Any pruning algorithm is based on a given measure $R(.)$ used to evaluate the convenience of retaining the branch T_t over pruning it, where the criterion is of type: prune node t if $R(t) \leq R(T_t)$. In general, such a measure considers the *complexity* or size of the tree i.e. the number of terminal nodes, and the *accuracy* i.e. error measures such as the error rate in classification trees and the mean square error of the predictions in regression trees. In the following, for sake of brevity we restrict our attention to simplifying classification trees. The idea of post-pruning was initially an innovation of the CART methodology (Breiman *et al.*, 1984). The algorithm is based on the *error-complexity measure* defined for the node t and for the branch T_t as

$$R_\alpha(t) = r(t)p(t) + \alpha, \tag{6}$$

$$R_\alpha(T_t) = \sum_{h \in |\bar{T}_t|} r(h)p(h) + \alpha|\bar{T}_t|, \tag{7}$$

where α is the penalty for complexity due to one extra terminal node in the tree, $r(t)$ is the error rate (the proportion of cases in node t that are misclassified), $p(t)$ is the proportion of cases in node t and $|\bar{T}_t|$ is the number of terminal nodes of T_t. Basically, the branch T_t should be pruned if $R_\alpha(t) \leq R_\alpha(T_t)$. Thus, using a down-top algorithm and a *training set*, the criterion is to prune each time the branch T_t that provides the lowest reduction in error per terminal node (i.e. the *weakest link*) as measured by $\alpha_t = \{R(t) - R(T_t)\}/\{|\bar{T}_t| - 1\}$. On the basis of the error-complexity measure $R_\alpha(.)$ a sequence of nested optimally pruned trees is generated pruning at each step the subtree with the minimum value of α_t. In the same framework of CART, Gelfand *et al.* (1991) provide an alternative procedure which iteratively optimizes the tree-growing and the pruning of classification trees. Variants to the CART pruning have been proposed in different contexts such as expert systems and artificial intelligence. In particular, Quinlan (1986, 1987, 1993) has developed some pruning methods for classification trees. The *reduced error pruning* directly and exclusively employs the test set to produce a sequence of pruned trees. The criterion is always to prune the node t if $R_{ts}(t) \leq R_{ts}(T_t)$ (where the subscript ts refers to the test set), choosing to prune at each step the branch with the largest difference. The down-top algorithm continues until no further pruning is possible, as the error rate would increase, and it ends with the smallest subtree with the minimum error rate with respect to the test set. Instead, the *pessimistic error pruning* uses a top-down pruning algorithm and produces only *one pruned tree* on the basis of the training set. The idea is to worsen the estimate of the error rate on the training set by applying the continuity correction for the Binomial distribution given by 0.5. This results in the *corrected* error rate $R^*(t) = r(t)p(t) + 0.5/n(t)$ for the node t, and the similarly defined rate $R^*(T_t)$ for the branch T_t, using a CART-like notation. Again, the branch T_t should be pruned if $R^*(t) \leq R^*(T_t)$, or alternatively if it is less than one standard error more than the corrected measure for its branch as in the *error based pruning* of C4.5 (Quinlan, 1993). The criterion employed in the pessimistic-error pruning can be viewed as a special case of the error-complexity pruning criterion when α is fixed to be equal to $0.5/n(t)$. Because of the top-down type algorithm only one pruned tree is identified whereas, using a down-top algorithm with the above criterion, a sequence of optimally pruned trees can be defined instead. Anyway, the continuity correction appears to be suitable only for the two class problem; a more general correction which is based on the number J of response classes is given by $(J-1)/J$ (Cappelli & Siciliano, 1997).

In decision tree induction, accuracy refers to the predictive ability of the decision tree to classify/predict an independent set of test data. In the particular case of classification trees the error rate, as measured by the number of incorrect classifications that a tree makes on the test data, is a crude measure since it does not reflect the accuracy of predictions for different classes within the data. In other words, classes are not equally likely, and those with few cases are usually predicted badly.

In this respect, the critical-value pruning of Mingers (1989b) represents an alternative pruning algorithm, as it does not rely on the error rate to define the set of pruned trees. In fact, this method specifies a critical value for the measure used in the partitioning criterion and prunes those nodes that do not reach the critical value for any node within their branch. The larger the critical value selected, the greater the degree of pruning and the smaller the resulting pruned tree; a set of optimally pruned trees can be generated by increasing the critical values. Similarly, Cappelli, Mola & Siciliano (1998) provide a pruning algorithm based on the *impurity-complexity measure* as an alternative to the error-complexity measure of CART. In particular, the error rate can be replaced by any impurity measure which takes account of the number of classes and the distribution of the cases over the classes. This approach might be viewed as a sort of critical-value pruning based on a very general *accuracy-complexity measure.*

3.2 Selection of decision rules

Several studies of empirical comparisons among pruning methods (Mingers, 1989b; Malerba *et al.*, 1993) have proved that all pruning methods consistently reduce the size of the tree while improving its accuracy. But there are also significant differences in the final decision rules in terms of both the number of terminal nodes and the error rate. Moreover, the performances of the simplification methods depend on the type as well as on the quality of data sets. As a matter of fact, there does not exist a best way to simplify trees as this depends on the definition of the domain of application and the criteria used to choose the best one. In fact, general methods for avoidance of overfitting might also amount to a form of *bias* rather than a statistical improvement of the prediction (Quinlan, 1986; Buntine, 1992; Schaffer, 1993).

Given a set of optimally pruned trees, a simple method to choose the optimal decision tree consists of selecting the one producing the minimum misclassification rate on an independent *test set* ($0 - SEE$ rule) or the smallest tree whose error rate on the test set is within one standard error of the minimum ($1 - SEE$ rule) (Breiman *et al.*, 1984). An alternative selection method consists of first defining a set of optimal decision trees provided by different pruning algorithms and then choosing the best rule according to some statistical criteria. Let T^* be the set of Q *optimal decision trees* resulting from Q different pruning methods. A simple selection procedure for choosing the best decision tree can be based on the misclassification rate to be minimized. Obviously, this criterion does not take into account the different number of terminal nodes of each optimal tree. A sophisticated procedure consists of summarizing the information given by each decision tree T_q^* by means of a table which cross-classifies the terminal nodes of the tree with the response classes. This table describes the conditional distributions of the cases over the response classes within each terminal node; it provides the estimates of the conditional probabilities $\pi(j|t_h)$ where t_h denotes the h-th terminal node for $h \subset H$. Ideally, for the best decision tree these conditional distributions should be internally homogeneous and externally heterogeneous, i.e. for each h it holds that $\pi(j|t_h) = 1$. Therefore, the problem of choosing the best decision tree can be converted into the problem of evaluating the predictability power of each optimal decision tree or equivalently the departure from independence in the table. As an example, the best decision tree can be chosen as the one that maximizes the corrected Akaike criterion for each tree T_q^*:

$$AIC^*(T_{q^*}^*) = max_{q \in Q} \; AIC^*(T_q^*). \tag{8}$$

for $AIC^*(T_q^*) = G^2(T_q^*) - 2df$, where $G^2(T_q^*)$ is the likelihood ratio statistic for testing the hypothesis of independence and df is the number of degrees of freedom. This index allows comparisons among trees with a different number of terminal nodes. The higher the value of the index the better the predictive power of the partition given by the terminal nodes of the decision tree (for applications see Cappelli & Siciliano, 1997). In order to describe the table associated with the q-th optimal decision rule, the latent budget model can also be considered (see also Section 2.4). In particular, for each optimal tree the conditional probability distributions $\pi(j|t_h)$ of the J response classes given the H terminal nodes can be decomposed according to the latent budget model (similar to (5)) with $G \leq min(H, J)$ latent budgets chosen such that the model fits the data. The final number G indicates the number of groups that can be considered to amalgamate the response classes. This is particularly convenient for the *prevalence estimation* of the prior probabilities of the classes and also their size in the population (Hand, 1996).

3.3 Compromise of decision rules

An alternative strategy to the selection of one final tree consists of the definition of a *compromise* or *consensus* using a set of trees rather than the single one. One approach, known as *tree averaging*, classifies a new object by averaging over a set of trees using a set of weights (Oliver & Hand, 1995). This procedure requires the definition of the set of trees (as it is impractical to average over every possible pruned tree), the calculation of the weights, and the independent data set to classify. Buntine (1992) defines a Bayesian approach to the estimation of the weights as well as the *path set* of trees as a subset of the set of all pruned trees. Oliver & Hand (1995) instead define the *fanny set* and the *extended set* which are not subsets of the set of pruned trees. Another proposal is provided by the set of *optimal decision trees* obtained by different simplification methods.

A more general strategy of compromise consists of creating a *consensus tree* among the optimal decision trees. When a decision tree is used to classify a case, a path of conditions is followed from the root node to one of its leaves. A path of conditions can be regarded as a *production rule* defined as the *conjunction* of splitting variables such as for example, in binary trees, $if \bar{s}_1 \wedge s_2 \wedge \bar{s}_4 \wedge s_9$ *then class* j where s_t and $\bar{s}_t$ denote respectively answers *yes* or *no* to the question induced by the splitting variable at node t. The *disjunction* of the production rules which provide the same class j, for $j = 1 \in J$, defines the *classification rule* for class j. Therefore, a decision tree can be regarded as a collection of classification rules. Two types of compromise can be proposed to combine different decision trees (Cappelli & Siciliano, 1997): the *coarser decision tree* is defined by applying the disjunction operator to the collections of classification rules associated with each optimal decision tree; the *finer decision tree* is defined by applying the conjunction operator to the collections of classification rules associated with each optimal decision tree. The smaller the differences in terms of both size and accuracy between the coarser and the finer tree (being the extremes of the set of the optimal trees) the higher the degree of *reliability* of the tree model induction.

4 Conclusion

This paper has reviewed some recent results in the field of segmentation and trees induction. Exploratory trees *versus* decision trees can be viewed as two important steps of a new paradigm of analysis called *semiparametric trees* where statistical modelling can fruitfully be applied for tree-growing as well as for tree-validation and assessment.

References

Aluja-Banet, T. & Nafria, E. (1998). Robust Impurity Measures in Decision Trees. In: *Data Science, Classification, and Related Methods* (ed. C. Hayashi *et al.*), 207-214. Tokyo: Springer.

Breiman L., Friedman J. H., Olshen R. A. & Stone C. J. (1984). *Classification and Regression Trees.* Wadsworth, Belmont CA.

Buntine, W. L. (1992). Learning Classification Trees. *Statistics & Computing*, **2**, 63-73.

Buntine, W. & Niblett, T. (1992). A Further Comparison of Splitting Rules for Decision-Tree Induction. *Machine Learning*, **8**, 75-85.

Cappelli, C. & Siciliano, R. (1997). On Simplification Methods for Decision Trees. In: *Proceedings of the NGUS-97 Conference*, 104-108.

Cappelli, C., Mola, F. & Siciliano, R. (1998). An Alternative Pruning Method Based on the Impurity-Complexity Measure. *COMPSTAT98 Proceedings in Computational Statistics*, in press.

Cestnik, B. & Bratko, I. (1991). On Estimating Probabilities in Tree Pruning. In: *Proceedings of the EWSL-91*, 138-150.

Clark, L.;A. & Pregibon, D. (1992). Tree-Based Models. *Statistical Models in S* (ed. J.M. Chambers & T. J., Hastie), 377-420. Belmont CA Wadsworth and Brooks.

Clark, P. & Niblett, T. (1989). The CN2 Induction Algorithm. *Machine Learning*, **3**, 261-283.

Gelfand, S. B., Ravishankar, C. S. & Delp, E. J. (1991). An Iterative Growing and Pruning Algorithm for Classification Tree Design. *IEEE Transactions on Pattern Analysis and Machine Intelligence*, **13**, *2*, 163-174.

Hand, D. J. (1996). Classification and Computers: Shifting the Focus. In: *COMPSTAT96 Proceedings in Computational Statistics* (ed. A. Prat), 77-88. Heidelberg: Physica Verlag.

Hand, D. J. (1997). *Construction and Assessment of Classification Rules.* New York: J. Wiley and Sons.

Hastie, T. & Tibshirani, R., (1990), *Generalized Additive Models.* London: Chapman & Hall.

Ho T. B., Nguyen, T. D. & Kimura, M. (1998). Induction of Decision Trees Based on the Rough Set Theory. In: *Data Science, Classification, and Related Methods* (ed. C. Hayashi *et al.*), 215-222. Tokyo: Springer.

Klaschka, J., Siciliano, R. & Antoch, J. (1998). Computational Enhancements in Tree-Growing Methods. In: *Proceedings of the IFCS Conference* (ed. A. Rizzi), in press.

Lauro, N. C. & D'Ambra, L. (1984). L'Analyse Non Symmetrique des Correspondances. In: *Data Analysis and Informatics III* (ed. E. Diday *et al.*), 433-446. Amsterdam: North Holland.

Malerba, D., Esposito, F. & Semeraro, G. (1996). A Further Comparison of Simplification Methods for Decision-Tree Induction. *Learning from Data: AI and Statistics* (ed. V. Fisher & H. J. Lenz), 365-374. Heidelberg: Springer Verlag.

Mingers, J. (1989a). An Empirical Comparison of Selection Measures for Decision Tree Induction. *Machine Learning*, **3**, 319-342.

Mingers, J. (1989b). An Empirical Comparison of Pruning Methods for Decision Tree Induction. *Machine Learning*, **4**, 227-243.

Mola, F. & Siciliano, R. (1992). A Two-Stage Predictive Splitting Algorithm in Binary Segmentation. In: *Computational Statistics* (ed. Y. Dodge & J. Whittaker), **1**, 179-184. Heidelberg: Physica Verlag.

Mola, F. & Siciliano, R. (1994). Alternative Strategies and CATANOVA Testing in Two-Stage Binary Segmentation. In: *New Approaches in Classification and Data Analysis* (ed. E. Diday *et al.*), 316-323. Heidelberg: Springer Verlag.

Mola, F & Siciliano, R (1997). A Fast Splitting Procedure for Classification trees. *Statistics and Computing*, **7**, 208-216.

Mola, F. & Siciliano, R. (1998). Visualizing Data in Tree-Structured Classification. In: *Data Science, Classification, and Related Methods* (ed. C. Hayashi *et al.*), 223-230. Tokyo: Springer.

Mola, F., Klaschka, J. & Siciliano, R.(1996). Logistic Classification Trees. In: *COMPSTAT '96* (ed. A. Prat), 165-177. Heidelberg: Physica Verlag.

Mola, F., Davino, C., Siciliano, R. & Vistocco, D. (1997). Use and Overuse of Neural Networks in Statistics. In: *Proceedings of the NGUS-97 Conference*, 57-68.

Oliver, J. J. & Hand, D. J. (1995). On Pruning and Averaging Decision Trees. In: *Machine Learning: Proceedings of the* 12*th International Workshop*, 430-437.

Quinlan, J. R. (1987). Simplifying Decision Tree. *International Journal of Man-Machine Studies*, **27**, 221-234.

Quinlan, J. R. (1993). *C4.5: Programs for Machine Learning.* Morgan Kaufmann, San Mateo, CA.

Russel R. (1993). Pruning Algorithms - A Survey. *IEEE Transactions on Neural Networks*, **4**, 5, 740-747.

Schaffer, C. (1993). Overfitting Avoidance as Bias. *Machine Learning*, **10**, 153-178.

Sethi, I. K. (1990). Entropy Nets: from Decision Trees to Neural Networks. *Proceedings of the IEEE*, **78**, *10*, 1605-1613.

Siciliano, R. (1998). Latent Budget Trees for Multiple Classification. *Proceedings of the IFCS Classification Group of SIS.* Heidelberg: Springer Verlag, in press.

Siciliano, R. & Mola, F. (1994). A Recursive Partitioning Procedure and Variable Selection. In: *COMPSTAT94 Proceedings in Computational Statistics* (ed. R. Dutter & W. Grossmann), 172-177, Heidelberg: Physica Verlag.

Siciliano, R. & Mola, F. (1996). A Fast Regression Tree Procedure. In: *Statistical Modelling. Proceedings of the* 11*th International Workshop on Statistical Modelling* (ed. A. Forcina *et al.*), 332-340. Perugia: Graphos.

Siciliano, R. & Mola, F. (1997). Multivariate Data Analysis and Modelling through Classification and Regression Trees. *Computing Science and Statistics* (ed. E. Wegman & S. Azen), **29**, *2*, 503-512. Interface Foundation of North America, Inc.: Fairfax.

Siciliano, R. & Mola, F. (1998a). On the Behaviour of Splitting Criteria for Classification Trees. In: *Data Science, Classification, and Related Methods* (ed. C. Hayashi *et al.*), 191-198. Tokyo: Springer.

Siciliano, R. & Mola, F. (1998b). Ternary Classification Trees: a Factorial Approach. In: *Visualization of categorical data* (ed. J. Blasius & M. Greenacre). New York: Academic Press.

Siciliano, R., Mooijaart, A. & van der Heijden, P. G. M. (1993). A Probabilistic Model for Nonsymmetric Correspondence Analysis and Prediction in Contingency Tables. *Journal of the Italian Statistical Society*, **2**, *1*, 85-106.

Taylor, P. C. & Silverman, B. W. (1993). Block Diagrams and Splitting Criteria for Classification Trees. *Statistics & Computing*, **3**, 147-161.

Exploring Time Series Using Semi- and Nonparametric Methods

Dag Tjøstheim

Department of Mathematics, University of Bergen, 5007 Bergen, Norway

1 Introduction

Exploratory procedures are essential ingredients at various stages of time series model building. For example, since linearity represents a great simplification, a linearity test ought to be implemented at an early stage. If linearity *is* rejected, one possibility is to search for a parametric model among such classical models as the threshold, the exponential autoregressive, the bilinear, or the more flexible class of smooth transition models (cf. Granger & Teräsvirta, 1993). If instead a nonparametric approach is followed, some simplification would usually be required, among other things, to avoid the curse of dimensionality. Selecting significant lags and modelling those lags by additive models represent one such simplification. I will indicate briefly both the testing for and estimation of such models.

The aim of much of time series analysis is to build a model where the residuals of the model are independent identically distributed (iid) random variables. If the residuals are not iid, this may give cause to re-evaluating and re-estimating the model. I will mention several nonparametric tests for independence.

It should be stressed that all of our nonparametric techniques are based on the assumption of stationarity. In applications in economics, for example, series are often thought to be nonstationary, and a first step consists of transforming them, using first differences or other devices, to obtain stationarity. Very recently an attempt has been made (Karlsen & Tjøstheim, 1998) to extend nonparametrics to classes of nonstationary processes (which include the random walk case).

All of the specification tests emphasized in this paper can be put into a common nonparametric framework, and demonstrating and clarifying this framework in a variety of different situations is the main purpose of the talk. The key idea of our procedure is to construct two statistics, which estimate the same quantity under H_0 – e.g linearity or independence – but different quantities under the alternative hypothesis H_A. Moreover, a distance function is introduced measuring the distance between the statistics. The null hypothesis is rejected if a large value of the distance functional occurs. The critical value is derived from the distribution of the distance functional under H_0. This distribution can either be constructed from asymptotic theory, often using a U-statistic argument, or from a randomization device. For small and moderate sample sizes it is our experience that the randomization argument gives a much better approximation than the asymptotics.

It should be made clear at the outset that our tests are designed to be used more as exploratory tools than as formal tests. Therefore we are not so concerned with formal power properties and guarding against pathological worst case alternatives. Such problems are difficult and challenging but not within the scope of this presentation. Our goal has rather been to construct exploratory devices which would be of direct use in a number of commonly encountered situations in model building.

This brief paper is to a large degree based on the more extensive survey paper Tjøstheim (1998). The reader interested in more details and more references should consult that paper. It should be noted that the word "semiparametric" is not used in a precise meaning, but rather to signify that the methods described to some extent include a parametric component; most often used as a means of comparison to a nonparametric quantity.

2 Linearity tests

Estimating and analysing linear models are simple tasks compared to the same for nonlinear models. It is therefore important to decide early in the modelling process whether a linear or a nonlinear model should be entertained. This fact constitutes the motivation for linearity tests. Such tests can roughly be divided into the classes of parametric and nonparametric tests. Many parametric tests can be thought of as being special cases of Lagrange multiplier tests. These have been emphasized by Luukkonen *et al.* (1988a & b), Granger & Teräsvirta (1993) and Teräsvirta (1994).

2.1 Nonparametric tests

The first nonparametric tests were proposed in the spectral domain. These tests originated with Subba Rao & Gabr (1980) and were improved by Hinich (1982).

An informal and much used exploratory technique in regression and time series analysis is to construct plots of the conditional mean and the conditional variance of the dependent variable Y given an explanatory variable X_k. The results to be presented in the remainder of this section could be seen as an attempt to quantify and formalize this much used looking–at–plot procedure. Essentially it amounts to computing the statistical fluctuations of the plots expected under the null hypothesis of linearity. My presentation will be based almost exclusively on Hjellvik & Tjøstheim (1995,1996) and Hjellvik *et al.* (1998), but see also Härdle & Mammen (1993) for a more general point of view.

For reasons having to do with the curse of dimensionality we look at one lag at a time; i.e. we estimate the conditional mean $M_k(x) = E(X_t \mid X_{t-k} = x)$

and the conditional variance $V_k(x) = \text{var}(X_t \mid X_{t-k} = x)$ nonparametrically and compare them with the linear regression of X_t on X_{t-k} and the corresponding residual variance; the comparison being made via test functionals now to be introduced.

If, with no restriction, we assume that $\{X_t\}$ is zero-mean, then the linear regression of X_t on X_{t-k} is given by $\rho_k X_{t-k}$ where $\rho_k = \text{corr}(X_t, X_{t-k})$, and the squared difference linearity test functional (cf. Hjellvik & Tjøstheim 1995,1996) is defined by

$$L(M_k) = \int (M_k(x) - \rho_k x)^2 w(x) dF(x) \tag{2.1}$$

where w is a weight function and F is the cumulative distribution function of X_t.

The conditional variance can be treated likewise using the functional

$$L(V_k) = \int (V_k(x) - \sigma_k^2)^2 w(x) dF(x) \tag{2.2}$$

where $\sigma_k^2 = (1 - \rho_k^2)\text{var}(X_t)$ is the residual variance in a linear regression of X_t on X_{t-k}. Actually, following the practice of ARCH modelling the $L(V_k)$-functional will be applied to the residuals $\{e_t\}$ of a linear AR (or ARMA) model fit. One can also look at aggregated functionals and functionals based on derivatives.

2.2 Estimation and asymptotic theory

There are two issues involved; i.e. that of estimating individual functions such as $M_k(x), V_k(x)$ and that of estimating the integrals of these functions. The integrals can be estimated using the empirical mean, so that, if n is the number of observations,

$$\widehat{L}(M_k) = \frac{1}{n} \sum_t \{\widehat{M}_k(X_t) - \widehat{\rho}_k X_t\}^2 w(X_t)$$

and

$$\widehat{L}(V_k) = \frac{1}{n} \sum_t \{\widehat{M}_k(\widehat{e}_t) - \widehat{\rho}_{e,k} \widehat{e}_t\}^2 w(\widehat{e}_t),$$

where $\widehat{e}_t$ are the residuals from a linear AR($\widehat{\text{p}}$) fitting; i.e.

$$\widehat{e}_t = X_t - \widehat{a}_1 X_{t-1} - \cdots - \widehat{a}_{\widehat{p}} X_{t-\widehat{p}}.$$

Turning our attention to the estimates of M_k, V_k, these were estimated by ordinary kernel estimates in Hjellvik & Tjøstheim (1995,1996), whereas local polynomial estimates (cf. Fan & Gijbels, 1995) were employed in Hjellvik *et al.* (1998).

To increase computational speed a cubic spline algorithm can be introduced (Hjellvik & Tjøstheim, 1995) in the computation of $\widehat{L}(M_k)$ and $\widehat{V}(M_k)$. The computation is done in three steps. First, kernel estimates of $M_k(x)$ and $V_k(x)$ for 20 equidistant x-values are computed for

$$\max\{-3\text{sd}(X_t), X_{\text{min}}\} \leq x \leq \min\{3\text{sd}(X_t), X_{\text{max}}\},$$

where sd$(\cdot)$ denotes empirical standard deviation. Second, standard cubic splines are used to interpolate for other x-values. Finally, the observations $\{X_t, t = 1, \ldots, n\}$ and corresponding residuals $\{\widehat{e}_t\}$ are inserted in the interpolated expressions for $\widehat{M_k}(\cdot)$ and $\widehat{V_k}(\cdot)$, respectively, and $\widehat{L}(M_k)$ and $\widehat{L}(V_k)$ are computed as in the above formulae.

This three-step algorithm was tested against the pure kernel estimation algorithm for several test situations and it gave very similar results.

Proofs of weak consistency and asymptotic normality are presented in Hjellvik *et al.* (1998). The proofs of the distributional results are based on $U-$statistic arguments and are quite intricate.

Unfortunately, the finite sample distribution is not close to that predicted by asymptotic theory unless n is very large. We think that the reason for the bad approximation is that, unlike a standard parametric setting, the next order terms in the Edgeworth expansion for these kind of functionals are very close to the leading normal approximation term. It cannot be ruled out that for each case there *exists* a bandwidth giving a fairly accurate approximation for a fixed moderate sample size, but in practice, when the truth is unknown, unlike in a simulation experiment, it would be difficult to find such a bandwidth.

2.3 A randomization/bootstrap approach to testing

The results described in the previous sub-section mean that for small and moderate sample sizes the asymptotic distribution ought not be used to construct the null distribution of the functionals. An alternative is to create the null distribution by randomizing or by bootstrapping the residuals

$$\widehat{e}_t = X_t - \sum_{1=1}^{\widehat{p}} \widehat{a}_i X_{t-i}$$

from the best linear autoregressive (or ARMA) fit to $\{X_t\}$. Randomized/bootstrapped values $\widehat{L}^*(\cdot)$ of the functional in the null situation are created by inserting in the expression for $\widehat{L}(\cdot)$ randomized/bootstrapped linear versions

$$X_t^* = \sum_{i=1}^{\widehat{p}} \widehat{a}_i X_{t-i}^* + e_t^*$$

of $\{X_t\}$. By taking a sufficiently large number of bootstrap replicas $\{e_t^*\}$ of $\{e_t\}$, in this way we can construct a null distribution for $\widehat{L}(\cdot)$.

The randomization/bootstrap approach applied to the test functionals $L(\cdot)$ has been evaluated on a large group of examples, both simulated and real data. It has also been compared to parametric tests and the bispectrum test. It performs well even for small sample sizes. We refer to Hjellvik *et al.* (1998).

3 Selecting significant lags

Selecting the order of a linear autoregressive model has been a research topic that has been extensively worked on starting from Akaike's (1969) path-breaking paper. If the data at hand fail to pass a linearity test of the kind described in the preceding section, then this issue takes on new importance, since if nonparametric analysis should be used subsequently, the curse of dimensionality would effectively preclude having many lags in the model. Actually the emphasis of the problem shifts. It is not so much determining the order of the model as determining a few main lags which capture the essential features of the data.

The gist of the problem is as follows: For a given collection of data $\{X_1, \ldots, X_n\}$ select a set of lags $i_1, \ldots, i_p$ where $i_p \leq L$, some upper limit, such that the mean square prediction error

$$E\{X_t - M(X_{t-i_1}, \ldots, X_{t-i_p})\}^2$$

is minimised. Here

$$M(X_{t-i_1}, \ldots, X_{t-i_p}) = E(X_t \mid X_{t-i_1}, \ldots, X_{t-i_p})$$

is the optimal least squares nonlinear predictor of X_t based on $X_{t-i_1}, \ldots, X_{t-i_p}$. In practice $E\{X_t - M(X_{t-i_1}, \ldots, X_{t-i_j})\}^2$ must be estimated for each subset of indices $\{i_1, \ldots, i_j\}$ and there are several difficulties (cf. Tjøstheim, 1998). Important progress has been made lately by Tschernig & Yang (1997).

4 Additive modelling and additivity tests

If the data do not pass a linearity test, major simplifications can still be obtained if the data can be modelled additively; i.e. if X_t can be represented to a good approximation as

$$X_t = \sum_{j=1}^{p} f_j(X_{t-i_j}) + e_t.$$

With no loss of generality we will write $f_i(X_{t-i})$ for a typical term of the sum in the sequel. The traditional way of estimating the f's nonparametrically has been by backfitting (Hastie & Tibshirani, 1990). This is a widely available and useful method. A disadvantage is that asymptotic theory is difficult to work out. Significant progress has been made in work by Opsomer & Ruppert (1997) and Linton *et al.* (1997). In this presentation I will put the emphasis on another method, which is now known as marginal integration. This method was to my knowledge first proposed in a paper by Auestad & Tjøstheim (1991). It was extended and treated more formally in Tjøstheim & Auestad

(1994). Independently the technique was discovered by Newey (1994), and Linton & Nielsen (1995). Subsequently the method has been applied to a number of new areas, and especially Oliver Linton has sought to demonstrate its potential. A rigorous account in the time series case is given in Masry & Tjøstheim (1997). See also Fan *et al.* (1995).

The idea behind marginal integration is very simple. In the model

$$X_t = c + \sum f_i(X_{t-i}) + e_t, \tag{4.1}$$

it is well-known that in general $E(X_t \mid X_{t-i}) \neq f_i(X_{t-i})$. However,

$$E(X_t \mid X_{t-1}, \ldots, X_{t-i}) = c + \sum f_i(X_{t-i}).$$

By integrating over the joint marginal distribution $F_{(k)}$ of $X_{t-1}, \ldots, X_{t-k-1}, X_{t-k+1}, \ldots, X_{t-p}$; i.e. by taking the expectation (not conditional!) of $M(X_{t-1}, \ldots, X_{t-i}) = E(X_t \mid X_{t-1}, \ldots, X_{t-i})$ with respect to these variables and with X_{t-k} fixed at x_k, and using the additivity, the marginal integrator (projector) at x_k is given by

$$P(x_k) = c + c_1 + f_k(x_k),$$

where $c_1 = \sum_{i \neq k} E\{f_i(X_i)\}$.

The quantity $P(x_k)$ can be estimated by

$$\widehat{P}(x_k) = \frac{1}{n} \sum_t \widehat{M}(X_{t-1}, \ldots, X_{t-k+1}, x_k, X_{t-k-1}, \ldots, X_{t-p}).$$

The marginal integrators $P(X_k)$ were originally intended (Auestad & Tjøstheim, 1991; Tjøstheim & Auestad, 1994) mainly as a means of looking for functional shapes of the additive components, so that subsequently parametric models could be built for these. An example in the time series regression case taken from Masry & Tjøstheim (1997) is given in Figure 1. It concerns a small simulation experiment for the first order system

$$Y_{t+1} = 0.5Y_t + 0.5X_{t+1}^2 + 0.5Z_{t+1}[\{1 + \exp(Z_{t+1})\}^{-1} - 0.5] + e_{t+1}$$

$$X_{t+1} = 0.5X_t + \varepsilon_{t+1}$$

$$Z_{t+1} = 0.5Z_t + \eta_{t+1}$$

where $\{e_t\}$, $\{\varepsilon_t\}$, and $\{\eta_t\}$ are generated as independent processes consisting of Gaussian variables with zero mean and variance one. The $\{X_t\}$, $\{Y_t\}$, and $\{Z_t\}$ processes were subsequentley adjusted so that they have zero mean (already the case for the $\{X_t\}$ and $\{Z_t\}$ processes) and unit variance. The marginal integration estimates of Y_{t+1} on Y_t, X_{t+1}, and Z_{t+1} based on the kernel estimate of

$$M(x, y, z) = E(Y_{t+1} | X_{t+1} = x, Y_t = y, Z_{t+1} = z)$$

were computed. Plots of $\widehat{P}_X(x)$, $\widehat{P}_Y(y)$, and $\widehat{P}_Z(z)$ for 10 independent realizations and for $n = 500$ and $n = 2000$ are given in Figure 1.

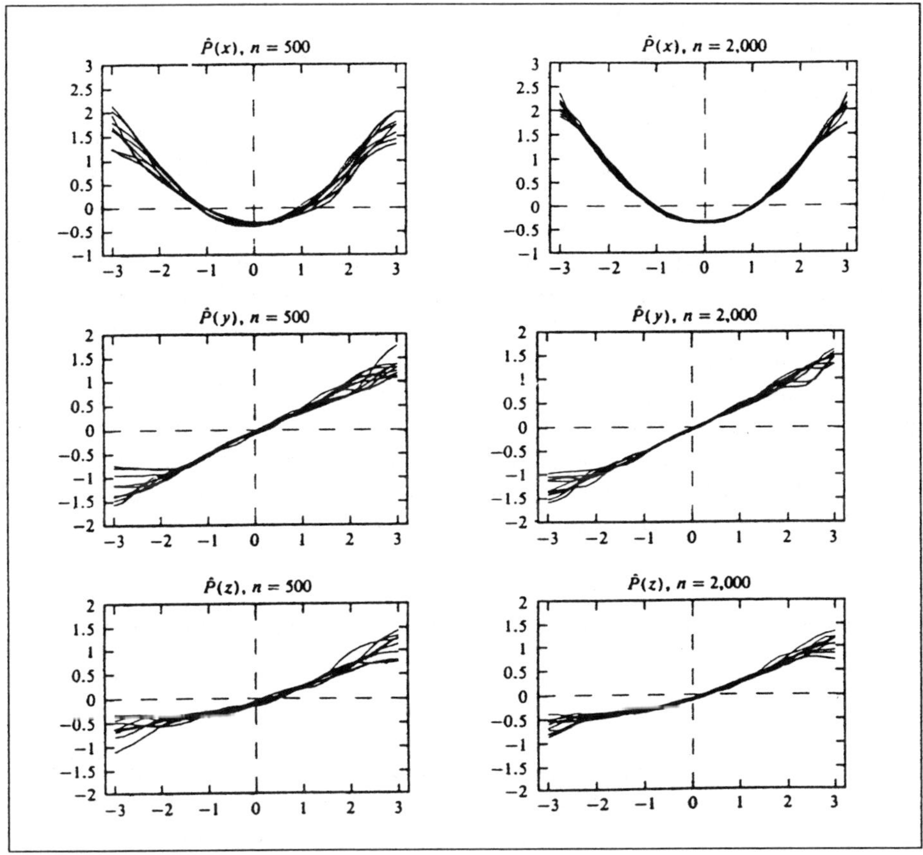

Fig. 1. Plots of $\widehat{P}_X(x)$, $\widehat{P}_Y(y)$, and $\widehat{P}_Z(z)$

The plots clearly reveal the linear dependence on Y_t and the nonlinear x-dependence and z-dependence on X_{t+1} and Z_{t+1}. Further, the character of the nonlinearity is also indicated. This can be used in a subsequent parametric or semiparametric analysis. The fact that $\widehat{P}(0) < 0$ in Figure 1 comes from the adjustment made so that $\{Y_t\}$ has a zero mean.

If there are not too many components, fairly accurate estimates of the components $f_k(\cdot)$ in (4.1) can be obtained, and they could be used as an end result of the analysis. Moreover, the estimates could be used as input estimates in a combined backfitting/marginal integration scheme. In an interesting paper Linton (1997) has obtained a certain type of optimality for this case.

The marginal integration method has been improved by using local polynomials, combination with back-fitting and by computational devices to speed up the procedure, but the methodology still needs to be more extensively tested to determine its true potential. A recent comparison between backfitting and marginal integration has been undertaken in Sperlich *et al.* (1997). Software for the marginal integration procedure can be found in the Xplore package developed at the Humboldt University in Berlin.

4.1 Tests of additivity

In principle, additivity tests can be built up quite analogously to the linearity test presented in Section 2, as mentioned already in Tjøstheim & Auestad (1994). One can compare the additive model estimate with the full nonparametric estimate. Work along these lines has so far only been set up in a quite general regression setting with independent sets of explanatory variables (cf. Linton & Gozalo, 1996b; and Sperlich *et al.* 1998). They look at regression situations with a vector of explanatory variables $\underline{X} = [X_1, \dots, X_p]$ and a response variable Y, where the observations $\underline{X}_1, \dots, \underline{X}_n$ are assumed to be iid. Linton & Gozalo (1996b) and Sperlich *et al.* (1998) essentially propose estimating $M(\underline{x}) = E(Y \mid \underline{X} = \underline{x})$ nonparametrically, via the Nadaraya-Watson estimator or using local polynomials. Denoting the corresponding estimator by $\widehat{M}(\underline{x})$, then the components f_j are estimated by marginal integration in an additive model fit to obtain

$$\tilde{M}\underline{x}) = \widehat{c} + \sum \widehat{f}_i(x_i).$$

The two estimates are subsequently compared using e.g. a functional of type

$$\widehat{A}(M) = n^{-1} \sum_{j=1}^{n} \{\widehat{M}(\underline{X}_j) - \tilde{M}(\underline{X}_j)\}^2 w(\underline{X}_j).$$

In an attempt to avoid the curse of dimensionality and to obtain a more finely tuned test, in Sperlich *et al.* (1998) the alternative hypothesis is more restricted. It is assumed that $M(\underline{x})$ is either additive or, if not, can at most have second order interactions, so that

$$M(\underline{x}) = c + \sum_i f_i(x_i) + \sum_{j<i} f_{ij}(x_i, x_j).$$

In this way one may get an indication as to which pairs of components are pairwise additive and which are not. Suitably normalized Sperlich *et al.* (1998) get a test where pairwise additivity in X_j and X_i is rejected if

$$\frac{1}{n} \sum_t \{\widehat{f}_i(X_{i,t}) + \widehat{f}_j(X_{j,t}) - \widehat{f}_{i,j}(X_{i,t}, X_{j,t})\}^2 w(X_{i,t}, X_{j,t})$$

is large enough. An alternative test can be based on the second order mixed derivatives of $M(\underline{x})$. Both tests are implemented sequentially, looking at one interaction term at a time and removing them according to the size of the P-values. Bootstrap arguments can be used roughly as in Section 2.3. The procedure is quite slow computationally if there are many potential interactions.

5 Tests of independence

Independence tests can be built in much the same spirit as additivity and linearity tests. There are other approaches, of course, for example the BDS test (Brock *et al.*, 1987) based on the correlation integral of chaos theory.

Independence between stochastic variables is defined in terms of distribution functions or – when they exist – density functions. Both distribution functions and density functions can be estimated nonparametrically.

The problem of measuring dependence between two random variables X and Y can be formulated as a problem of measuring the distance between the two bivariate distributions $F_{X,Y}$ and $F_X \otimes F_Y$, where $F_{X,Y}$, F_X and F_Y are the joint and marginal distribution functions, respectively, and where $F_X \otimes F_Y(x,y) = F_X(x)F_Y(y)$. Conventional distance measures between two distribution functions F_1 and F_2 are the Kolmogorov – Smirnov distance and the Cramer – von Mises type distance.

These distance functions can be taken as measures of dependence, and it is easily seen that they fit into the general framework of specification testing outlined in the introduction. An & Cheng (1990) have used the Kolmogorov – Smirnov distance in connection with a linearity test of theirs. It could be converted into an independence test, but apart from this, as far as I know, all the work pertaining to measuring dependence and testing of independence has been done in terms of the Cramer – von Mises distance. More details and some examples are given in Skaug & Tjøstheim (1993) and Hong (1997).

It is clear that this way of testing independence can be extended to testing of conditional independence. Linton & Gozalo (1996a) has explored this possibility.

5.1 Measures and tests based on density functions

For two random variables X and Y having a joint density $p_{X,Y}$ and marginals p_X and p_Y we measure the degree of dependence by $\Delta(p_{X,Y}, p_X \otimes p_Y)$ where Δ now is a distance measure between two bivariate density functions. The Hellinger distance and the Kullback-Leibler distance are examples of distance measures that have been used.

Estimates may be obtained by replacing integrals by empirical averages of functions of kernel density estimates. An asymptotic theory can be developed, but basing a test on such a theory may be hazardous as the real level will typically deviate substantially from the nominal level. This is amply demonstrated in Skaug & Tjøstheim (1996) where sometimes the level was twice that of the nominal level for a sample size of 100 observations. Permutation tests produce tests with exact levels in this situation.

The procedures described so far are general purpose tests, and for special classes of alternatives it is possible to find tests with better properties. In fact such tests were suggested in Hjellvik & Tjøstheim (1996) as special cases of the linearity test.

6 Nonparametrics and nonstationarity

Parametric models can be estimated for certain types of nonstationarity, and for the so-called unit root processes, for example, one obtains super efficiency of parameter estimates. Everything that has been done in this presentation has been under the implied assumption of stationarity, but this is not an absolute prerequisite for conducting a nonparametric analysis. Actually, there is a need, e.g. in econometrics, for models that are both nonstationary and nonlinear, and nonparametric methods could be viable tools for exploring such models.

Karlsen & Tjøstheim (1998) have singled out the class of null recurrent continuous state space Markov chains as a class of processes which are nonstationary and in general nonlinearly generated. It is possible to build up a theory of nonparametric estimation for this class. Both the conditional mean, the conditional variance and the invariant density can be estimated. The key idea is to use the splitting chain technique, where the chain is decomposed into a sequence of independent parts according to its regeneration points. A tail condition on the recurrence time distribution plays a vital role in the development of the asymptotic properties.

References

Akaike, H. (1969) Fitting autoregressions for predictions. *Ann. Inst. Statistist. Math.*, **21**, 243-247.

An, H-Z. & Cheng, B. (1990). A Kolmogorov – Smirnov type statistic with application to test for normality in time series. *Inter. Stat. Rev.*, **59**, 287-307.

Auestad, B. & Tjøstheim, D. (1991). Functional identification in nonlinear time series. In: *Nonparametric Functional Estimation and Related Topics* (ed. G. Roussa), 493-507. Amsterdam: Kluwer Academic.

Brock, W.A., Dechert, W.D. & Scheinkman, J.A. (1987). A test for independence based on the correlation dimension. Preprint, Department of Economics, University of Wisconsin.

Fan, J. & Gijbels, I. (1995). *Local Polynomial Modeling and Its Application. Theory and Methodologies.* London: Chapman and Hall.

Fan, J., Härdle, W. & Mammen, E. (1998). Direct estimation of low dimensional components in additive models. *Ann. Statist.*, in press.

Granger, C.W.J. & Teräsvirta, T. (1993). *Modelling Nonlinear Economic Relationships.* Oxford: Oxford University Press.

Härdle, W. & Mammen, E. (1993). Comparing nonparametric versus parametric regression fits. *Ann. Statist.*, **21**, 1926-1947.

Hastie, T. & Tibshirani, R. (1990). *Generalized Additive Models*, London: Chapman and Hall.

Hinich, M. (1982). Testing for Gaussianity and linearity of a stationary time series. *J. Time Ser. Anal.*, **3**, 169-176.

Hjellvik, V. & Tjøstheim, D. (1995). Nonparametric tests of linearity for time series. *Biometrika*, **82**, 351-368.

Hjellvik, V. & Tjøstheim, D. (1996). Nonparametric statistics for testing linearity and serial independence. *J. Nonparametric Stat.*, **6**, 223-251.

Hjellvik, V., Yao, Q. & Tjøstheim, D. (1998). Linearity testing using polynomial approximation. *J. Stat. Plan. and Inf.*, in press.

Hong, Y. (1998). Testing for pairwise serial independence via the empirical distribution function. *J. Roy. Stat. Ser. B*, in press

Karlsen, H. & Tjøstheim, D. (1998). Nonparametric estimation in null recurrent time series models. Preprint, Department of Mathematics, University of Bergen.

Linton, O. (1997). Efficient estimation of additive nonparametric regression models. *Biometrika*, **84**, 469-473.

Linton, O. & Gozalo, P. (1996a). Conditional independence restrictions: testing and estimation. Preprint, Cowles Foundation, Yale University, New Haven, Connecticut.

Linton, O. & Gozalo, P. (1996b). Testing additivity in generalized nonparametric regression models. Preprint, Cowles Foundation, Yale University, New Haven, Connecticut.

Linton, O. & Nielsen, J.P. (1995). A kernel method of estimating structured nonparametric regression based on marginal integration. *Biometrika*, **82**, 93-101.

Linton, O., Mammen, E. & Nielsen, J.P. (1997). The existence and asymptotic properties of a backfitting projection algorithm under weak conditions. Preprint, Cowles Foundation, Yale University, New Haven Connecticut.

Luukkonen, R., Saikkonen, P. & Teräsvirta, T. (1998a). Testing linearity in univariate time series. *Scand. J. Statist.*, **15**, 161-175.

Luukkonen, R., Saikkonen, P. & Teräsvirta, T. (1988b). Testing linearity against smooth transition autoregression, *Biometrika*, **75**, 491-499.

Masry, E. & Tjøstheim, D. (1997). Additive nonlinear ARX time series and projection estimates. *Econometric Theory*, **13**, 214-252.

Newey, W.K. (1994). Kernel estimation of partial means. *Econometric Theory*, **10**, 233-253.

Opsomer, J.D. & Ruppert, D. (1997). Fitting a bivariate additive model by local polynomial regression. *Ann. Statist.*, **25**, 186-211.

Skaug, H.J. & Tjøstheim, D. (1993). A nonparametric test of serial independence based on the empirical distribution function. *Biometrika*, **80**, 591-602.

Skaug, H.J. & Tjøstheim, D. (1996). Testing for serial independence using measures of distance between densities. In: *Athens Conference on Applied Probability and Time Series*, Vol. II, in memory of E.J. Hannan. Springer Lecture Notes in Statistics **115** (ed. P.M. Robinson & M. Rosenblatt), 363-377.

Sperlich, S., Linton, O. & Härdle, W. (1997). A simulation comparison between integration and backfitting methods of estimating separable nonparametric regression models. Report Humboldt Sonderforschungsbereich 373, Humboldt-Universität zu Berlin, Wirtschaftswissenschaftliche Fakultät.

Sperlich, S., Tjøstheim, D. & Yang, L. (1998). Nonparametric estimation and testing of interaction in additive models. Preprint, Sonderforschungbereich 373, Humboldt-Universität zu Berlin, Wirtschaftswissenschaftliche Fakultät.

Subba Rao, T. & Gabr, M.M. (1980). A test for linearity of stationary time series. *J. Time Ser. Anal.*, **1** 145-158.

Teräsvirta, T. (1994). Specification, estimation and evaluation of smooth transition autoregressive models. *J. Amer. Statist. Assoc.*, **89**, 208-218.

Tjøstheim, D. (1998). Nonparametric specification procedures for time series. In: *Asymptotics, Nonparametrics and Time Series* (ed. S. Ghosh), in press. New York: Marcel Dekker.

Tjøstheim, D. & Auestad, B. (1994). Nonparametric identification of nonlinear time series: Projections. *J. Amer. Statist. Assoc.*, **89**, 1398-1409.

Tschernig, R. & Yang, L. (1997). Nonparametric lag selection in time series. Report Humboldt Sonderforschungsbereich 373, Humboldt-Universität zu Berlin, Wirtschaftswissenschaftliche Fakultät.

Time Series Forecasting by Principal Component Methods

Mariano J. Valderrama[1], Ana M. Aguilera[1] and Juan C. Ruiz-Molina[2]

[1] Department of Statistics and Operations Research, University of Granada, 18071-Granada, Spain
[2] Department of Statistics and Operations Research, University of Jaén, 23071-Jaén, Spain

Abstract. On the basis of Functional Principal Component Analysis (FPCA), two forecasting approaches for time series are developed in this paper. The first one uses weighted multiple linear regression among principal components whereas the second one applies Kalman filtering on approximate state-space models. The forecasting performance of both methods is discussed on a real financial time-series.

Keywords. Functional principal components, weighted least-squares linear regression, state-space models, Kalman filtering

1 Introduction to FPCA

Functional Principal Component Analysis (FPCA) is a generalization for stochastic processes of the multivariate PCA where the sample data are now curves instead of the vectors of the classic multivariate analysis. An interesting perspective on the analysis of functional data can be seen in Ramsay & Dalzell (1991) and in the recent book by Ramsay & Silverman (1997).

In order to introduce the theoretical framework of FPCA, let us consider a second order and quadratic mean continuous random process, $\{X(t) : t \in [T_1, T_2]\}$, whose sample functions have squares integrable over $[T_1, T_2]$.

The *ith* principal component (p.c.) associated with the process $\{X(t)\}$ in the interval $[T_1, T_2]$ is given by

$$\xi_i = \int_{T_1}^{T_2} (X(t) - \mu(t)) f_i(t) dt, \tag{1}$$

where f_i, called the *ith* principal factor, is the normalized eigenfunction corresponding to the *ith* largest eigenvalue λ_i of the covariance kernel $C(t,s)$, and $\mu(t)$ is the mean function of the process. That is, f_i and λ_i are the solution to the following second kind of integral equation:

$$\int_{T_1}^{T_2} C(t,s) f_i(s) ds = \lambda_i f_i(t) \quad t \in [T_1, T_2]. \tag{2}$$

The principal components defined above have the same optimal properties as in the finite case. That is, ξ_i is the normalized generalized linear combination of the process variables having maximum variance, λ_i, out of all generalized linear combinations which are uncorrelated with $\{\xi_j\}_{j=1}^{i-1}$. Thus, the variance explained by the *ith* principal component is $V_i = \lambda_i/V$, with $V = \sum_i \lambda_i$ as the total variance of the process in the interval $[T_1, T_2]$.

Then, the process has the following principal component expansion:

$$X(t) = \mu(t) + \sum_{i=1}^{\infty} \xi_i f_i(t) \quad t \in [T_1, T_2]. \tag{3}$$

The representation (3) truncated in the m-th term is the best m-dimensional linear model for the process $\{X(t)\}$ in the least-squares sense with $\sum_{i=1}^{m} \lambda_i/V$ as the variance explained by this linear model.

2 Weighted principal component forecasting

Aguilera *et al.* (1996b) proposed an adaptation of functional Principal Component Prediction (PCP) models for forecasting a continuous-time series in a future interval from discrete-time observations in the past. This functional approach allows the forecasting of a time-series from unequally spaced time values (omitted or missing observations) which is one of its main advantages if we take into account that the literature on this general class of problem is actually sparse and complex. Moreover, PCP models do not impose restrictive hypotheses on the stochastic process generating the data and not only allow us to forecast a continuous-time stochastic process in a future interval but also its reconstruction between the discretization time points in the past.

The PCP models are obtained as an extension of principal component regression of multiple responses (MPCR) to forecast an infinite set of responses (the process variables in the future) from an infinite set of predictors (the process variables in the past). These models are based on linear regression of the principal components associated with the process in the future against its principal components in the recent past.

In order to build a PCP model for predicting the process in a future interval $[T_3, T_4]$ in terms of its evolution in a past interval $[T_1, T_2]$ $(T_2 \leq T_3)$, let us denote by (f_i, ξ_i) and (g_j, η_j) the principal factors and components associated with the process in the past and the future intervals, respectively.

Then a PCP model for the process in the future (Aguilera *et al.*, 1997) is given by

$$\tilde{X}^q(s) = \mu(s) + \sum_{j=1}^{q} \tilde{\eta}_j^{p_j} g_j(s), \quad s \in [T_3, T_4], \tag{4}$$

where $\tilde{\eta}_j^{p_j}$ is the least-squares linear estimator for the p.c. η_j against the first p_j p.c.'s of the process in the past. That is,

$$\tilde{\eta}_j^{p_j} = \sum_{i=1}^{p_j} \frac{E[\eta_j \xi_i]}{\lambda_i} \xi_i = \sum_{i=1}^{p_j} \beta_i^j \xi_i. \tag{5}$$

The model defined by (4) will be denoted by $\mathrm{PCP}(q; p_1, \ldots, p_q)$.

In order to use a PCP model to forecast a time series $\{x(t) : t > T\}$ from the time S (S> T),

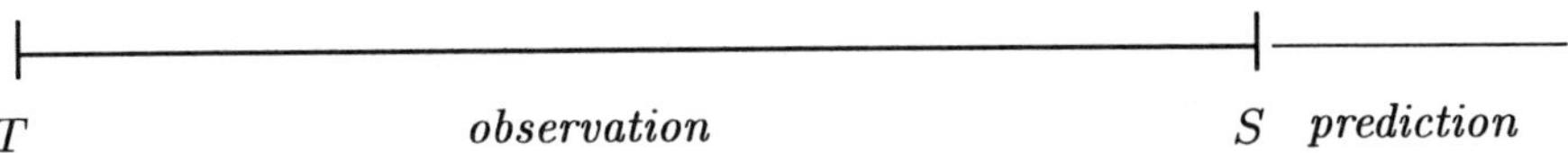

we propose to cut the series into periods of length h (h >0) (see Figure 1.1) and define the following process by rescaling:

$$\{X_w(t) = x(wh + t) : t \in (T, T + h]; \quad w = 0, 1, \ldots\}. \tag{6}$$

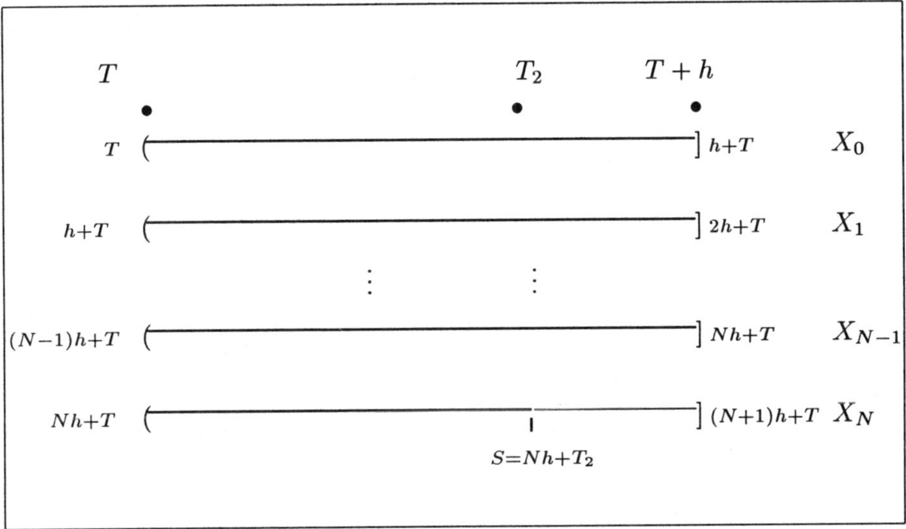

Fig. 1.1. Sample paths obtained after cutting the original series

Then, we construct a PCP model to forecast the process $\{X(t)\}$ in a future interval $[T_3, T_4]$ in terms of its evolution in a past interval $(T_1, T_2]$ with $T \leq T_1 < T_2 \leq T_3 < T_4 \leq T + h$. Finally, we propose to use this model to forecast the original series $x(t)$ in the interval $[Nh + T_3, Nh + T_4]$ for a large N. The selection of the past and future intervals is determined by the forecasting period previously fixed by the design.

Once we have defined the PCP model, the next step is to estimate it from discrete-time observations of the process sample paths. The natural estimators of the principal factors from a set of independent sample paths are the eigenfunctions of the sample covariance kernel. That is, the sample principal factors in the interval $[T_1, T_2]$ are the solutions to the integral equation

$$\int_{T_1}^{T_2} \hat{C}(t, s)\hat{f}_i(s)ds = \hat{\lambda}_i \hat{f}_i(t), \quad t \in [T_1, T_2]. \tag{7}$$

In order to give higher importance to recent observations the following weighted estimation of the sample covariance function proposed in Aguilera *et al.* (1996b) will be used:

$$\hat{C}(t,s) = \frac{N}{(N-1)S_N} \sum_{w=0}^{N-1} P_w (X_w(t) - \bar{X}(t))(X_w(s) - \bar{X}(s)), \tag{8}$$

where $\bar{X}$ is the weighted estimation of the mean defined as

$$\bar{X}(t) = \frac{1}{S_N} \sum_{w=0}^{N-1} P_w X_w(t), \tag{9}$$

and P_w is the weight for the sample-path w with $S_N = \sum_{w=0}^{N-1} P_w$.

Moreover, the approximation of the sample principal factors from discrete-time values is performed by means of cubic spline interpolation of the sample paths between the observed data in terms of the basis of cubic B-splines (Aguilera *et al.*, 1996a).

3 Kalman filtering on approximate state-space models

On the basis of the previous paper by Ruiz-Molina *et al.* (1995), we propose a flexible state-space model with minimal dimension, in the class of time-invariant models, for estimating a time series.

The starting point is the principal component expansion (3) truncated in the fifth term where the principal factors f_i are approximated by using the Rayleigh-Ritz method with a basis of trigonometric functions. This approximated representation for the process is given by

$$X_5(t) = c_1 + \sqrt{2}\{\cos(2\pi t)c_2 + \sin(2\pi t)c_3 + \cos(4\pi t)c_4 + \sin(4\pi t)c_5\} + \epsilon(t) \tag{10}$$

where c_j are correlated random variables with zero mean and variances that are linear combinations of the approximated principal values, and $\epsilon(t)$ is the approximation random error.

By differentiating twice both terms in equation (10), and applying the general decomposition theorem (Kailath, 1980), the following system representation is obtained:

$$\dot{\mathbf{z}}(t) = \begin{pmatrix} 0 & 1 \\ -4\pi^2 & 0 \end{pmatrix} \mathbf{z}(t) + \begin{pmatrix} 0 \\ 1 \end{pmatrix} w(t) \tag{11}$$

$$y(t) = (1\ 0)\,\mathbf{z}(t) + v(t) \tag{12}$$

where $\mathbf{z}(t) = \left(X_5(t), \dot{X}_5(t)\right)'$, $v(t)$ is a zero-mean Gaussian white noise satisfying $E\left[v(t)v(s)\right] = r\delta(t-s)$,$(r > 0)$, and $w(t) = 4\pi^2\epsilon(t) + \ddot{\epsilon}(t)$ can be considered as a zero-mean Gaussian white noise satisfying $E\left[w(t)w(s)\right] = q\delta(t-s)$, $(q \geq 0)$. This last assumption is not restrictive and it has been proved to be acceptable by means of several simulation studies.

The linear system given by equations (11) and (12) is uniformly completely observable and controllable and so the optimal filter is uniformly asymptotically stable and there exists a constant solution for the minimum error variance equation in the steady-state case.

In order to get a more flexible model we introduce the parameter θ_1 in the system matrix instead of the term $-4\pi^2$ and rewrite the model given by (11) and (12) in the following form that will be called Approximate State-Space (ASS) model:

$$\dot{\hat{\mathbf{z}}}(t) = \begin{pmatrix} 0 & 1 \\ \theta_1 & 0 \end{pmatrix} \hat{\mathbf{z}}(t) + \begin{pmatrix} \theta_2 \\ \theta_3 \end{pmatrix} \nu(t) \tag{13}$$

$$y(t) = (1\,0)\,\hat{\mathbf{z}}(t) + \nu(t). \tag{14}$$

where the innovation $\nu(t)$ is also a Gaussian white noise and $(\theta_2, \theta_3)'$ is the Kalman gain matrix.

In the estimation procedure such a model will be discretized by using a sampling interval with unit amplitude. The method used is the prediction error method (PEM) that determines estimates of the parameters by minimizing

$$\sum_{t=1}^{N} \nu^2(t).$$

Then, the Kalman filter is applied on the estimated ASS model to provide forecasts for the data. The validation of the models is based on Akaike's Final Prediction Error (FPE) which is given by

$$FPE = \frac{1 + n/N}{1 - n/N} \times V$$

where n is the total number of estimated parameters, N is the length of the data record and V is the loss function (quadratic fit).

4 Modelling and forecasting a financial time series

The principal component approaches described above are now applied to model and forecast the evolution of a financial time series in Madrid Stock Market.

4.1 Data description

The evolution of stock prices along the time can be modelled as a sample path of a continuous time stochastic process so that the sample information about stock markets is a set of curves.

For this application we have randomly chosen the Spanish *Banco Bilbao Vizcaya* (BBV). In order to forecast the stock prices of BBV during the last five weeks of 1997 we have weekly observations recorded at the close of the Thursday's trading from the begining of 1992 (three hundred and twelve values).

It is known that direct statistical analysis of stock prices is difficult because consecutive prices are highly correlated and the variances of prices increase with time. Consequently it is more covenient to analyse changes in prices taking into account that results for changes, for example a forecast, can easily be used to give appropriate results for prices. In this paper price changes are measured by means of the weekly returns studied in economics research (Taylor, 1986). If we denote by $X(t)$ the continuous time stochastic process that models the stock price of a certain bank at time t, the weekly returns are given by the stochastic process

$$R(t) = \frac{X(t) - X(t-1)}{X(t-1)},$$

so that a model for weekly returns can be used to forecast stock prices.

4.2 Estimating a PCP model for the weekly returns

In order to build a PCP model for predicting the weekly returns of BBV in the last five weeks of 1997 we have cut the return series into periods of amplitude $h = 13$ weeks (see Figure 1.1) and considered two different intervals in each of them: $[1, 8]$ and $[9, 13]$ as the past and the future intervals, respectively. In each interval we have weekly observations of twenty four sample paths corresponding to the evolution of the return process. That is, we have twenty four replicates of an eight dimensional vector as the past and twenty four replicates of a five dimensional vector as the future. In order to estimate a PCP model we will use only the first twenty three sample paths in each interval and the remainder observations will be used to measure the accuracy of the forecasts provided by the model.

Firstly, the FPCA of the process $R(t)$ has been estimated in the past and the future intervals by using exponential weights $P_w = \exp(w)$. The percentages of accumulated variance explained by the p.c.'s in each period appear in Table 4.1. As the variance explained by the first two p.c.'s in the future is greater than a 96% of the total variance, the PCP model will be constructed in terms of these 2 p.c.'s.

Table 4.1. Percentage of accumulated variance explained by the weighted FPCAs

p.c.	[1,8]	[9,13]
1	93.02	78.93
2	97.96	96.96
3	99.58	99.33
4	99.84	99.93
5	99.95	100.00
6	99.99	100.00
Total variance		
	0.00443	0.00277

Secondly, Table 4.2 shows the estimated linear correlations between the p.c.'s in the two intervals. After using stepwise regression for selecting the p.c.'s in the past that will be entered in the linear models (5) as predictors for each of the two first p.c.'s in the future, the following PCP model for the weekly return process has been derived:

$$\begin{aligned} \text{PCP(2;1,2)}: \tilde{R}^2(s) &= \bar{R}(s) + \tilde{\eta}_1^1 \hat{g}_1(s) + \tilde{\eta}_2^1 \hat{g}_2(s) \\ \tilde{\eta}_1^1 &= 0.7090\hat{\xi}_1 \\ \tilde{\eta}_2^2 &= -1.2102\hat{\xi}_2 - 1.4427\hat{\xi}_3 \end{aligned}$$

Table 4.2. Linear correlations between the p.c.'s in the future and the past

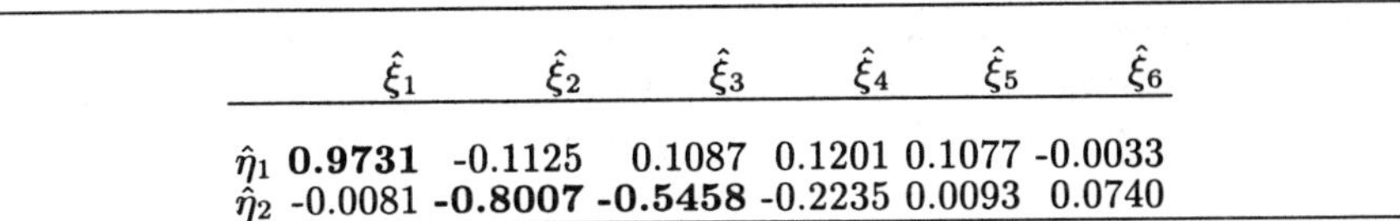

	$\hat{\xi}_1$	$\hat{\xi}_2$	$\hat{\xi}_3$	$\hat{\xi}_4$	$\hat{\xi}_5$	$\hat{\xi}_6$
$\hat{\eta}_1$	**0.9731**	-0.1125	0.1087	0.1201	0.1077	-0.0033
$\hat{\eta}_2$	-0.0081	**-0.8007**	**-0.5458**	-0.2235	0.0093	0.0740

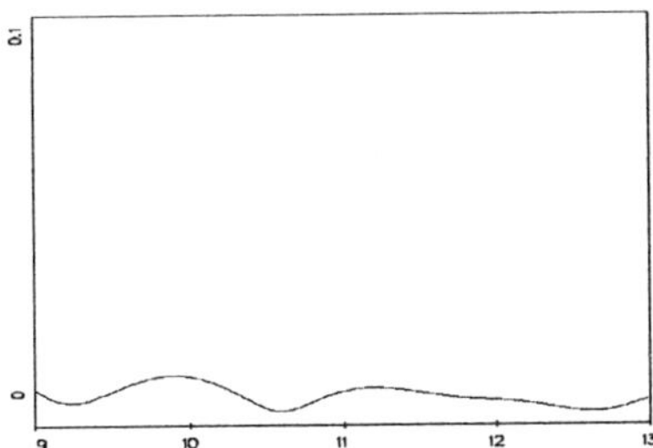

Fig. 4.1. MSE committed by PCP(2;1,2) model

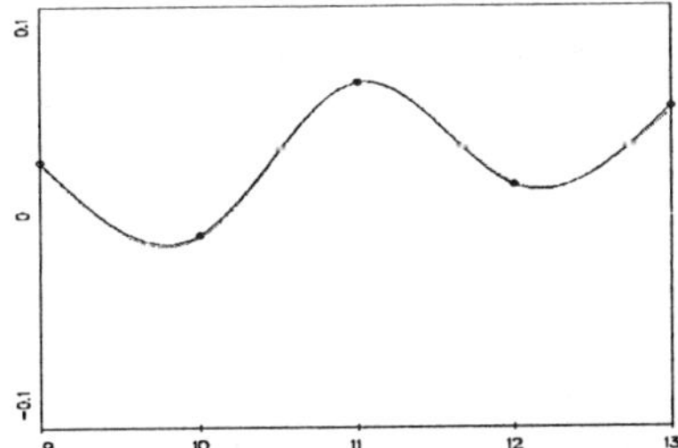

Fig. 4.2. Scatter plot, cubic spline interpolation (solid curve) and its smoothing by the exponential weighted PCP(2;1,2) model (dotted curve) for the last five weeks of the twenty third return curve

Figure 4.1. represents the Mean-Square Error (MSE) generated by this PCP model, given by

$$MSE^2(s) = \frac{N}{(N-1)S_N} \sum_{w=0}^{N-1} P_w \left(IR_w(s) - \tilde{R}^2(s) \right)^2 \quad s \in [9, 13], \qquad (15)$$

where $IR_w(s)$ is the cubic spline interpolating to the return sample path w between the observed values in the future interval.

The smoothing given by this PCP model for the last five weeks of the twenty third sample path is drawn in Figure 4.2 and superposed with the natural cubic spline interpolation of the corresponding real weekly returns.

To perform all the computations we have developed the statistical system SMCP^2 (see Figure 4.3) which consists of a set of libraries, coded in Turbo Pascal by using Object Oriented Programming, and two executable programs: PCAP and REGRECOM. The SMCP^2 program provides accurate and fast estimation of PCP models for large data sets.

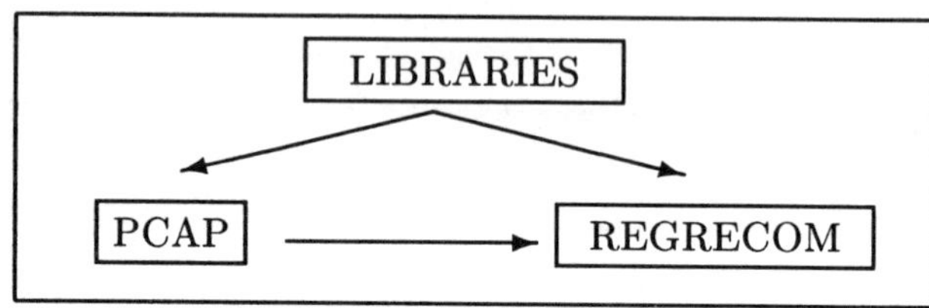

Fig. 4.3. SMCP^2 statistical software

4.3 Estimating an ASS model for the weekly returns

We are now going to estimate the system given by (13) and (14) to predict by Kalman filtering, the weekly return process in the last five weeks of 1997 from the three hundred and eight weekly values starting at the begining of 1992.

The estimates of the parameters in (13) are shown in Table 4.3. The residual analysis and the Akaike's FPE have shown the goodness of the estimated model. The software used was Matlab 4.2 and the System Identification and Statistics Toolboxes.

Table 4.3. Estimation of parameters for the ASS model

	Parameter		
	θ_1	θ_2	θ_3
Estimate	-0.3592	-0.0060	0.0079
Standard error	0.0139	0.0174	0.0105
$\hat{\sigma}_\nu^2 = 0.0016$		FPE = 0.0166	

4.4 Forecasting performance

In order to evaluate the forecasting performance of the above estimated PCP and ASS models, they have been applied to predict the returns of BBV bank in the last five weeks of 1997. The following AR(1) model has also been adjusted to the return series as the most adequate Box-Jenkins model:

$$R(t) = 0.00595 - 0.19183R(t-1) + a(t)$$

where $a(t)$ is a white noise with estimated variance $\hat{\sigma}^2 = 0.00156$.

As a measure of the prediction accuracy of these models we have considered the standard Mean-Square Forecasting Error (MSFE) given by

$$MSFE = \sqrt{\frac{1}{5}\sum_{i=1}^{5}(R(t) - \tilde{R}(t))^2},$$

where $\tilde{R}(t)$ denotes the estimated value. The results are shown in Table 4.4.

Table 4.4. Real and forecasted BBV weekly returns in the last five weeks of 1997

		Last five weeks of 1997				
	MSFE	1	2	3	4	5
Real		0.01537	-0.02487	0.03991	-0.01173	0.04423
PCP(2;1,2)	0.01465	-0.00746	-0.01721	0.05750	-0.00519	0.03235
ASS	0.03208	0.01170	0.01121	0.00682	0.00005	-0.00674
AR(1)	0.02916	-0.00103	0.00729	0.00569	0.00600	0.00594

The stock prices for the last five weeks of 1997 have been derived by means of the formula $X(t) = X(t-1)(R(t)+1)$ from the weekly returns previously computed. The real stock prices, their forecasts as well as the corresponding MSFE are shown in Table 4.5.

Table 4.5. Real and forecasted BBV stock prices in the last five weeks of 1997

		Last five weeks of 1997				
	MSFE	1	2	3	4	5
Real		4625	4510	4690	4635	4840
PCP(2;1,2)	58.5064	4521	4443	4698	4674	4825
ASS	108.1129	4608	4660	4692	4692	4660
AR(1)	97.9622	4550	4583	4610	4638	4665

Furthermore, the weekly return prediction given by the PCP(2;1,2) model in the last five weeks of 1997 is displayed in Figure 4.4 and superposed with the natural cubic spline interpolation of the corresponding real weekly returns.

From the previous analysis of the adjusted models we can conclude that the ASS and AR(1) models have a similar forecasting performance whereas the PCP model reduces to a half the MSFE and improves the predictions taking into account the long-term trend of the series.

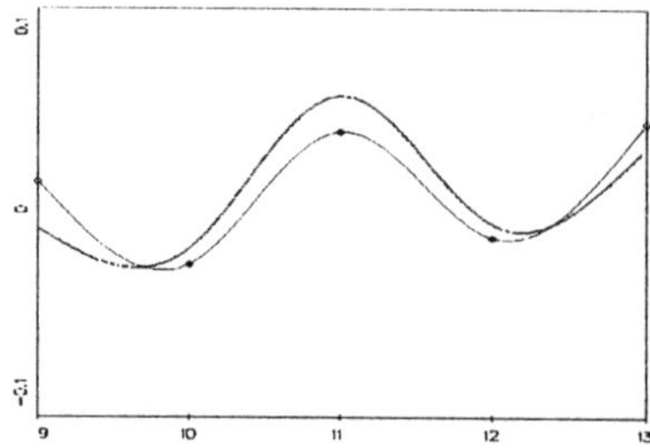

Fig. 4.4. Scatter plot, cubic spline interpolation (solid curve), and weekly return forecasting given by PCP(2;1,2) (dotted curve) for the last five weeks of 1997

Acknowledgements

This research was supported by Dirección General de Enseñanza Superior, Ministerio de Educación y Cultura, Spain, under project PB96-1436.

References

Aguilera, A.M., Gutiérrez, R., Ocaña, F.A. & Valderrama, M.J. (1995). Computational approaches to estimation in the principal component analysis of a stochastic process. *Applied Stochastic Models and Data Analysis*, **11**, 279-299.

Aguilera, A.M., Gutiérrez, R. & Valderrama, M.J. (1996a). Approximation of estimators in the PCA of a stochastic process using B-splines. *Commun. Statist.-Simula.*, **25**, 671-690.

Aguilera, A.M., Ocaña, F.A. & Valderrama, M.J. (1996b). On a weighted principal component model to forecast a continuous time series. In: *COMPSTAT96 Proceedings in Computational Statistics* (ed. A. Prat), 169-174, Heidelberg: Physica-Verlag.

Aguilera, A.M., Ocaña, F.A. & Valderrama, M.J. (1997). An approximated principal component prediction model for continuous time stochastic processes. *Appl. Stoch. Models Data Anal.*, **13**, 61-72.

Kailath, T. (1980). *Linear Systems.* New Jersey: Prentice Hall.

Ramsay, J.O. & Dalzell, C.J. (1991). Some tools for functional data analysis (with discussion). *J. R. Statist. Soc. B*, **53**, 539-572.

Ramsay, J.O. & Silverman, B.W. (1997). *Functional Data Analysis.* New York: Springer-Verlag.

Ruiz-Molina, J.C., Valderrama, M.J. & Gutiérrez, R. (1995). Kalman filtering on approximative state-space models. *Journal of Optimization Theory and Applications*, **84**, 415-431.

Taylor, S.J. (1986). *Modelling Financial Time Series.* Chichester: Wiley.

Contributed Papers

A Simulation Study of Indirect Prognostic Classification

N. M. Adams[1], D. J. Hand[1] and H. G. Li[2]

[1] Department of Statistics, The Open University, Walton Hall, Milton Keynes, MK7 6AA, UK.

[2] Department of Mathematics, The University of Essex, Wivenhoe Park, Colchester, Essex, CO4 3SQ, UK.

1 Introduction

Prognostic classification refers to identifying the class to which an object is likely to belong in the future. In many situations the classes are defined indirectly via some intermediate variables whose values are not known at present and only become known in the future. An example of this framework is provided in the classification of degrees at our university. The degree class is determined from continuous assessment work and examination scores, using deterministic rules. The standard classification approach (the 'direct' approach) would predict the degree grade directly from initial information, such as age, high school performance and so on. An alternative approach, previously unexplored, is to predict the intermediate variables and then apply the classifying rule set. We term this 'indirect prognostic classification', and describe the method in Section 2.

Problems where the classes are defined in terms of intermediate variables often occur in the social and behavioral sciences. Li & Hand (1997) give a detailed analysis of such a problem occurring in the prediction of bank account status, but the comparison of direct and indirect classifiers in that case was confounded by the high degree of noise in the data. The purpose of this paper is to compare direct classification with indirect classification in the controlled context of a simulation study, where we can investigate the factors influencing the relative performance of the methods.

2 Indirect prognostic classification

The indirect prognostic framework has the following structure. An observation has feature vector $\mathbf{x}$, consisting of the initial information. At some point later the intermediate variables, upon which the classification is based, are observed. These can either be *continuous intermediate variables*, $\mathbf{y}$, or *categorical intermediate variables*, denoted $\mathbf{w}$. The object we ultimately wish to predict is the class variable, z, which is defined structurally in terms of the $\mathbf{y}$ and $\mathbf{w}$. Observed $\mathbf{y}$ variables are partitioned into $\mathbf{w}$ variables, using thresholds provided by the rule set. To illustrate with the example of degree grades, we have feature vector $\mathbf{x}$, consisting of the students' initial information and an intermediate vector $\mathbf{y}$, the students' scores on examination and continuous assessment. The $\mathbf{w}$ variables emerge after the application of cut-off scores to continuous assessment and examination scores.

For two class situations, to which we restrict this paper, the direct approach to this problem would usually obtain a classification by applying a threshold to a model for $p(1|\mathbf{x})$, the posterior probability of an object with feature vector $\mathbf{x}$ belonging to class 1. However, such approaches ignore the structural relationship between the class variable z and the intermediate variables $\mathbf{y}$. The approach we consider here is to predict the $\mathbf{y}$ or $\mathbf{w}$ variables from the $\mathbf{x}$ variables and then take advantage of the known structure of the problem to obtain a classification z. We restrict attention to modelling the relationship between the feature space and the intermediate space with linear models. In the case of predicting the $\mathbf{y}$ variables we use linear regression, and in the case of predicting the $\mathbf{w}$ variables we use logistic regression. We assume we have available a design set on which both $\mathbf{x}$ and $\mathbf{y}$ have been measured. For new cases, of course, only $\mathbf{x}$ is measured.

In this paper we describe two simulation studies. The first is intended to be extremely demanding for conventional classification methodology, while the second is closer to real problems that have an indirect structure. Theoretical discussion of indirect models, as well as comparison with direct models in real data applications, is given in Hand, Li & Adams (1998).

3 Simulation 1

This simulation employs a set of ten binary predictor variables, $\mathbf{x}$, and ten binary intermediate variables, $\mathbf{w}$. Each intermediate variable is correlated with one of the predictor variables, and the extent of this relationship is the same for each pair: in each case $p(w_i = x_i) = \tau$, where τ is a factor in the simulation. The class indicator, z, takes the value 0 when the sum of the intermediate variables is even, and the value 1 otherwise. We can imagine this as forming a multidimensional representation of a chequerboard pattern. This is an extremely ambitious problem because we use a training sample of only 20 observations to learn the highly nonlinear decision rule for a population of $2^{10} = 1024$ possible patterns. There is no need to employ a test set, because we can simply evaluate classifier performance over the known true distributions.

We used logistic discrimination as the direct classifier in this example. We do not expect this to perform well because of both the nonlinearity of the decision surface and the small sample size. The indirect classifier predicts $\hat{w}_i = 0$ if more than half of the training observations with $x_i = 0$ have $w_i = 0$, with a similar rule for $\hat{w}_i = 1$. Finally, the class is predicted by seeing if the sum of the $\hat{w}_i$ is even or odd.

For $\tau = 0.78, 0.80, \ldots, 0.98$, we took 100 random training samples and measured the true error rate of direct and indirect models, using the known distributions. Logistic regression is able to make no progress on this data at all, and essentially achieves 50% error rate, regardless of the magnitude of τ. In contrast, the indirect model does well, with an average error rate that depends on τ as shown in Figure 1. (Note that if the indirect model predicts one or more variables incorrectly, the error rate is exactly 0.5, about the same as logistic regression. Fortunately, however, especially for large τ, this is a rare occurrence.)

Because of the small size of the training data relative to the complexity of the problem, flexible methods, such as neural networks and decision trees, are unable to capture the nonlinearity of the problem.

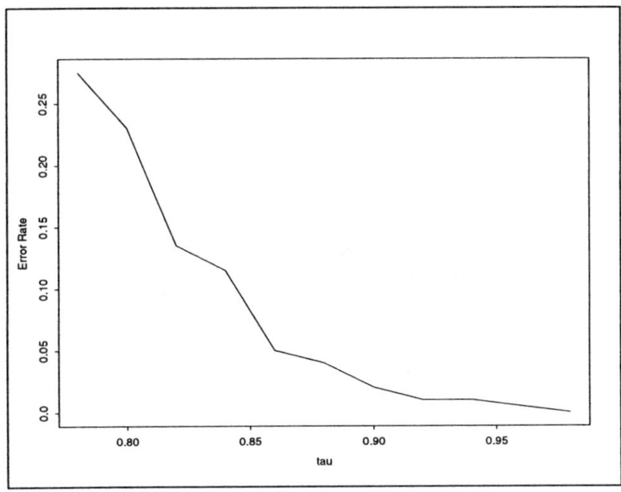

Fig. 1. Average error rate of the indirect model

4 Simulation 2

The second simulation involves a two class problem with two features x_1 and x_2, each sampled from a standard normal distribution. The continuous intermediate variables y_1 and y_2 are defined as

$$\begin{aligned} y_1 &= u_{11}x_1 + u_{12}x_2 + \epsilon_1 \\ y_2 &= u_{21}x_1 + u_{22}x_2 + \epsilon_2 \end{aligned}$$

where the ϵ_j are independent and $\epsilon_j \sim N(0, \sigma^2)$, and

$$u_{i1}^2 + u_{i2}^2 = 1$$

for $i = 1, 2$. The choice of this constraint on the $\mathbf{u}$ vectors simplifies the multiple correlation between y_j and $\mathbf{x}$ to

$$R^2 = \frac{1}{1 + \sigma^2}.$$

Objects belonging to class 1 are defined as having both y_1 and y_2 greater than a threshold T. The shape of the optimal decision boundary in x space is determined by the angle between the two $\mathbf{u}$ vectors: $\theta = u_{11}u_{21} + u_{12}u_{22}$. Thus we can picture the bivariate y space as being partitioned into quadrants with objects falling in one quadrant defined as class 1, and all others as class 0. The left frame of Figure 2 presents the y space representation of a sample of simulated data, with the partitioning thresholds included. The right frame gives the x space representation of the sample, with class 0 represented by crosses and class 1 represented by squares. The optimal decision boundary in $\mathbf{x}$ space is plotted as a dotted line. The decision surface in $\mathbf{x}$ space is piecewise linear and the classes overlap due to the nonzero value of σ.

Our simulation involves four parameters: σ^2, the strength of the relationship between $\mathbf{y}$ and $\mathbf{x}$; θ, which determines the shape of the decision boundary; T,

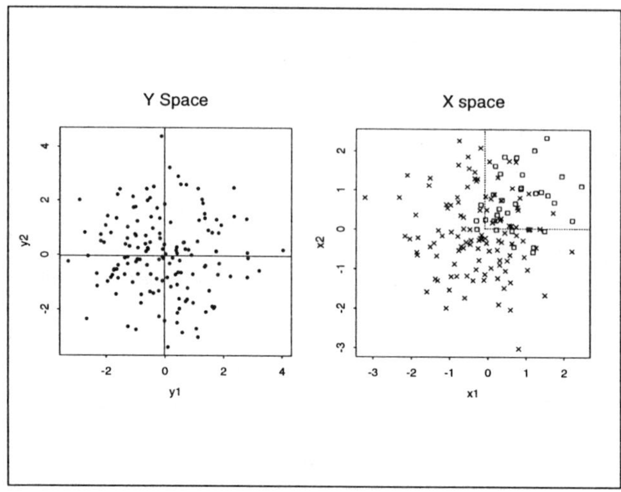

Fig. 2. A sample of $n = 150$ observations from simulation 2, with $\theta = 90^o$, $T = 0$ and $\sigma^2 = 1.0$; this value of σ^2 is chosen for illustration

the position of the threshold on the $\mathbf{y}$ variables; and n, the training sample size. In this study we examined $\sigma^2 \in [0.01, 0.09]$, corresponding to $R^2 \in [0.9174, 0.9901]$. Three shapes of decision boundary were used, one linear ($\theta = 0^o$) and two nonlinear ($\theta = 90^o, 135^o$). The partitioning threshold T took two values, 0 and –0.5. (The latter value of T is chosen to prevent the prior for class 1 from becoming very small). Finally, we took $n = 50$ and $n = 150$. Twenty replications were performed for each factor combination.

4.1 Classifiers

For this simulation we employed two indirect and two direct classifiers. The indirect classifiers used linear regression to predict $\mathbf{y}$ from $\mathbf{x}$, and logistic regression to predict $\mathbf{w}$ from $\mathbf{x}$. The direct classifiers were logistic discrimination, and a multilayer neural network, both of which model $p(1|\mathbf{x})$. The neural network approach is a nonlinear generalisation of logistic discrimination, which uses the optimisation criterion

$$E = \sum_{i=1}^{n} z_i \log \hat{f}(\mathbf{x}_i) + (1 - z_i) \log(1 - \hat{f}(\mathbf{x}_i)) + \lambda g(\mathbf{w})$$

where $\hat{f}$ is the neural network output, $g(\mathbf{w}) = \sum w_i^2$, the sum of squared weights, is a penalty function, and λ is a parameter that controls the flexibility of the model. This method of introducing bias to neural networks is usually referred to as weight decay. Two difficulties with the automatic application of neural networks (especially in the context of simulations) are choice of λ and choice of the number of hidden nodes. The second problem is less critical than the first, since providing we have sufficient hidden nodes to yield a sufficiently flexible mapping, the weight decay will control the complexity of the fit. λ is estimated on line in conjunction with estimating the network weights, using the *evidence approximation* of Mackay (1992), which is essentially an empirical Bayes procedure. Our simulations involve choosing a single network, defined as 'best' according to the evidence procedure, from a variety of network architectures and random starting points.

5 Results

Figure 3 shows the simulation results for a sample of size 50, with $\theta = 0$ and $T = 0$. The left panel of Figure 3 plots the mean of the 20 simulation results for each of the four classifiers, while the right panel plots the variance. Note

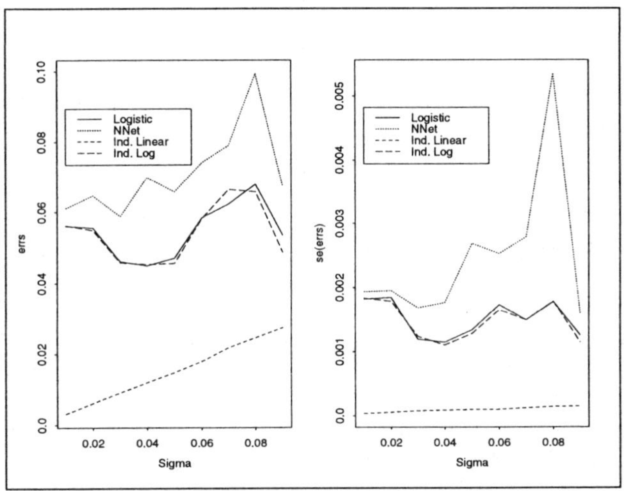

Fig. 3. Results for simulation 2, with $n = 50$, $\theta = 0$ and $T = 0$

that the degradation of performance in error rate (and variance) is linear for the indirect linear regression model. This reflects the fact that the indirect linear regression model is essentially the correct model, and σ^2 is sufficiently small to allow the model to perform well. It means that for large enough σ^2 the indirect model will perform worse than the other models, even though it is the correct model. The simulation results for other factor combinations are broadly similar. For nonlinear decision boundaries, the performance of logistic regression degrades markedly. Neural network are usually outperformed by the other classifiers and also exhibit higher variance.

Figure 4 shows the results for a sample of size $n = 150$, with $T = -0.5$, $\theta = 90$ and $\sigma^2 = 0.05, 0.06, \ldots, 0.09$. This Figure is typical of the simulation results for this sample size. This example involves a nonlinear decision surface and consequently logistic regression does not perform well. Neural networks do much better, on average, with $n = 150$, but still exhibit high variance. Indirect logistic models continue to outperform neural networks (while exhibiting lower variance), although the difference is appreciably smaller.

6 Conclusion

Both here, and in the applications described in Li & Hand (1997) and Hand, Li & Adams (1998), we have shown that indirect models can have advantages

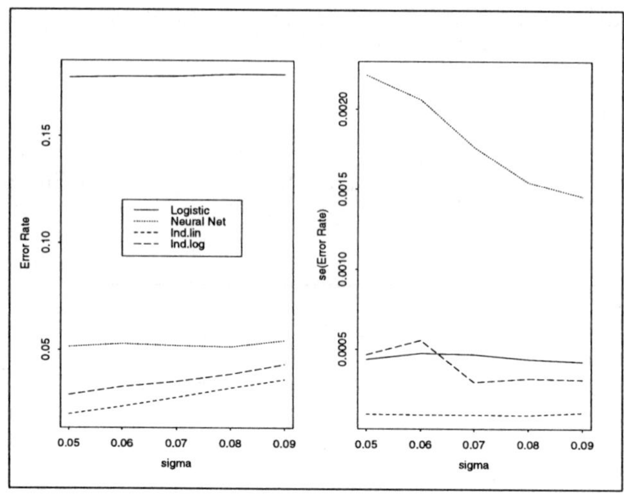

Fig. 4. Results for simulation 2, with $n = 150$, $\theta = 90$ and $t = -0.5$

over direct models. These advantages are most apparent in small sample situations, where classifiers capable of modelling a nonlinear boundary fail because of insufficient data. A secondary benefit is the lower variance observed for the indirect linear regression model, due to its small number of parameters. We might suppose that for very large samples a neural network model can capture the same function as the indirect model, and thereby achieve the same performance. In addition, of course, there are the complex and time consuming tasks of model estimation and selection, which are significantly less severe for indirect models.

Acknowledgments

The work of Niall Adams on this project was supported by grant GR/K55219 from the EPSRC, under its "Neural networks - the key questions" initiative. The work of the third author was funded by grant number R022250001 from the ESRC under the Realising Our Potential Award scheme. Thanks are due to Brian Ripley for making available his S-plus neural network routines.

References

Hand, D.J., Li, H.G. & Adams, N.M. (1998). Supervised classification with structured class definitions. Technical Report, Department of Statistics, The Open University.

Li, H.G. & Hand, D.J. (1997). Direct versus indirect credit scoring classifications. Technical Report, Department of Statistics, The Open University.

Mackay, D.J.C. (1992). The evidence framework applied to classification networks. *Neural Computation*, **4**, 720–736.

Model Search: An Overview

Herman J. Adèr[1] and Dirk J. Kuik[1] and David Edwards[2]

[1] Faculty of Medicine, Department of Epidemiology and Biostatistics. Vrije Universiteit, Amsterdam.

[2] Statistics Dept., Novo Nordisk, Bogsvaerd, Denmark

Keywords. Parallel model search, loglinear modelling, regression analysis, user interaction, geographic information systems

1 Introduction

Since standard software does not incorporate the full spectrum of model search techniques, more sophisticated methods are not often used in practice. This paper aims to give a rundown of what is possible in this area and indicates some ideas for applications not previously thought of.

We start with an overview of search procedures commonly used in regression analysis and log linear modelling. After that, in Section 2.2, the approach proposed by Edwards & Havránek will be described. This was first developed for graphical and hierarchical loglinear models and later generalized to arbitrary model families (Edwards & Havránek, 1987). Even for simple regression problems, the EH procedure offers more insight than the usual stepwise procedures since it produces an overview of the accepted and rejected models.

Section 4 summarizes two implementations of this procedure.

Finally, other applications will be mentioned. After an application in the field of rule selection in Artificial Intelligence, we discuss possibilities and consequences of allowing analyst interaction during the search process; an application to modelling in Geographic Information Systems will also be sketched.

2 Model search procedures

Most applied statistics is based on frequentist, model-based inference. It is not surprising, therefore, that the choice of the most appropriate model for a research question given a collected data set is a problem commonly encountered. A range of approaches to solve this problem has been proposed.

The basic approach to model search could be called: the *complete procedure*. All terms are defined by the user and some estimation procedure is executed. The results are found in one session. The model may be re-estimated with some terms replaced or left out. For more intricate models than the linear, in many statistical packages this is still the only way (SPSS for all GLM, BMDP for intrinsically non-linear models).

The most widely used automated approach is *stepwise* model search, using significance tests. From a computational point of view, this approach consists of two components: (i) a working model, and (ii) criteria for changing the working model into an 'adjacent' model. Terms are now introduced or omitted from the working model using ordinary significance tests. Alternative variants are *backward* search: starting with the full model, non-significant terms are successively omitted; and *forward* search: starting from the simplest model, significant terms are successively included. The stepwise procedures are simple to implement and have a natural flavour in the realm of classical statistics where hypothesis testing is a central activity. However, as formalized procedures they have many disadvantages, and have been widely criticized. For example: in non-orthogonal designs the individual tests at the different stages are dependent, but the dependency structure is model dependent and complex, so that the repeated sample properties of the procedure are intractable.

Another common approach to model search is to find the model that optimizes some appropriate criterion. For example, in multiple regression various criteria can be used, including the mean square error, Mallows C_p or R^2.

Two criteria, AIC and BIC, have been proposed as a basis for model selection. Akaike's Information Criterion (Akaike, 1974) was derived from information-theoretic arguments. Later, Schwarz (1978) proposed a modification of this, the Bayesian Information Criterion: this was derived as a simple approximation, asymptotically, to the posterior probability of the model. Of these approaches, those based on the AIC and BIC criteria seem to be preferable, in so far that they have some theoretical justification, whereas the approaches based on significance tests appear to be purely heuristic.

2.1 Procedures that allow for user interaction

Statisticians tend to differ in opinion on the proper way to account for knowledge originating from the research domain to which the data relate. Some of them take the position that solely automated search methods are inappropriate: instead, they advocate methods that allow interaction with the analyst during model search. During a stepwise process the analyst has the opportunity to indicate whether a model proposed by the search software should be accepted or not. At each step a set of 'interesting' terms is indicated by the user. From this set, the analyst/researcher chooses one for inclusion or omission from the model. Grounds for this choice are found in the subject matter field: for instance, if a diastolic blood pressure is more interesting than a systolic, and the latter is but a little bit more significant, precedence is given to the former. We call this approach the *Interactive stepwise approach.*

2.2 The EH model search algorithm

Edwards & Havránek (1987) describe an algorithm to search a model space and find a subset that is optimal in some predefined way.

Consider a set of models $\mathcal{M}$ on which a non-strict partial order $\prec$ is defined.

The problem is to find a partition $\mathcal{M} = A \cup R, A \cap R = \emptyset$, in which A is the set of accepted models and R the set of rejected models.

Edwards & Havránek assume *coherence* of $\mathcal{M}$. This means that for models $m_1, m_2 \in \mathcal{M}$:

$$m_1 \prec m_2, m_1 \in A \Rightarrow m_2 \in A, \qquad m_1 \prec m_2, m_2 \in R \Rightarrow m_1 \in R \tag{1}$$

In other words: if a model m is accepted, then models that are comparable and larger than m in the partial order are accepted, and if a model m is rejected, models that are smaller than m are also rejected.

As an example, assume that $\prec$ represents hierarchical model inclusion. Now, let $m_1 = (011100), m_2 = (011101)$ be two models of A and let m_1 be accepted. Since $m_1 \prec m_2$ (all variables of m_1 also occur in m_2), m_2 is also accepted. Note that $\prec$ is a partial order since not all models can be compared in this way.

For each $S \subset \mathcal{M}$ a *maximal* set and a *minimal* set are defined:

$$\max(S) = \{s \in S | s \prec t \Rightarrow t \notin S\}, \quad \min(S) = \{s \in S | t \prec s \Rightarrow t \notin S\} \tag{2}$$

In the sequel, we assume $\prec$ to represent model inclusion as in the example above. In this case, a smaller model is more parsimonious than a model that is larger. In common parlance we call this model 'simpler'.

The technique essentially uses the concepts of *a-dual* and *r-dual* of a set $S \in \mathcal{M}$. The a-dual of S(notation: $D_a(S)$) contains the most simple models in $\mathcal{M}$, that are not more parsimonious than any model of S. If S coincides with the rejected models, $D_a(S)$ contains the simplest models that conceivably may be accepted. A similar definition is given of $D_r(S)$. If S corresponds to the accepted models, then $D_r(S)$ includes the least parsimonious models that may conceivably be rejected.

By T we indicate the set of models not yet accepted or rejected during the iterative construction of A and R . It can be proven that

$$\max(T) = D_r(A) \setminus R, \qquad \min(T) = D_a(R) \setminus A \tag{3}$$

Figure 1 conveys an impression of what may go on in an iteration step. In the left panel, four models have not yet been assigned to the subset of either rejected or accepted models. After the iteration, in the right panel two more models have been rejected, while one is accepted. One model is as yet undecided and has to be assigned in future iterations.

3 Applications

3.1 Loglinear models

From the point of view of the present article, loglinear models differ from regression models in several ways: firstly, they often include multiple response variables, and secondly, they always include interactions between variables.

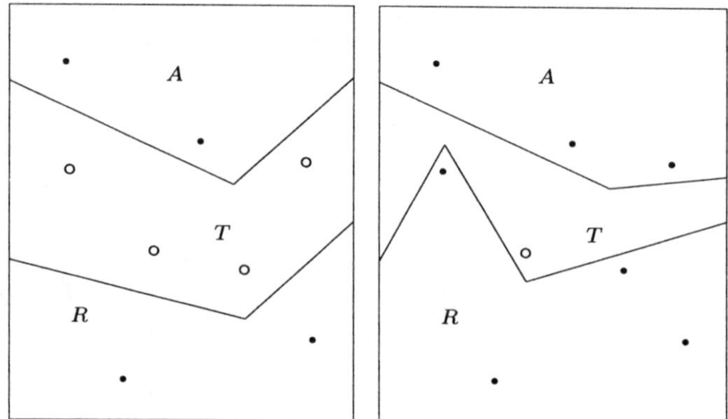

Fig. 1. Before and after an iteration in EH model search.
A: Accepted models; *R*: Rejected models; *T*: Yet undetermined

In many cases, regression models can be simply characterized by the presence or absence of covariates, thereby generating a binary lattice of models, and allowing simple stepwise model selection. In contrast, hierarchical loglinear models are subject to a constraint that higher order interaction terms only make sense when all the corresponding lower order terms also are in the model. This model family constitutes a distributed lattice. However, a subclass, the graphical loglinear models, that are characterized through the presence or absence of two-factor interaction terms (corresponding to edges in the interaction graph) constitute a binary lattice, and, again, a simple stepwise approach can be adopted for this family (Edwards, 1995).

3.2 Regression models

To search a large model space using the EH algorithm, two ingredients are essential: an appropriate goodness-of-fit measure and a threshold value to discriminate between rejected and accepted models.

As we saw in Section 2, in the case of multiple linear regression analysis AIC or BIC is used as a goodness-of-fit measure. In this case, it turns out to be possible to find a threshold that is independent from the number of predictors in the model.

For logistic regression analysis, the 'Deviance' D is used to assess goodness-of-fit:

$$D = -2\ln\left\{\frac{\text{(likelihood of the current model)}}{\text{(likelihood of the saturated model)}}\right\} \tag{4}$$

With D a summary measure is available to determine a rough accepted/rejected threshold: When the number of data patterns in the sample is J and the number of terms of the fitted model is p, then D is asymptotically χ^2-distributed with $J-p$ degrees of freedom under the assumption that the model is correct (Hosmer & Lemeshow, 1989, pages 138-139). The value $J-p$ is an estimate of the expected value of D and can be used as an upperbound for acceptance: models with lower D are unlikely to be rejected.

This measure has several drawbacks, one of them being that it is dependent on the number of terms of the model to be fitted. Therefore, the procedure adopted in the implementation is slightly more elaborate: In cases where the threshold function fails, significance tests on D-differences are used to compare hierarchically ordered models.

For Cox's regression a similar approach can be used since the definition of the Deviance is the same.

4 Implementations

MIM (Edwards, 1995) implements the algorithm for graphical modelling, including a broad family of statistical models for discrete and continuous variables. Hierarchical loglinear models, graphical Gaussian models, graphical association models and standard MANOVA models are included as special cases. The program offers facilities to define and manipulate models, display their independence graphs, test for comparison of models, and estimate incomplete data or latent variables using the EM-algorithm.

An interesting aspect of the EH procedure is the possibility to search different branches of the search tree in parallel (Havránek, 1992). This has been applied by Adèr in an implementation of multiple and logistic regression analysis (Adèr, Kuik & van Rossum, 1996). This implementation makes use of an empty shell that allows for parallelism.

5 Other possibilities

5.1 Rule learning in artificial intelligence

Mitchell (1997) describes an approach similar to the EH algorithm in a completely different field: *Learning rules from training instances.* Let V indicate a set of `if-then-rules` called a *Version Space.* On the basis of incoming training examples, rules in V are divided into two classes: appropriate and inappropriate (as to the description of the training set read so far). The approach has as a difference with the EH algorithm that no use is made of the a-dual and r-dual concepts. Furthermore, it is difficult to see how a Goodness-of-Fit measure should be formulated in this case.

On the other hand, from this example it follows that the EH algorithm is relevant to the blossoming field of Case-Based reasoning. It may eventually offer a tool in Artificial Neural Network learning where it could be applied to preprocess and optimize training sets.

5.2 Modelling rate data in geographical information systems

Consider rate data from various geographical areas. A model could set the rates for certain sets of contiguous areas to be equal. The full model would have different rates for each area, and the null model would set all the rates equal. Applying the EH algorithm to this kind of data would result in finding the simplest models for a given data set, and, in the end, in smoothing the data.

5.3 Allowing for user interaction during model search

The possibility to allow user interaction during the model search can be implemented in various ways and at different stages of the model search process.

There are three possible sets the user may want to change at each iteration, namely: (a) the set of undetermined models, (b) the set of accepted models and, (c) the set of the rejected models.

Basically, the EH algorithm just stores the minimal accepted and maximal rejected models, and proceeds by fitting the minimal or maximal undetermined models. But any other (undetermined) models can be fitted instead of the minimal or maximal undetermined models, for instance models chosen by the user. If the user wants to force one of the undetermined models either to be rejected or to be accepted, there is no problem. However, if one of the other sets is involved, for example, when the model (m_1) has already been accepted and is now reallocated to the set of rejected models, we have a 'contradiction', namely a pair of models $m_2 \prec m_1$ such that m_1 is rejected and m_2 is accepted. To remove the contradiction, we have to revise the set of rejected and accepted models. As long as we do this coherently, user interaction is feasible even in the EH procedure.

References

Adèr, H.J., Kuik, D.J. & van Rossum, H. (1996). Parallel model selection in Logistic Regression Analysis. In: *COMPSTAT96 Proceedings in Computational Statistics* (ed. A. Prat), 163-168. Barcelona: Physica-Verlag.

Akaike, H. (1974). A New Look at the Statistical Model Identification. *IEEE AC*, **19**, 716-723.

Edwards, D. (1995). *Introduction to Graphical Modelling.* New York: Springer.

Edwards, D. & Havránek, T. (1987). A Fast Model Selection Procedure for Large Families of Models. *Journal of the American Statistical Association*, **82** (397), 205-213.

Havránek, T. (1992). Parallelization and Symbolic Computation Techniques in Model Search. In: *SoftStat '91. Advances in Statistical Software* (ed. F. Faulbaum), 219-227. Stuttgart-New York: Gustav Fischer Verlag.

Hosmer, D.W. & Lemeshow, S. (1989). *Applied Logistic Regression.* New York - Chichester - Brisbane - Toronto - Singapore: John Wiley & Sons.

Mitchell, T. (1977). Version spaces: a candidate elimination approach to rule learning. In: *Proceedings Fifth International Joint Conference on Artificial Intelligence IJCAI-77.* Pittsburgh, PA.

Schwarz, G. (1978). The Bayesian Information Criterion. *Ann. Statist.*, **6**, 461-464.

Speeding up the Computation of the Least Quartile Difference Estimator

José Agulló

Departamento de Fundamentos del Análisis Económico,
Universidad de Alicante, E-03080, Alicante, Spain

Abstract. We propose modified p-subset algorithms for computing the least quartile difference and least trimmed difference estimates in a multiple linear regression model.

Keywords. Computation, robust regression

1 Introduction

In this paper we consider the multiple linear regression model

$$y_i = \mathbf{x}_i^T \beta + \alpha + \epsilon_i, \qquad i = 1, \ldots, n \tag{1}$$

where y_i is the dependent variable, β is a $(p-1)$-vector of slope parameters, α is an intercept term, $\mathbf{x}_i$ is a $(p-1)$-vector and ϵ_i is a random error. We suppose that the errors ϵ_i are independent and identically distributed. Our interest is to estimate the slope parameter β. For any β and α we denote residuals as $r_i(\beta, \alpha) = y_i - \mathbf{x}_i^T \beta - \alpha$. We also use the quantities $r_i(\beta) = y_i - \mathbf{x}_i^T \beta$ which do not depend on α.

The most well-known estimator of β is the *least squares* (LS) estimator. The LS estimator is optimal in several situations but it is severely affected by outliers. To get a reliable outlier detection and estimation, a high breakdown point estimator should be used. The most popular high breakdown estimator is the *least median of squares* (LMS) estimator (Rousseeuw, 1984; Rousseeuw & Leroy, 1987), defined by

$$(\hat{\beta}_{\text{LMS}}, \hat{\alpha}_{\text{LMS}}) = \text{argmin}_{(\beta,\alpha)} \{|r_i(\beta, \alpha)|, 1 \leq i \leq n\}_{h:n},$$

where $h = [(n+p+1)/2]$. (The subscript $h{:}\,n$ denotes the hth order statistic out of the indicated subset of size n.) $\hat{\beta}_{\text{LMS}}$ has excellent robustness properties: It has asymptotically a 50% breakdown point, and nearly minimizes the maximum bias curve in the class of all the estimates which depend only on the residuals. However, its rate of convergence is $n^{1/3}$ and its gaussian efficiency is asymptotically zero.

Recently, Croux, Rousseeuw & Hössjer (1994) proposed the *least quartile difference* (LQD) estimator:

$$\hat{\beta}_{\text{LQD}} = \text{argmin}_\beta QD_n(r_1(\beta), \ldots, r_n(\beta)),$$

where

$$QD_n(r_1, \ldots, r_n) = \{|r_i - r_j|; 1 \leq i < j \leq n\}_{\binom{h}{2}:\binom{n}{2}}.$$

Another recent proposal is the *least trimmed difference* (LTD) of Stromberg, Hawkins & Hössjer (1995), defined as

$$\hat{\beta}_{\text{LTD}} = \text{argmin}_{\beta} TD_n(r_1(\beta), \ldots, r_n(\beta)),$$

where

$$TD_n(r_1, \ldots, r_n) = \sum_{l=1}^{\binom{h}{2}} \{(r_i - r_j)^2; 1 \leq i < j \leq n\}_{l:\binom{n}{2}}.$$

For $h = [(n + p + 1)/2]$ the LQD and LTD estimators have a 50% breakdown point. Moreover, their distributions are asymptotically gaussian with the usual rate of convergence $n^{1/2}$. The LQD has a gaussian efficiency of 67% and the LTD of 66%. However, the major drawback to the widespread use of LQD or LTD is the relative high cost in computation time.

The basic procedure for computing high breakdown estimators is the p-subset algorithm (Rousseeuw & Leroy, 1987). This algorithm consists of minimizing the objective function over trial estimates corresponding to exact fits of subsets of size p (out of the n available observations). For each trial estimate, all residuals are computed and then the objective function is evaluated. The trial estimate that minimizes the objective function gives an approximate estimate. The p-subsets considered can be either all possible such sets, or a random subsample of them. The time needed to compute the objective function determines the total computation time, because the remaining steps of the algorithm are the same for all estimators.

It is well-known that the objective function of LMS can be computed in $O(n)$ time, but it requires $O(n \log n)$ time if the intercept term is adjusted as described by Rousseeuw & Leroy (1987, p. 201). Croux & Rousseeuw (1992) develop an $O(n \log n)$-time algorithm for computing QD_n. However, Rousseeuw & Croux (1992) observe that an important drawback of TD_n is that it needs $O(n^2)$ time. Nevertheless, in the next section we propose a procedure that computes TD_n in $O(n \log n)$ time.

If QD_n and TD_n are computed by means of efficient algorithms, then the p-subset algorithms for LQD and LTD have time complexities of the same order as the p-subset algorithm for LMS. However, the computation times for LMS, LQD and LTD can be very different due to the constant factors involved. For instance, the empirical results in Croux & Rousseeuw (1994) indicate that LQD requires several times more computer time than LMS.

In this paper we propose p-subset algorithms to compute the LQD and LTD regression estimates in nearly the same computer time as the LMS. Section 2 proposes a fast algorithm to calculate the objective function TD_n. In Section 3 we construct modified p-subset algorithms for computing the LQD and LTD estimates. An empirical comparison of the p-subset algorithms is described in Section 4. Finally, Section 5 contains the conclusions.

2 Computation of TD

At first sight it seems that the computation of QD_n and TD_n requires the calculation of the $\binom{n}{2}$ differences among residuals. This would imply $O(n^2)$ time. However, Croux & Rousseeuw (1992) propose an efficient algorithm that computes QD_n in only $O(n \log n)$ time. This algorithm first sorts the residuals so that the values r_i appear in nondecreasing order, and then considers the set $D = \{d_{ij} \mid d_{ij} = r_j - r_i, 1 \le i < j \le n\}$. This set contains all the elements above the main diagonal of the following matrix:

$$\begin{pmatrix} 0 & r_2 - r_1 & \dots & r_i - r_1 & \dots & r_j - r_1 & \dots & r_n - r_1 \\ & & & & & & & \vdots \\ & & & & & & & \vdots \\ & & & 0 & \dots & r_j - r_i & \dots & r_n - r_i \\ & & & & & & & \vdots \\ & & & & & & & \vdots \\ & & & & & & & 0 \end{pmatrix}$$

Notice that all the elements in D are nonnegative. Moreover, the elements of each row (column) of the above matrix are in nondecreasing (nonincreasing) order. Throughout the execution of the algorithm two arrays Lb and Rb define a partition of D in the three following subsets:

$D_L = \{d_{ij} \mid i = 1, \dots, n,\ i + 1 \le j < Lb(i)\}$,
$D_C = \{d_{ij} \mid i = 1, \dots, n,\ Lb(i) \le j \le Rb(i)\}$,
$D_R = \{d_{ij} \mid i = 1, \dots, n,\ Rb(i) < j \le n\}$.

The candidates to be QD_n are in D_C. The elements in D_L are smaller than QD_n and the elements in D_R are greater than QD_n. The cardinal of D_L is denoted by L. Initially $D_C = D$, so that $Lb(i) = i + 1$, and $Rb(i) = n$. The algorithm iteratively refines the set D_C by comparing the elements in D_C with a partition element λ. After each iteration the algorithm determines whether QD_n is smaller, equal or greater than λ. If $QD_n < \lambda$ ($QD_n > \lambda$), then D_R (D_L) is enlarged by those elements discarded from D_C for which $d_{ij} \ge \lambda$ ($d_{ij} \le \lambda$). If $QD_n = \lambda$, then the elements in D_C for which $d_{ij} < \lambda$ ($d_{ij} > \lambda$) are transferred to D_L (D_R). As partition element λ, the algorithm chooses the weighted median of the median elements of each row of D_C (where the weights are the number of elements in each such row). The refinement continues until: a) λ coincides with QD_n, or b) the size of D_C is smaller than or equal to n. In case b) the algorithm calculates QD_n selecting the $(\binom{h}{2} - L)$-smallest element in D_C.

Next we propose a procedure to compute TD_n in $O(n \log n)$ time. Consider the last partition of D found by Croux & Rousseeuw's algorithm. To compute TD_n we note that $TD_n = S_1 + S_2$, where S_1 is the sum of squares of the $\binom{h}{2} - L$ smallest elements in D_C, and S_2 is the sum of squares of the elements in D_L. The calculation of S_1 is carried out in the following way, according to the cases above. In case a), $S_1 = (\binom{h}{2} - L) \times QD_n^2$, because all the elements in D_C are equal to QD_n. In case b), S_1 is obtained as a by-product in the selection process of the $(\binom{h}{2} - L)$-smallest element of D_C. Note that this selection process necessarily identifies the $\binom{h}{2} - L - 1$ elements in D_C that are smaller than or equal to QD_n. To compute S_2 we note that D_L can be described as

$$D_L = \{d_{ij} \mid j = 1, \dots, n,\ Tb(j) \le i < j\},$$

where

$$Tb(j) = \min\{i \mid \sum_{l=1}^{i} Lb(l) - Lb(l-1) \ge j\}, \qquad j = 1, \dots, n,$$

with $Lb(0) = 1$ and $Lb(n) = n+1$. Let $s_0 = 0$ and $s_i = s_{i-1} + r_i, i = 1, \ldots, n$. Since the elements in D_L have the form $d_{ij} = r_j - r_i$ and $d_{ij}^2 = r_i^2 + r_j^2 - 2r_ir_j$, the sum of the squares of the elements in D_L is

$$S_2 = \sum_{i=1}^{n-1} r_{i+1}^2(Lb(i) - i - 1) + \sum_{i=2}^{n} r_i^2(i - Tb(i)) - 2\sum_{i=1}^{n-1} r_i(s_{Lb(i)-1} - s_i).$$

The calculation of the array Tb from the array Lb can be carried out in $O(n)$ time as follows:

```
j := 1
do i = 1, n
      k := 0
      while (k < Lb(i) - Lb(i - 1))
            Tb(j) := i
            j := j + 1
            k := k + 1
      endwhile
enddo
```

Notice that, given the last partition of D found by Croux & Rousseeuw's algorithm, the proposed procedure computes TD_n in $O(n)$ time. Since Croux & Rousseeuw's algorithm requires $O(n \log n)$ time, the objective function TD_n also requires $O(n \log n)$ time.

Remark. During the process of partition of D it is possible to obtain lower bounds for TD_n. Indeed, every time that the partition element λ implies a modification of subset D_L it can be shown that:

$$TD_n \geq S_2 + (\binom{h}{2} - L) \times \lambda^2, \tag{2}$$

where S_2 denotes the sum of squares of the L elements in the modified subset D_L.

3 Modified p-subset algorithms

In this section we propose modified p-subset algorithms for LQD and LTD. These algorithms avoid the calculation of the objective function for most of the trial estimates. Moreover, they obtain the same approximation as the usual algorithm in a significantly smaller computation time.

We consider first the calculation of the LQD estimate. Let ω^* be the smallest QD_n found so far. For each trial estimate, the modified algorithm sorts the residuals and then computes $C = \#\{(i,j) \ : \ |r_i - r_j| < \omega^*, 1 \leq i < j \leq n\}$. If $C < \binom{h}{2}$, the trial estimate does not provide an improvement of the currently smallest objective value, and the computation of QD_n is unnecessary. Otherwise, the trial estimate provides an improvement and the computation of QD_n is carried out taking ω^* as first partition element of D. In this way, all elements of D that are greater than or equal to ω^* are expediently discarded from D_C.

Assuming that the r_i, $i = 1, \ldots, n$, are in nondecreasing order, the computation of C can be carried out as follows:

$j := 1; C = -n * (n-1)/2$
  do $i = 1, n-1$
  while $(j < n)$ **and** $(r_{j+1} - r_i < \omega^*)$
    $j := j + 1$
  endwhile
  $C := C + j$
enddo
return C

Let us consider the computation of LTD. Let τ^* be the currently smallest TD_n, and ω^{**} be the QD_n for the currently best trial estimate. For each trial estimate, the modified algorithm starts the computation of TD_n taking ω^{**} as first partition element of D. Each time that the subset D_L is modified, the algorithm computes the lower bound for TD_n given by (2). If this lower bound is greater than or equal to τ^*, then the trial estimate does not improve the currently smallest objective value, and the next trial estimate in list is considered. In this case the full computation of TD_n is unnecessary.

Notice that the above modifications are applicable not only if the algorithm examines all p-subsets but also if it only examines a random subset of p-subsets.

4 Empirical comparison

In order to compare the computation times of LMS, LQD and LTD estimates we performed an empirical experiment. We calculated the average computation time for the estimation of regression coefficients over 10 gaussian simulated samples of size n for the model (1) with $p = 2$. We applied an exhaustive inspection of all p-subsets. LQD and LTD were computed using both the usual p-subset algorithm and the modified p-subset algorithm described in Section 3. Figure 1 shows the average computation time (in seconds on a Pentium Pro 200 MHz) versus the sample size.

The results suggest that the modified p-subset algorithms not only save a considerable amount of computer time but also that they require similar computer time as the computation of the LMS estimate. For example, for $n = 200$ the modified p-subset algorithm for LQD (LTD) is about nine (six) times faster than the usual algorithm. Moreover, the modified algorithm for LQD requires nearly the same computer time as the LMS. However, the computation of the LTD needs about 40% more time than the LMS.

5 Conclusions

We have proposed modified p-subset algorithms for computing the LQD and LTD estimators. The new algorithms obtain the same approximations as the usual p-subset algorithm in significantly less computer time. A simulation experiment suggests that the modified algorithms compute the LQD and LTD in nearly the same computer time as the LMS. This fact eliminates the main drawback of the LQD and LTD estimators. Since the LQD estimator approximately minimizes the maximum bias curve in the class of regression estimators based only on differences of residuals (Berrendero & Romo, 1996) and its maximum bias curve is only a bit greater than that of the LMS, the LQD estimator must be preferred.

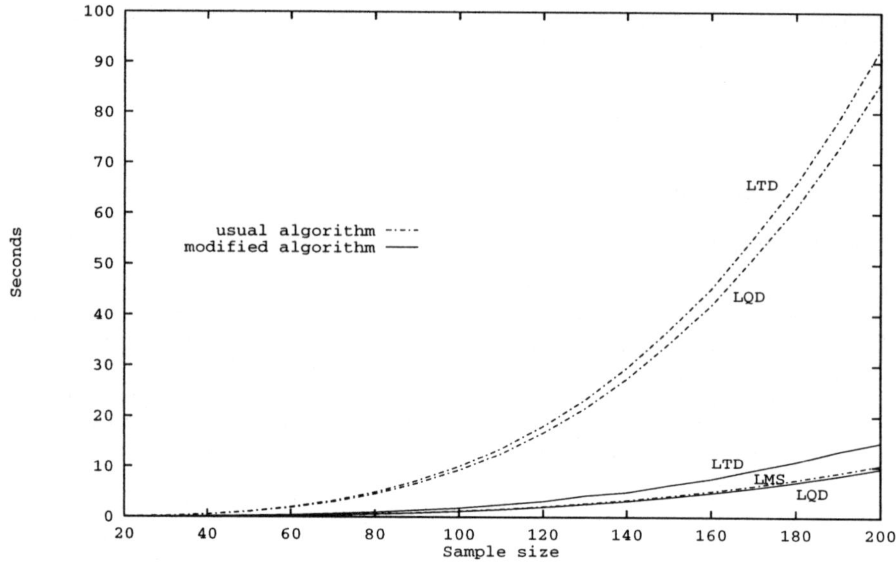

Fig. 1. Computation times of LMS, LQD and LTD estimates for $p = 2$

Acknowledgements

We would like to thank an anonymous referee and the editor for useful comments. This work has been supported by Projects PB96-0339 and PB92-0342 of DGICYT, Spain.

References

Berrendero, J.R. & Romo, J. (1996). Stability under contamination of robust regression estimators based on difference of residuals. Working paper 96-64, Statistics and Econometric Series 18, Universidad Carlos III de Madrid.

Croux, C. & Rousseeuw, P.J., (1992). Time-Efficient Algorithms for Two Highly Robust Estimators of scale. In: *Computational Statistics,* Vol. 1, 411–428. Heidelberg: Physica Verlag.

Croux, C. & Rousseeuw, P.J., (1994). High breakdown regression by minimization of a scale estimator. In: *COMPSTAT 1994: Proceedings in Computational Statistics,* 245–250. Heidelberg: Physica Verlag.

Croux, C., Rousseeuw, P.J. & Hössjer, O. (1994). Generalized S-estimators. *Journal of the American Statistical Society,* **89**, 1271–1281.

Rousseeuw, P.J. (1984). Least median of squares regression. *Journal of the American Statistical Society,* **79**, 871–880.

Rousseeuw, P.J. & Croux, C. (1992). Explicit scale estimators with high breakdown point. In: *L_1-Statistical Analysis and Related Methods,* 77–92. Amsterdam: North-Holland.

Rousseeuw, P.J. & Leroy, A. (1987). *Robust regression and outlier detection.* New York: Wiley.

Stromberg, A.J., Hawkins, D. & Hössjer, O. (1995). The Least Trimmed Differences Regression Estimator and Alternatives. Tech. report 1995:26, Dept. Mathematical Statistics, Lund University.

Piece-wise Detection of Spinal Scoliosis

A. J. Baczkowski[1] and X. Feng[2]

[1] Department of Statistics, University of Leeds, Leeds, UK
[2] Neural Computing Research Group, Department of Computer Science and Applied Mathematics, Aston University, Birmingham, UK

Abstract. In this paper we present kernel-based methods for modelling local deformities of the spine arising from scoliosis. This extends the work of Mardia *et al.* (1996a) who derived a global measure of spinal deformation.

Keywords. Kernel methods, scoliosis, shape, spine

1 Introduction

Scoliosis is a deformity of the spinal column which gives the spine an apparent curvature when it is viewed from the front. The normal spine lies in a plane, the lateral or front-back plane, and scoliosis can thus be regarded as a three-dimensional deformity; see for example Graf & Mouilleseaux (1990). Although it is possible to perform corrective surgery for the condition, clinicians need to be able to assess its nature and severity.

With radiography it is possible to obtain different clinical measurements of the spine such as the Cobb angle (Cobb, 1948), the degree of vertebral rotation (Bunnell, 1985), and the degree of lateral deviation and tilt of the vertebrae (Drerup & Hierholzer, 1992).

To overcome the obvious drawbacks of using even low-dose radiographic techniques, a recently introduced procedure projects a series of white-light fringes onto the back of a patient; see Curran & Groves (1990). Analysis of the observed pattern of light fringes by this Quantec spinal measurement system allows the three-dimensional shape of the spine to be reconstructed.

Using the output from the Quantec system, Mardia *et al.* (1996a) present several simple statistics for assessing whether the patient under study suffers from scoliosis. These statistics are global averages of shape information over the length of the spine and, while they allow a decision about whether scoliosis is present or not, and indicate the overall degree of scoliosis, they do not reveal information about the precise location of the deformity.

In this paper a kernel-based method is used to determine the degree of deformation at each point along the spine, thus giving clinicians local information whilst also retaining the modelling simplicity of the global measure of Mardia *et al.* (1996a).

2 Global model of spinal deformation

The Quantec system gives the locations $\{x_i, y_i, z_i\}$, for $i = 1, \ldots, n$, at points along the spinal prominence, with x representing the distance along the spine measured from top to bottom, y the distance along the front-back axis, and z the distance along the left-right axis. Typically $n = 300$ for the Quantec system. The y- and z-values are centred to have mean zero. Mardia *et al.* (1996a) proposed using the simple model,

$$x_i = i, \quad y_i = \alpha_i cos\theta + \epsilon_i, \quad z_i = \alpha_i sin\theta + \delta_i,$$

where ϵ_i and δ_i are mutually independent Gaussian variables with zero mean and common variance σ^2, and $i = 1$ corresponds with the top of the spine. They derived estimates for α_i at each location i and also for the global average θ, which represents the departure of the spinal plane from the $z = 0$ lateral plane. Geometrically, the α_i represent the perpendicular distances of the data values from the mean spinal line.

Expressions for the mean and variance of these parameter estimates can be derived, though it should be noted that the number of parameters increases with the number of observations n; see Neyman & Scott (1948).

This simple model is quite successful at highlighting abnormal individuals.

Example 1

For the normal spine *srh9a* we obtain the estimate $\hat{\theta} = 9.7°$ while the abnormal spine *ds7* gives $\hat{\theta} = -59.0°$. Figure 1 shows plots of y against x and z against x for these two individuals.

Note that the Quantec system is only able to resolve y distances to 0.5mm. and z distances to 1mm. Figure 1 clearly shows this problem of resolution, as witnessed by the periodic large jumps in the y- and more noticeably the z-values.

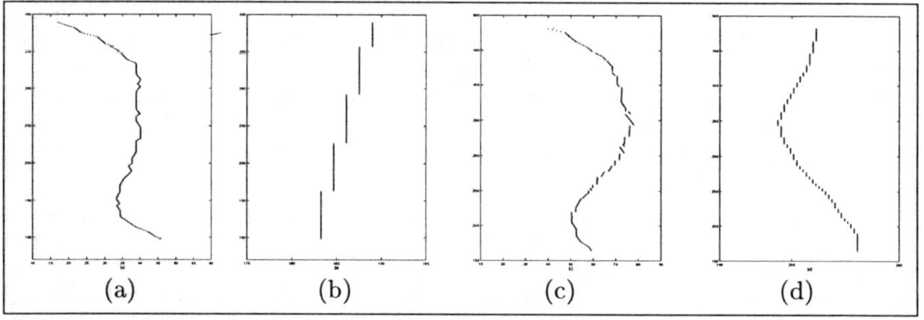

Fig. 1. Projections of profiles of normal spine *srh9a* and abnormal spine *ds7* (not to same scale). *srh9a*: (a) y vs x, (b) z vs x. *ds7*: (c) y vs x, (d) z vs x

Although $\hat{\theta}$ gives a useful global shape measure for crudely classifying normal and abnormal spines, it does not give information about the location of any abnormality. A further problem is that the estimates $\hat{\alpha}_i$ are highly inter-correlated which hinders their interpretation.

In this paper we present models which attempt to estimate α_i and the local deviation θ_i from the lateral plane at each location i using only those data values within a window of bandwidth m centred on i. The mean and variance of the parameter estimates can be easily derived. Although parameter estimates at adjacent locations are correlated, it is possible to split the spine into non-overlapping segments; for example, corresponding to small groupings of the individual spinal vertebrae. The parameter estimates obtained from these non-overlapping segments will be independent.

3 Model 1

Consider the simple model,

$$y_j = \alpha cos\theta + \epsilon_j, \quad z_j = \alpha sin\theta + \delta_j,$$

for values $j = 1, \ldots, m$, located in a small region of the spine, where ϵ_j and δ_j are as before. Maximum likelihood estimates for α and θ are given by,

$$\hat{\alpha} = \overline{y} cos\hat{\theta} + \overline{z} sin\hat{\theta} = \sqrt{\overline{y}^2 + \overline{z}^2} \quad \text{and} \quad \tan\hat{\theta} = \overline{z}/\overline{y},$$

where $\overline{y}$ and $\overline{z}$ are the local means of y_j and z_j respectively, each averaged over m values. The local variance σ^2 is estimated by $\frac{1}{2m}(s_{yy} + s_{zz})$, where $s_{yy} = \sum_{j=1}^{m}(y_j - \overline{y})^2$ and $s_{zz} = \sum_{j=1}^{m}(z_j - \overline{z})^2$.

For large m the variances of these parameter estimates satisfy,

$$Var[\hat{\alpha}] \approx \frac{\sigma^2}{m}, \quad Var[\hat{\theta}] \approx \frac{\sigma^2}{m\alpha^2}, \quad Var[\hat{\sigma}^2] \approx \frac{\sigma^4}{m}.$$

Plots showing the local estimates of α and σ^2 for each location i along the spine can be produced. The values of α provide a smoothed measure of the mean distance of the spinal prominence from the mean line. The variance estimates σ^2 allow the assumption of constant variance to be checked.

Example 2

Figure 2 shows the results of fitting this model to the normal spine *srh9a* and the abnormal spine *ds7* of Example 1. We have used a bandwidth $m = 40$ points which corresponds approximately to three vertebrae. The estimates of θ are given for each set of m points centred at locations i, with $i = 20, 40, 60, \ldots$. Thus the estimates $\hat{\theta}$ are not strictly independent here.

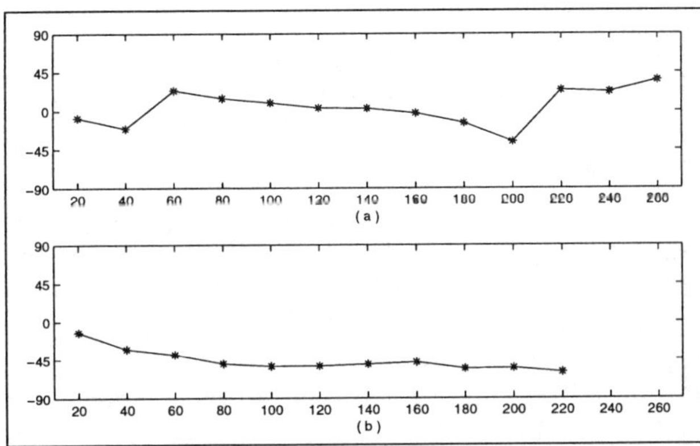

Fig. 2. Plot of $\hat{\theta}$ against index i: (a) normal spine *srh9a*, (b) abnormal spine *ds7*

Figure 2a shows that the $\hat{\theta}$ values for the normal spine *srh9a* are scattered about zero. There are some small local departures from zero but the individual is clinically "normal". The averaged $\hat{\theta}$ value is 8.3°, close to the global measure in Example 1.

For the abnormal spine *ds7* it is seen that the $\hat{\theta}$ values steadily decrease for this individual as one moves down the spine (i increasing). The averaged $\hat{\theta}$ value is −47.3°, again close to the corresponding global measure determined in Example 1.

4 Model 2

Consider now modifying the global model of Section 2 by estimating the parameters α_j and θ for a subset of values $j = 1, \ldots, m$ centred on a location i. Denote these estimates as $\alpha_{j(i)}$ and θ_i.

In model 1 of Section 3, the θ and α were allowed to vary with i while being locally fixed constants as j varied. Here θ is again a constant for all j for a given location i, while the α parameters are allowed to vary with j.

The presentation described above is equivalent to placing a rectangular kernel function $w(i,j)$ at location i taking values unity if $\mid i-j \mid < \frac{1}{2}m$ and zero otherwise. More generally other forms of kernel function might be used.

For simplicity we assume that the variance σ^2 is constant. The y_i and z_i are again centred for $i = 1, \ldots, n$.

Suitable weighting of the log-likelihood (Wand & Jones, 1995, p165) yields maximum likelihood estimates $\hat{\theta}_i$ and $\hat{\alpha}_{j(i)}$ satisfying,

$$tan2\hat{\theta}_i = \frac{2S_{yz}}{S_{yy} - S_{zz}},$$

and

$$\hat{\alpha}_{j(i)} = y_j cos\hat{\theta}_i + z_j sin\hat{\theta}_i,$$

where $S_{yy} = \sum_{j=1}^{n} y_j^2 w(i,j)$ is a measure of the local variability of the y-values, and similarly $S_{zz} = \sum_{j=1}^{n} z_j^2 w(i,j)$, and $S_{yz} = \sum_{j=1}^{n} y_j z_j w(i,j)$.

A local consistent estimate of the variance, σ_i^2 say, measured about the point i, is given by

$$\tilde{\sigma}_i^2 = \frac{\sum_{j=1}^{n} t_{j(i)}^2 w(i,j)}{\sum_{j=1}^{n} w(i,j)},$$

where the residual distance of (x_j, y_j, z_j) from the local best-fitting plane is $t_{j(i)} = y_j sin\hat{\theta}_i - z_j cos\hat{\theta}_i$ and can be regarded as an approximate Gaussian random variable with zero mean and variance σ_i^2. With a rectangular window it follows that $\tilde{\sigma}_i^2$ is just the mean of $t_{j(i)}^2$ over the window bandwidth.

To check on the departures of the spine from the lateral $z = 0$ plane we can plot $\hat{\theta}_i$ as a function of i. Similarly, plots of $\tilde{\sigma}_i^2$ against i can be used to monitor the variance. The values i can be positioned sufficiently far apart to ensure independence of the parameter estimates, based on non-overlapping spinal segments.

For $w(i,j) = 1$ and large values of the bandwidth m it is easily shown that the local estimates $\hat{\theta}_i$ becomes close to the global estimate $\hat{\theta}$ of Section

2. Note also that a pooled estimate of α_i can be obtained using a weighted average of the $\hat{\alpha}_{i(j)}$.

Example 3

We again use a rectangular window with $m = 40$. Figure 3 shows $\hat{\theta}_i$ plotted against the index i for the two individuals *srh9a* and *ds7* of Example 1.

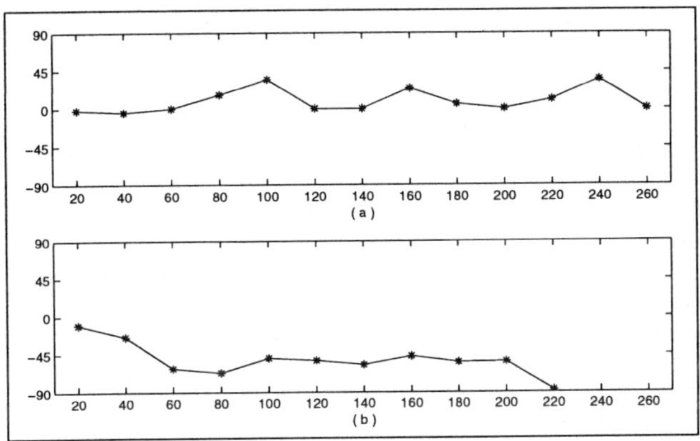

Fig. 3. Plot of $\hat{\theta}_i$ against index i: (a) normal spine *srh9a*, (b) abnormal spine *ds7*

Figure 3a shows that $\hat{\theta}_i$ for the normal spine is generally close to zero with the mean value of $\hat{\theta}_i$ being 11.0°. For the abnormal spine *ds7*, Figure 3b shows a similar decreasing behaviour to Figure 2b, with overall mean of the $\hat{\theta}_i$ being −42.8°. This individual exhibits an approximately constant twist along the length of the spine.

5 Extending models 1 and 2

A simple modification to models 1 and 2 is to allow ϵ_j and δ_j to have variances σ_y^2 and σ_z^2 respectively. For model 1 the parameter estimates are as given in Section 3. For model 2 the simplicity of the solutions of the parameter estimates is removed. We now have,

$$\hat{\alpha}_{j(i)} = \frac{y_j \hat{\sigma}_z^2 cos\hat{\theta}_i + z_j \hat{\sigma}_y^2 sin\hat{\theta}_i}{\hat{\sigma}_z^2 cos^2\hat{\theta}_i + \hat{\sigma}_y^2 sin^2\hat{\theta}_i},$$

with $t = tan\hat{\theta}$ the solution of the quadratic equation

$$S_{yz}(vt^2 - 1) + t(S_{yy} - vS_{zz}) = 0,$$

where $v = \hat{\sigma}_y^2/\hat{\sigma}_z^2$, and

$$\hat{\sigma}_y^2 = \frac{\sum_{j=1}^n w(i,j)(y_j - \hat{\alpha}_{j(i)} cos\hat{\theta}_i)^2}{\sum_{j=1}^n w(i,j)}$$

and

$$\hat{\sigma}_z^2 = \frac{\sum_{j=1}^{n} w(i,j)(z_j - \hat{\alpha}_{j(i)} sin\hat{\theta}_i)^2}{\sum_{j=1}^{n} w(i,j)}$$

These equations can be solved iteratively.

6 Discussion

The results from using the kernel procedure can be used to give both a global measure of abnormality if m is large, corresponding to the methodology of Mardia *et al.* (1996a), as well as giving an indication of the locality and degree of the spinal deformity if m is small.

Our analyses have been based upon the raw centred data. Inspection of Figure 1 suggests that it may be useful to smooth the output from the Quantec system prior to performing a data analysis.

We have experimented with both Gaussian and rectangular windows and several different choices of the window bandwidth m.

This paper has been motivated by the simple model of Mardia *et al.* (1996a) but has developed a local rather than a global measure of deformity. Other approaches could be considered. For example, Mardia *et al.* (1996b) sought to measure the geometrical torsion at all points along the spine. A Markov chain model for the spinal profile might also be used to assess deformity; see Ripley & Sutherland (1990).

References

Bunnell, W.P. (1985). Vertebral rotation: a simple method of measurement on routine radiographs. *Orthopaedic Transplant*, **9**, 114.

Cobb, J.R. (1948). Outline for the study of scoliosis. *The Academy of Orthopaedic Surgeons Instructional Course Lectures*, **5**, 261-275.

Curran, P. & Groves, D. (1990). An accurate, fast and cost effective method for the measurement of body shape and the assessment of spinal deformity. In: *Proceedings of the 6th International Symposium on Surface Topography and Spinal Deformity*, 24-30, Estoril.

Drerup, B. & Hierholzer, E. (1992). Evaluation of frontal radiographs of scoliotic spines - part II: relation between lateral deviation, lateral tilt and axial rotation of vertebrae. *Journal of Biomechanics*, **25**, 1443-1450.

Graf, H. & Mouilleseaux, B. (1990). *3-D Analysis of the Lumbar Scoliosis in Adults.* Leeds: Thackray Orthopaedic Monograph Series.

Mardia, K.V., Baczkowski, A.J., Feng, X., & Millner, P.A. (1996a). A study of three-dimensional curves. *Journal of Applied Statistics*, **23**, 139-148.

Mardia, K.V., Baczkowski, A.J., & Feng, X. (1996b). Detection of scoliosis by modelling geometric torsion. In: *Proceedings in Image Fusion and Shape Variability Techniques*, 212-213. Leeds: Leeds University Press.

Neyman, J. & Scott, E.L. (1948). Consistent estimates based on partially consistent observation. *Econometrica*, **16**, 1.

Ripley, B.D. & Sutherland, A.I. (1990). Finding spiral structures in images of galaxies. *Philosophical Transactions of the Royal Society of London, Series A*, **332**, 477-485.

Wand, M.P. & Jones, M.C. (1995). *Kernel Smoothing.* London: Chapman & Hall.

A Comparison of Recent EM Accelerators within Item Response Theory

F. Bartolucci, A. Forcina and E. Stanghellini

Dipartimento di Scienze Statistiche, Università di Perugia, Italy
Via A. Pascoli, 1, C.P. 1315 Succ.1, 06100 Perugia (I)

Abstract. We use a latent class model to analyse the results of an aptitude test in Statistics assigned to a group of students at our University. We implement various algorithms for maximum likelihood estimation of the parameters (the EM algorithm, two accelerators of the EM algorithm introduced recently and a plain Fisher-scoring algorithm) and compare their relative performance on a real and a simulated data set.

Keywords. Item response theory, latent class model, EM accelerator

1 Introduction

The Item Response Theory models are typically used to analyse the results of an aptitude test assigned to a group of students. Usually, within these models, the ability of the examinee is considered a *unobservable* or *latent trait* and a parameter for each examinee is used to denote his position on this *latent trait*. Moreover one or more parameters are used to describe the features of each test item.

The first and the simplest item response model is that of Rasch (1961) which is referred to dichotomous scored responses and uses just one parameter for each item. Another possibility to analyse item responses is to assume that the examinee's ability is an unobservable categorical variable instead of a continuous one. The resulting model is therefore a latent class model (see e.g. Bartholomew, 1987 or Heinen, 1996). Examples of the use of such models for Item Response analysis can be found, for instance, in Lindsay *et al.* (1991).

Maximum likelihood estimates of the parameters of this model can easily be performed through the EM algorithm (Dempster *et al.*; 1977, Wu, 1983). The nice properties of such an algorithm are its simplicity and insensitiveness to the starting values. However, one limitation of the EM algorithm is its slow convergence which becomes dramatic as the number of parameters increases. Applications of such an algorithm to latent class in the described context can be prohibitive, as the number of observed items, and, therefore, of parameters, is quite high.

Several attempts have been made to improve the convergence rate of the EM algorithm. Jamshidian & Jennrich (1997) provide a review of such methods and propose two new accelerators based on quasi-Newton methods called QN1 and QN2. The first is a pure accelerator and the second is a hybrid one.

In this paper we implement and compare the relative performance of the EM algorithm with that of the two accelerators suggested by Jamshidian & Jennrich (1997) and a plain Fisher-scoring algorithm in the context of an examination session in Statistics at our University. Additional comparisons are based on a simulated data set. Finally, a result due to Lang (1992) concerning the information matrix for log-linear models with missing data is used to provide standard errors of the estimates in the EM based algorithms.

2 The latent class model

The latent class model implies that examinees are drawn from an overall population divided into C categories of a latent factor representing *ability* so that subjects within the same class are homogeneous and that, given the latent class, the answers to different questions by the same subject are independent.

More formally, the probability that a random selected subject from the class c answers the set of J questions with the response vector $\mathbf{r}$ is:

$$P(\mathbf{r}|c) = \prod_j \lambda_{j|c}^{r_j} \left(1 - \lambda_{j|c}\right)^{1-r_j}$$

where $\lambda_{j|c}$ is the probability to observe a correct answer to item j for a member of class c. Consequently, if we denote by π_c the probability to draw randomly a subject from class c, we have that

$$P(\mathbf{r}, c) = \pi_c P(\mathbf{r}|c). \tag{1}$$

Therefore, once the parameter estimates are available, we can also estimate for each examinee the posterior probability that (s)he belongs to latent class c given the response vector $\mathbf{r}$

$$P(c|\mathbf{r}) = P(\mathbf{r}, c) / \sum_c P(\mathbf{r}, c).$$

These estimated probabilities may be used to assess the performance of each examinee, for instance by computing posterior expectations of suitable scores assigned to the level of ability corresponding to the various latent classes.

Let $m_{\mathbf{r},c}$ be the expected frequency in the cell $(\mathbf{r}, c)$ of the contingency table *Response* $1 \times \cdots \times$ *Response* $J \times$ *Latent Class* with $J+1$ dimensions and sC cells, with $s = 2^J$. A log-linear model for $m_{\mathbf{r},c}$ may easily be written down: it contains all the two-way interactions between each response and the latent variable. If $\{n_{\mathbf{r},c}\}$ are the cell counts of this table, we can observe only the margins $n_{\mathbf{r}} = \sum_c n_{\mathbf{r},c}, \forall \mathbf{r}$, hence the problem reduces naturally to one of incomplete data.

In a more general context, Lang (1992) dealt with log-linear models with incomplete data as follows. Assume that the vector $\boldsymbol{x}$ of observations follows a log-linear model with $\boldsymbol{m} = \mathrm{E}(\boldsymbol{x})$ and $\log(\boldsymbol{m}) = \boldsymbol{Z\beta}$, where $\boldsymbol{Z}$ is the appropriate design matrix and $\boldsymbol{\beta}$ is a vector of unknown parameters. Assume also that observations are available only on a suitable margin of the multi-way table which can be written in the form $\boldsymbol{y} = \boldsymbol{Lx}$ where $\boldsymbol{L}$ is a matrix of ones and zeros with at least a 1 in each row and at most a 1 in each column. The basic latent class model considered here is thus a special case with $\boldsymbol{L} = \boldsymbol{I}_s \otimes \mathbf{1}_C^T$.

2.1 A Fisher-scoring algorithm

The log-likelihood of the observed data is simply

$$l = \boldsymbol{y}^T \log(\boldsymbol{n}) - \mathbf{1}_s^T \boldsymbol{n} + constant$$

where $\boldsymbol{n} = \boldsymbol{L}\boldsymbol{m}$. The first derivative and the expected second derivative with respect to $\boldsymbol{\beta}$ are easily computed by the chain rule and we have

$$\boldsymbol{g} = \frac{\partial L}{\partial \boldsymbol{\beta}} = \boldsymbol{Z}^T \mathrm{diag}(\boldsymbol{m}) \boldsymbol{L}^T \mathrm{diag}(\boldsymbol{n})^{-1} (\boldsymbol{y} - \boldsymbol{n})$$

and

$$\boldsymbol{H} = -E[\frac{1}{\partial \boldsymbol{\beta}} (\frac{\partial L}{\partial \boldsymbol{\beta}})^T] = \boldsymbol{Z}^T \mathrm{diag}(\boldsymbol{m}) \boldsymbol{L}^T \mathrm{diag}(\boldsymbol{n})^{-1} \boldsymbol{L} \mathrm{diag}(\boldsymbol{m}) \boldsymbol{Z}.$$

With these ingredients, a Fisher-scoring algorithm at each step updates the starting value $\boldsymbol{\beta}_0$ to $\boldsymbol{\beta}_0 + \boldsymbol{H}^{-1}\boldsymbol{g}$; however a simple line search is inserted to avoid instability. We also found that efficient starting values may be obtained as follows: first we compute $\boldsymbol{m}_0$ by splitting each observed frequency uniformly (apart from a small random perturbation) among the C latent categories and then compute $\boldsymbol{\beta}_0$ as ordinary least squares from $\log(\boldsymbol{m}_0)$.

2.2 EM algorithm

The EM algorithm is very often used to compute the m.l.e. whenever the observed data can be considered as incomplete. In the context of latent class models, the two steps specify into the following:

E-step: compute $\hat{\boldsymbol{m}} = E(\boldsymbol{x} \mid \boldsymbol{y}, \boldsymbol{\beta})$=diag$(\boldsymbol{m})\{[diag(\boldsymbol{n})^{-1}\boldsymbol{y}] \otimes \mathbf{1}_C\}$

M-step: find the m.l.e. $\tilde{\boldsymbol{\beta}}$ by fitting a log-linear model on $\hat{\boldsymbol{m}}$ using $\boldsymbol{Z}$ like design matrix.

As already mentioned, the main drawback of this algorithm is the slow convergence rate. Several accelerators of the EM algorithm have been proposed. Here we focus on the QN1 and QN2 accelerators recently proposed by Jamshidian & Jennrich (1997).

2.2.1 QN1 accelerator

Briefly, the QN1 accelerator searches for the vector of parameters $\hat{\boldsymbol{\beta}}$ such that $\tilde{\boldsymbol{g}}(\hat{\boldsymbol{\beta}}) = \mathbf{0}$ via a quasi-Newton method which uses an approximated matrix $\mathbf{A}$ instead of the real Jacobian matrix of $\tilde{\mathbf{g}}(\boldsymbol{\beta})$. We introduce the EM step as $\tilde{\boldsymbol{g}} = \tilde{\boldsymbol{\beta}} - \boldsymbol{\beta}$. Starting from an initial value for $\boldsymbol{\beta}$, $\tilde{\boldsymbol{g}} = \tilde{\mathbf{g}}(\boldsymbol{\beta})$ and $\mathbf{A} = -\mathbf{I}$, this accelerator consists of the following steps:

i) Compute $\Delta\boldsymbol{\beta} = -\mathbf{A}\tilde{\boldsymbol{g}}$ and $\Delta\tilde{\boldsymbol{g}} = \tilde{\boldsymbol{g}}(\boldsymbol{\beta} + \Delta\boldsymbol{\beta}) - \tilde{\boldsymbol{g}}$;

ii) Sum to $\mathbf{A}$ the adjustment matrix:

$$\Delta\mathbf{A} = \frac{(\Delta\beta - \mathbf{A}\Delta\tilde{g})\Delta\beta^T \mathbf{A}}{\Delta\beta^T \mathbf{A}\Delta\tilde{g}};$$

iii) Let $\boldsymbol{\beta}$ equal to $\boldsymbol{\beta} + \Delta\boldsymbol{\beta}$, $\tilde{\boldsymbol{g}}$ equal to $\tilde{\boldsymbol{g}} + \Delta\tilde{\boldsymbol{g}}$ and check for convergence.

2.2.2 QN2 accelerator

In short, the QN2 accelerator directly maximizes the incomplete data log-likelihood $l(\boldsymbol{\beta})$ via a different quasi-Newton method. In order to implement this accelerator, we need to compute the gradient vector of $l(\boldsymbol{\beta})$, which we denote by $\mathbf{g}(\beta)$. Starting from the initial values $\boldsymbol{\beta}$, $\boldsymbol{g} = \boldsymbol{g}(\boldsymbol{\beta})$, $\tilde{\mathbf{g}} = \tilde{\mathbf{g}}(\boldsymbol{\beta})$ and $\mathbf{S} = \mathbf{0}$, this accelerator consists of the following steps:

i) Compute $\boldsymbol{d} = -\tilde{\boldsymbol{g}} + \mathbf{S}\boldsymbol{g}$;
ii) Let $\Delta\boldsymbol{\beta} = \alpha\boldsymbol{d}$ where α is the value which maximizes $l(\boldsymbol{\beta} + \alpha\boldsymbol{d})$;
iii) Compute $\Delta\boldsymbol{g} = \boldsymbol{g}(\boldsymbol{\beta} + \Delta\beta) - \boldsymbol{g}$ and $\Delta\tilde{\boldsymbol{g}} = \tilde{\boldsymbol{g}}(\boldsymbol{\beta} + \Delta\beta) - \tilde{\boldsymbol{g}}$;
iv) Sum to $\mathbf{S}$ the adjustment matrix:

$$\Delta\mathrm{S} = \left(1 + \frac{\Delta\boldsymbol{g}^T\Delta\boldsymbol{\beta}^*}{\Delta\boldsymbol{g}^T\Delta\boldsymbol{\beta}}\right)\frac{\Delta\boldsymbol{\beta}\Delta\boldsymbol{\beta}^T}{\Delta\boldsymbol{g}^T\Delta\boldsymbol{\beta}} - \frac{\Delta\boldsymbol{\beta}^*\Delta\boldsymbol{\beta}^T + \left(\Delta\boldsymbol{\beta}^*\Delta\boldsymbol{\beta}^T\right)^T}{\Delta\boldsymbol{g}^T\Delta\boldsymbol{\beta}}$$

where $\Delta\boldsymbol{\beta}^* = -\Delta\tilde{\boldsymbol{g}} + \mathbf{S}\Delta\boldsymbol{g}$;
v) Let $\boldsymbol{\beta} = \boldsymbol{\beta} + \Delta\boldsymbol{\beta}$, $\boldsymbol{g} = \boldsymbol{g} + \Delta\boldsymbol{g}$ and $\tilde{\boldsymbol{g}} = \tilde{\boldsymbol{g}} + \Delta\tilde{\boldsymbol{g}}$ and check for convergence.

3 An example

To compare the above algorithms, we fitted a latent class model with $C = 2$, to a data set of responses of 106 students to a 4-item test. The observed frequencies are (ordered lexicographically with r_1 running slowest):

2 1 0 1 4 1 2 1 3 2 3 3 20 18 17 28.

The model has a good fit with a deviance of 4.1663 on 6 d.f.. The following table gives the estimated probabilities of a correct answer given the latent class:

	Latent class	
Item	1	2
1	0.6782	0.9983
2	0.7282	0.9282
3	0.3272	0.6213
4	0.3001	0.6358

Notice that the probability of a correct answer is always larger for the second latent class which, clearly, refers to the most capable students. Moreover, while in the Rash model the ordering of items according to difficulty would be constrained to be the same for the two classes, here the easiest item is n. 2 for the first and n. 1 for the second class. The estimated probability for a subject to be in the second class, π_2, is 0.6516.

4 Implementation and relative performance

The algorithms have been implemented in MATLAB. The M-step of the EM algorithm is performed via the Iteratively Weighted Least Square (IWLS, McCullagh & Nelder, 1989); the same algorithm has also been used in both the accelerators. For the QN1 algorithm, having noticed that the log-likelihood could occasionally decrease in some steps, we inserted a univariate search of the kind used by Jamshidian & Jennrich (1997) for QN2. This search consists in finding, via a Newton-Raphson method, the value of α which maximises

$l(\boldsymbol{\beta}+\alpha\Delta\boldsymbol{\beta})$. This modification forces the log-likelihood not to decrease. Both accelerators have been further modified so that the actual step may be either a plain EM step or an accelerated one depending on which gives the largest increase in the log-likelihood.

Our starting values are based on a preliminary rule for splitting y_r, the number of subject with response pattern $\boldsymbol{r}$, among the latent classes. In our implementation we assume that latent classes are ordered from less to most capable and that the larger the total number of correct answers $\sum r_j$, the larger the probability of belonging to a more capable latent class. This rule combined with a slight random perturbation seems to work very well in the examples we have analysed.

To evaluate the relative efficiency of the various algorithms, we first tried to find a threshold which was a reasonably close approximation from above to the deviance at the m.l.e. (of the order, say, 10^{-4}). This is a compromise in the sense that requiring an unnecessarily high level of accuracy would have been unfair to the EM algorithm which is reasonably fast at getting near the m.l.e. but very slow in the final stages.

Results on the computing time of the various algorithms relative to the computing time of the EM algorithm, are given in the table below. In addition to the data in the example above denoted as A, we simulated a 6 item test on 500 subjects by a Rash model. This is denoted as B_2 and B_3 respectively according to whether the fitted model has 2 or 3 latent classes. To reduce dependence on the starting value, these times are the average in 10 trials.

Algorithm	A	B_2	B_3
QN1	1/15	1/8	1/10
QN2	1/30	1/8	1/10
Fisher-scoring	1/50	1/15	1/14

5 Discussion

The experiments above seem to indicate that all the alternatives to the EM algorithms can improve the computing time considerably. However, further investigations would be necessary to assess whether these improvements depend on the number of items and the number of assumed latent classes.

For instance the advantage of QN2 relative to QN1 seems to vanish with a larger number of items. Moreover our experience is that QN1 is more reliable and less sensitive to starting values.

Clearly the EM algorithm is the less sensitive to starting values. On the other hand, the Fisher-scoring algorithm is very vulnerable to inappropriate starting values in the sense that it is then likely to converge towards a point on the boundary of the parameter space. For the data sets analysed in this paper our starting rule always seems to provide good starting values even for the Fisher-scoring algorithm. However, we do not feel confident to recommend unsupervised use of this algorithm.

A final comment concerns the performance of the Iterative Proportional Fitting (IPF) relative to IWLS in the M step. Apparently the first one is slower than the second one when the number of classes is small, but it becomes competitive when this number increases.

References

Bartholomew, D.J. (1987). *Latent Variable Models and Factor Analysis.* New York: Oxford University Press.

Dempster, A.P., Laird, N.M. & Rubin, D.B. (1977). Maximum Likelihood from Incomplete Data via the EM Algorithm. *Journal of the Royal Statistical Society Series B*, **39**, 1-22.

Heinen, T. (1996). *Latent Class and Discrete Latent Trait Models; Similarities and Differences.* Thousand Oakes, CA: Sage.

Jamshidian, M. & Jennrich, R.J. (1997). Acceleration of the EM Algorithm by using Quasi-Newton Methods. *Journal of the Royal Statistical Society Series B*, **59**, 569-587.

Lang, J.B. (1992). Obtaining the observed information matrix for the Poisson log linear model with incomplete data. *Biometrika*, **79**, 405-407.

Lindsay, B., Clogg, C.C. & Grego, J. (1991). Semiparametric Estimation in the Rash Model and Related Exponential Response Models, Including a Simple Latent Class Model for Item Analysis. *Journal of the American Statistical Association*, **86**, 96-107.

McCullagh, P. & Nelder, J.A. (1989). *Generalized Linear Models: Second Edition.* London: Chapman and Hall.

Rasch, G. (1961). On General Laws and the Meaning of Measurement in Psychology. *Proceedings of the IV Berkeley Symposium on Mathematical Statistics and Probability*, **4**, 321-333.

Wu, C.F.J. (1983). On the convergence properties of the EM algorithm. *Annals of Statistics*, **11**, 95-103.

Gröbner Basis Methods in Polynomial Modelling

R.A. Bates, B. Giglio, E. Riccomagno and H.P. Wynn

Department of Statistics, University of Warwick, Coventry, CV4 7AL, UK

1 Introduction

The Gröbner basis (G-basis) method in the design of experiments was introduced by Pistone & Wynn (1996) and followed up by several strands of work one in particular addressing real practical applications: Holliday, Pistone, Riccomagno & Wynn (1997). This paper continues this latter series.

The authors are particularly interested in complex experimental design and modelling situations which arise in industry. Part of this work arises in computer experiments under an EU grant, Brite-Euram III: $(CE)^2$, Computer Experiments for Concurrent Engineering. The objective of that project is to use statistical models to "emulate" large scale computer code which is typically expensive, in computer time, to run.

These applications have common aspects. The experimental designs are often non-standard. Sometimes the industrial partners are *discouraged* by the authors from using standard factorial designs but encouraged towards space-filling designs which allow more complex modelling. For such situations the G-basis methods have been found to be useful.

The basic output from a G-basis run is a saturated set of modelling terms for a polynomial model guaranteed to be estimable. That is to say the regression model is guaranteed to be non-singular. The suggestion is that the method is combined with some new strategies for modelling which exploit this benefit. In addition, certain more general considerations arise. The method exposes the possibility, clearly already known from factorial design methods, that there are several polynomial models fitting the same data. In computer experiments where, theoretically at least, there are no errors, this means that there are several different models *interpolating* the same data.

2 The G-basis method

Since the G-basis method is explained at some length in the other papers mentioned we only give a short description.

The main idea is to consider an experimental design in m factors as a zero-dimensional polynomial ideals. That is to say the design points are considered as the zeros of a set of polynomial equations.

The algebraic setting is the ring of all polynomials in m indeterminates where each indeterminate corresponds to a statistical factor and every polynomial is associated to the deterministic part of a linear regression statistical model. That is the models are of the form $Y(x) = \mu(x) + \epsilon(x)$ where μ is the deterministic part of the model and ϵ is a centered random error. The mean, μ is a polynomial determined with the G-basis method.

A design is represented by the ideal of all polynomials interpolating the design points. We call this the *design ideal.* Such an ideal is described by a finite basis (see Hilbert's basis theorem in Cox, Little & O'Shea, 1997), that is a system of polynomial equations whose solution set is given by the design points. Different systems of equations can have the same design as a solution set. That is, there are different bases of the same ideal. G-bases are particular bases of an ideal depending on a so-called term-ordering. A term-ordering is a total ordering on the set of all monomials. A monomial corresponds either to a linear term in the statistical model or to an interaction. The choice of a term-ordering allows us to define the leading term of a polynomial as the largest element in the polynomial with respect to the term-ordering. Among the properties of a term-ordering we recall that of satisfying the divisibility condition: for the following monomials $x^\alpha, x^\beta, x^\gamma$ if $x^\alpha < x^\beta$ then $x^\alpha x^\gamma < x^\beta x^\gamma$. In this paper we mainly use the `tdeg` term-ordering and the `plex` term-ordering. For the technical definitions we refer to Cox, Little & O'Shea. Here we simply note the following. The `plex` term-ordering is the lexicographic term-ordering over an initial ordering on the factors, so that one factor is always favoured over the others. The `tdeg` term-ordering takes in account the total degree of a monomials, for example $x_1 x^2 <_{\texttt{tdeg}} x_1^2 x_2^2$, and compares monomials of the same total degree in a lexicographic manner $x_1 x_2^2 <_{\texttt{tdeg}} x_1^2 x_2$ when we assume $x_2 < x_1$.

As an example of design ideal consider the design union of the 2^2-full factorial design at levels ± 1 and the central point $(0,0)$. The (reduced) G-basis of the design ideal with respect to the $\texttt{tdeg}(x_1 > x_2)$ is given by the system formed by the three polynomials $x_1^2 - x_2^2 = 0$, $x_2^3 - x_2 = 0$ and $x_1 x_2^2 - x_1 = 0$. The definition of a G-basis is as follows. The set $G = \{g_i : i = 1, \ldots, t\}$ is a G-basis for the ideal I with respect to the term-ordering, τ, if and only if the ideal generated by the leading terms of the g_i's $(i = 1, \ldots, t)$ is equal to the ideal generated by the leading terms of the elements of I.

After defining the G-basis for the design ideal I we consider the quotient space of all polynomials modulo the design ideal. Such a space is a vector space and is spanned over the coefficient set by the monomials that are not divided by the leading terms of the G-basis of I. This vector space basis is called Est and is such that the design matrix for the design points and for the monomials in Est is invertible. Notice that Est has as many elements as there are distinct design points. In the previous example the leading terms are x_1^2, x_2^3 and $x_1 x_2^2$ and thus we have $Est = \{1, x_1, x_2, x_2^2, x_1 x_2\}$.

Given a polynomial (model) p we can decompose it over the G-basis $G = \{g_j, j = 1, \ldots, t\}$ of a design ideal as

$$p(x) = \sum_{j=1}^{t} s_j(x) g_j(x) + r(x)$$

where the support of $r(x)$ (that is the monomials in $r(x)$) is a subset of Est. The above division operation can be performed with respect to any set of polynomials, F. But when F is a G-basis, G, the remainder r is unique given the term-ordering, τ.

3 Interpolation: the non-error case

The traditional study of confounding or aliasing depends on the underlying algebra and is essentially independent of distributional assumptions on the errors. In Pistone & Wynn (1996) it was pointed out that the G-basis method addresses the general problem of confounding. Two polynomials are confounded if they interpolate the same data. The G-basis method gives some kind of partial classification of such cases, as follows. Different monomial orderings, in general, give different models, *Est*. Since these models are saturated they are exact interpolators. Hence we will have two interpolators for the same data.

Here is a simple artificial example. In two dimensions consider the following design $D = \{(0,1),(1,4),(2,2),(3,6),(4,0),(5,5),(6,3)\}$ and the output variable $Y = (3.2, 4.3, 5.2, 1.2, 3.4, 5.4, 8.0)$. The τ_1=$\mathtt{tdeg}(x_1 < x_2)$ term-ordering gives the following polynomial model

$$Est_{\tau_1} = \{1, x_1, x_2, x_1^2, x_1x_2, x_2^2, x_1^3\}$$

with $\hat{\theta}_{\tau_1} = (0.7534, -0.3784, 2.9606, 0.4615, -0.0262, -0.5140, 0.0504)$ while τ_2=$\mathtt{tdeg}(x_2 < x_1)$ gives

$$Est_{\tau_2} = \{1, x_2, x_1, x_2^2, x_1x_2, x_1^2, x_2^3\}$$

with $\hat{\theta}_{\tau_2} = (1.6219, 1.8346, 0.4033, -0.2326, 0.0458, 0.0103, -0.0240)$.

The issue, now, is to choose between competing interpolators. This issue is real in large computer experiments in which the emulators really do, typically, interpolate the simulators they are modelling at the selected input sites. Smoothness is a criterion which has typically been used to choose between different models and the suggestion here is based on a simple such measure.

Let $Est_\tau = \{x^\alpha : \alpha \in L\}$ be the model derived under a particular monomial ordering τ by the G-basis method. Here $\#(L) = n$ (where n is the sample size) and we are in the exact interpolation case. For a data vector, Y, and design, D, define the design matrix as $X = \{x^\alpha\}_{x \in D, \alpha \in L}$. Then the fitted model is $p(x) = \sum_{\alpha \in L} \hat{\theta}_\alpha x^\alpha$ where $\theta = \{\theta_\alpha\}_{\alpha \in L} = X^{-1}Y$, where vectors and matrices have been compatibly indexed. The proposition is to use a simple measure of smoothness at each point. Thus define the Hessian of $p(x)$ as $H = \{\frac{\partial^2 p(x)}{\partial x_i \partial x_j}\}$ and an overall measure of curvature at x: $trace(H^\alpha)^{\frac{1}{\alpha}}$, where typically we have $\alpha = 2$, in which case $trace(H^2) = \sum_{i,j}(\frac{\partial^2 p(x)}{\partial x_i \partial x_j})^2$. The measure of smoothness used here is, then,

$$\phi = (\sum_{x \in D} trace(H^2))^{\frac{1}{2}}$$

This is simpler than the more standard form in smoothing theory when $\sum_{x \in D}$ is replaced by the integral over a design region. It takes advantage of the use of space filling experimental designs which cover the design region more uniformly. The computations are straightforward for

$$\frac{\partial^2 p(x)}{\partial x_i \partial x_j} = \sum_{\alpha \in L} \hat{\theta}_\alpha \frac{\partial^2 x^\alpha}{\partial x_i \partial x_j}$$

The result is that ϕ^2 is a quadratic form in $\hat{\theta}$: $\phi^2 = \hat{\theta}^T Q_1 \hat{\theta}$ where Q_1 is a nonnegative definite matrix which only depends on the design and the term-ordering. For the above example we have respectively $\phi_{\tau_1} = 3.1574$ and $\phi_{\tau_2} = 2.4995$. The strategy is in simple form to consider the value of ϕ as τ ranges over a range of monomial orderings and pick the interpolator which minimizes ϕ.

Table 1. A 3-factor 16-point design

p	500	500	500	800	200	500	500	500	350	350	650	650
b	396	403	404	402	402	645	151	405	248	550	249	555
a	8	13	3	8	8	8	8	8	5	5	5	5
Y	271.4	268.9	282.8	266.2	297.5	295.1	262.2	269.4	274.8	291.1	266.9	285.4

p	350	350	650	650
b	252	551	254	550
a	11	11	11	11
Y	261	276.5	263.1	280.4

4 Modelling strategies

Several different kinds of modelling strategies are available and we discuss one which combines measures of smoothness with more standard procedures.

In the standard regression case a wide variety of methods is available. An important class is covered by model choice criteria such as Mallow's C_p or the Akaike criterion (AIC).

The G-basis method can be combined with these methods using one or more of the following

1. prior screening with a model with all linear terms: $\theta_0 + \sum_i \theta_i x_i$
2. use of Step 1. above (or another method) to suggest an initial ordering for the G-basis method
3. running stepwise regression based on the starting Est_τ from the G-basis method
4. using a combination of a smoothing criteria, ϕ and a model order choice criterion.
5. evaluation of the criteria in Step 4. over a wide choice of different τ (and hence Est).

A model Est_τ has the so-called "order ideal" property. If a monomial term x^α is in Est_τ then the term x^β which divides x^α is also included: $\beta \leq \alpha$ in the usual entry-wise ordering. This has some benefits in cutting down on the number of possible submodels and to some extent reflects statistical practice. In Caboara, Pistone, Riccomagno & Wynn (1997) the collection of all Est_τ for a particular design D as τ ranges over all monomials is referred to as the "fan" of the design and individual Est_τ called "leaves" (see also Mora & Robbiano, 1988). This lead to the use of the letter L above.

Despite the restriction to order ideals the class of possible models is large. It is useful to restrict stepwise regression to models of this kind but a special "walk"on leaves needs to be performed to search the fan in a sensible way. This can either be done *(i)* algebraically on the set of all monomial ordering τ or *(ii)* by simply moving between order ideal models —whether or not they are obtainable as an Est_τ for some τ.

5 Case study

This case study is part of a much larger engine modelling exercise and used only for illustration. The data consist of an output at a particular load/speed setting. We consider the design in the three factors a, b, p with 16 distinct design points given in Table 1. The last row of Table 1 gives the output, Y.

Table 2. ANOVA table and t values

	Value	Std. Error	t value	$Pr(>\|t\|)$
1	275.7937	1.8023	153.0219	0.0000
ap^2	−75.9247	16.2835	−4.6627	0.0007
p^3	32.6955	6.6669	4.9042	0.0005
ba	−4.9431	2.6133	−1.8915	0.0852
ap	53.7587	12.2695	4.3815	0.0011

Residual standard error: 7.209 on 11 degrees of freedom
Multiple R-Squared: 0.7259
F-statistic: 7.283 on 4 and 11 degrees of freedom, the p-value is 0.004043
Scale factor: 60

The following two interpolators are obtained with the G-basis method with respect to the plex and the tdeg term-ordering respectively and the same initial ordering $p > b > a$

$$Est_{\texttt{plex}} = \{1, a, a^2, a^3, a^4, b, ba, ba^2, b^2, b^2a, b^2a^2, b^3, b^3a, b^3a^2, b^4, p\}$$
$$Est_{\texttt{tdeg}} = \{1, a, b, p, a^2, ba, pa, b^2, pb, p^2, a^3, ba^2, pa^2, b^2a, pba, b^3\}$$

Notice that the plex model is of fourth degree in a and b and includes p only in the linear term. The tdeg model is of third degree in a and b and includes all the linear and second order terms. These model structures are typical of tdeg and plex, that is tdeg includes first all the lower order terms and plex includes first as many as possible powers of the lower factor. The two saturated models differ on four terms. These two monomial orderings are taken for illustration, more general strategies are mentioned in Section 6.

The next step is to fit the two models to the data, Y, in Table 1. We obtain the following two exact interpolators

$$\begin{aligned} Fit_{\texttt{plex}} &= 246714.0 + 21349.93a - 12034.73a^2 + 1356.805a^3 - 42.52327a^4 \\ &-3170.615b + 260.2529ba - 17.46151ba^2 + 10.98772b^2 \\ &-.3984503b^2a + .02822986b^2a^2 - .01743636b^3 + .000120693b^3 \\ &a - .000010456886b^3a^2 + .000010735636b^4 - .05216667p \\ Fit_{\texttt{tdeg}} &= -5143.543 + 1240.072a + 20.44767b + .06642581p - 103.1794a^2 \\ &-2.462621ba - .08776544pa - .02739751b^2 + .0003155914pb \\ &+.0001309597p^2 + 3.873335a^3 + .01785048ba^2 + .006525472pa^2 \\ &+.0027319016b^2a - .00002858413pba + .0000047474536b^3 \end{aligned}$$

Next we compute the curvatures of the above two fits (in the non-error case) that is the ϕ values in Section 3. They are $\phi_{\texttt{plex}} \simeq 17374.9$ and $\phi_{\texttt{tdeg}} \simeq 257.6$ respectively. From this result we choose the tdeg model. Next we perform a step-wise regression over the chosen model and reduce the model to the following five terms

$$1, ap^2, p^3, ba, ap \tag{1}$$

The ANOVA table based on Splus is given in Table 2. To allow model parameters to be compared roughly we centered and scaled the factors. (Note the choice of the scale factor in the Splus routine AIC is 60). The fit for model (1) is

$$245.5918 - 0.0001ap^2 + 0.0001p^3 - 0.0026ba + 0.0327ap$$

and its curvature is $\phi \simeq 1.3$.

6 Advanced implementation issues

In the previous sections we suggested selecting models obtained by the algebraic procedure for identifiability by choosing the term-ordering empirically. A total degree term-ordering such as `tdeg` favours lower order terms, while a lexicographic term-ordering, such as `plex`, favours the first indeterminates. It is possible to consider the models obtained by varying all the term-orderings and thus to look at all the models that the design identifies according to the G-basis method, namely the full fan. A recent algorithm to move between leaves of a fan using linear algebra techniques is due to Collart, Kalkbrener & Mall (1997) and is called *Gröbner walk.* They consider two different term-orderings and move on a path between them by updating the G-basis. Theoretically it is possible, but computationally infeasible, to compute the full fan of a given design.

Alternatively, without using G-basis techniques one could consider all the models with an order ideal structure (that is satisfying the divisibility condition) and with as many terms as design points. This step is computationally as prohibitive as the computation of the full fan.

A third option, favoured by the authors, is to work in the space of models which are order ideals switching between identifiable models evaluating the model choice/smoothness criteria at each iteration using a global search procedure. The relation between changing the model rather than the term-ordering is the subject of current research with an aim of unifying the approaches and speeding computation.

References

Caboara, M., Pistone, G., Riccomagno, E. & Wynn, H. P. (1997). The fan of an experimental design. *Annals of Statistics* (submitted).

Caboara, M. & Robbiano, L. (1997) Families of Ideals in Statistics *Proc. ISSAC '97*, ACM, 404-417.

Collart, S., Kalkbrener, M. & Mall, D. (1997) Converting bases with the Gröbner walk *J. Symb. Comput.*, **24**, 465-469.

D. Cox, J. Little & D. O'Shea (1997). *Ideal, Varieties, and Algorithms*, second edition. New York: Springer-Verlag.

Holliday, T., Pistone, T., Riccomagno & Wynn H. P. (1997). The Application of Algebraic Geometry to the Analysis of Designed Experiments: a case study. *Computational Statistics*, in press.

Mora, T. & Robbiano, L. (1988). The Gröbner fan of an ideal *J. Symb. Comput.*, **6**, 183-208.

Pistone, G. & Wynn, H. P. (1996). Generalised confounding with Gröbner bases. *Biometrika*, **83**, 653-666.

Analysis of High Dimensional Data from Intensive Care Medicine

Marcus Bauer[1], Ursula Gather[1] and Michael Imhoff[2]

[1] Department of Statistics, University of Dortmund, D-44221 Dortmund, Germany

[2] Surgical Department, Community Hospital, Beurhausstrasse 40, D-44137 Dortmund, Germany

Abstract. As high dimensional data occur as a rule rather than an exception in critical care today, it is of utmost importance to improve acquisition, storage, modelling, and analysis of medical data, which appears feasable only with the help of bedside computers. The use of clinical information systems offers new perspectives of data recording and also causes a new challenge for statistical methodology. A graphical approach for analysing patterns in statistical time series from online monitoring systems in intensive care is proposed here as an example of a simple univariate method, which contains the possibility of a multivariate extension and which can be combined with procedures for dimension reduction.

Keywords. Clinical information systems, decision support, high dimensional time series, online monitoring, phase space reconstruction

1 Introduction

Increasing technical possibilities in online recording of complex data structures produce manifold challenges for statistical methods. For instance, the use of clinical information systems (CIS) in intensive care medicine makes it possible to report online, simultaneously, and automatically up to 2000 physiological variables, laboratory data, device parameters etc. Even senior physicians may not be able to develop a systematic response to any problem involving more than seven variables (Miller, 1956). To allow for a more differentiated approach to therapy and computer aided clinical decision making, it seems necessary to develop tools for a suitable bedside decision support.

The patient data are multivariate time series. Thus modelling and analysis of the underlying dynamic is a central task, which should be solved to obtain tools for a decision support. In recent years much progress has been made in multivariate time series analysis, but for really high dimensional data, such as intensive care data, traditional multivariate time series methods fail because of the so-called "curse of dimensionality" (Friedmann, 1994).

Therefore, we see the necessity of new methodological approaches, which are able to extract the important information from high dimensional data and which work automatically and with fast algorithms. One should not underestimate or disregard that the results of a statistical online analysis have to be readable in an easy manner, such that physicians and nurses are able to recognize ad hoc the extracted information on the state of the patient.

In Section 2 we describe the data acquisition and storage with a CIS. The resulting demands for statistical methodology are formulated in Section 3 and a simple procedure for a graphical analysis of time series data as a tool for an ad hoc decision support is proposed in Section 4.

2 Data acquisition and storage

As the patient record is one of the most important tools in intensive care therapy, one has to give attention to acquisition and storage of these records. More and more devices with integrated microprocessors are in use at surgical intensive care units for monitoring patients and for therapeutic interventions. The use of CIS is unavoidable for processing the enormous data floods. In a clinical evaluation carried out at a major German surgical intensive care unit, a clinical information system was run for six years. Experience has shown that a well configured and well maintained CIS improves dramatically the quality of therapy and care (Imhoff, 1992, 1995).

The CIS is based on a network of autonomous Unix workstations, one for each bed. Bedside devices (such as monitors, ventilators, etc.) are connected locally via serial interfaces. All patient data are stored on the local hard disk at the bedside and simultaneously mirrored onto a second workstation within the network. An administrative data server is used for administration of the network and the CIS, and may serve as a communication hub with central data services like the Hospital Information System. For undisturbed data analysis the patient record is transferred into a secondary SQL (Sybase SQL server) and exported into standard statistical software (SPSS, SAS).

Thus, it is guaranteed that most of the data relevant for patient monitoring is recorded in a regular, reliable, and correct way and hence the technical requirements for using statistical methods or tools for bedside decision support are fulfilled.

3 Challenges for statistical methods

The necessity of systematic research on statistical analysis of complex data structures has been repeatedly pointed out during the past two decades (see for example Tukey, 1977; Michie, 1994). Modern statistical methods and new areas like neural networks, statistical assistant systems, projection pursuit, "data mining" etc. are concerned with modelling, analysing and visualizing complex data structures. The need for appropriate methods and the existence of some of them is partly due to the increasing speed and capacity of computers. But in situations with really high dimensional data, which we often find in the lifesciences, the usefulness of the above-mentioned procedures is limited because the computational effort exceeds any possible computational power (Huber, 1993). Another fundamental problem is that in order to fill the highdimensional sample space one needs very large sample sizes, which are seldom given in praxis.

This is especially true in the context of intensive care medicine. Here the challenge for statistical methods is to develop new types of methods for data analysis, covering the following features:

- ability to deal with multivariate / high dimensional time series
- allowing for individual patient monitoring
- designed for online-monitoring data
- ability of pattern identification
- implementation with fast algorithms
- allowing for simple interpretation.

To derive such methods, it is necessary to combine parsimonious model building with (automatic) procedures for dimension reduction on the existing computational basis. Another aspect, which has to be considered is robustness. (In intensive care medicine this is especially important for pattern identification.)

4 A new approach to pattern recognition for physiological variables from online monitoring systems

To fix ideas, let us be concerned in the following with the special task of analysing univariate intensive care online-monitoring data. We give an example of a simple univariate method, which satisfies the principles of parsimony, robustness and an ad hoc visualization and interpretation of the results.

One basic purpose of clinical monitoring of patients is to develop tools for automatic detection of qualitative patterns like outliers, level changes and trends in physiologic data. Different mathematical approaches exist, in particular in the framework of statistical time series analysis, and have been implemented in experimental and commercial software packages today. Mainly two approaches are pursued, i.e. procedures based on ARIMA-models (batch processing), see for example Imhoff *et al.* (1997), and state space models (sequential processing) following Smith & West (1983) as well as Daumer *et al.* (1996). It is not worked out yet how we can profit from these approaches to get a basis for future bedside online multivariate time series analysis. The reason for this is that existing methods have actually been constructed for monitoring univariate variables and their multivariate extensions fail because of the "curse of dimensionality". Further, the possibilities for online-monitoring are partly restricted.

Here, we discuss a new graphical approach for pattern recognition in univariate time series based on phase space reconstruction. Let $\{y_t\}_{t\in\{1,\dots,N\}}$ be a time series. Takens (1981) considered the set of m-dimensional vectors, the

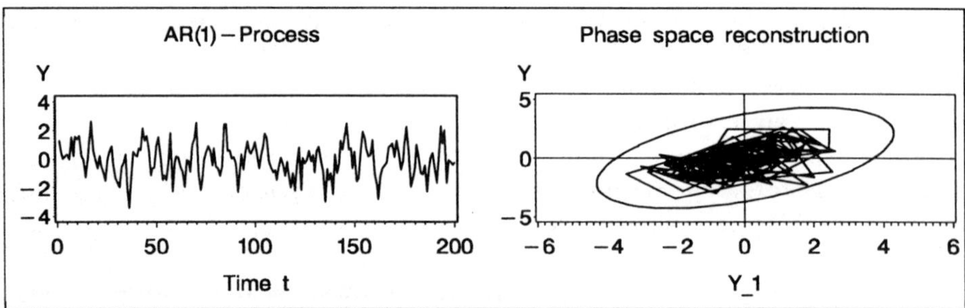

Fig. 1. Realization of an AR(1)-process $Y_t = 0.5Y_{t-1} + \epsilon_t$, $\epsilon_t \sim N(0,1)$, $t \in I\!N$ and its phase space reconstruction; the phase space vectors form an elliptic cloud

components of which are the time delayed observations of this time series:

$$\mathbf{y}_t := (y_{t+(m-1)T}, \ldots, y_{t+2T}, y_{t+T}, y_t)', \quad \mathbf{y}_t \in I\!R^m$$

with $T, m \in I\!N \backslash \{0\}$, and $t = 1, \ldots, N - (m-1)T$. The time delay is denoted by T and m is called the embedding dimension. Thus, the univariate time series is transformed into an m-dimensional space, the so-called phase space. The set $\{\mathbf{y}_t \mid t = 1, \ldots, N - (m-1)T\}$ forms the phase space reconstruction. The phase space retains the properties of the state space, the axis of which are all variables, which characterize the dynamic. A mathematical justification of this approach is given in Takens (1981).

Analytical methods based on phase space reconstruction, which have been developed in theoretical physics to find properties of nonlinear dynamics, assume large sample sizes. Furthermore the data come from carefully controlled physical experiments. In the analysis of biological and ecological systems, we often have small sample sizes and random errors and moreover empirical data rather than data from controlled experiments. Thus the exact topological results of Takens (1981) are not longer valid for stochastic systems.

Nevertheless the concept of phase space reconstruction can be used for stochastic processes. In Figure 1 an observation of an AR(1)-process with Gaussian error terms and its 2-dimensional phase space reconstruction ($m = 2$) is depicted for $T = 1$. The chronological observations are combined in order to show the movement through space. The dependence structure can be clearly recognized by the elliptic form of the vector cloud. Typical disturbances of a time series like outliers, level shifts and trends can be visualized by phase space reconstructions, too. In Figure 2a an outlier is inserted in a simulated AR(1)-process and in Figure 2c a level change at time point 112 is added. The outlier at time point 152 arises in the phase space vectors $\mathbf{y}_t$ and $\mathbf{y}_{t+1}$, such that these vectors extrude from the regular observations (Figure 2b). Similar, all observations occuring after the level shift lie outside the original ellipse and form a new one (Figure 2d). Such features are often found in variables of intensive care data, for typical examples see Figure 3.

First attempts to model and monitor linear physiologic time series with

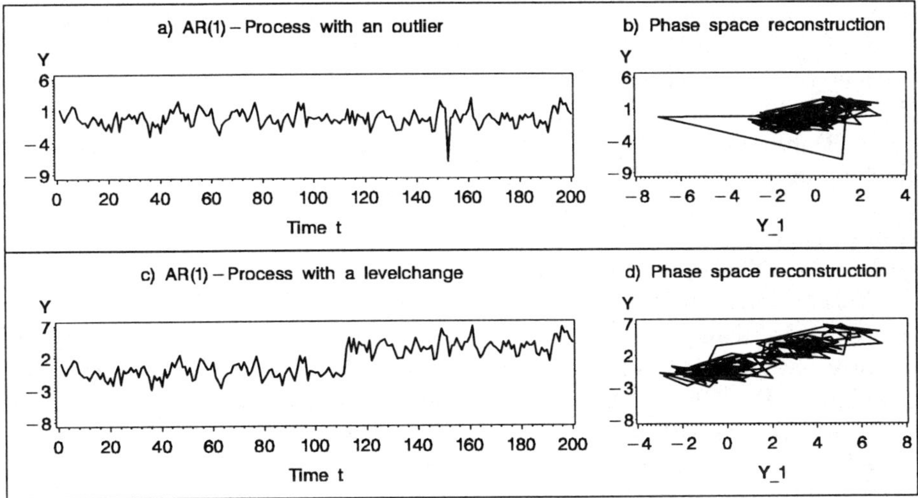

Fig. 2. Simulated time series and phase space reconstruction of an AR(1)-process with outlier and with level shift

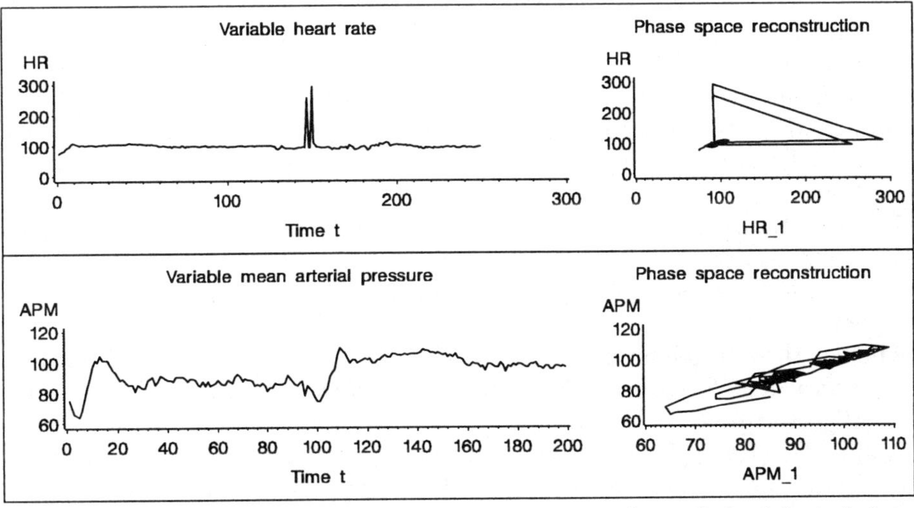

Fig. 3. Typical time series and phase space reconstructions of physiological data, variables 'heart rate' and 'mean arterial pressure'

phase space reconstructions are done by Bauer (1997), proposing a robust and automatic procedure with low computational effort for identification of special patterns in time series. Although, up to now, this pattern identification procedure has been succesfully applied to controlled clinical studies, the general use in automatic online monitoring remains for future developments. A generalization to multivariate time series and an extension in connection with procedures for dimension reduction in the situation of high dimensional data is under current research. The combination of this method with procedures for dimension reduction is necessary. A detailed analysis of the dependencies between the observed variables is a further fundamental task when modelling high dimensional data. Here the use of graphical models for multivariate time series (Dahlhaus, 1996) or the combination of methods from statistics and machine learning (Morik *et al.*, 1994) seem promising.

References

Bauer, M. (1997). *Identification of outliers and interventions in online monitoring data.* Ph.D. (in German), University of Dortmund.

Daumer, M., Falk, M. & Beyer, U. (1996). Online monitoring using multiprocess Kalman filtering. Discussion paper 54 of SFB 386, TU Munich.

Dahlhaus, R. (1996). *Graphical interaction models for multivariate time series.* Technical Report, University of Heidelberg.

Friedmann, J.H. (1994). An overview of predictive learning and function approximation. In: *From Statistics to Neural Networks* (ed. V. Cherkassky, J.H. Friedmann & H. Wechsler), 1-61. Berlin: Springer.

Huber, P.J. (1993). Projection pursuit and robustness. In: *New Directions in Statistical Data Analysis and Robustness* (ed. S. Morgenthaler, E.M. Ronchetti & W.A. Stahel), 139-146. Basel: Birkhäuser.

Imhoff, M. (1992). Acquisition of ICU data: concepts and demands. *International Journal of Clinical Monitoring and Computing*, **9**, 229-237.

Imhoff, M. (1995). A clinical information system (CIS) in intensive care: how to make it work? *8th European Congress on Intensive Care Medicine*, Athens.

Imhoff, M., Bauer, M., Gather, U. & Löhlein, D. (1997). Statistical pattern detection in univariate time series of intensive care online-monitoring data. *10th European Congress on Intensive Care Medicine*, Paris, September 1997.

Michie, D., Spiegelhalter, D.J. & Taylor, C.C. (1994). *Machine Learning, Neural and Statistical Classification, Artificial Intelligence.* New York: Ellis Horwood.

Miller, G. (1956). The marginal number seven, plus or minus two: Some limits to our capacity for processing information. *Psychol. Rev.*, **63**, 81-97.

Morik, K., Potamias, G., Moustakis, V.S. & Charissis, G. (1994). Knowledgeable learning using MOBAL: a medical case study. *Applied Artificial Intelligence*, **8** (4), 579-592.

Smith, A.F.M. & West, M. (1983). Monitoring renal transplants: an application of the multi-process Kalman filter. *Biometrics*, **39**, 867-878.

Takens, F. (1981). Detecting strange attractors in turbulence. In: *Dynamic Systems and Turbulence* (ed. D.A. Rand & L.S. Young), 366-381. New York: Springer-Verlag.

Tukey, J.W. (1977). *Exploratory data analysis.* Reading, MA: Addison-Wesley.

Fitting Non-Gaussian Time Series Models

Michael A. Benjamin, Robert A. Rigby and Mikis D. Stasinopoulos

STORM Research Centre, University of North London, London N7 8DB, UK

Abstract. We consider a class of Generalized Autoregressive Moving Average (GARMA) models which extend the univariate Gaussian ARMA time series model to a flexible model for non-Gaussian time series data. Estimation of the model is carried out using an iteratively reweighted least squares algorithm. The model is demonstrated by its application to a time series data set.

Keywords. ARMA model, exponential family, GARCH model, generalized linear model, iteratively reweighted least squares, time series

1 Introduction

This paper considers the problem of extending Gaussian ARMA time series models to a non-Gaussian framework. Two approaches to this problem were originally described by Cox (1981) as observation driven and parameter driven models. Several methods have been used for the parameter driven (or state space) modelling approach (see Shephard & Pitt, 1997, for references). Typically these models tend to require complicated estimation techniques and occasionally crude approximations.

This paper focuses on observation driven models. In the models developed here, the conditional distribution of the dependant variable at time t, given the previous data is modelled by an Exponential Family distribution. Previously, Zeger & Qaqish (1988) developed autoregressive Exponential Family models, in particular autoregressive Poisson and Gamma models, while Li (1994) developed a moving average version of the model.

In this paper we extend the work of the authors Zeger & Qaqish (1988) and Li (1994) giving rise to Generalized Autoregressive Moving-Average (GARMA) Models. In Section 2 of this paper GARMA models are defined. Section 3 describes the model fitting algorithm and its implementation. Model comparison and inference is considered in Section 4.

The GARMA model can be used on a variety of time dependent responses which also have time dependent covariates. For example, count data with a conditional Poisson or Binomial distribution or continuous data with a conditional Gamma distribution (e.g. the volatility in a GARCH model).

The algorithm implemented here uses the GLIM statistical package although other packages with a weighted least squares algorithm such as S-PLUS could implement this fitting algorithm. Section 5 illustrates the methodology with an example. Conclusions are given in Section 6.

2 GARMA model definition

The conditional distribution for each observation y_t for $t = 1, \ldots, n$ given the previous information set $\boldsymbol{D}_t = \{\boldsymbol{x}_t, \ldots, \boldsymbol{x}_1, y_{t-1}, \ldots, y_1, \mu_{t-1}, \ldots, \mu_1\}$ is assumed to belong to the same Exponential Family, i.e.

$$f(y_t \mid \boldsymbol{D}_t) = exp\left\{\frac{y_t\theta_t - b(\theta_t)}{\varphi} + c(y_t, \varphi)\right\} \tag{1}$$

where θ_t and φ are the canonical and scale parameters respectively with $b(.)$ and $c(.)$ being specific functions which define the particular Exponential Family. The notation is the same as for Generalized Linear Models (GLMs) with independent observations (McCullagh & Nelder, 1989), but here conditional rather than marginal distributions are modelled.

As with standard GLMs the mean $\mu_t = E(y_t \mid \boldsymbol{D}_t) = b'(\theta_t)$ is related to the linear predictor, η_t, by a twice differenciable one-to-one monotonic function g which is called the link function (McCullagh & Nelder, 1989). Unlike the standard GLM linear predictor where $\eta_t = \boldsymbol{x}_t'\boldsymbol{\beta}$ and $\boldsymbol{\beta}' = (\beta_1, \beta_2, \ldots, \beta_r)$, here there is an additional component τ_t which allows autoregressive moving-average components to be included in the predictor. A general model for μ_t is given by $g(\mu_t) = \eta_t = \boldsymbol{x}_t'\boldsymbol{\beta} + \tau_t$, with

$$\tau_t = \sum_{j=1}^{p} \phi_j \mathcal{A}(y_{t-j}, \boldsymbol{X}_{t-j}, \boldsymbol{\beta}) + \sum_{j=1}^{q} \theta_j \mathcal{M}(y_{t-j}, \mu_{t-j}) \tag{2}$$

where $\mathcal{A}$ and $\mathcal{M}$ are functions representing the autoregressive and moving-average terms respectively, and $\boldsymbol{\phi}' = (\phi_1, \ldots, \phi_q)$ and $\boldsymbol{\theta}' = (\theta_1, \ldots, \theta_p)$ are the autoregressive and moving average parameters respectively. The moving-average error terms $\mathcal{M}$ could for example be deviance residuals, Pearson residuals, residuals measured on the original scale (i.e. $y_t - \mu_t$) or as below residuals on the predictor scale (i.e. $g(y_t) - \eta_t$).

Model (2) is too general for practical application, so here we take the following flexible and parsimonious submodel which includes many well known special cases,

$$g(\mu_t) = \eta_t = \boldsymbol{x}_t'\boldsymbol{\beta} + \sum_{j=1}^{p} \phi_j\{g(y_{t-j}) - \boldsymbol{x}_{t-j}'\beta\} + \sum_{j=1}^{q} \theta_j\{g(y_{t-j}) - \eta_{t-j}\} \tag{3}$$

Equations (1) and (3) together define the GARMA(p, q) model. The parameters $\boldsymbol{\beta}$, $\boldsymbol{\phi}$ and $\boldsymbol{\theta}$ are estimated by maximum likelihood as discussed in Section 4. For certain functions g it may be necessary to replace y_{t-j} with y_{t-j}^* in equation (3) to avoid the non-existence of $g(y_{t-j})$ for certain values of y_{t-j}. The form of y_{t-j}^* depends on the particular function g and is defined for specific cases in Section 3.

A Box-Cox power transformation (Box & Cox, 1964) may provide a suitable flexible form for g in equation (3), giving rise to a Box-Cox transformation of both sides (TBS) model. The optimal value for the power parameter λ may be obtained from its profile likelihood.

3 Estimation

The GARMA model fitting procedure performs maximum likelihood estimation (MLE) using iteratively reweighted least squares (IRLS).

Let the model parameters to be estimated be denoted $\boldsymbol{\gamma}' = (\boldsymbol{\beta}', \boldsymbol{\phi}', \boldsymbol{\theta}')$. These parameters are estimated by maximum likelihood. The log likelihood of the data $\{y_{m+1}, \ldots, y_n\}$ conditional on the first m observations $\{y_1, \ldots, y_m\}$ and on $\eta_t = g(y_t)$ for $t = 1, 2, \ldots, i$ where $i = \max(p, q)$ and $m \geq \max(p, q)$ is given by $l = \sum_{t=m+1}^{n} \log f(y_t | \boldsymbol{D}_t)$.

Hence from equation (1) the score function $U(\boldsymbol{\gamma}) = \frac{\partial l}{\partial \boldsymbol{\gamma}} = \left(\frac{\partial l}{\partial \boldsymbol{\beta}'}, \frac{\partial l}{\partial \boldsymbol{\phi}'}, \frac{\partial l}{\partial \boldsymbol{\theta}'}\right)'$ is given by

$$U(\boldsymbol{\gamma}) = \sum_{t=m+1}^{n} \frac{\partial \mu_t}{\partial \eta_t} \frac{\partial \eta_t}{\partial \boldsymbol{\gamma}} \frac{1}{v_t} (y_t - \mu_t) \tag{4}$$

where $v_t = Var(y_t \mid \boldsymbol{D}_t) = \varphi V(\mu_t) = \varphi b''(\theta_t)$, (McCullagh & Nelder, 1989). [Note that since the predictor is non linear in the parameters, $\frac{\partial \eta_t}{\partial \boldsymbol{\gamma}}$ is obtained recursively using (9) below.]

A Fisher Scoring Algorithm procedure is used to maximise the conditional log-likelihood function, l. The current estimate $\boldsymbol{\gamma}^k$ of $\boldsymbol{\gamma}$ is updated by the iterative procedure,

$$\boldsymbol{\gamma}^{(k+1)} = \boldsymbol{\gamma}^{(k)} + \alpha I^{-1}(\boldsymbol{\gamma}^{(k)}) U(\boldsymbol{\gamma}^{(k)}) \tag{5}$$

where $I(\boldsymbol{\gamma}) = -E\left(\frac{\partial^2 l}{\partial \boldsymbol{\gamma} \partial \boldsymbol{\gamma}'}\right) = \frac{1}{\varphi} \frac{\partial \eta'}{\partial \boldsymbol{\gamma}} \boldsymbol{W} \frac{\partial \eta}{\partial \boldsymbol{\gamma}'}$ denotes the Fisher information matrix, $\boldsymbol{W} = \text{diag}(w_t)$ where w_t is given in equation (7) below and $0 < \alpha \leq 1$ is the step length.

The score function $U(\boldsymbol{\gamma})$ and the Fisher information matrix $I(\boldsymbol{\gamma})$ are in the same form as those used in the Generalized Linear Model except that the usual explanatory variables $\boldsymbol{x}_j$ are replaced by $\frac{\partial \eta}{\partial \boldsymbol{\gamma}_j}$ for all j. Hence the fitting can be achieved using an adjusted dependent variable regression i.e. by iteratively reweighted least squares (IRLS), Green (1984).

The algorithm proceeds as follows:

(i) Given $\boldsymbol{\gamma}^{(k)}$ calculate a new $\boldsymbol{\eta}^{(k)}$, $\boldsymbol{\mu}^{(k)}$, $\left(\frac{\partial \boldsymbol{\eta}}{\partial \boldsymbol{\mu}}\right)^{(k)}$, $V(\boldsymbol{\mu})^{(k)}$, $\left(\frac{\partial \boldsymbol{\eta}}{\partial \boldsymbol{\gamma}}\right)^{(k)}$, $\boldsymbol{w}^{(k)}$ and $\boldsymbol{z}^{(k)}$, where the values $z_t^{(k)}$ and $w_t^{(k)}$ of the adjusted dependent variable $\boldsymbol{z}^{(k)}$ and weights $\boldsymbol{w}^{(k)}$ are constructed from the equations

$$z_t = \frac{\partial \eta_t}{\partial \boldsymbol{\gamma}'} \boldsymbol{\gamma} + \alpha (y_t - \mu_t) \frac{\partial \eta_t}{\partial \mu_t} \tag{6}$$

$$w_t^{-1} = \left(\frac{\partial \eta_t}{\partial \mu_t}\right)^2 V(\mu_t) \tag{7}$$

for $t = m+1, \ldots, n$.

(ii) Estimate $\boldsymbol{\gamma}^{(k+1)}$ by regressing $\boldsymbol{z}^{(k)}$ on $\left(\frac{\partial \boldsymbol{\eta}}{\partial \boldsymbol{\gamma}'}\right)^{(k)}$ with weights $\boldsymbol{w}^{(k)}$,

$$\boldsymbol{\gamma}^{(k+1)} = I^{-1}(\boldsymbol{\gamma}^{(k)}) \left(\frac{\partial \eta'}{\partial \boldsymbol{\gamma}}\right)^{(k)} \boldsymbol{w}^{(k)} \boldsymbol{z}^{(k)} \tag{8}$$

(iii) Update k to $k+1$ and repeat steps (i) and (ii) until the parameter estimates (or deviance) converges.

Within step (i) the predictor $\eta^{(k)}$ is calculated by *recursion* using equation (3) while the derivatives $\left(\frac{\partial\eta}{\partial\gamma}\right)^{(k)}$ are calculated with respect to the regression, autoregressive and moving average parameters respectively and are obtained from the following *recursive* equations for $t > i$,

$$
\begin{aligned}
\frac{\partial\eta_t}{\partial\beta_s} &= x_{ts} - \sum_{j=1}^{p}\phi_j x_{t-j,s} - \sum_{j=1}^{q}\theta_j\frac{\partial\eta_{t-j}}{\partial\beta_s} && \text{for } s = 1,2,\ldots,r \\
\frac{\partial\eta_t}{\partial\phi_s} &= \{g(y_{t-s}) - x'_{t-s}\beta\} - \sum_{j=1}^{q}\theta_j\frac{\partial\eta_{t-j}}{\partial\phi_s} && \text{for } s = 1,2,\ldots,p \qquad (9) \\
\frac{\partial\eta_t}{\partial\theta_s} &= \{g(y_{t-s}) - \eta_{t-s}\} - \sum_{j=1}^{q}\theta_j\frac{\partial\eta_{t-j}}{\partial\theta_s} && \text{for } s = 1,2,\ldots,q
\end{aligned}
$$

Note that when there are no moving average components no recursion is necessary to evaluate the new η and its partial derivatives in equations (3) and (9). When fitting moving average components, we need to fix initial values for η_t and its derivatives for $t = 1,2,\ldots,i$. Initial values for η_t are fixed by using $\eta_t = g(y_t)$ while initial values for the derivatives are taken to be zero, both for $t = 1,2,\ldots,i$.

4 Model inference

We can test between two nested GARMA models H_0 and H_1 with fitted deviances $\hat{D}_0/\varphi = -2\hat{l}_0$ and $\hat{D}_1/\varphi = -2\hat{l}_1$ and total number of fitted parameters κ_0 and κ_1 respectively. Assuming the Exponential Family scale parameter φ is known, the likelihood ratio statistic for testing between the models is $\Lambda = (\hat{D}_0 - \hat{D}_1)/\varphi$ which has an approximate Chi-squared distribution with $(\kappa_1 - \kappa_0)$ degrees of freedom under H_0. If φ is not known we base our test on $(\hat{D}_0 - \hat{D}_1/\hat{\varphi}(\kappa_1 - \kappa_0)$ which has an approximate F-distribution with $(\kappa_1 - \kappa_0, n - m - \kappa_1)$ degrees of freedom, where the scale papameter φ is estimated using the generalised Pearson statistic,

$$\hat{\varphi}_P = \frac{1}{n - m - \kappa_1}\sum_{t=m+1}^{n}(y_t - \hat{\mu}_t)^2/V(\hat{\mu}_t) \qquad (10)$$

5 Illustrative example

An important special case of the GARMA model is a reparameterised form of the Generalised Autoregressive Conditional Heteroscedastic (GARCH) model introduced by Bollerslev (1986). Assume $\{\epsilon_t, t = 1,2,\ldots,n\}$ is a process with conditional distribution $\epsilon_t|\boldsymbol{D}_t \sim N(0, h_t)$ where

$$h_t = \boldsymbol{x}'_t\boldsymbol{\beta} + \sum_{j=1}^{p}\phi_j(\epsilon^2_{t-j} - \boldsymbol{x}'_{t-j}\boldsymbol{\beta}) + \sum_{j=1}^{q}\theta_j(\epsilon^2_{t-j} - h_{t-j}) \qquad (11)$$

Now if we let $y_t = \epsilon_t^2$, then y_t has a conditional Gamma distribution, i.e.

$$y_t|\boldsymbol{D}_t \sim h_t\chi_1^2 \equiv Ga(\mu_t, \varphi) \tag{12}$$

with mean $\mu_t = E(y_t|\boldsymbol{D}_t) = h_t$ and fixed scale parameter $\varphi = 2$. Hence the model for y_t is a Gamma GARMA(p,q) model with identity function g and scale parameter φ. The above model defines a GARMA-GARCH(p,q) model. The family of GARMA-GARCH(p,q) models is a reparameterised form of the standard family of GARCH models.

Model (11) allows time dependent explanatory variables to also influence h_t, the conditional variance of ϵ_t. The standard GARCH model, however sets $\boldsymbol{x}_t'\boldsymbol{\beta} = \beta_0$ for all t. In which case the GARMA-GARCH parameterisation has the advantage that the autocorrelation function and partial autocorrelation function of the squared process $\{y_t\} = \{\epsilon_t^2\}$ cut of at lags q and p for the GARMA-GARCH$(0,q)$ and GARMA-GARCH$(p,0)$ respectively.

We can use the GARMA-GARCH model to model the volatility of the Standard and Poor 500 Index (SP_t) from January 1988 to December 1992. The data consisting of 1,265 observations is plotted in figure 1(a). The returns, r_t, are calculated by $10 \times \Delta \log SP_t$ and then centred about their sample mean, giving $\epsilon_t = r_t - \bar{r}_t$. The series volatility, $y_t = \epsilon_t^2$ was plotted against t in figure 1(b) and modelled by the GARMA-GARCH given by (11) and (12), with constant mean $\boldsymbol{x_t}'\boldsymbol{\beta} = \beta_0$ for all t.

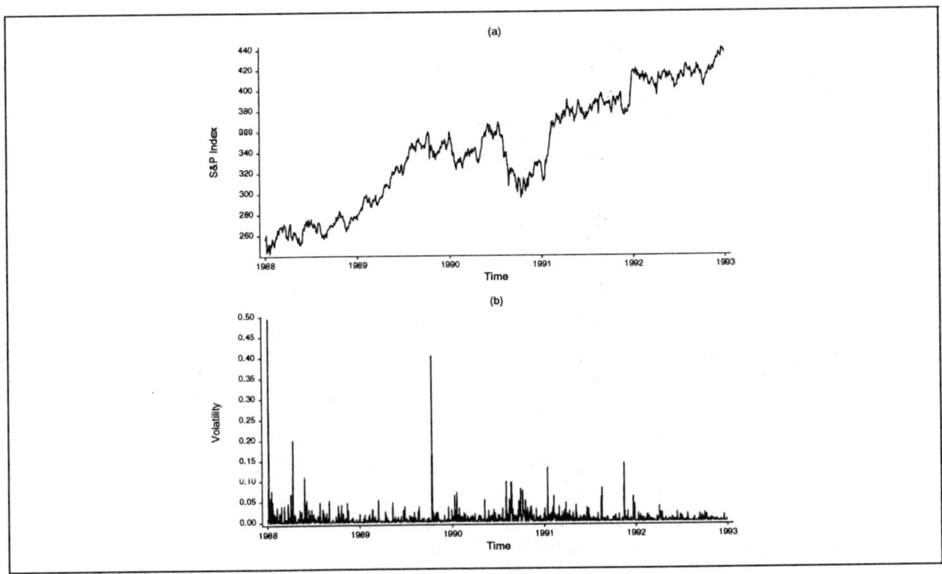

Fig. 1. Standard & Poor 500 Index 1988-1992: (a) S&P Index vs Time (b) Volatility vs Time

We consider choosing suitable values for p and q to eliminate the autocorrelation structure in y_t. The algorithm outlined in Section 3 was used to fit the GARMA-GARCH model to the data for $\max(p,q) \leq 3$ and the resulting fitted GLIM deviances, conditional on the first $m = 20$ observations, are shown in table 1.

	ARMA(p,q) Model			
	MA(0)	MA(1)	MA(2)	MA(3)
AR(0)	4154.8	4140.8	4135.0	4135.0
AR(1)	4140.2	4069.6	4069.4	4064.2
AR(2)	4134.4	4069.4	4033.5	4017.8
AR(3)	4134.4	4064.2	4018.0	4006.9

Table 1. Fitted GLIM deviances for GARMA-GARCH models

Using a fixed scale parameter of $\varphi = 2$ and a forward selection procedure using χ^2 tests for choosing between adjacent entries in the table led to a chosen model with order $p = q = 3$. However, as the value of $\hat{\varphi}$ calculated using the generalised Pearson statistic in (10) showed a departure from the value 2, F-tests described in Section 4 (with an estimate of the scale parameter $\hat{\varphi}_p$ from the $p = q = 3$ model) were used as a guide to fitted model selection leading to the choice of $p = q = 2$.

6 Conclusion

We have proposed a class of Generalised Autoregressive Moving Average (GARMA) models which extend univariate ARMA models to a non-Gaussian situation. The maximum likelihood fitting algorithm has been shown to be within the iteratively reweighted least squares (IRLS) framework. The fitted likelihood is used for model comparison.

Acknowledgements: This work was supported in part by a research grant from the Economic and Social Research Council.

References

Bollerslev, T. (1986). Generalised Autoregressive Conditional Heteroskedasticity. *Journal of Econometrics*, **31**, 304-327.

Box, G. E. P. & Cox, D. R. (1964). An Analysis of Transformations *Journal of the Royal Statistical Society*, **26**, 211-243.

Cox, D. R. (1981). Statistical Analysis of Time Series: Some Recent Developments. *Scandinavian Journal of Statistics*, **8**, 93-115.

Green, P. (1984). Iterative Reweighted Least Squares for Maximum Likelihood Estimation, and some Robust and Resistant Alternatives *Journal of the Royal Statistical Society, B*, **46**, 149-192.

Li, W. K. (1994). Time Series Models Based on Generalised Linear Models: Some Further Results. *Biometrics*, **50**, 506-511.

McCullagh, P. & Nelder, J.A. (1989). *Generalized Linear Models* (2nd ed.), London, Chapman and Hall.

Shephard, N. & Pitt, M. K. (1997). Likelihood Analysis of Non-Gaussian Measurement Time Series. *Biometrika*, **84**, 653-667.

Zeger, S. L. & Qaqish, B. (1988). Markov Regression Models for Time Series: A Quasi-likelihood Approach. *Biometrics*, **44**, 1019-1032.

Using Singular Value Decomposition in Non-Linear Regression

Philip Brain

IACR-Long Ashton Research Station, Department of Agricultural Sciences, University of Bristol, Long Ashton, Bristol BS41 9AF, UK

Abstract. The explicit form of a non-linear model may not be known. However, in some cases, it is known that there is an underlying non-linear relationship, which varies from individual to individual by means of differing scale parameters; when the form of the underlying relationship is known, this is known as a *parallel curve analysis*. This analysis can be extended to fit general functions (*general parallel curves*) that are only specified at a set of x-values. Such models can be fitted using Singular Value Decomposition. Two examples of the use of *general parallel curves* are presented. The first involves a study of wheat growth, the second, the detection of pesticide interactions.

Keywords. AMMI, SHMM, general parallel curves, synergism, growth curves

1 Introduction

Non-linear equations are widely used for describing results from biological experiments. These non-linear equations can be used to compare treatments, commonly using *parallel curve analysis* (as used in Genstat, Genstat committee, 1993). If we denote the jth observation of the dependent variable for the ith treatment by y_{ij} and assume that the responses with the corresponding independent variable are denoted by x_{ij}, the parallel curve model assumes that

$$y_{ij} = a_i + \sum_{k=1}^{r} b_{ik} f_k(x_{ij}; c) \qquad (1)$$

where, a_i and b_{ik} are linear parameters which vary with treatment i, $\mathbf{c}$ is a vector of non-linear parameters not varying with treatment, and $f_k(.)$ are a set of known functions.

The general framework above allows each treatment to be measured at a different set of x-values. However, it is often the case that the measurements are made on different treatments at the same set of values of the independent variable, so x_{ij} in equation (1) reduces to x_j, i.e. independent of treatment, and we will develop the theory assuming that this is so.

Sometimes the equation of the component response curves are not known, or we choose not to define them to detect deviations from current models. In either case the functions $f_k(.)$ are then not known explicitly. I define *generalised parallel curves* as unknown component response curves which are unaffected by treatment. The functions $f_k(.)$ in equation 1 are then defined only at the observed values of x_j, so can be represented in their general form as f_{jk}, (which corresponds to the value of the kth component function at x_j). Equation 1 then reduces to

$$y_{ij} = a_i + \sum_{k=1}^{r} b_{ik} f_{jk} \tag{2}$$

which is the equation for *generalised parallel curves.*

2 Estimation and inference

2.1 Estimating the parameters of the generalised parallel curve

If we assume that there are n treatments and each individual is measured at c values of x, the observed responses can then be stored in an $n \times c$ matrix $\boldsymbol{Y}$, where the jth column contains the observed responses for each treatment at the jth value of the independent variable, and the ith row contains the observed responses for the ith treatment at each value of the independent variable. If we further denote the $n \times r$ matrix of coefficients b_{ik} by $\boldsymbol{B}$, the $c \times r$ matrix of coefficients f_{jk} by $\boldsymbol{F}$, and the $n \times 1$ matrix of a's by $\boldsymbol{A}$, then equation 2 reduces to:

$$\boldsymbol{Y} = \boldsymbol{A}\,\boldsymbol{1}^T + \boldsymbol{B}\,\boldsymbol{F}^T$$

where, $\boldsymbol{1}$ is a $n \times 1$ matrix of 1's. This is an AMMI model (Gauch, 1988) and the parameters can be estimated using Singular Value Decomposition (SVD) on the suitably centred data matrix $\boldsymbol{Y}$.

2.2 Inference - deciding on the number of component curves

The full SVD analysis of $\boldsymbol{Y}$ (centred or uncentred) will explain all the variability but the data matrix can be approximated by one, two, or more components of the SVD, in a similar fashion to stepwise linear regression. However, testing the decrease in the residual variance in this way can give spurious results. Mandel (1971) showed that in the SVD of a random matrix the first component will explain a significant proportion of the variability if the degrees of freedom based on changes in the number of fitted parameters are used to calculate mean deviances. He estimated the effective degrees of freedom for a series of combinations and these can be used to test significances. The Mandel d.f. are used for testing the number of components required.

2.3 Detecting lack of fit of specified functions

Once the number of components has been decided, it is possible to estimate the variance, using the residual sum of squares after fitting these components, together with the appropriate degrees of freedom. If specific functions are now fitted, it is possible to detect lack of fit by testing the increase in the residual variance, as compared with fitting the general parallel curves.

3 Examples of the application of generalised parallel curves

3.1 Analysing wheat plant heights over the growing season

In an experiment on the development of wheat (Lovell, *pers. comm.*), plant height was measured over time by making repeated measures on a sample of 10 plants, from each of two cultivars (Avalon and Riband) on 19 dates, with records for each plant being made on the same set of dates. Plots of plant height *versus* accumulated day degrees for a typical plant showed that the growth curve appeared to consist of two components. It was not clear what analytic response curve could be fitted to the growth curves, so *generalised parallel curves* were fitted to obtain more information on the form of the underlying growth components. Carrying out a Singular Value Decomposition on the 10 × 19 matrix of plant heights for Avalon gave the Latent Roots (Λ^2) presented in Table 1, where it can be seen that two parallel curve components are required to describe the growth curves. These are presented in Figure 1.

Table 1. Results of a full SVD of the matrix of plant heights for Wheat cv. Avalon.

Latent Root Number	Variance accounted for (Λ^2)	Nominal d.f.	(Mandel's d.f.)	F-test (using Mandel d.f. *vs* LR's 4-10 pooled)	
1	265162.71	28	(48.09)	5502.879	***
2	159.37	26	(36.37)	4.373	***
3	41.82	24	(28.56)	1.461	NS
4-10 (Pooled)	77.11	112	(76.98)		
Total		190	(190)		

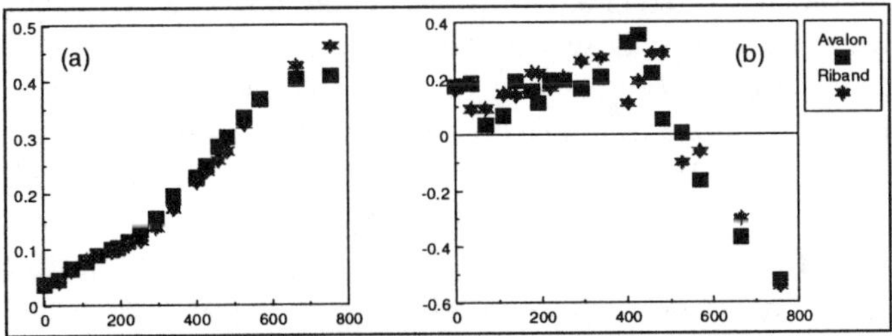

Fig. 1. Estimates of **(a)** component 1 and **(b)** component 2 of the *generalised parallel curves* for wheat cultivars Avalon and Riband (1995)

For the two cultivars there is a typical growth curve, which is similar (Figure 1a), indicating that there are two parts to the growth stage. The second component growth curve (Figure 1b) shows that the growth curve differs from plant to plant by an initially increasing linear trend, which reaches a maximum at 400 day-degrees, before decreasing linearly. No further components were needed. A further analysis would specify the growth functions as specific equations, when lack of fit could be tested.

3.2 Detecting interactions between mixtures of fungicides

Investigations of the behaviour of mixtures of pesticides have been carried out for many years with the aim of finding *synergists*, compounds which enhance each others activity, and *antagonists*, which reduce activity. To detect synergism / antagonism in a mixture of pesticides a measure of "independence", or "non-interaction" is needed, and as noted by Finney (1971) one such is *independent action*. The *independent action* definition of non-interaction forms the basis of the Colby test (Colby, 1967), which is used to detect interactions for regulatory purposes.

The Colby test has two major practical problems. Firstly it only detects interactions at a single dose of each of the two compounds, so may detect differing types of response at different doses. The other drawback is that there may be, in practice, an implicit lower limit to the response. If this is not taken into account, false antagonism is likely to be detected by the Colby test; an extreme example is obtained if we assume the Control response is 100, and the minimum possible response is 5. In this case, a high dose of both compounds 1 and 2 would give observed responses of 5. The Colby definition of non-interaction would then predict that the response for the mixture would be 0.25, considerably lower than the observed value of 5, and, thus, falsely indicating antagonism.

A series of experiments were carried out to detect interactions between fungicides (Kendall, Hollomon & Stormonth, 1994). The results for a typical experiment are presented in Table 2; the experiment consisted of 3 replicates, and the residual mean square is 0.0094 on 72 d.f. The standard error of the difference between observed and fitted is 0.056. If we denote the observed response for dose i of fungicide 1 (d_{1i}) and dose j of fungicide 2 (d_{2j}) by y_{ij}, the fitted control response by C, the effect of dose i of fungicide 1 alone by F_{1i} and the effect of dose j of fungicide 2 alone by F_{2i} (with a zero dose of either compound having an effect of 1), then the Colby test predicts

$$y_{ij} = C\, F_{1i}\, F_{2j}$$

in the absence of an interaction. We note that the Colby test in the form given above may give incorrect conclusions if there is a non-zero response to high doses; however, this flaw can readily be corrected by adding another parameter to the equation above to give

$$y_{ij} = C\, F_{1i}\, F_{2j} + D \qquad (3)$$

which is the modified Colby test presented by Brain & Davies (1995).

Theoretically, both F_{1i} and F_{2j} will be well predicted by a dose-response curve (a logistic curve *vs* log(dose)), so this could be used to predict the response surface in the absence of an interaction. However, in practice, the dose-response curve may not be appropriate and the constraint imposed by an inappropriate equation may induce a false interaction or mask a real one. Thus, although there is a good reason to suppose that the response surface can be modeled by a smooth function, a general function is more likely to detect interactions.

Table 2. Observed response (turbidity reading) on *S. Tritici.* at different dose levels. There was one significant difference from the modified Colby test, highlighted in **bold**.

		Dose of Cyproconazole (μg/ml)					
		0.000	0.060	0.320	1.600	8.000	40.000
Dose of Flusilazole (μg/ml)	0.000	0.60	0.83	0.88	0.77	0.87	0.28
	0.130	0.66	0.73	0.75	0.88	0.86	0.23
	0.640	0.57	0.81	0.82	**1.03**	0.86	0.29
	3.200	0.67	0.82	0.86	0.84	0.97	0.24
	16.000	0.74	0.98	0.80	0.91	1.03	0.29
	80.000	0.26	0.31	0.26	0.32	0.32	0.29

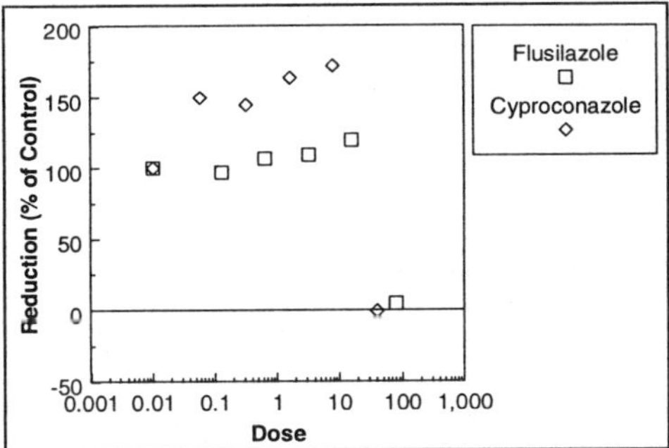

Fig. 2. Estimated dose response curves for the two fungicides. The residual mean deviance (lack of fit) is 0.01682 on nominally 16 d.f. The lack of fit test is significant at the 5% probability level, so there is some evidence of interaction between the compounds

When comparing equations (2) and (3), it is apparent that (3) is a special case of (2), with one component parallel curve. The "slopes" of the relationships in equation (2) are themselves point estimates of a function, rather than varying from individual to individual, and the individual-varying intercept is now constant. Equation (3) has been used in genotype × environment studies and is known as the Shifted Multiplicative Model (SHMM). Thus, we can readily estimate the parameters of the equation and obtain best-fit point estimates of the dose-response curves under the assumption of independence. The residual sum of squares between the observed and predicted responses can be tested for significance against the variability between replicates; the presence of significant lack of fit of equation (3) implies that there is significant evidence of interactions between the two compounds.

The results of fitting the SHMM to the results in Table 2, in the form of the point-estimated dose-response curves for two fungicides are shown in Figure 2. There is significant evidence of interaction between the compounds (significant lack of fit); the combination causing this is shown in Table 2 and is probably spurious. The dose-response curves in this case are clearly not described by the "standard" dose-response curve. Describing the interaction by a response surface based on dose-response curves would give obvious lack of fit, but this would be a combination of the lack of fit of the curve and the presence of an interaction.

4 Conclusions

Multiplicative models have long been used in plant breeding studies, where they enable environmental differences to be accounted for in a sensible way; in this context, the fitted parameters correspond to qualitative factors. In this paper, I have shown that they can have a valuable role in regression analysis, where *generalised parallel curves* can provide an appropriate model. In this context, the parameters of multiplicative models provide point estimates of non-parametric, intrinsically continuous, functions. These can give indications of appropriate parameteric functions which can be then fitted and then allow lack of fit for these specific models to be tested.

References

Brain, P. & Davies, J. (1995). Detecting interactions between mixtures of compounds using the shifted multiplicative model (SHMM). *Aspects of Applied Biology,* **41**, 87-94.

Colby, S.R. (1967). Calculating synergistic and antagonistic responses of herbicide mixtures. *Weeds,* **15**, 20-22.

Cornelius, P.L., Seyedsadr, M. & Crossa, J. (1992). Using the shifted multiplicative model to search for "separability" in crop cultivar trials. *Theoretical & Applied Genetics*, **84**, 161-172.

Finney, D.J. (1971). *Probit analysis. (3rd edition).* Cambridge University Press.

Gauch, H.G. (1988). Model selection and validation for yield trials with interaction. *Biometrics,* **44**, 705-716.

Genstat Committee (1993). *Reference Manual (Genstat 5 Release 3).* Oxford: Oxford University Press.

Kendall, S.J., Hollomon, D.W. & Stormonth, D.A. (1994). Towards the rational use of triazole mixtures for cereal disease control. *Proceedings of Brighton Crop Protection Conference - Pests and Diseases* - 1994, 549-556.

Mandel, J. (1971). A new analysis of variance model for non-additive data. *Technometrics*, **13**, 1-18.

Nachit, M.M., Nachit, G., Ketata, H., Gauch, H.G. & Zobel, R.W. (1992). Use of AMMI and linear regression models to analyze genotype-environment interactions in Durum-wheat. *Theoretical & Applied Genetics,* **83**, 597-601.

A Modelling Approach for Bandwidth Selection in Kernel Density Estimation

Mark J Brewer

Department of MSOR, University of Exeter, Exeter, EX4 4QE, UK

Abstract. A new procedure is proposed for bandwidth selection in univariate kernel density estimation. Rather than concentrate on minimising some criterion based upon the mean integrated square error (MISE), which depends directly on the (unknown) true density, we build a model for the data and use sampling methods to make inferences about the bandwidths. The model is Bayesian, and it is noted that it allows for systematic adjustment for subjective changes in smoothness of the density estimate.

Keywords. Kernel density estimation, Markov chain Monte Carlo, cross-validation

1 Introduction

This paper considers a new procedure for the selection of bandwidths in univariate kernel density estimation. The procedure differs from previous bandwidth selection methods in that it does not set out directly to minimise some criterion related to the mean integrated square error (MISE)—see Jones *et al.* (1996), Silverman (1986) and Wand & Jones (1995) for example. Instead, we formulate a model which represents a cross-validated likelihood function. The bandwidth is then treated as a parameter of the model to be estimated.

Let $f(x)$ be the unknown true density function. Then the (fixed) kernel density estimator based on a sample $\{x_1, x_2, \ldots, x_n\}$ can be written

$$\hat{f}(x) = \frac{1}{n}\sum_{i=1}^{n} K_\lambda\left(x - x_i\right), \tag{1}$$

where the kernel function K is a symmetric density function and λ is a bandwidth which controls the smoothness of $\hat{f}(x)$. This bandwidth is *global* in the sense that it is constant over i. The choice of λ is well-known to be crucial, and of more importance than the choice of kernel K. Choosing λ usually involves a trade-off between smoothness and bias of the density estimate. For full discussion of the many available bandwidth selectors, see Jones *et al.* (1996) or Park & Turlach (1992) for example.

Section 2 describes the basic model used for modelling the data and (global) bandwidth. The model is based around a cross-validated representation of the data. In Section 3, possible ways of estimating this global bandwidth are discussed. The successful methods revolve around Monte Carlo routines (whether Markov chain Monte Carlo—MCMC—or not). Section 4 then applies the model and methods to real data.

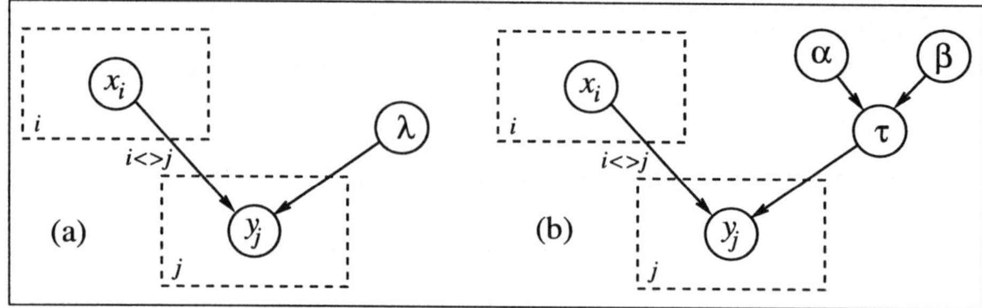

Fig. 1. Graphs of models for the global bandwidth—(a) non-Bayesian, (b) Bayesian

2 A Model for Bandwidth Selection

Consider expression (1) for the fixed kernel density estimator. We would like to be able to use this expression to formulate a model for the bandwidth λ. It is tempting to think of the kernel estimator as a mixture of (say) Normals, with λ^2 as the constant variance over all mixture components. In mixture modelling however, the usual aims are to estimate the number of components and to allocate data points to those components, while in kernel estimation these quantities are fixed. In addition, only one point of data is assigned to each mixture component, occurring at the mode. Trying to estimate the variance (or λ itself), by methods such as those of Diebolt & Robert (1994), is clearly not possible, as the relevant integral will have infinite value.

In kernel density estimation, one common way around this problem is to use cross-validation. Each point of data is assumed to have originated from the kernel density based on all the other observations. We build a model based around this representation of the data: assume we have univariate data x_i, $i = 1, 2, \ldots, n$ and *equivalently* y_j, $j = 1, 2, \ldots, n$ so that $x_i = y_j$ when $i = j$. Having two occurrences of the data in this way allows us to define a sensible graphical model; consider initially the graph of Figure 1 (a), having joint density

$$f(\boldsymbol{y}, \boldsymbol{x}, \lambda) = \prod_{j=1}^{n} f(y_j \mid \{x_{-j}\}, \lambda) = \prod_{j=1}^{n} \frac{1}{n-1} \sum_{\substack{i=1 \\ i \neq j}}^{n} K_\lambda \left(y_j - x_i\right), \qquad (2)$$

where $\{x_{-j}\}$ is the set of observations excluding x_j. We take the functions K to be Normal throughout. From this, we create a Bayesian model, modelling (as standard) the precision $\tau = 1/\lambda^2$ with a Gamma prior; then the graph corresponds to that of Figure 1 (b). The joint density of the whole Bayesian model is then written (substituting τ for λ where necessary)

$$f(\boldsymbol{y}, \boldsymbol{x}, \tau, \alpha, \beta) = f(\tau \mid \alpha, \beta) \prod_{j=1}^{n} f(y_j \mid \{x_{-j}\}, \tau). \qquad (3)$$

3 Estimation of the bandwidth

Estimation of λ from our Bayesian model involves simply finding the expectation from (3) given the data and prior parameters. We could also consider the usual cross-validated likelihood as a scaled density function, and if we are prepared to do this we would need an expectation from (2) representing a "non-Bayesian" model. Hence we require expectation of

$$f(\tau \mid \boldsymbol{y}, \boldsymbol{x}, \alpha, \beta) = c_1 f(\tau \mid \alpha, \beta) \prod_{j=1}^{n} f(y_j \mid \{x_{-j}\}, \tau) \tag{4}$$

$$f(\lambda \mid \boldsymbol{y}, \boldsymbol{x}) = c_2 \prod_{j=1}^{n} f(y_j \mid \{x_{-j}\}, \lambda) \tag{5}$$

for the Bayesian and non-Bayesian models respectively, where c_1 and c_2 are appropriate constants.

More usually, one would wish to maximise (5) to obtain a pseudo-maximum likelihood estimate (LCV) of the bandwidth (see Silverman, 1986, Section 3.4.4). Since (5) is positively skew, the LCV will be smaller than $E(\lambda)$ here. In fact, for a data set of size n, it can be shown (by reversing the order of the product and summation at (2)) that this function is a mixture of $(n-1)^n$ functions of Gamma form. For the Bayesian model, the difference between the LCV and expected value will depend on the prior distribution for τ.

We would like to estimate λ by evaluating relevant integrals of functions (4) and (5), but this is an onerous task—the package Maple, for example, was unable to perform the calculations (either analytically or numerically) for more than a handful of data points. Hence we consider various sampling methods for estimating λ.

Rejection Sampling

To avoid numerical underflow, it is clear that we should work with logs of expressions at (4) and (5) (cf. log-likelihoods). To use rejection sampling to estimate λ therefore, we require an envelope function g (of τ or λ as appropriate) such that $\log f - \log g \leq M$ for some constant M. Choosing g to be Gamma seems sensible; M is chosen via maximisation of $\log f - \log g$, and this is not trivial. Once found, we sample U from $U(0,1)$, τ or λ from g, and accept if $\log U \leq \log f - \log g - M$.

Metropolis-Hastings Sampling

Alternatively, we can use Metropolis-Hastings (M-H) sampling to generate (correlated) observations from a Markov chain (see Brewer *et al.*, 1996) Note that again we use logs during calculation. We define a proposal distribution q (taken to be Gamma here; for the Bayesian model, to generate a new value τ_{t+1} given τ_t, we sample τ' from q and calculate a where

$$a = \begin{cases} \min\{R, 0\} & \text{if } f(\tau_t \mid \boldsymbol{y}, \boldsymbol{x}, \alpha, \beta) q(\tau_t, \tau') > 0, \\ 0 & \text{if } f(\tau_t \mid \boldsymbol{y}, \boldsymbol{x}, \alpha, \beta) q(\tau_t, \tau') = 0. \end{cases} \tag{6}$$

and $R = \log f(\tau' \mid \boldsymbol{y}, \boldsymbol{x}, \alpha, \beta)) + \log q(\tau', \tau_t) - \log f(\tau_t \mid \boldsymbol{y}, \boldsymbol{x}, \alpha, \beta) - \log q(\tau_t, \tau')$. We set $\tau_{t+1} = \tau'$ if, for $U \sim U(0,1)$, $\log U \leq a$, else we set $\tau_{t+1} = \tau_t$.

Auxiliary Variable Sampling

The auxiliary variable sampling method (AV) is applicable for the Bayesian model only. This method is adapted from Besag & Green (1993), and the version below is a more efficient version than that in Brewer *et al.* (1996).

Given the posterior distribution for τ at (4), we sample a proposed new value τ' from the prior on τ, $f(\tau \mid \alpha, \beta)$. Then, for each of the n remaining

	Method	Code	Bandwidth λ	Standard Error
	Rule-of-thumb	ROT	4.789	N/A
	Plug-in	PI	3.895	N/A
	Likelihood CV	LCV	4.069	N/A
Non-Bayes:	Rejection	NBR	4.303	0.031
	M-H	NBMH	4.401	0.031
Bayes:	Rejection	BR	4.097	0.029
	M-H	BMH	4.175	0.030
	AV	BAV	4.145	0.030

Table 1. Bandwidth estimates for Lean Body Mass data

product terms from (4) in turn we generate u_j for an auxiliary variable U_j from $U(0, f(y_j \mid \{x_{-j}\}, \tau))$, *until* the condition $f(y_j \mid \{x_{-j}\}, \tau') \geq u_j$ fails. Upon failure, we start again with a new τ', but if the condition holds for all j then we set $\tau_{t+1} = \tau'$. Since we consider only the conditional densities, there is no need to use logs for this method.

4 Application

The methods are now applied to the "Lean Body Mass" variable of the Australian Institute of Sports data from Cook & Weisberg (1994). Table 1 shows values of the bandwidth selected using different methods. The rule-of-thumb bandwidth (ROT) is given by Silverman (1986); the plug-in estimate (PI) originates from Sheather & Jones (1991) and is based upon minimising asymptotic MISE. The Likelihood CV method (LCV) is simply that mentioned earlier of maximising (5) over λ. The results from the sampling methods described above are for samples of size 1000, and for the M-H and AV methods this meant taking every 20th observation from 20000 to reduce autocorrelations. For the Bayesian model, a Ga(0.001,0.001) prior was used for τ. Figure 2 illustrates some of the curves for bandwidths from Table 1.

Jones *et al.* (1996) also study this data set for various bandwidth selectors: there they conclude that only the plug-in estimate suggests the data to be of a bimodal nature and illustrate that since the data consist of values for both male and female athletes, an assumption of bimodality is reasonable. If one accepts that a rôle for kernel density estimation is the identification of (numbers of) modes, then one can argue that the plug-in estimate has performed well, but that there is little to choose between most of the methods.

We note that the non-Bayes models will give bandwidth estimates higher than the LCV values. The bandwidths from the Bayes models will depend on the prior; the vague prior used will result in smaller bandwidths than the non-Bayesian model. While the estimation of λ by taking expectation at (5) may be considered inappropriate, the value of the likelihood at these points has generally been found to be within 2 or 3 percent of the maximum value.

The sampling methods used to estimate λ from the models here all give similar results, but at differing costs. For rejection sampling, one has to find a suitable envelope, perform function maximisation to obtain M, and then obtain samples. While this will produce independent observations, and hence rejection sampling will be the preferred method of a competent statistician, the tasks of deciding on the envelope and function maximisation will be non-trivial to a non-statistician. For M-H sampling, one has to ascertain a sensible proposal distribution q at (6), and this may not be an automatic choice. The process of generating observations for these two methods will be much quicker

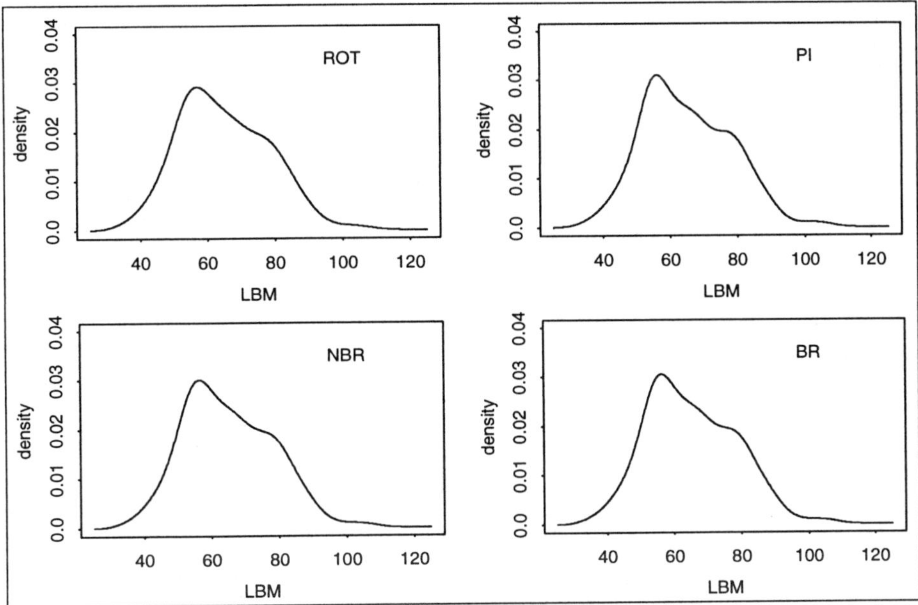

Fig. 2. Kernel density estimates of the Lean Body Mass data

Method	Code	Bandwidth λ	Standard Error
Rejection	BRa	3.750	0.028
M-H	BMHa	3.682	0.026
AV	BAVa	3.725	0.026

Table 2. Bandwidth estimates given Ga(1,0.1) prior on τ for Lean Body Mass data

than for the AV procedure, but note that this latter method requires merely the data as input. For this reason, a non-statistician requiring an automatic choice of bandwidth may prefer the AV selector.

Now suppose that we wish to obtain a kernel density which has (in MISE terms) less bias than those resulting from the bandwidths in Table 1, i.e. we would like the bandwidth to be *smaller*. For the data, using the informative prior Ga(1,0.1) results in the bandwidths of Table 2; two curves are shown in Figure 3. The resulting curves suggest slightly more strongly the existence of a mode around LBM=75. However, attempting to highlight this mode by reducing the bandwidth further will also affect the rest of the function—in this case, a (spurious?) mode around LBM=105 would appear.

5 Discussion

The modelling approach for bandwidth selection has been shown to be feasible and does not "require" the true unknown density. The modelling approach also lends itself to selection of automatic *variable* bandwidths; see Brewer (1998) for full details. An example of an adaptive kernel density is shown in Figure 4, along with a plot of the variable bandwidth values. As can be seen, the mode at LBM=75 has been highlighted but without the cost of vastly increasing variance or introduction of spurious modes. Also, the adaptive estimate has successfully smoothed out the step at LBM=105.

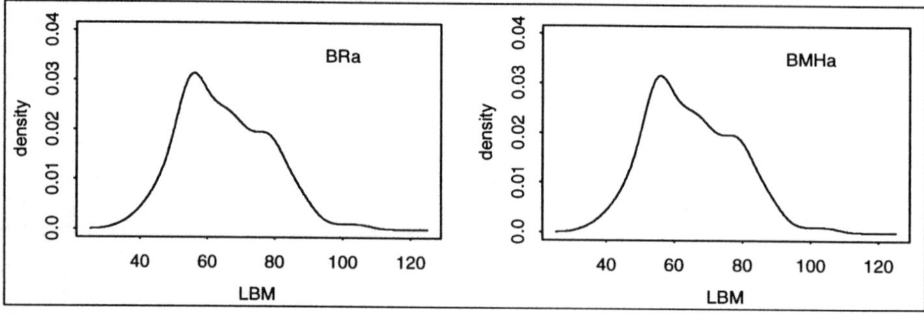

Fig. 3. Kernel estimates with informative prior

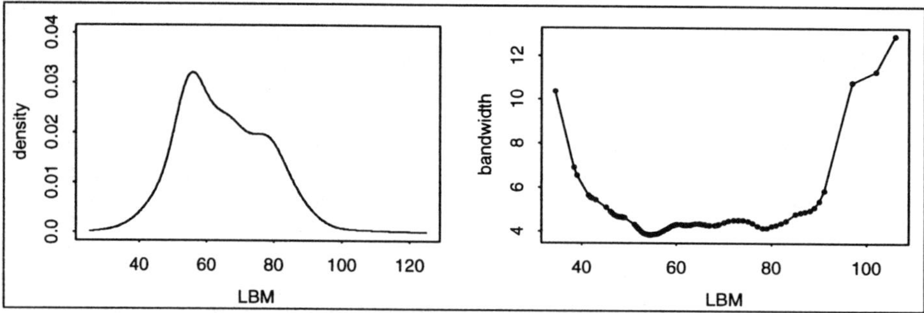

Fig. 4. Variable kernel density and bandwidth values

References

Besag, J. & Green, P.J. (1993). Spatial Statistics and Bayesian Computation. *Journal of the Royal Statistical Society*, B **55**, No. 1 25–38.

Brewer, M.J. (1998). A Model-Based Approach for Variable Bandwidth Selection in Kernel Density Estimation. Technical report, URL `http://msor0.ex.ac.uk/Staff/MJBrewer/research/98_1.ps` .

Brewer, M.J., Aitken, C.G.G. & Talbot, M. (1996). A Comparison of Hybrid Strategies for Gibbs Sampling in Mixed Graphical Models. *Computational Statistics and Data Analysis*, **21**, 343–365.

Cook, R.D. & Weisberg, S. (1994). *An Introduction to Regression Graphics.* New York: Wiley.

Diebolt, J. & Robert, C.P. (1994). Estimation of Finite Mixture Distributions through Bayesian Sampling. *Journal of the Royal Statistical Society*, B **56**, No. 2 363–375.

Jones, M.C., Marron, J.S. & Sheather, S.J. (1996). A Brief Survey of Bandwidth Selection for Density Estimation. *Journal of the American Statistical Association*, **91**, No. 433 401–407.

Park, B.-U., and Turlach B.A. (1992). Practical Performance of Several Data-Driven Bandwidth Selectors. *Computational Statistics*, **7**, 251–285.

Sheather, S.J. & Jones, M.C. (1991). A Reliable Data-Based Bandwidth Selection Method for Kernel Density Estimation. *Journal of the Royal Statistical Society*, B **53**, No. 3 683–690.

Silverman, B.W. (1986). *Density Estimation for Statistics and Data Analysis.* London: Chapman and Hall.

Wand, M.P. & Jones, M.C. (1995). *Kernel Smoothing.* London: Chapman and Hall.

A New Method for Cross-Classification Analysis of Contingency Data Tables

Sergio Camiz [1] and Jean-Jacques Denimal [2]

[1] Dipartimento di Matematica, Università "La Sapienza", Roma; Université des Sciences et Technologies, Lille. e-mail:camiz@mat.uniroma1.it.

[2] Université des Sciences et Technologies, Lille. e-mail:denimal@alea.univ-lille1.fr

Abstract. For the interpretation of a two-way contingency table cross-classification, based on two hierarchies, new indexes are developed that are able to evaluate the relative contribution of nodes of either hierarchy to the nodes of the other hierarchy or partitions. Indexes are based on the group's inertia projection onto the orthogonal basis associated with the explanatory hierarchy, in a vector space with χ^2 metrics. An application based on the study of family supplies in Eastern Lombardia is given.

Keywords. Contingency tables, hierarchical classification, cross-classification

1. Introduction

A two-way contingency table may be so large that its dependence structure may not be easily readable. Reduction of the number of both rows and columns through classification, giving a reduced dimensional table, may help in its understanding, provided that the resulting cross-classification may be analyzed with suitable tools, in particular able to explain either one-way partition through the other. Orlóci (1978) and Feoli & Orlóci (1979) propose the *analysis of concentration* for evaluating vegetation table *sharpness*: a kind of correspondence analysis based on a normalised table, a task possible only when considering presence-absence data, as used by Camiz (1993, 1994) for the investigation of the *best* cross-classification of a vegetation table. Even in its best application, the association between groups is detected only through the usual correspondence analysis practice, namely the inspection of scatter plots, with the addendum that the position of points in one partition must be considered with respect to all other partition points, so that the association between classes of the two sets is not directly revealable. Govaert (1984) developed techniques based on Diday's (1970) *dynamical clouds*, whose results depend on the *a priori* choice of the suitable number of groups of each partition. Greenacre (1988) proposes a method based on Hirotsu (1983) statistics to find optimal cut-points for both hierarchies, obtained independently through Ward (1963) criterion. Benzécri *et al.* (1980) decompose node inertia on canonical bases, whereas Lebart *et al.*(1979) propose statistical methods to outline each group's *typical variables*, i.e. those whose occurrence is significantly higher or lower than the expected value, using a hypergeometric distribution. This does not allow a direct comparison of classifications, since the items used for the explanation of a group are not considered to form a hierarchy; also, they may be the same for two different groups of the same partition, but at different frequency levels. Nevertheless, it is a useful method to gain deep insight into a classification structure.

The proposed technique, initially developed by Denimal (1997), is based on any double ascending hierarchical classification performed on both rows and columns of a two-way contingency table. The derived indexes evaluate the relative contribution of nodes of one hierarchy both to the nodes of the other hierarchy and to partitions. In this way relations are found between both hierarchies and partitions, helping in interpretation and graphical representation of results.

2. The method

Given a two-way contingency table K for two sets of characters I and J, we consider two hierarchies H^I, H^J built on I and J, respectively, using Ward's (1963) criterion. With these hierarchies, two bases are associated on vector spaces $\mathbb{R}^{|J|}$ and $\mathbb{R}^{|I|}$ respectively, each vector (but one) representing a node in the corresponding hierarchy. On such vectors lines belonging to the two different merging clusters have opposite coordinates. Being orthogonal, these bases allow us to decompose the squared distance between each group centroid and the grand centroid. Thus, *indexes* may be established, to interpret hierarchy and partitions on either set in terms of the hierarchy on the other.

Let a generic element of K be k_{ij}, $i \in I, j \in J$. As usual, marginal and grand totals may be written as $k_{i\cdot} = \sum_{j\in J} k_{ij}$, $k_{\cdot j} = \sum_{i\in I} k_{ij}$, $k_{\cdot\cdot} = \sum_{(i,j)\in I\times J} k_{ij}$. Given two subsets $p \subseteq I$, $q \subseteq J$, we denote by k_p, k_q, and k_{pq} the partial sums $k_p = \sum_{i\in p} k_{i\cdot}$, $k_q = \sum_{j\in q} k_{\cdot j}$, and $k_{pq} = \sum_{(i,j)\in p\times q} k_{ij}$. If we set these in the correspondence analysis frame, where each element $i \in I$ is represented by a point $(k_{ij} / k_{i\cdot})_{j\in J} \in \mathbb{R}^{|J|}$ with mass $k_{i\cdot}/k_{\cdot\cdot}$ (and analogous formulae hold for j), the centroids of all subsets p and q have representations given by $c_p^I = (k_{pj} / k_p)_{j\in J} \in \mathbb{R}^{|J|}$ and $c_q^J = (k_{iq} / k_q)_{i\in I} \in \mathbb{R}^{|I|}$ respectively, where c^I and c^J are the grand centroids, i.e. the marginals. If we provide the vector spaces $\mathbb{R}^{|I|}$ and $\mathbb{R}^{|J|}$ with the χ^2 metrics, a basis *associated to hierarchy* H^J $(e^0, e^1, \ldots, e^{|J|-1})$ for $\mathbb{R}^{|J|}$ (an analogous one may be given for $\mathbb{R}^{|I|}$) may be given by:

$$\forall j \in J, \quad e^0_j = k_{\cdot j}/k_{\cdot\cdot}$$

$$\forall m = (q_1, q_2) \in H^J, \quad e^m_j = \begin{cases} k_{\cdot j}/k_{q_1} & \text{if } j \in q_1 \\ -k_{\cdot j}/k_{q_2} & \text{if } j \in q_2 \\ 0 & \text{otherwise} \end{cases} \tag{1}$$

with q_1 and q_2 the two subsets that merge at the H^J m-th node. This basis is orthogonal (Benzécri *et coll.*, 1973-82; Weiss 1978; Cazes, 1984).

To interpret the H^I hierarchy in terms of H^J, the squared distance $\| c^I_p - c^I \|^2$ of the H^I branch or group p from the centroid c^I is partitioned according to the elements of the hierarchy H^J associated basis, and we have

$$\delta^I_p = \| c^I_p - c^I \|^2 = \sum_{m=(q_1,q_2)\in H^J} \Delta_{p,(q_1,q_2)} = \sum_{m=(q_1,q_2)\in H^J} \frac{k_{q_1} k_{q_2}}{k_{q_1} + k_{q_2}} k_{\cdot\cdot} \left(\frac{k_{pq_1}}{k_p k_{q_1}} - \frac{k_{pq_2}}{k_p k_{q_2}} \right)^2 \tag{2}$$

As a consequence, for each group p of hierarchy H^I, indexes $i^I_{p,(q_1,q_2)} = \dfrac{\Delta^I_{p,(q_1,q_2)}}{\delta^I_{p,(q_1,q_2)}}$ may be calculated for the most significant nodes $m = (q_1, q_2) \in H^J$. In particular, indexes i, summing to 1, measure the relative contribution of the hierarchy nodes in the explanation of group's distance from the centroid.

As the squared distance δ^I_p is defined in χ^2 metric, the quantity $k_p \|c^I_p - c^I\|^2$ may be interpreted as the χ^2 statistic in the frame of the classical test of fit, adjusting the observed distribution $c^I_p = (k_{pj} / k_p), j \in J$ to the theoretical $c^I = (k_{.j} / k_{..}) j \in J$. Let us denote $p_j = k_{.j} / k_{..}$ and $n_j = k_{pj}, j \in J$. Assuming that p_j is the probability of belonging to the cluster j, we know that $(n_1, n_2, .., n_{|J|})$ has a multinomial distribution with parameters k_p and $p_j, j \in J$. Since $\mathbb{R}^{|J|}$ is provided with the χ^2 metric, whose (diagonal) elements are $(1 / p_j), j \in J$ and with the orthogonal basis given by (1), the difference $c^I_p - c^I$ is decomposed in this basis as:

$$c^I_p - c^I = \sum_{m=1}^{|J|-1} \frac{u_m e^m_j}{\|e^m_j\|} \tag{3}$$

It was proved by Weiss (1978) that variables u_m satisfy the equations: $E(u_m) = 0$, $cov(u_m, u_{m'}) = 0$ if $m \neq m'$, $var(u_m) = 1/k_p$. The difference $c^I_p - c^I$ is normally distributed in $\mathbb{R}^{|J|}$, the variables $\sqrt{k_p} u_m$, $m \in (1, |J| - 1)$ are independent and have a normal $N(0,1)$ distribution. It follows that $k_p \delta^I_{p,(q_1,q_2)} = k_p u_m^2$, where $m = (q_1, q_2)$ is asymptotically distributed as χ^2 with 1 degree of freedom. It is thus possible to calculate the significance of the influence of each H^J node on both H^I nodes and partition groups.

If an H^J node is composed by branches q_1 and q_2, it is evident that $\Delta_{p,(q_1,q_2)}$ depends on the difference $f_{pq_1} = \dfrac{k_{pq_1}}{k_p k_{q_1}}$ and $f_{pq_2} = \dfrac{k_{pq_2}}{k_p k_{q_2}}$, which may be used to check the association of both branches with group p of the H^I hierarchy. If $f_{pq_i} > 1$, the association between groups p and q_i is positive, and negative if $f_{pq_i} < 1$. The elements k_{pq} have hypergeometric distributions with mean $\mu = k_p k_q / k_{..}$ and standard deviation $\sigma = \sqrt{\dfrac{(k-k_p)(k-k_q)k_p k_q}{k^2(k-1)}}$. Thus, the standardised variable $s_{pq} = (k_{pq} - \mu) / \sigma$ may be used as a test-value and a grey-scale may be built on it. This allows a graphical representation of the association strength, something close to the shadow matrices of Sneath & Sokal (1962) and the graphical representations of Benzécri *et al.* (1973-82: I, 82 and 365). For our purposes, a five-step grey-scale was defined, based on the four cutpoints in the s_{pq} distribution -5σ, -3σ, $+3\sigma$, $+5\sigma$. Analogous indexes may be calculated for the groups q of the H^J hierarchy.

3. Application to Eastern Lombardia family supplies data

The data consist in a contingency table of 592 different family supplies crossed with the 641 Lombardia region eastern municipalities. The aim of the exploratory analysis (Badioli *et al.*, 1997) was to classify the municipalities according to the different structure of supplies. The two hierarchies considered were obtained through Ward's (1963) method applied to Euclidean distance among items, computed on their coordinates on the first two correspondence analysis axes (Benzécri *et coll.*, 1973-82; Lebart *et al.*, 1984). In this study the first axis opposes major cities, actual supply centres for the area, to smallest mountain sites, due to opposition between high rank supplies (*lawyers, architects, Insurances*) and those present nearly everywhere, like *primary schools* and *general food.* The second axis opposes supplies relative to *tourism*, like *hotels, campings, restaurants, etc.*, with all the others, so that touristic centres are set apart on the factor plane. The chosen partitions gave an 8 × 6 cross-classification, that explains around 7% of total table variation. The attempt to explain the municipality groups through the distribution of supplies was very hard to perform, based on the Lebart *et al.* (1979) technique, since too many items were involved. The technique previously proposed was then used, through a Fortran program written on purpose. This allows us to show results in a very synthetical way: Figure 1 represents the reduced cross-classification table, whose rows are municipality groups and columns the supply ones.

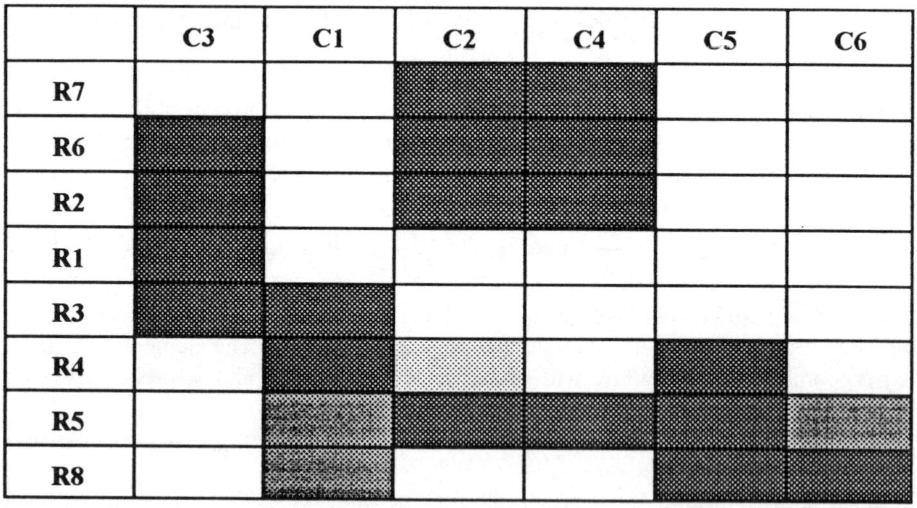

	C3	**C1**	**C2**	**C4**	**C5**	**C6**
R7						
R6						
R2						
R1						
R3						
R4						
R5						
R8						

Fig 1. The association among groups of municipalities (rows) and groups of supplies (columns) in the 8×6 cross-classification of Eastern Lombardia family supplies data.

In the figure the white cells represent very low frequencies, i.e. negative association: then, considering the dark grey cells, a strong association is visible between the two partitions. The first two supply groups contain supplies of general diffusion (C3: *general food, car repair, churches*, etc.; C1: *clothes, shoeshops, hospitals*, etc.), the second two represent most of the tourism supplies (C2: *night clubs, restaurants, thermal resources,* etc.; C4: *hotels, camping grounds, ski-lifts*, etc.), and the last correspond to highest rank supplies (C5: *travel agencies, supermarkets, physicians,* etc.; C6: *theatres, bookshops, financial supplies*, etc.). With this characterisation of supply groups, it is possible to describe the municipality groups. The distribution of supply groups shows that a *rank* structure of municipalities may easily be identified, following the high-low position of rows on the table, going from the poorest country / mountain areas through to the richest cities, whereas the touristic structure is somehow independent. In fact, the tourism supplies are present mainly in the mountain areas, very poor in common supplies, by the lakes shores, and in the main cities, since there both hotels and restaurants support the work activities too.

4. Conclusions

Specifically suited for the interpretation of a cross-classification, the usefulness of the method is proved even when the interpretation of only one classification is important, provided that it may be explained by the other hierarchy. In addition, when cross-classification is an important issue, the method is helpful in detecting the deepest levels where either partition may be explained through the other (Camiz & Denimal, 1998). Thus it may help in detecting the optimal cross-classification too.

It must be emphasized that, even if it may be considered as a standalone technique, as proposed by Denimal (1997), the method is totally independent from the way hierarchies were built, since it is based only on their existence. So, one may choose for any situation the hierarchy that considered as most appropriate, and then use this method for the interpretation of results.

In the example of Eastern Lombardia supplies, the graphical representation of association through shadowing was effective in showing very synthetically the different municipality structure, which was most difficult to see through the alternative techn-iques. It is evident that further investigation concerning most interesting supply levels and their differences in the municipality groups may be achieved only through specific techniques, such as Lebart *et al.* (1979). Its use as a second step, involving quantitative aspects of only a selected number of items of interest, seems more appropriate.

Some development directions are now under investigation: the use of the procedure in order to improve structuring of vegetation data tables, very close to the Bertin (1977) graphical representation of data, and the generalisation to other cases, such as three-way contingency data tables, individual × character, and individual × variable tables.

Acknowledgements

The first author was supported in this work by Italian Consiglio Nazionale delle Ricerche, in the frame of Progetto Finalizzato Trasporti 2, grant n. PFT2 96.00121.PF74, whose scientific manager is Prof. Silvana Stefani, and by La Sapienza University 60% personal grant.

References

Badioli, M., Camiz, S. & Stefani, S. (1997). La localizzazione dei servizi alle famiglie nella Lombardia Orientale. In: *Atti del 3° Convegno Nazionale del Progetto Finalizzato Trasporti 2*. Roma, Consiglio Nazionale delle Ricerche, on Cd-Rom.

Benzécri, J.P. *et coll.* (1973-82). *L'Analyse des Données*, 2 vol. Paris: Dunod.

Benzécri, J.P., Lebeaux, M.O. & Jambu, M. (1980). Aides à l'interprétation en classification automatique. *Cahiers de l'Analyse des Données*, **5**(1), 101-123.

Bertin, J. (1977). *La graphique et le traitement graphique de l'information.* Paris: Flammarion.

Camiz, S. (1993). Computer Assisted Procedures for Structuring Community Data. *Coenoses*, **8** (2), 97-104.

Camiz, S. (1994). A Procedure for Structuring Vegetation Tables. *Abstracta Botanica*, **18** (2), 57-70.

Camiz, S. & Denimal, J.J. (1998). Interpretation of a Cross-Classification: a New Method and an Application. *IFCS-98 Sixth International Conference*, in press.

Cazes, P. (1984). Les correspondances hiérarchiques. CEREMADE, Université Paris-Dauphine, *Les Cahiers de Mathématiques de Décision.*

Diday, E. (1970). *La méthode des nuées dynamiques et la reconnaissance des formes*. Paris: IRIA.

Denimal, J.J. (1997). Aides à l'interpretation mutuelle de deux hiérarchies construites sur les lignes et les colonnes d'un tableau de contingence. *Revue de Statistique Appliquée*, 4.

Feoli, E. & Orlóci, L. (1979). Analysis of concentration and detection of underlying factors in structured tables. *Vegetatio*, **40**, 49-54.

Govaert, G. (1984). *Classification croisée.* Université de Paris VI, Thèse d'état.

Greenacre, M. (1988). Clustering the Rows and Columns of a Contingency Table. *Journal of Classification*, **5**, 39-51.

Hirotsu, C. (1983). Defining the Pattern of Association in Two-way Contingency Tables. *Biometrika*, **70**, 579-589.

Lebart, L., Morineau, A. & Fénelon, J.P. (1979). *Traitement des données statistiques.* Paris: Dunod.

Lebart, L., Morineau, A. & Warwick, K. (1984). *Multivariate Descriptive Statistical Analysis.* New York: John Wiley and Sons.

Orlóci, L. (1978). *Multivariate Analysis in Vegetation Research.* The Hague: Junk.

Sneath, P.H.A. & Sokal, R.R. (1962). Numerical Taxonomy. *Nature*, **193**, 855-860.

Ward, J.H. (1963). Hierarchical Grouping to Optimize an Objective Function. *Journal of the American Statistical Association,* **58**, 236-244.

Weiss, M.C. (1978). Décomposition hiérarchique du chi-deux associée à un tableau de contingence à plusieurs entré. *Revue de Statistique Appliquée,* **26**(1), 23-33.

Bayesian Analysis of Overdispersed Count Data with Application to Teletraffic Monitoring

Olivier Cappé [1], Randal Douc [1], Eric Moulines [1] and Christian Robert [2]

[1] Ecole Nationale Supérieure des Télécommunications Dpt. Signal – CNRS URA 820, 46 Rue Barrault, 75634 Paris Cedex 13, France
[2] Laboratoire de Statistique, CREST, INSEE, Paris, France

Keywords. Teletraffic, segmentation, classification, overdispersed count data

1 Introduction

Traffic modelling of modern telecommunication networks is a concern of major importance. Potential applications of efficient traffic models would be numerous including dynamic bandwidth allocation, network dimensioning or statistical multiplexing... As of today however, the characterization of network traffic is still an open and challenging statistical issue. It has first been reported by several authors that traditional traffic models (homogeneous Poisson model) are largely inappropriate for the arrival processes measured on most types of network connections, and especially for wide-area networks such as the internet. Moreover, it has been demonstrated that many teletraffic data sets exhibit nonstandard characteristics such as heavy tailed distribution or long range dependence. Finally, the huge size of the data sets involved imposes severe computational constraints on the analysis methodology.

In this contribution, we focus on the task of modelling aggregated count data on medium time scales (a few seconds to a few minutes). The motivation for using aggregated data is twofold, first it is a good solution to cope with the dimensionality problem by selecting the time-resolution (the aggregation interval) depending on the total duration of the data to be analyzed, second, it makes it possible to deal with the different types of available teletraffic data (block-packet counts - number of packets received during a time interval, or byte data - size of the transferred data).

The two traces shown on Figure 1 are typical example of the data sets considered in this paper. Perhaps the most striking feature of Figure 1 is the fact that both traces look pretty much similar: the persistence of the burstyness of the data whatever the duration of the interval used for aggregation is a consequence of the long-range memory effect. For a fixed time scale however, the main feature of the data is the presence of abrupt changes. If longer durations of traffic (several hours) were to be analyzed, it would also be necessary to take into account long-term trends and/or drifts, but these effects can be neglected for the data considered here.

In order to summarize the information contained in traces such as those shown on Figure 1, we consider models that have a strong segmentation and classification potential. Segmentation means finding contiguous regions of the data set that can be considered as homogeneous, whereas classification aims at identifying various levels of activity in the data. Typical statistical models for these two purposes would be a changepoint model (segmentation) or a mixture model (classification). For teletraffic data we need a model that is somewhere in between these two situations since the traces exhibit both a strong temporal structure and a "clustered" behaviour (for instance, with repeated bursts of comparable level).

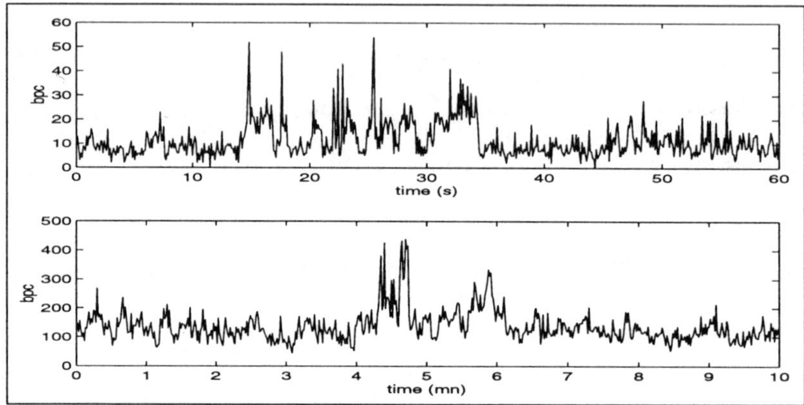

Fig. 1. Top: Block packet counts with 0.1 second aggregation, spanning one minute of traffic. Bottom: Block packet counts with 1 second aggregation, spanning ten minutes of traffic (both traces involve 600 data points)

2 Choice of the model

Many of the methods proposed for modelling count data such as Arjas & Heikkinen (1997) for a semiparametric model, Green (1995) for a change-point model, Robert & Titterington (1997) for a parametric model, are based on the hypothesis that the data has, at least locally, a Poisson distribution. This assumption is clearly ruled out here because of the aggregation that has the effect of increasing the variability of the data compared to the Poisson distribution. An alternative distribution which has been used in many disciplines involving overdispersed count data, such as accident statistics or market research, is the negative binomial. The use of the negative binomial in this context can be understood since it corresponds to the distribution of a Poisson mixture with Gamma mixing distribution. The negative binomial is thus most suited in cases where the observed data averages several sources with different mean rates of activity.

We will use the following parameterization of the negative binomial distribution

$$\mathrm{Neg-Binomial}(\mathrm{n}|\kappa,\pi) = \binom{n+\kappa-1}{\kappa-1}\pi^{\kappa}(1-\pi)^{\mathrm{n}} \quad \text{for } \mathrm{n}\in\mathbb{N}$$

where $\pi \in (0,1)$ and $\kappa > 0$ are both treated as continuous parameters.

From a computational point of view, a very attractive model for data such as that displayed on Figure 1 is the hidden Markov model, or HMM (MacDonald & Zucchini, 1997; Robert & Titterington, 1997). HMMs have been successfully used in many applications involving classification and segmentation, with a computational complexity (be it for estimation or prediction) that is moderate. For a negative binomial HMM, it is relatively straightforward to obtain maximum likelihood estimates of the parameters using the

ECM algorithm (details of the implementation are omitted here since this is not the main topic of the contribution). Our experience with the application of the negative binomial HMM model to aggregated teletraffic data can be summarized as follows: *(1)* The marginal distribution of the data is well modelled when using a negative binomial HMM with a small number of states (2 to 4), except for the top 1-2% quantile for which the data is generally more heavy tailed than predicted by the model. *(2)* All components of the HMM corresponds to negative binomial distributions that are largely overdispersed compared to the Poisson distribution, and the overdispersion is more pronounced for states that with high mean value. *(3)* The segmentation obtained with such a model is satisfying in many respects, but raises some doubts concerning the Markovian assumption: the real data is systematically more "bursty" than implied by the model.

We thus propose a model that retains the main ingredients of the negative binomial HMM while relaxing the Markovian assumption which seems unrealistic in practise. We choose to use a Bayesian model specification together with a Markov chain Monte Carlo estimation technique which is based on the reversible jumps methodology proposed by Green (1995).

3 Model description

The proposed model relies on an implicit (non-observable) splitting of the observation interval into successive *segments* separated by *boundaries*, where each segment is linked to a *class* by means of a *label.* Note that unless otherwise specified the word "label" refers only to the segment labels and not to the classification of single data points (which we should refer to as indicators). With this choice of the parameterization, the segment boundaries and labels will be treated as variable dimension data. Using the indicators rather than the segment label would have the advantage of fixing the size of the data to be simulated but with the disadvantage of making prior specification cumbersome. For the sake of the simplicity of the exposition, we will only describe the sampler used when the number of classes is considered to be fixed and known. Techniques that allow for a variable number of classes are very similar to those used for the simulation of segments but with the difference that they also imply the simulation of variable dimension continuous parameters.

Let $\{n_t\}_{t=1,\dots,T}$ ($n_t \in \mathbb{N}$) denote the observed data. Given the parameters of the model, the counts n_t are assumed to be conditionally independent with marginal distribution

$$n_t \sim \mathrm{Neg-Binomial}(\kappa_{\mathrm{m}}, \pi_{\mathrm{m}})$$

where m is the label associated with the segment that contains t, that is such that $b_k \leq t < b_{k+1}$, where b_K are the segments boundaries. The a priori structure of the model is the following: *Number of Segments* – $p(K) \propto \exp(K\tau_K)$, with $\tau_K \in (0,1)$. This prior is intended to be less informative, in a context where the number of segments may become important, than the Poisson prior proposed by Green (1995). *Segment boundaries* – They are set a priori uniformly with the convention that $b_1 = 1$ and $b_{K+1} = T+1$ and the constraint that $b_k > b_{k-1}$ for $k = 2, \dots, K$, and thus $p(b_2, \dots, b_K|K) = (T-1)!/[(T-(K-1))!(K-1)!]$. *Segment label* – Given K and M, all valid label sequences (ie. such that no adjacent segments share the same label) are a priori equally likely: $p(l_1, \dots, l_K|K, M) = M(M-1)^{K-1}$. *Parameters of each class* – For the type of model considered here, specifying a proper prior for the parameters of each class is an absolute requirement (see e.g. Robert & Titterington, 1997). We use an independent prior of the form

$$p(\kappa_1, \pi_1, \ldots, \kappa_M, \pi_M) = M! \prod_{m=1}^{M} \mathrm{Gamma}(\kappa_\mathrm{m}|\alpha_\kappa, \beta_\kappa)\mathrm{Beta}(\pi_\mathrm{m}|\alpha_\pi, \beta_\pi)$$

on all configurations of the parameters such that $\pi_1 > \pi_2 > \ldots > \pi_m$. This ordering constraint is imposed following Richardson & Green (1997) so as to avoid label switching artifacts during the simulation (although there are other approaches to this issue as pointed out by several discussants of the paper by Richardson & Green - see also Robert & Titterington, 1997, on this point). The beta distribution was chosen since it is the conjugate prior for π whereas the reasons of the choice of the gamma distribution for κ will be made clearer in Section 4.3.

4 Sampling technique

In this section, we briefly describe the characteristics of the MCMC sampler highlighting only the parts that are less conventional.

4.1 Updating the segment boundaries

Following Stephens (1994), the boundaries are updated using a Gibbs sweep which modifies each segment boundary b_k conditionally to the previous and next ones b_{k-1} and b_{k+1}. The conditional probabilities for the boundary position are obtained by an exhaustive enumeration whose computational cost is linear in the size of the segment thanks to the conditioning on the previous and next boundary. Chib (1997) proposes a novel technique based on Markovian modelling which would make it possible to sample in block the segment boundaries. The method of Chib is however not directly amenable to cases where the number of segments varies.

4.2 Updating the segment labels

Conditionally to the segment boundaries, the label are drawn in block. The computation of the conditional probabilities can be done independently for all segments, but the simulation has to take into account the constraint that no successive labels can be identical.

4.3 Updating the parameters of each class

The parameters $\pi_1, \ldots, \pi_M$, ares sampled from the full conditional according to

$$\pi_m | \cdots \sim \mathrm{Beta}(\kappa_\mathrm{m}\bar{\mathrm{N}}^{(\mathrm{m})} + \alpha_\pi, \bar{\mathrm{S}}^{(\mathrm{m})} + \beta_\pi)\mathbf{I}_{(\pi_{\mathbf{m+1}}, \pi_{\mathbf{m-1}})}$$

where $\bar{N}^{(m)}$ is the number of data points classified within class m (or more precisely, the sum of the duration of all segments whose label is m) $\bar{S}^{(m)}$ denotes the sum of these points.

For the parameters κ_m, a one-at-a-time Metropolis-Hastings strategy is used with a gamma proposal distribution that is tuned to match the mode and the log-curvature around the mode of the exact conditional distribution. The full conditional distribution for κ_m is

$$p(\kappa_m|\cdots) \propto (\kappa_m)^{\alpha_\kappa - 1} \left\{ \prod_{r=1}^{\bar{R}^{(m)}} (\kappa_m + r - 1)^{\bar{C}_r^{(m)}} \right\} \mathrm{e}^{-[\beta_\kappa + \bar{\mathrm{N}}^{(\mathrm{m})} \log(1/\pi_\mathrm{m})]\kappa_\mathrm{m}}$$

where $\bar{R}^{(m)}$ denotes the maximum value of the data points in class m, and $\bar{C}_r^{(m)}$ are the rank statistics, ie. the number of points belonging to class m greater or equal to r. The distribution above does not belong to any standard distribution family, however it appears that in practise it can be approximated with great accuracy by a gamma distribution. First, the mode is found, starting from the moment estimate $\kappa_m = [\bar{S}^{(m)}/\bar{N}^{(m)}]\pi_m/(1-\pi_m)$, using a few Newton steps. Note that in this case the Newton algorithm is globally convergent since $\log p(\kappa_m|\cdots)$ is strictly concave as soon as $\bar{C}_r^{(m)} > \alpha_\kappa - 1$, which is the case whenever one at least of the observations allocated to the mth class is different from zero. Then, a gamma distribution with matched mode and log-curvature is used as the proposal for a Metropolis-Hastings procedure. The acceptance ratio associated to this proposal kernel is usually found to be greater than 98%.

4.4 Creating or removing segments

We now come to more elaborate moves which will have the effect of modifying the number of segments K. The technique used is substantially simpler than that needed in most of the examples given by Green (1995) since all the variable dimension data is of discrete type (as far as K only is concerned). There is however one potential difficulty with the constraint imposed on the sequence of labels that no label should be repeated. The following simple solution was found to be relatively efficient:

Let $(K+1)$ denote the current number of segments, we first consider merging two consecutive segments and we will denote with a prime sign the quantities pertaining to the lower dimension K. We proceed as follows:

1. Select the segment merging move with probability P_M.
2. Choose a segment k in $\{1,\ldots,K\}$ (merging will be performed on the segments numbered k and $k+1$).
3. Choose the label l'_k of the merged segment in $\{1,\ldots,M\}$.

The move is systematically rejected if $l'_k = l_{k-1}$ or $l'_k = l_{k+2}$.

For ease of presentation, we describe the reverse move assuming that the current number of segments is K and quantities denoted with a prime sign will now refer to the highest dimension $(K+1)$. The move consists in splitting a segment in two. The following steps are in order:

1. Select the Split-2 move with probability P_S.
2. Choose a segment k in $\{1,\ldots,K\}$ with probability $(b_{k+1}-b_k-1)/(T-K)$ (note that longer segments are thus favoured and that length one segments which can't be split are avoided).
3. Position the new sub-segment boundary b'_{k+1} in $\{b_k+1,\ldots,b_{k+1}-1\}$.
4. Choose the two new sub-segment labels independently in $\{1,\ldots,M\}$.

The move is systematically rejected if any two successive labels in the sequence $(l_{k-1}, l'_k, l'_{k+1}, l_{k+1})$ are identical. All choices are random with uniform probabilities (except for the choice of the segment in the Split move).

The corresponding proposal ratio for the split move is found to be P_S/P_M $K/(T-K)/M$. This solution is not optimal in the sense that the proposal mechanism itself may result in "blank" moves (i.e. moves for which the chain stays in its present state), but it seems sufficient in practise. Attempts to optimize this simple scheme did not significantly improve the sampler behaviour in the case of the discrete variable dimension data. Note that the above procedure cannot be applied any more when $M = 2$, for which a more specialized scheme is needed (the case $M = 2$ is very special because there are only two valid label sequences whatever the number of segments). As usual the acceptance ratio combines the likelihood and prior ratios with the proposal ratio (see Green, 1995, for details). Figure 2 displays a typical example of application of the above analysis procedure to an aggregated traffic trace.

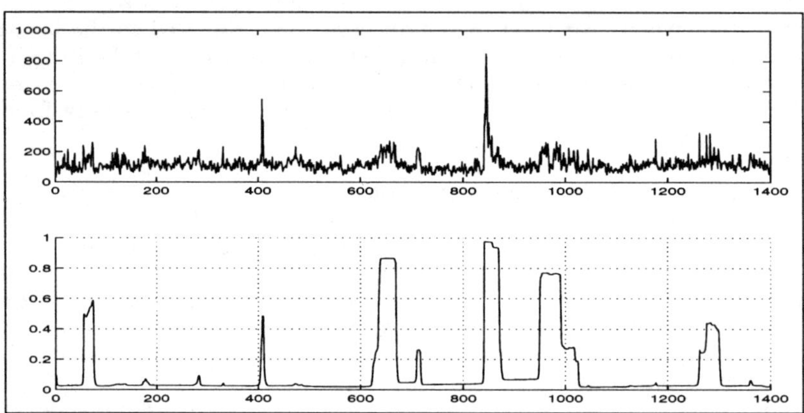

Fig. 2. Top: Block packet counts with 1 second aggregation. Bottom: Classification result - posterior probability for the class with highest mean count (M=3, 85 000 sampler sweeps)

References

Arjas, E. & Heikkinen, J. (1997). An algorithm for nonparametric Bayesian estimation of a Poisson intensity. *Computational Statistics*, **12**, 385-402.

Chib, S. (1997). Estimation and comparison of multiple changepoint models. Preprint.

Green, P. J. (1995). Reversible jump Markov chain Monte Carlo computation and Bayesian model determination. *Biometrika*, **82**, 711-732.

MacDonald, I. L. & Zucchini, W. (1997). *Hidden Markov models and other models for discrete-valued time series.* London: Chapman & Hall.

Paxson, V. & Floyd, S. (1995). Wide-Area Traffic: The Failure of Poisson Modeling. *IEEE/ACM Transactions on Networking*, **3**, 226-244.

Richardson, S. & Green, P. J. (1997). On the Bayesian analysis of mixture with an unknown number of components. *J. R. Statist. Soc. B*, **59**, 731-792.

Robert, C. P. & Titterington, M. (1997). Reparameterisation strategies for hidden Markov models and Bayesian approaches to maximum likelihood estimation. *Statistics and Computing*, in press.

Stephens, D. A. (1994). Bayesian retrospective multiple changepoint identification. *Appl. Statist.*, **43**, 159-178.

An Alternative Pruning Method Based on the Impurity-Complexity Measure

Carmela Cappelli, Francesco Mola and Roberta Siciliano

Dipartimento di Matematica e Statistica, Università di Napoli Federico II

Abstract. This paper provides a new pruning method for classification trees based on the impurity-complexity measure. Advantages of the proposed approach compared to the error-complexity pruning method are outlined showing an example on a real data set.

Keywords. Classifier, error rate, complexity measure

1 Introduction

The problem of developing a classifier from a set of examples can be successfully faced using nonparametrical methods based on classification tree procedures. These involves two main stages: *creating* the *maximal tree* by dividing recursively a set of N cases (*training set*) into two subsets according to a splitting rule; *pruning* the tree, i.e. removing the branches which overfit the data and therefore explain the training set rather than the investigated phenomenon. The pioneer pruning procedure has been introduced by Breiman *et al.* (1984) with the CART (Classification And Regression Trees) methodology. This pruning method is based on the *error*-complexity measure which takes into account the misclassification rate to evaluate the classification accuracy. In this paper we propose an alternative pruning method based on the *impurity*-complexity measure. The idea is to evaluate the accuracy by means of the impurity as measured by the Gini index of heterogeneity rather than by the misclassification rate. The latter in fact, opposed to the former, does not depend on the distribution of cases over the classes as well as on the number of classes.

In the following, after a brief presentation of the CART pruning procedure, the proposed pruning method with its main properties are described. An application on a real data is finally shown.

2 Error-complexity pruning

CART pruning method as well as other pruning methods proposed in literature (for a review, see Mingers, 1989; Esposito *et al.*, 1993) do not inspect all the possible subtrees of the maximal tree, but they produce either a single subtree or a sequence of optimally pruned subtrees employing a criterion which is always based on the misclassification rate defined at any node t as:

$$r(t) = 1 - max_j \; p_j(t), \tag{1}$$

where $p_j(t)$ is the proportion of cases at node t belonging to class j. In particular, the CART pruning considers both the aspects of the accuracy (*error rate*) and the complexity (*number of terminal nodes*) of the tree by introducing the *error-complexity* measure. This is defined for any node t and for the subtree T_t branching from t as:

$$R_\alpha(t) = R(t) + \alpha, \tag{2}$$

$$R_\alpha(T_t) = R(T_t) + \alpha|\bar{T}_t|, \tag{3}$$

where $R(t) = r(t)p(t)$ (with $p(t) = N_t/N$) and $R(T_t) = \sum_{h \in H_t} r(h)p(h)$ are the weighted error rates at node t and at subtree T_t respectively, H_t is the set of terminal nodes of T_t having cardinality $|\bar{T}_t|$ (i.e. the number of leaves of the subtree) and α is a sort of penalty for complexity. Note that the error rate which appears in the (2) as well as in the (3) is the *resubstitution estimate* of the true error rate, i.e. the error rate produced on the training set which is a rather optimistic estimate. As long as:

$$R_\alpha(t) > R_\alpha(T_t), \tag{4}$$

the subtree T_t has an error-complexity measure smaller than the node t, and therefore it will be kept. But as α increases the two measures tend to become equal making useless retaining the subtree that will be cut. The critical value of α can be found by solving the above inequality, getting to:

$$\alpha_t = \frac{R(t) - R(T_t)}{|\bar{T}_t| - 1}, \tag{5}$$

so that α, named *the complexity parameter*, gives the reduction in error per terminal node. The algorithm is in two stages: first, a sequence of nested subtrees $T_{max} \supset T_{(1)} \supset \ldots \supset T_{(k)} \supset \ldots \supset \{t_1\}$ is generated pruning at each step the subtree branching from the node with the minimum value of α, which is called *weakest link*; second, a final tree is selected as the one producing the minimum misclassification rate on an independent *test set*. This criterion is referred as $0 - SE$ rule. As an alternative criterion Breiman *et al.* suggest selecting the smallest tree whose error rate on the test set is within one standard error of the minimum. This criterion, named $1 - SE$ rule, in most cases, results in the selection of a very small tree, showing a bias toward overpruning (Cappelli & Siciliano, 1997).

3 Impurity-compexity pruning

The starting point of our proposal is a drawback of the error rate: the error rate is a rough measure that, being based on the modal class, does not take into account the distribution of cases among classes as well as the number of response classes. As a result, it might not be a suitable pruning criterion in multiclass problems. As an alternative criterion, we propose to replace the error rate in the CART procedure with the impurity as measured by the Gini index of heterogeneity defined at any node t as:

$$i(t) = 1 - \sum_{j=1}^{J} p_j^2(t). \tag{6}$$

Both the impurity and the error rate vary from 0 to $(J-1)/J$, but the impurity is never lower than the error rate: $i(t) \geq r(t)$. The higher the number of classes the higher the impurity for an equal error rate, in other words, the Gini index is very sensitive to changes in the number of classes as well as to changes in the distribution of cases among the response classes. Therefore, for each internal node t and its subtree T_t, we define the *impurity complexity* measures:

$$I_\beta(t) = I(t) + \beta, \tag{7}$$

$$I_\beta(T_t) = I(T_t) + \beta|\bar{T}_t| \tag{8}$$

where $I(t) = i(t)p(t)$ and $I(T_t) = \sum_{h \in H_t} i(h)p(h)$. The decrease in impurity per terminal node, then, is given by:

$$\beta_t = \frac{I(t) - I(T_t)}{|\bar{T}_t| - 1}, \tag{9}$$

so that, in our approach, at each iteration, the weakest link is identified by the minimum decrease in impurity per terminal node. Equation (9) can also be written as:

$$\beta_t = \frac{1}{|\bar{L}_t|} \sum_{l \in L_t} \Delta i(s^*, l)p(l), \tag{10}$$

where, for each non terminal node l of T_t, in the set L_t having cardinality $|\bar{L}_t| = |\bar{T}_t| - 1$ [1], $\Delta i(s^*, l)$ is the decrease in impurity induced by the best split s^* which maximizes:

$$\Delta i(s, l) = i(l) - [i(2l)\frac{p(2l)}{p(l)} + i(2l+1)\frac{p(2l+1)}{p(l)}], \tag{11}$$

with $p(l) = N_l/N$ and $p(2l)/p(l) = N_{2l}/N_l$ [2]. Notice that:

$$I(t) - I(T_t) = I(t) - [I(T_{2t}) + I(T_{2t+1})]. \tag{12}$$

The equivalence between equation (9) and equation (10) is verified by definition in the case of a branch of depth one, i.e. $|\bar{T}_t| = 2$, equation (12) being equal to $\Delta i(s^*, t)p(t)$. For $|\bar{T}_t| = 3$ and, without loss of generality, let $|\bar{T}_{2t}| = 2$ (i.e. node $2t + 1$ is terminal), then

$$I(T_{2t}) = i(2t)p(2t) - \Delta i(s^*, 2t)p(2t), \tag{13}$$

so that, equation (12) can be rewritten as:

$$I(t) - I(T_t) = \Delta i(s^*, t)p(t) + \Delta i(s^*, 2t)p(2t) = \sum_{l \in L_t} \Delta i(s^*, l)p(l), \tag{14}$$

where L_t covers the non terminal nodes of the subtree T_t, namely t and $2t$. By induction equation (14) follows for any $|\bar{T}_t|$.

An important property of β_t is provided by its relation to the chi-square distribution under the hypothesis of independence. This relation results from the equivalence between $\Delta i(s^*, t)$ and the numerator of the sample version of

---

[1] This relation holds for strictly binary trees where the number of internal nodes is equal to the number of leaves minus one.

[2] In binary trees it is convenient to number nodes as follows: node l is split into node $2l$ and $2l + 1$.

the Goodman and Kruskal τ (Mola & Siciliano, 1997), as well as from the χ^2 approximation of the τ sample version as shown by Light & Margolin (1971). In our notation, we can write:

$$\Delta i(s^*, l)N_l \sim i(l)\frac{N_l}{N_l - 1}\frac{\chi^2_{(J-1)}}{J-1}. \quad (15)$$

For a conservative testing procedure, $i(l)$ can be replaced by its maximum which is equal to $(J-1)/J$, moreover $N_l/(N_l-1)$ can be approximated by 1, yielding, after some simplifications, the following result:

$$\chi^2_e = JN\beta_t \sim \chi^2_{(J-1)}, \quad (16)$$

where χ^2_e is the empirical chi-square. This allows us to assign a p-value to each subtree of the sequence, so that, the final tree can be selected on the basis of a significance level.

4 An application on a real data set

The main features of the Impurity-Complexity Pruning (ICP) compared to the Error-Complexity Pruning (ECP) are shown in an application on a real data set. This consists of 286 graduates at the Faculty of Economics of the University of Naples over the period 1986 – 1989. These data relate eight categorical variables which are: *final score*, *sex*, *place of residence*, *age*, *high school diploma*, *study plan*, *time needed to graduate*, *thesis subject*. The response variable is the final score. As the response variable is ordinal, it has been coded first into two classes (*low*, *high*) and then into three classes (*low*, *medium*, *high*) providing two data sets, denoted respectively by *Grad2* and *Grad3*. Each data set has been randomly divided into two subsets, a training set (70%) and a test set (30%). The maximal trees have 42 and 55 terminal nodes respectively. In Tables 1 – 4 are reported for each data set the descriptions of the sequences of pruned subtrees provided by the two methods [3]. Note that the empirical value of the χ^2 can be associated only with the impurity-complexity pruned trees. In order to allow a comparison between the performances of the two methods, we have considered the CART selection rules, namely $0-SE$ ($T^*_{(k)}$, trees) and $1-SE$ ($T^{**}_{(k)}$ trees).

Table 1. *Grad2* data set: sequence of pruned subtrees, ECP method

Trees	Number of terminal nodes	α_{min}	weakest link	$\hat{R}(T)$
$T_{(1)}$	34	0.0013	52	30.7
$T^*_{(2)}$	30	0.0025	25	30.7
$T_{(3)}$	23	0.0029	111	32.1
$T_{(4)}$	20	0.0033	54	31.1
$T_{(5)}$	14	0.0050	55	33.6
$T_{(6)}$	10	0.0075	24	34.6
$T_{(7)}$	4	0.0100	7	31.6
$T^{**}_{(8)}$	2	0.0250	3	34.0
$\{t_1\}$	1	0.1100	1	46.0

[3] Both the methods have been implemented in the MATLAB environment.

Table 2. *Grad2* data set: sequence of pruned subtrees, ICP method

Trees	Number of terminal nodes	β_{min}	weakest link	χ^2_e	$\hat{R}(T)$
$T_{(1)}$	33	0.0012	105	0.458	31.6
$T_{(2)}$	32	0.0016	218	0.611	31.6
$T_{(3)}$	30	0.0023	98	0.879	31.7
$T_{(4)}$	28	0.0031	100	1.184	30.7
$T_{(5)}$	26	0.0033	220	1.261	30.7
$T_{(6)}$	25	0.0040	444	1.528	30.7
$T_{(7)}$	22	0.0048	111	1.834	32.1
$T_{(8)}$	21	0.0056	101	2.139	33.1
$T_{(9)}$	20	0.0061	52	2.330	33.2
$T_{(10)}$	19	0.0062	50	2.368	32.2
$T_{(11)}$	18	0.0064	110	2.445	32.2
$T_{(12)}$	16	0.0068	54	2.598	31.2
$T_{(13)}$	15	0.0070	5	2.674	31.2
$T_{(14)}$	14	0.0074	49	2.827	31.2
$T_{(15)}$	12	0.0075	27	2.865	29.0
$T^{*}_{(16)}$	11	0.0092	48	3.514	29.0
$T_{(17)}$	9	0.0095	13	3.629	30.1
$T_{(18)}$	8	0.0098	25	3.744	30.1
$T_{(19)}$	7	0.0128	2	4.889	30.2
$T^{**}_{(20)}$	5	0.0133	12	5.081	31.6
$T_{(21)}$	4	0.0155	6	5.921	33.9
$T_{(22)}$	3	0.0184	7	7.029	33.9
$T_{(23)}$	2	0.0385	3	14.707	34.0
$\{t_1\}$	1	0.0580	1	22.041	46.0

Table 3. *Grad3* data set: sequence of pruned subtrees, ECP method

Trees	Number of terminal nodes	α_{min}	weakest link	$\hat{R}(T)$
$T_{(1)}$	39	0.0025	392	45.6
$T_{(2)}$	36	0.0033	8	43.6
$T_{(3)}$	28	0.0038	28	43.1
$T^{*}_{(4)}$	14	0.0043	6	37.2
$T_{(5)}$	12	0.0050	15	37.7
$T_{(6)}$	9	0.0067	29	39.9
$T^{**}_{(7)}$	7	0.0100	5	39.1
$\{t_1\}$	1	0.0200	1	45.0

As a general observation, the ICP method makes a larger number of iterations than the ECP method, therefore it produces a broader variety of pruned subtrees. In both the applications, the proposed approach has found a better tree with respect to the error rate. Moreover, the empirical values of the χ^2 associated with the ICP sequences allow a different selection based on the choice of a significance level. For example, when choosing a significance level equal to 0.1 only a few trees are significant, these are: trees from $T_{(13)}$ up to $\{t_1\}$ for *Grad*2 and trees from $T_{(20)}$ up to $\{t_1\}$ for *Grad*3.

5 Concluding remarks

In this paper an alternative pruning method for classification trees has been introduced. The main properties of this method are as follows: it is based on an accuracy measure that, as opposed to the classical error rate, takes account of the distribution of cases among classes as well as of the number of classes; it allows a *statistical* selection of the final tree by means of the associated empirical values of χ^2; it offers a broader choice of pruned subtrees.

Table 4. *Grad3* data set: sequence of pruned subtrees, ICP method

Trees	Number of terminal nodes	β_{min}	weakest link	χ^2_e	$\hat{R}(T)$
$T_{(1)}$	49	0.0012	113	0.691	46.0
$T_{(2)}$	48	0.0014	790	0.806	46.0
$T_{(3)}$	47	0.0017	205	0.979	46.0
$T_{(4)}$	46	0.0023	475	1.325	46.2
$T_{(5)}$	45	0.0024	100	1.382	46.2
$T_{(6)}$	44	0.0026	1578	1.498	45.2
$T_{(7)}$	43	0.0027	50	1.555	45.2
$T_{(8)}$	42	0.0031	56	1.786	46.2
$T_{(9)}$	41	0.0036	785	2.074	46.2
$T_{(10)}$	40	0.0042	237	2.419	44.1
$T_{(11)}$	39	0.0045	30	2.592	44.1
$T_{(12)}$	38	0.0052	118	2.995	44.1
$T_{(13)}$	36	0.0053	17	3.053	42.1
$T_{(14)}$	32	0.0057	115	3.283	42.1
$T_{(15)}$	15	0.0059	6	3.398	38.7
$T_{(16)}$	12	0.0066	4	3.802	37.0
$T^*_{(17)}$	11	0.0070	57	4.032	35.6
$T_{(18)}$	9	0.0072	29	4.147	37.7
$T_{(19)}$	8	0.0076	15	4.378	38.2
$T_{(20)}$	7	0.0082	28	4.723	38.1
$T^{**}_{(21)}$	6	0.0093	5	5.357	37.3
$T_{(22)}$	5	0.0120	14	6.912	43.6
$T_{(23)}$	4	0.0153	2	8.813	47.6
$T_{(24)}$	3	0.0222	7	12.787	44.6
$T_{(25)}$	2	0.0245	3	14.112	44.8
$\{t_1\}$	1	0.0260	1	14.976	45.0

Acknowledgement
Research supported by MURST 60%.

References

Breiman, L., Friedman, J. H., Olshen, R. A. & Stone, C. J. (1984). *Classification and Regression Trees.* Wadsworth, Belmont CA.

Cappelli, C. & Siciliano, R. (1997). On simplification methods for decision trees. In: *Proceedings of the Ngus-97 Conference*, 104-108.

Esposito, F., Malerba, D. & Semeraro, G. (1993). Decision tree pruning as a search in the state space. In: *Machine Learning: ECML-93, Lecture Notes in Artificial Intelligence* (ed. P. Bradzil), 165-184. Berlin: Springer Verlag.

Light,R.J. & Margolin, B.H. (1971). An analysis of variance for categorical data. *Journal of the American Statistical Association*, **66**, 534-544.

Mingers, J. (1989). An empirical comparison of pruning methods for decision tree induction. *Machine Learning*,**4**, 227-243.

Mola, F & Siciliano, R (1997). A Fast Splitting Procedure for Classification trees. *Statistics and Computing*, **7**, 208-216.

Bayesian Inference for Mixture: The Label Switching Problem

Gilles Celeux
Inria Rhône-Alpes, Montbonnot 38330 France

Abstract. A K-component mixture distribution is invariant to permutations of the labels of the components. As a consequence, in a Bayesian framework, the posterior distribution of the mixture parameters has theoretically $K!$ modes. This fact involves possible difficulties when interpreting this posterior distribution. In this paper, we discuss the problem of labelling and we propose a simple and general clustering-like tool to deal with this problem.

Keywords. MCMC algorithm, labelling latent structure, k-means algorithm

1 The labelling problem

Bayesian analysis of mixtures received practical implementation since the emergence of Markov Chain Monte Carlo (MCMC) methods. Notable references for MCMC implementation of Bayesian analysis of mixtures are Lavine & West (1992), Diebolt & Robert (1994), Bensmail *et al.* (1997) and Richardson & Green (1997). It is striking that all the authors neglect or minimize the label switching problem occurring in Bayesian analysis of mixtures. This contribution is aimed to analyze this problem in different contexts and to propose a clustering procedure to overcome it.

A K-component mixture distribution is invariant to permutations of the labels $k = 1, ..., K$ of the components. As a consequence, the posterior distribution of the mixture parameters has (theoretically) $K!$ intrinsic modes which are symmetric when using a prior distribution that is indifferent to the ordering of the components. Indeed, this posterior distribution is not interpretable in its own terms and there is a need to isolate one of its $K!$ parts related to an unique labelling. A label switching occurs as some labels of the mixture components permute. This is unimportant when the goal is just to find the maximum likelihood estimate of the parameters since one estimate is just as good as any other obtained through a label switching. But, the effect of label switching is of great importance when we are concerned with Bayesian inference. Thus, avoiding the effects of label switching is desirable for a truly Bayesian inference especially when the mixture model is considered in a clustering aim.

The paper is organized as follows. In Section 2, we analyze the possible effects of label switching and review different typical situations, not reduced to the mixture context, where label switching is likely to occur. In Section 3, we discuss the drawbacks of labelling by ordering some component characteristics. In Section 4, we present a simple cluster analysis procedure to deal with the label switching problem in any situation. It is illustrated with numerical experiments on a mixture of exponential distributions.

2 Impact of the labelling problem

To illustrate our ideas, we consider Gaussian mixtures in this section, but the labelling problem arises for any mixture of distributions or more generally for any model with a finite hidden structure. The mixture distribution

$$f(x|\theta) = \sum_{k=1}^{K} \pi_k \phi(x_i|\mu_k, \Sigma_k), \tag{1}$$

where $\phi(\cdot|\mu, \Sigma)$ is the multivariate normal density with mean μ and covariance matrix Σ, $\pi = (\pi_1, \ldots, \pi_K)$ is a vector of group mixing proportions such that $\pi_k \geq 0$ and $\sum \pi_k = 1$, and $\theta = (\pi_1, \ldots, \pi_K; \mu_1, \ldots, \mu_K; \Sigma_1, \ldots, \Sigma_K)$, is invariant to permutations of the labels $k = 1, \ldots, K$ of the components.

Before presenting ways to control label switching, it is of interest to examine how the posterior distribution is in practice.

For well-separated components, label switching will rarely occur, especially in an informative setting, when using a MCMC algorithm. The approximate posterior distribution $\Pi(\theta|\mathbf{x})$ will be unimodal and thus meaningfully summarized by posterior modes or means.

For poorly separated components, label switching is likely to occur, especially in a non or weakly informative context, when using MCMC methods. The approximate posterior distribution will present more or less unequal peaks and may produce a confused picture of the distribution of interest. Moreover it can lead to poor estimates of the parameters. For instance, the estimates $\hat{\mu}_k$ derived from the mean of the posterior distribution $\Pi(\theta|\mathbf{x})$, $\mathbf{x} = (x_1, \ldots, x_n)$ denoting a sample from $f(x|\theta)$, are approximately equal to $\mu = \sum_{\ell=1}^{K} \pi_\ell \mu_\ell$ for any $k, 1 \leq k \leq K$ when the labels are a priori equally likely and treated totally symmetrically.

In some circumstances such as, for instance, estimating a multivariate Gaussian mixture with well-separated components in a clustering purpose, the labelling problem is of little practical importance, but there are many situations for which it is more acute. We can cite:

- Mixture of exponential distributions (Gruet *et al.*, 1998) which by their very nature are quite overlapping near 0. Such mixtures are useful for instance for modelling lengths of stays data (Millard & McClean, 1993).
- Full Bayesian inference for mixtures varying number of components via reversible jump samplers (Richardson & Green, 1997): relabelling is in order each time two components are merged into one and one component split into two.
- Competing risk problems for lifetime data. In those problems, the actual failure time T is $T = \min(T_1, \ldots, T_K)$ where T_ℓ, $1 \leq \ell \leq K$, denotes latent failure times, modelled, for instance, with Weibull distributions. And, the label switching problem is likely to occur when no strong prior information on the latent components is available (see Bacha *et al.*, 1998).

3 Forcing a unique labelling

If the labels are a priori equally likely, a simple idea to deal with the labelling problem when running a MCMC algorithm is to force an unique labelling by putting some constraints on the parameters to avoid any label switching. For univariate Gaussian mixtures, Richardson & Green (1997) experimented with several constraints. Most often, they consider in a natural way that the means

μ_k are in increasing order: a MCMC move is accepted if and only if the ordering of the means is unchanged. They also considered other constraints such as ordering the variances σ_k^2. They noticed that the most appropriate labelling depends on the mixture at hand and that it is advisable to postprocess the MCMC run according to different choices of labels in order to get the clearest picture of the component parameters. It is the first consequence of this way of labelling: As shown in the Section 4.3.1 of Richardson & Green (1997), the chosen constraints can influence greatly the shape of the posterior density of parameter estimates, and, as illustrated by Stephen (1997), it can happen that any constraint leads to a misleading representation of the mixture. Moreover the ordering to identify the labels requires prior indifference to the labelling before ordering. Finally, it is very difficult, if possible, to achieve in any meaningful way a choice of constraints for multivariate mixtures...

For those reasons, we do not think that forcing constraints for an unique labelling is a good way to circumvent the label switching problem. We think that a better way to deal with this problem is to use clustering-like procedures like the one presented in the next section.

4 A clustering procedure

We think that a good way to deal with the labelling problem consists of permuting the sample of the model parameters derived from the MCMC algorithm by using a clustering procedure. Stephens (1997) proposed such a procedure in a full Bayesian perspective. His procedure consists of relabelling the model parameters sample in such a way to get a unimodal marginal distribution. For the specific case of univariate Gaussian mixture with conjugate prior distributions, it consists of a clustering-like algorithm relabelling the sample to get a good fit to some member of the natural conjugate family for the parameters. This procedure works well, but can be numerically demanding and it is restricted to the limited framework of Bayesian analysis of latent structure models with conjugate prior distributions. For instance, it can not be used for non Bayesian stochastic algorithms as the stochastic EM algorithm (see Celeux *et al.*, 1996). Moreover, it is an *off-line* algorithm needing stocking huge arrays of data.

In the same spirit, we present now a solution which has several advantages. It is quite simple, not specific to Bayesian analysis with conjugate prior distributions or to the mixture context, and it does not need stocking any data. Roughly speaking, it is simply a sequential version of the k-means algorithm on the normalized sequence of the simulated parameters by the stochastic algorithm at hand.

Our procedure works as follows. Let $\theta^1, \ldots, \theta^M$ be the sequence of p dimensional vector parameters generated through a MCMC method or any other stochastic algorithm, p denoting the number of parameters of the mixture to be analyzed. (Eventually, this sequence is cleared from values obtained during a burn-in period.) We initiate our *on-line* k-means type algorithm with $K!$ clusters in the following way, from the first m vectors $\theta^1, \ldots, \theta^m$. (Typically, we take $m = 100$.)

For $i = 1, \ldots, p$ the variance of each coordinate θ_i is computed

$$(s_i^{[0]})^2 = \frac{1}{m} \sum_{j=1}^{m} (\theta_i^j - \bar{\theta}_i)^2$$

where $\bar{\theta}_i = 1/m \sum_{j=1}^m \theta_i^j$.

The initial reference centre is defined as $\bar{\theta}_1^{[0]} = \bar{\theta} = 1/m \sum_{j=1}^{m} \theta^j$. Then, the $K! - 1$ other centres $\theta_2^{[0]}, \ldots, \theta_{K!}^{[0]}$ are deduced from $\theta_1^{[0]}$ by permuting the labelling of the mixture components. From this initial position, the rth iteration of the sequential procedure consists of two steps:

- the vector θ^{m+r} is assigned to the cluster $j\star$ minimizing the normalized squared distance

$$||\theta^{m+r} - \bar{\theta}_j^{[r-1]}||^2 = \sum_{i=1}^{p} \frac{(\theta_i^{m+r} - \bar{\theta}_{ij}^{[r-1]})^2}{(s_i^{[r-1]})^2}, \; j = 1, \ldots, K!, \qquad (2)$$

  $\bar{\theta}_{ij}^{[r-1]}$ denoting the ith coordinate of the jth current centre. If $j\star \neq 1$, the coordinates of the vector θ^{m+r} are permuted to ensure that $j\star = 1$ in order to reestablish the initial labelling.
- the $K!$ centres and the p normalizing coefficients s_i^2 are updated. The first centre becomes

$$\bar{\theta}_1^{[r]} = \frac{m+r-1}{m+r} \bar{\theta}_1^{[r-1]} + \frac{1}{m+r} \theta^{m+r}$$

  and the $K! - 1$ other centres are derived from $\bar{\theta}_1^{[r]}$ by permuting the labelling of the mixture components.
  The updated variances are, for $i = 1, \ldots, p$

$$(s_i^{[r]})^2 = \frac{m+r-1}{m+r} (s_i^{[r-1]})^2 + \frac{m+r-1}{m+r} (\bar{\theta}_i^{[r-1]} - \bar{\theta}_i^{[r]})^2 + \frac{1}{m+r} (\theta_i^{m+r} - \bar{\theta}_i^{[r]})^2.$$

Remarks: The normalization of the distances (2) makes the procedure independent of scale transformation on the mixture parameters. The choice of m is not really sensitive. It must be large enough to ensure that $\bar{\theta}_1^{[0]}$ is a reasonable crude approximation of the posterior mean of θ, but not too large to avoid that a label switching litters this crude approximation. We illustrate

Table 1. Posterior means of the mixture parameters from the three Gibbs sampler GIBBS, CONST and CLUST.

method	π_1	λ_1	π_2	λ_2
GIBBS	0.51	0.80	0.49	0.80
CONST	0.85	0.86	0.15	0.34
CLUST	0.69	1.11	0.31	0.49

our procedure with the Bayesian analysis of a mixture of two exponential distributions. We generated a sample $x_1, \ldots, x_n$ of size $n = 100$ of the following distribution

$$f(x|\theta) = \sum_{k=1}^{2} \pi_k \lambda_k \exp(-\lambda_k x), \; x > 0,$$

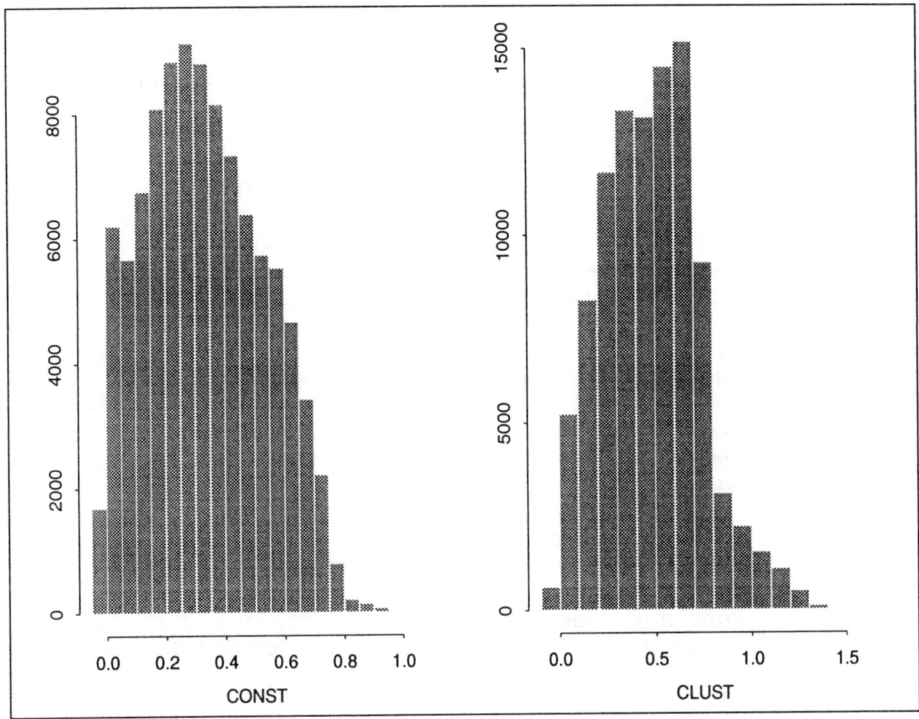

Fig. 1. Marginal posterior distribution of λ_2 for the algorithms CONST and CLUST

with $\lambda_k > 0$, $k = 1, 2$, are the scale parameters of the exponential distributions and $\theta = (\pi_1 = .5, \pi_2 = .5, \lambda_1 = 1.0, \lambda_2 = .5)$.

The considered conjugate prior distribution for θ was a Dirichlet distribution $\mathcal{D}(1,1)$ for the proportions π_1 and π_2 and a Gamma distribution $\mathcal{G}(.5,.5)$ for the scale parameters λ_1 and λ_2. Let z_i be the unknown integer $(1 \leq z_i \leq K)$ identifying the component label generating observation x_i for $i = 1, \ldots, n$. From an initial position $\theta^{[0]}$, we consider the following Gibbs sampling implementation:

1. For $i = 1, \ldots, n$, simulate $z_i^{[r]} | x_i, \theta^{[r]}$ according to its posterior distribution $(t_{ik}^{[r]}, k = 1, 2)$ conditional on $\theta^{[r]}$, namely
$$t_{ik}^{[r]} = \frac{\pi_k^{[r]} f(x_i | \theta_k^{[r]})}{\sum_j \pi_j^{[r]} f(x_i | \theta_j^{[r]})},$$
and for $k = 1, 2$ compute $n_k^{[r]} = \text{card}\{i / z_i^{[r]} = k\}$ and $\bar{x}_k^{[r]} = \sum_{i/z_i^{[r]}=k} x_i$.
2. For $k = 1, 2$, simulate $\pi_k^{[r+1]} \sim \mathcal{D}(1 + n_1^{[r]}, 1 + n_2^{[r]})$.

3. Simulate $\lambda_1^{[r+1]} \sim \mathcal{G}\,(.5+n_1^{[r]}, .5+\bar{x}_1^{[r]})$ and $\lambda_2^{[r+1]} \sim \mathcal{G}\,(.5+n_2^{[r]}, .5+\bar{x}_2^{[r]})$.

Concerning maximum likelihood inference, the EM algorithm initiated with the true parameter values gave the following estimates $p_1 = 0.52$, $\lambda_1 = 1.27$, $p_2 = 0.48$ and $\lambda_2 = 0.51$ which are quite good according to the high overlap of the mixture components.

We ran the Gibbs sampler with a burn-in period of 100000 iterations and took the next $M = 100000$ values of θ generated by the Markov chain as a sample of its posterior distribution. Table 1 gives the posterior means obtained with the raw Gibbs sampler described above (GIBBS), the Gibbs sampler working with the constraint $\lambda_1 > \lambda_2$ applied each iteration (CONST) and the Gibbs sampler incorporating our relabelling procedure (CLUST). Because of the small number (36) of visits to each labelling in this run, the figures given for GIBBS are subject to large Monte Carlo variability.

From this table, it is apparent that, as we expected, GIBBS confuses the scale parameters of the exponential distributions (And subsequently, it gives artificially good estimates of the equal proportions.) whereas, on the contrary, CONST tends exaggerating the differences between these scale parameters and poorly estimates the proportions. CLUST provides good estimates of the scale parameters, but overestimates the proportion p_1. It is worth noting that maximizing the likelihood provides better estimates.

Figure 1 displays the histograms of the scale parameter λ_2 for the sequence of 100000 recorded iterations of the algorithms CONST and CLUST. They are different because the left hand one is the posterior distribution from a Bayesian analysis based on an ordered prior distribution and the right hand one is the result of a post-processing procedure designed to interpret the output of an analysis using an unordered prior distribution. This procedure seems sensible: For instance the mean of the posterior distribution derived from CLUST provides more reliable parameter estimates.

References

Bacha M., Celeux G., Idée E., Lannoy A. & Vasseur D. (1998). *Estimation de modèles de durées de vie fortement censurées.* Paris : Eyrolles.

Bensmail, H., Celeux, G. , Raftery, A. E. & Robert, C. P. (1997). Inference in Model-Based Cluster Analysis. *Statistics and Computing*, **7**, 1-10.

Celeux G., Chauveau D. & Diebolt J. (1996). Stochastic versions of the EM algorithm: an experimental study in the mixture case. *Journal of Statistical Computation and Simulation*, **55**, 287-314.

Diebolt, J. & Robert, C. P. (1994). Bayesian estimation of finite mixture distributions. *Journal of the Royal Statistical Society* B, **56**, 363-375.

Gruet M. A., Philippe A., & Robert C. P. (1998) Estimation of the number of components in a mixture of exponential distributions. Preprint Crest.

Lavine, M. & West, M. (1992). A Bayesian method for classification and discrimination. *The Canadian Journal of Statistics*, **20**, 451-461.

Millard P. & McClean S. (1993) Patterns of length of stay after admission in geriatric medicine : an event history approach. *The Statistician*, **42**, 263-274.

Richardson, S. & Green, P. J. (1997) On Bayesian analysis of mixtures with an unknown number of components. (with discussion) *Journal of the Royal Statistical Society* B, **59**, 731-792.

Stephens, M. (1997) Discussion of the Richardson & Green article. *Journal of the Royal Statistical Society* B, **59**, 768-769.

Simulation of Multifractional Brownian Motion

Grace Chan[1] & Andrew T. A. Wood[2]

[1] Department of Statistics and Actuarial Science, University of Iowa, Iowa City, IA 52242–1419, USA
[2] Department of Mathematical Sciences, University of Bath, Bath, BA2 7AY, UK

Abstract. Multifractional Brownian motion (MFBm) is a generalization of Fractional Brownian motion (FBm) in which the Hurst parameter varies with time in a prescribed manner. In this paper we investigate an approximate method for simulating realizations of MFBm on a finite grid.

Keywords. Fractal dimension, Hurst parameter, kriging, long-range dependence, sample path roughness, stationary increments

1 Introduction

1.1 Review of FBm

FBm is a zero-mean self-similar Gaussian process $\{B_H(t)\}_{t\geq 0}$ with covariance function given by

$$cov\left\{B_H(s), B_H(t)\right\} = \frac{1}{2}\left\{|s|^{2H} + |t|^{2H} - |s-t|^{2H}\right\} \tag{1}$$

where $H \in (0,1)$ is known as the *Hurst parameter*. Note that $B_H(0) \equiv 0$ and $var\{B_H(t)\} = t^{2H}$. In the special case $H = 1/2$, FBm reduces to ordinary Brownian motion. An important property of FBm, which follows from (1), is that it has *stationary increments.*

FBm has proved to be of considerable interest in diverse fields such as Hydrology, Signal Processing and Financial Mathematics. Basic properties of FBm are given by Mandelbrot & Van Ness (1968) and Falconer (1990), and multivariate generalizations are considered by Adler (1981).

The Hurst parameter H plays a dual rôle: on the one hand, it determines the *roughness* of the sample path; and on the other, it determines the *long-range dependence* properties of the process. With regard to sample path roughness, H determines the fractal (or Hausdorff) dimension, D, of the (graph of the) sample path through the following simple formula: $D = 2 - H$. See Adler (1981) and Falconer (1990) for details.

For future reference, we now mention two distinct integral representations of FBm:

$$B_H(t) = C_1(H)\left\{\int_{-\infty}^{t} (t-s)^{H-(1/2)} dW(s) - \int_{-\infty}^{0} (-s)^{H-(1/2)} dW(s)\right\} \tag{2}$$

and

$$B_H(t) = C_2(H) \int_{-\infty}^{\infty} \{1 - \cos(st) - \sin(st)\}|s|^{-H-(1/2)} dW(s) \tag{3}$$

where $C_1(H)$ and $C_2(H)$ are chosen so that $var\{B_H(t)\} = |t|^{2H}$, and W is a pure *white noise* process, i.e. Brownian motion on $(-\infty, \infty)$. For (2), see Mandelbrot & Van Ness (1968, p.423). Representation (3) is (essentially) due to Hunt; see Mandelbrot & Van Ness (1968, p. 435) for closely related formulae.

1.2 Simulation of FBm

Suppose that we wish to simulate $\{B_H(t)\}$ at locations on a grid $\delta, 2\delta, \ldots, n\delta$, where the grid width δ is open to choice, and n, the number of grid points, is very large. Then, writing $t_j = j\delta$, we define the increments

$$Y_j = \delta^{-H}\left\{B_H(t_j) - B_H(t_{j-1})\right\} \qquad (j = 1, \ldots, n) \tag{4}$$

and proceed as follows:

Step 1: Simulate $Y_1, \ldots, Y_n$;
Step 2: For each j, calculate $B_H(t_j) = \delta^H(Y_1 + \ldots + Y_j)$.

It follows from (1) that the Y-sequence is stationary, and that the covariance between Y_j and Y_k is given by (9) with $H_u = H_v = H$. So Step 1 may be performed efficiently using the *circulant embedding* approach for simulating stationary Gaussian processes; see Section 4 for an outline of this method.

2 Multifractional Brownian motion

In some real datasets there is evidence that the roughness of the sample path varies with location. In such cases, a single number H (or D) may not provide an adequate global description of the roughness of the sample path and there is motivation for developing models which allow for varying roughness. Lévy-Véhel (1995) has considered such datasets in Image Analysis and Signal Processing contexts, and these led him to consider a generalization of FBm which he calls Multifractional Brownian motion (MFBm). A key feature of MFBm is that sample path roughness is described by a function $H(t)$ rather than just a single number. It makes sense to define the *local* fractal dimension of the sample path as $D(t) = 2 - H(t)$.

Lévy-Véhel's (1995) construction of MFBm is as follows: given a continuous function $H(t)$ strictly bounded by 0 and 1, define $X(H(t), t) \equiv B_{H(t)}(t)$ for $t \geq 0$ using (2). However, for present purposes, Lévy-Véhel's construction is inconvenient because the covariance of $B_{H_1}(t_1)$ and $B_{H_2}(t_2)$, obtained via (2), is rather complex. It turns out that, if we use (3) rather than (2), then the covariance of $B_{H_1}(t_1)$ and $B_{H_2}(t_2)$ is much simpler, and is given by

$$\frac{g(H_1, H_2)}{2}\{|t_1|^{H_1+H_2} + |t_2|^{H_1+H_2} - |t_1 - t_2|^{H_1+H_2}\} \tag{5}$$

where $g(H_1, H_2) = \{I(H_1)I(H_2)\}^{-1/2}\, I\{(H_1 + H_2)/2\}$ and

$$I(H) = \begin{cases} \dfrac{\Gamma(1-2H)}{H}\sin\left\{\dfrac{\pi}{2}(1-2H)\right\} & H \in (0, 1/2), \\ \pi & H = 1/2, \\ \dfrac{\Gamma\{2(1-H)\}}{H(2H-1)}\sin\left\{\dfrac{\pi}{2}(2H-1)\right\} & H \in (1/2, 1). \end{cases} \tag{6}$$

The derivation of (5) follows directly from (3). We omit the details, except to mention that (6) is obtained using formulae on p.422 of Gradshteyn & Ryzhik (1965). Note that when $H_1 = H_2$, $g(H_1, H_2) \equiv 1$. It would be interesting to know whether the differences between the MFBm's constructed via (2) and (3) are of any significance in statistical applications.

3 Simulation of MFBm

3.1 The algorithm

Suppose that we are given a (deterministic) time-varying Hurst function $H(t)$ and that we wish to simulate an MFBm $X(H(t), t) \equiv B_{H(t)}(t)$ at locations on a finite grid. Choose $0 < H_1 < \dots H_m < 1$, for example

$$H_u = u/(m+1) \qquad (u = 1, \dots, m). \tag{7}$$

These H_u's will be Hurst parameters for m correlated FBms whose joint covariance structure is determined by (5). Let δ, n and the t_j's be as in subsection 1.2 and, bearing (3) in mind, define increments

$$Y_{j,u} = \delta^{-H_u} \{B_{H_u}(t_j) - B_{H_u}(t_{j-1})\} \qquad (j = 1, \dots, n;\ u = 1, \dots, m) \tag{8}$$

and write $W_j = (Y_{j,1}, \dots, Y_{j,m})^T$. Our simulation algorithm may be specified in outline as follows:

Step 1 Simulate $\{Y_{j,u} : j = 1, \dots, n;\ u = 1, \dots, m\}$.
Step 2 For each j and u, calculate $B_{H_u}(t_j) = \delta^{H_u}(Y_{1,u} + \dots + Y_{j,u})$.
Step 3 For each j, "predict" $X(H(t_j), t_j)$ using some form of kriging based on the "observations" $\{B_{H_u}(t_j) : j = 1, \dots, n;\ u = 1, \dots, m\}$.

Using (5), it is seen that $\{W_j : j - 1, 2, \dots\}$ is a stationary vector valued Gaussian sequence with the covariance of $Y_{j,u}$ and $Y_{k,v}$ given by

$$\frac{g(H_u, H_v)}{2} \left\{ |j-k-1|^{H_u+H_v} + |j-k+1|^{H_u+H_v} - 2|j-k|^{H_u+H_v} \right\}. \tag{9}$$

Thus Step 1 can be performed efficiently using the circulant embedding approach for stationary *vector-valued* Gaussian processes; see Section 4. Since Step 2 is a matter of simple arithmetic, this leaves Step 3, which we now discuss.

3.2 Kriging neighbourhoods

The kriging in Step 3 is performed as follows: at each location $(H(t_j), t_j)$, we specify a set of neighbours $N_j - \{(v, k)\}$ and then predict $X(H(t_j), t_j)$ by

$$X(H(t_j), t_j) = \sum_{(v,k) \in N_j} \gamma_{v,k}^{(j)} B_{H_v}(t_k) \tag{10}$$

We shall use the two types of N_j specified in Figure 1 ("small" and "large").

If we adopt a Mean Squared Error (MSE) criterion then, using (5) to obtain all covariances below, the optimal choice of $\gamma^{(j)} = (\gamma_{v,k}^{(j)} : (v,k) \in N_j)$ is given by

$$\hat{\gamma}^{(j)} = \{cov(Z_j)\}^{-1} cov\left(Z_j, B_{H(t_j)}(t_j)\right) = A_j^{-1} B_j \quad \text{(say)} \tag{11}$$

Small

	t_{j-1}	t_j	t_{j+1}
H_{u+1}	•	•	•
$H(t_j)$		X	
H_u	•	•	•

Large

	t_{j-2}	t_{j-1}	t_j	t_{j+1}	t_{j+2}
H_{u+2}	•	•	•	•	•
H_{u+1}	•	•	•	•	•
$H(t_j)$			X		
H_u	•	•	•	•	•
H_{u-1}	•	•	•	•	•

Fig. 1. Kriging Neighbourhoods (X=target, o=neighbour)

where $Z_j = (B_{H_v}(t_k) : (v,k) \in N_j)$; and the optimal MSE is given by

$$MSE_j = var\left\{B_{H(t_j)}(t_j)\right\} - B_j^T A_j^{-1} B_j \quad (j = 1, \ldots, n). \tag{12}$$

If $(H(t_j), t_j)$ is close to the boundary of the simulation region then some of the $B_{H_v}(t_k)$ at neighbouring locations may not have been simulated. In such cases, we simply use those neighbouring locations which are available, with the obvious modifications to (10)–(12).

4 Simulation of stationary Gaussian processes

We briefly outline the *circulant embedding* approach for simulating stationary Gaussian processes. For further details, see Chan & Wood (1997a,b). This procedure can be used to perform Step 1 of the algorithms for simulating FBm and MFBm; see subsections 1.2 and 3.1 respectively.

Let $\ldots, Y_{-1}, Y_0, Y_1, \ldots$ be a stationary sequence of zero-mean scalar Gaussian variables with covariance function $\sigma_{i-j} = cov(Y_i, Y_j)$. Suppose that we wish to simulate $Y = (Y_1, \ldots, Y_n)^T \sim N_n(0, V)$. Note that $V = [v_{ij}]$ is a Toeplitz covariance matrix since $v_{ij} = \sigma_{i-j}$ depends only on $i - j$. For large n, the circulant embedding approach provides a fast, efficient and sometimes theoretically exact procedure.

We recall the definition of a circulant matrix: $C = [c_{ij}]$ is an $m \times m$ circulant matrix if $c_{ij} = a_k$ where $k = j - i \pmod m$ and $a_0, \ldots, a_{m-1}$ are arbitrary numbers. The simulation procedure is based on the following facts.

(a) Any symmetric $n \times n$ Toeplitz matrix V with first row $\sigma_0, \ldots, \sigma_{n-1}$ can be embedded in (i.e. is a submatrix of) the symmetric circulant matrix of even dimension $m \geq 2(n-1)$ whose first row $a_0, \ldots, a_{m-1}$ is given by $a_j = \sigma_j$ if $0 \leq j \leq m/2$ and $a_j = \sigma_{m-j}$ if $m/2 < j \leq m-1$.

(b) The unit eigenvectors $q_0, \ldots, q_{m-1}$ of a symmetric circulant matrix C have components $q_j(k) = m^{-1/2} \exp(-2\pi\sqrt{-1}jk/m)$, $j, k = 0, 1, \ldots, m-1$; and the corresponding eigenvalues are $\lambda_j = \sum_{k=0}^{m-1} a_k \exp(-2\pi\sqrt{-1}jk/m)$.

The main implication of point (b) is that, if C is a symmetric non-negative definite $m \times m$ circulant matrix, and $m = 2^g$ where g is an integer, then $G \sim N_m(0, C)$ can be simulated very efficiently via two applications of the Fast Fourier Transform. Moreover, if $Y \sim N_n(0, V)$ is the vector we want to simulate, and C is a non-negative definite circulant embedding for V, then $Y = (G_i, i = 1, \ldots, n)^T$ has exactly the required distribution. We should choose $m = 2^g \geq 2(n-1)$ to be as small as possible subject to the embedding matrix C in point (a) being non-negative definite; if it is not possible to find such an m, then Chan & Wood (1997a) suggest setting negative eigenvalues in C to zero, in which case the resulting procedure is approximate.

In the above discussion, we have only considered stationary scalar-valued Gaussian sequences; similar ideas can be used for vector-valued sequences (which we use for simulating MFBm) though the details are a little more involved; see `http://www.maths.unsw.edu.au/~grace` and Chan & Wood (1997b) for further details.

5 Numerical results

We now describe numerical results obtained using the simulation algorithm presented in the previous section. Three choices for $H(t)$ were considered:

linear: $H(t) = t$;
logistic: $H(t) = 0.3 + 0.3/[1 + \exp\{-100(t - 0.7)\}]$;
periodic: $H(t) = 0.5 + 0.49\sin(4\pi t)$.

We only report the results with $\delta = 1/n$, $n = 1000$, $m = 8$ and the H_u's chosen according to (7); other cases were broadly similar.

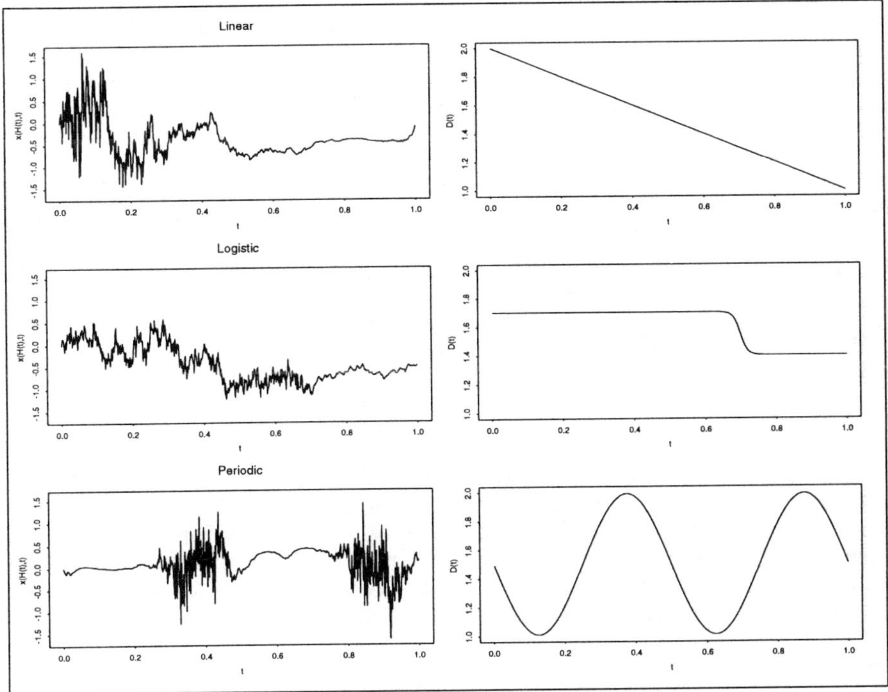

Fig. 2. Three examples of Multifractional Brownian motion

In Figure 2, each function on the left is a single realization of an MFBm, obtained using the small kriging neighbourhoods (see Figure 1); and each curve on the right is the corresponding local dimension function $D(t) = 2 - H(t)$. Observe that Figure 2 displays exactly the behaviour we would expect: the larger (smaller) the value of $D(t)$ on the right, the rougher (smoother) the sample path on the left.

In Figure 3, the MSE obtained from (12) is displayed in the linear and periodic cases. Note that, in both cases, the MSE is very small unless $H(t)$

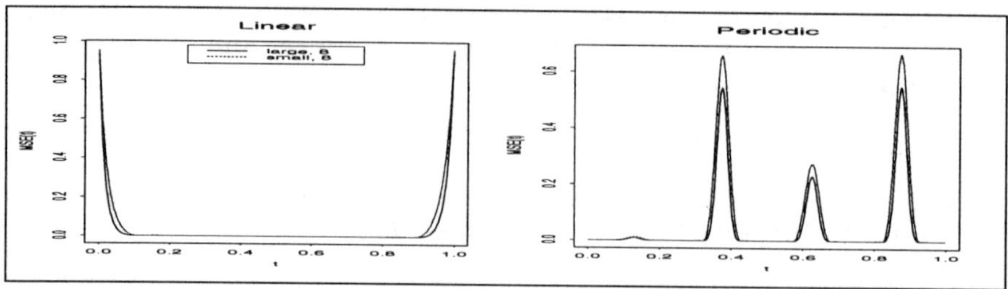

Fig. 3. Prediction mean squared errors

is "out of range" (i.e. unless either $H(t) < H_1 = 1/9$ or $H(t) > H_m = 8/9$). In the logistic case, $H(t)$ was always "within range" and the MSE was uniformly close to zero. In particular, the largest value of MSE_j in (12) in the logistic case was 3.2×10^{-5} (5×10^{-5}) when the large (small) kriging neighbourhoods were used. These MSE results suggest two broad conclusions: first, the importance of choosing H_1 and H_m so that $H(t)$ stays within range (i.e. $H_1 \leq H(t) \leq H_m$ for all t); and second, that the gain in distributional accuracy achieved by using the large instead of small kriging neighbourhoods is rather modest.

We also compared CPU timings for the small and large kriging neighbourhoods. The results indicate that if the small kriging neighbourhoods are used, then the proportional of total CPU time spent in Step 3 is very modest; but if the large kriging neighbourhoods are used, then this proportion is rather substantial. Bearing this and the comments in the previous paragraph in mind, and noting that the algorithm is somewhat easier to program when the small neighbourhoods are used, we suggest that the approach based on the small kriging neighbourhoods is to be preferred.

6 Conclusions

The results in Section 4 indicate that our algorithm for simulating MFBm using the small kriging neighbourhoods should have sufficient distributional accuracy for most practical purposes. However, note how important it is to ensure that $H(t)$ stays "within range" (see Section 5); this can always be arranged in practice, provided $H(t)$ is bounded away from 0 and 1. In future work, we hope to discuss statistical applications of MFBm.

References

Adler, R. (1981). *The Geometry of Random Fields.* New York: Wiley.

Chan, G. & Wood, A.T.A. (1997a). An algorithm for simulating stationary Gaussian random fields. *Applied Statistics* **46**, 171–181.

Chan, G. & Wood, A.T.A. (1997b). Simulation of stationary Gaussian vector fields. Submitted to *Statistics and Computing.*

Falconer, K.J. (1990). *Fractal Geometry: Mathematical Foundations and Applications.* New York: Wiley.

Gradshteyn, I.S. & Ryzhik, I.M. (1965). *Tables of Integrals, Series and Products.* New York: Academic Press.

Lévy-Véhel, J. (1995). Fractal approaches in signal processing. *Fractals*, **3**, 755–775.

Mandelbrot, B. B. & Van Ness, J. W. (1968). Fractional brownian motions, fractal noises and applications. *SIAM Review*, **10**, 422–437.

Simulating Categorical Data from Spatial Models

Trevor F. Cox[1] and Michael A.A. Cox[2]

[1] Department of Statistics, University of Newcastle, Newcastle upon Tyne, NE1 7RU, UK
[2] Department of Management Studies, University of Newcastle, Newcastle upon Tyne, NE1 7RU, UK

Abstract. Categorical data are simulated using random rotating hyperplanes superimposed on a spatial pattern of points in a d-dimensional space and also by random hyperspheres. These data can be used as a source for testing various statistical techniques. Their use in multidimensional scaling in particular is investigated.

Keywords. Categorical data, multidimensional scaling, rotating hyperplane process, spatial process

1 Introduction

Sibson *et al.* (1981), a paper in a series of three, studying the robustness of multidimensional scaling (MDS), simulate dissimilarities from an underlying spatial pattern of points. They use four models, the first of which uses a Poisson hyperplane process to split the space containing the points into two half-spaces a number of times. For every hyperplane those points in one half-space are allocated the value unity and those in the other half-space the value zero. The model thus generates binary data, the dimension of which is the number of hyperplanes. The Hamming distance is then used as a measure of dissimilarity, the model giving a dependence structure to the dissimilarities. This is important as most papers on the Monte Carlo testing of multidimensional scaling have dissimilarities simulated to be independent, see for example Stenson & Knoll (1969), Klahr (1969), Sherman (1972), Cox & Cox (1990).

Sibson *et al.* use another model to generate categorical data where centres of hyperspheres form a Poisson process and where the radii of the hyperspheres follow some distribution. Binary data are generated as with the hyperplane process but with points within spheres being allocated the value unity and points outside the value zero.

This paper extends these spatial models of Sibson *et al.* to simulate categorical data for use in a variety of multivariate techniques. There are two aspects to the simulation of the data, the initial pattern of points in a d dimensional space and then the division of the space into regions where all points in a particular region are allocated to the same category for the associated variable. Thus if a spatial pattern of n points is used, and then there are p separate divisions of the space by some mechanism, then an n by p data matrix of categorical data will be generated. These categorical data can then be augmented if required by some continuous data. For instance by the distance of each point from some origin, or by some measurement relating to the original spatial pattern of points, for example the intensity of the underlying spatial process at the particular point. The data can then be subjected to multivariate techniques, results of which can be related back to the original spatial pattern. This is important for some multivariate techniques such as multidimensional scaling since a measure of its validity is how well the technique can reconstruct spatial patterns of points from dissimilarities based on the distances between the points in the original spatial pattern.

2 Simulating the categorical data

A computer program was written to simulate the categorical data. Any d dimensional spatial pattern of points can be used as the starting configuration. The spatial patterns chosen were: an homogeneous Poisson process; a Poisson cluster process where cluster centres follow a Poisson process and then for each cluster, a random number of cluster members are randomly distributed about the cluster centre; a regular process where points at the vertices of a regular lattice are randomly displaced from their positions; an inhomogeneous Poisson process with intensity function chosen as a polynomial in the coordinates of the space. See Cressie (1993) for further details.

The methods for dividing the space to generate the categorical data were variations on a hypersphere process and a rotating hyperplane process. For the first hypersphere process, concentric hyperspheres are placed in the d-dimensional space with centre a random point from a Poisson process. The volume of the inner hypersphere is proportional to the probability p_1 for the first category, the volume between this first hypersphere and the second is proportional to p_2, and so forth. Points in the spatial pattern lying between the ith and $(i+1)$st hypersphere are allocated to category $i+1$. The radius of the outer hypersphere can either be fixed or random according to some distribution. Points in the region outside the final hypersphere are allocated to the final category. If the probability for this category is to be a particular value then the radius of the outer hypersphere needs to be set accordingly. Also, only random hyperspheres which lie totally in the sampling region can be used in this case. The second version of the process does not require the hyperspheres to be concentric, but has their centres as cluster members in a cluster of a Poisson cluster process. The third version has the hypersphere centres as the points in an inhibition process.

In the Poisson rotating hyperplane process a hyperplane is rotated sequentially $c-1$ times, about an axis lying in the hyperplane, through angles proportional to the probabilities for each category, each rotation forming a new hyperplane. All points between two consecutive hyperplanes are then allocated to the appropriate category.

For the rotating hyperplane process let the centroid of the plane to be rotated be given by $\mathbf{a}$. Let $\mathbf{n}_1$ be the unit normal vector to the plane. The random plane is actually generated with $\mathbf{a}$ as a realised point from a d-dimensional Poisson process and $\mathbf{n}_1$ as a unit vector with direction hyperspherically uniformly distributed. Let $\mathbf{m}$ be a random unit vector at $\mathbf{a}$ again hyperspherically uniformly distributed. Then $\mathbf{n}_1$ is rotated towards $\mathbf{m}$ which in turn rotates the plane about an axis orthogonal to $\mathbf{n}_1$ and $\mathbf{m}$. Only the half-plane bounded at the axis of rotation is used. If there are to be c categories for the categorical variable being simulated with associated probabilities p_j, then the unit vector $\mathbf{n}_1$ is rotated successively through angles $\theta_j = 2\pi p_j$ in turn towards $\mathbf{m}$ finishing back where it started. Each rotation gives rise to another unit normal vector defining another half-plane, a rotation of the original half-plane. Let these unit normals be denoted by $\mathbf{n}_1, \mathbf{n}_2, \ldots, \mathbf{n}_c$ with associated rotated half-planes, $P_1, P_2, \ldots, P_c$. Points in the spatial pattern which lie between the half-planes P_i and P_{i+1} are allocated to category i.

Consider the rotation of $\mathbf{n}_1$ to $\mathbf{n}_2$. Let $\mathbf{n}_2 = \alpha\mathbf{n}_1 + \beta\mathbf{m}$. Then

$$\mathbf{n}_2.\mathbf{n}_2 = \alpha^2 + \beta^2 + 2\alpha\beta\mathbf{n}_1.\mathbf{m} \quad \text{and} \quad \mathbf{n}_2.\mathbf{n}_1 = \alpha + \beta\mathbf{n}_1.\mathbf{m}$$

which gives

$$\beta = \sqrt{\frac{1 - \cos^2(\theta_1)}{1 - (\mathbf{n}_1.\mathbf{m})^2}}, \qquad \alpha = \cos(\theta_1) - \beta\mathbf{n}_1.\mathbf{m}.$$

The positive root is needed for $\theta_1 < \pi$ and the negative root for $\theta > \pi$. The normal $\mathbf{n}_2$ is then rotated to $\mathbf{n}_3$ and so forth.

To find which two half-planes enclose a particular point $\mathbf{x}$ of the spatial pattern, the angle ϕ_i between the unit vector $\mathbf{u} = (\mathbf{x} - \mathbf{a})/|\mathbf{x} - \mathbf{a}|$ and the plane P_i is found for $i = 1, \ldots, c$, where $\phi_i = \pi/2 - \cos^{-1}(\mathbf{u}.\mathbf{n}_i)$. Suppose each half-plane is now projected past the axis of rotation. This now splits the space into $2c$ regions each bounded by two successive half-planes. It is impossible to tell whether the (acute) angle ϕ_i is the angle between the original half-plane or its projection past the axis or rotation, and so the sequence of angles $\{\phi_i\}$ has to be considered. Let the sign of ϕ_i be denoted by + or − and then each region will have a unique sequence of + and − terms. For example consider half-plane P_i rotating around the axis of rotation. Then every point in regions encountered in the first 180° will have a positive value of ϕ_i. All regions in the next 180° will have a negative value of ϕ_i. This is illustrated in Figure 1 where there are four half-planes (lines) in a two-dimensional space together with their extensions, superimposed on a realisation of a Poisson process. Each point in a region has the same sequence of +'s and −'s. To find the correct allocation of a particular point $\mathbf{x}$, if the sequence $\{\phi_i\}$ commences with j positive values then the angle of $\mathbf{x}$ with the first half-plane is $\sum_{i=1}^{j-1} \theta_i + \phi_j$

while if the sequence commences with j negative values then the required angle is $\pi + \sum_{i=1}^{j-1} \theta_i - \phi_j$ and hence the category can be found.

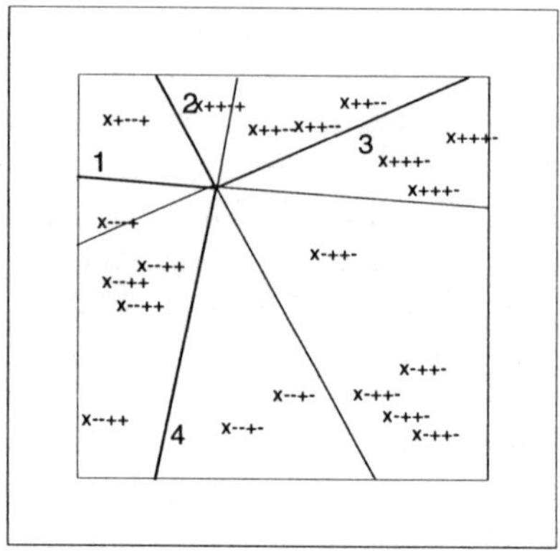

Fig 1. A rotating hyperplane (line) in a 2-dimensional space superimposed on a realisation from a Poisson process. Bold lines are the four half-planes; faint lines are their projections

3 Statistical analyses using the simulated data

Categorical data generated using the processes just described can be subjected to various statistical techniques such as contingency table analysis and its extension to log-linear models, cluster analysis and logistic discrimination and regression. Only a limited description of some results for multidimensional scaling can be given here.

Multidimensional scaling attempts to construct a configuration of points in a space, usually Euclidean, where each point represents an object or individual. The configuration is such that distances between points match as well as possible the original data in the form of dissimilarities measured between the objects or individuals. The dissimilarities can be measured in a variety of ways, see for example Cox & Cox (1994). For categorical data a dissimilarity using Gower's general similarity coefficient is a convenient measure, Gower (1971), where the dissimilarity between objects r and s based on p categorical variables is given by

$$\delta_{rs} = 1 - p^{-1} \sum_{i=1}^{p} s_{rsi}$$

where $s_{rsi} = 1$ if objects r and s share the same category on variable i, and zero otherwise.

This dissimilarity measure was used on various categorical data sets generated using the methods described previously. The dissimilarities were then subjected to multidimensional scaling and the degree to which the MDS configuration agreed with the original spatial pattern of points generating the categorical data was measured by the Procrustes statistic, see for example

Sibson (1978). A value of the statistic equal to zero implies a perfect fit, while values close to unity imply no fit at all. The fit of the distances in the MDS configuration to the dissimilarities is given by the STRESS.

Figure 2(i) shows three rotating Poisson hyperplanes (lines) in a realisation of a two dimensional Poisson process, giving rise to three categorical variables, the first with three categories with probabilities (1/3, 1/2, 1/3), the second with three categories with probabilities (1/6, 1/3, 1/2) and the third with two categories with probabilities (1/6, 5/6). Figure 2(ii) shows the results of using MDS on the categorical data generated. The configuration has been subjected to Procrustes analysis to match it up with the original configuration of points. The value of the STRESS was zero and that of the Procrustes statistic was 0.17. The points in the MDS configuration are more clustered but this is not surprising as there are only four possible values of the dissimilarity measure. It should be reported that in this example the starting configuration used by the MDS program was the original spatial configuration of points generating the data and it could be argued that this gives an unfair advantage. When a random starting configuration of points is used the points form more clusters of coincident points, again with zero STRESS. A problem of a zero STRESS MDS configuration is that often the points can be moved an apreciable amount within the configuration without changing the zero STRESS value.

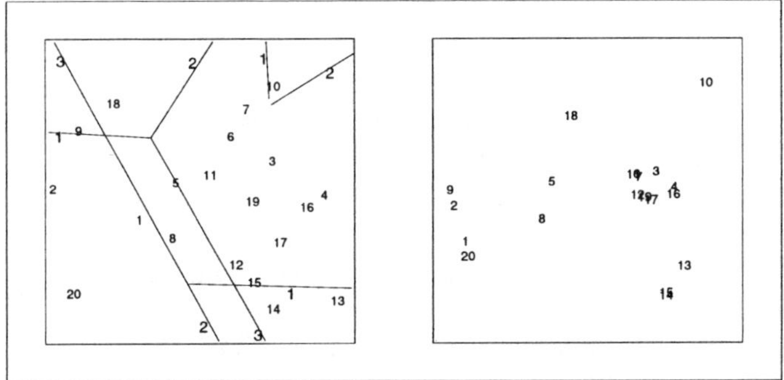

Fig. 2. Three rotating Poisson hyperplanes (lines) in a realisation of a two dimensional Poisson process together with the MDS configuration based on the generated categorical data

Table 1 shows Procrustes statistics for various MDS configurations matched to the original spatial patterns from which the categorical data were generated using random rotating hyperplanes. Various spatial patterns, number of dimensions, number of variables and number of categories were used. For each variable, equal probabilities for categories were chosen. The first row of the table shows the effect of changing the number of categories for a Poisson process. The Procrustes statistic tends to be larger when there are fewer categories. The second row shows the effect of changing the number of variables, with the Procrustes statistic being larger when there are fewer variables. The third row shows that the higher the dimension, the higher the value of the statistic. Similar results for regular and Poisson cluster processes for various numbers of variables and categories are also shown in the table.

Table 1. Procrustes statistics for matching MDS configurations to the original configurations generating the categorical data. Key: d-dimensions; n-number of points; v-variables; c-categories. Results are averaged over five simulations

Poisson process, 2d, 3v, n=30					
2c 0.279	3c 0.217	4c 0.152	5c 0.183	7c 0.142	9c 0.183
Poisson process, 2d, 3c, n=30					
2v 0.192	3v 0.177	4v 0.200	5v 0.178	7v 0.104	9v 0.120
Poisson process, 10v, 4c, n=50					
2d 0.118	3d 0.181	4d 0.309	5d 0.353	7d 0.437	9d 0.432
Regular process, 2d, 3v, n=45					
2c 0.244	3c 0.193	4c 0.191	5c 0.180	7c 0.168	9c 0.214
Regular processs, 2d, 3c, n=45					
2v 0.233	3v 0.193	4v 0.170	5v 0.194	7v 0.147	9v 0.134
Poisson cluster process, 2d, 3v, n=55					
2c 0.242	3c 0.194	4c 0.184	5c 0.248	7c 0.226	9c 0.183
Poisson cluster process, 2d, 3c, n=55					
2v 0.185	3v 0.214	4v 0.184	5v 0.169	7v 0.182	9v 0.158

References

Cox, T.F. & Cox, M.A.A. (1990). Interpreting stress in multidimensional scaling. *J. Statist. Comput. Simul.*, **37**, 211-223.

Cox, T.F. & Cox, M.A.A. (1994). *Multidimensional Scaling.* London: Chapman and Hall.

Cressie, N.A.C. (1993). *Statistics for Spatial Data, Revised Edition.* New York: Wiley.

Gower, J.C. (1971). A general coefficient of similarity and some of its properties. *Biometrics*, **27**, 857-874.

Klahr, D. (1969). A Monte Carlo investigation of the statistical significance of Kruskal's nonmetric scaling procedure. *Psychometrika*, **34**, 319-330.

Sherman, C.R. (1972). Nonmetric multidimensional scaling: a Monte Carlo study of the basic parameters. *Psychometrika*, **34**, 323-355.

Sibson, R. (1978). Studies in the robustness of multidimensional scaling: Procrustes statistics. *J. R. Statist. Soc.*, **B 40**, 234-238.

Sibson, R., Bowyer, A. & Osmond, C. (1981). Studies in the robustness of multidimensional scaling: Euclidean models and simulation studies. *J. Statist. Comput. Simul.* **13**, 273-296.

Stenson, H.H. & Knoll, R.L. (1969). Goodness of fit for random rankings in Kruskal's nonmetric multidimensional scaling. *Multivariate Behavioral Research*, **8**, 511-517.

Robust Factorization of a Data Matrix

Christophe Croux[1] and Peter Filzmoser[2]

[1] ECARE and Institut de Statistique, Université Libre de Bruxelles, Av. F.D. Roosevelt 50, B-1050 Bruxelles, Belgium
[2] Department of Statistics and Probability Theory, Vienna University of Technology, Wiedner Hauptstraße 8-10, A-1040 Vienna, Austria

Abstract. In this note we show how the entries of a data matrix can be approximated by a sum of row effects, column effects and interaction terms in a robust way using a weighted L_1 estimator. We discuss an algorithm to compute this fit, and show by a simulation experiment and an example that the proposed method can be a useful tool in exploring data matrices. Moreover, a robust biplot is produced as a byproduct.

Keywords. Alternating regressions, biplot, factor model, robustness

1 Introduction

Multivariate data can often be represented in the form of a data matrix whose elements will be denoted by y_{ij}, where $1 \leq i \leq n$ denotes the row index, and $1 \leq j \leq p$ the column index. Each entry in the data matrix is supposed to be the realization of a random variable

$$Y_{ij} = \mu_{ij} + \varepsilon_{ij}, \tag{1}$$

where μ_{ij} is the median value of each variable Y_{ij} and the residuals ε_{ij} are supposed to form a white noise. It is assumed that the values μ_{ij} can be decomposed as a sum of four terms:

$$\mu_{ij} = c + a_i + b_j + \sum_{l=1}^{k} \lambda_{jl} f_{il}, \tag{2}$$

with $k \leq p$. The constant c can be interpreted as an overall median, a_i as a row effect and b_j as a column effect. The last term represents the interaction between rows and columns and is factorized as the scalar product between a vector of *loadings* $\lambda_{j\cdot} = (\lambda_{j1}, \ldots, \lambda_{jk})^\top$ and a vector of *scores* $f_{i\cdot} = (f_{i1}, \ldots, f_{ik})^\top$. The above model is like the FANOVA model introduced by Gollob (1968), which combines aspects of analysis of variance and factor analysis. We are mainly interested in data matrices in which the rows represent individuals, and the column variables possibly represent different types of measurement. Therefore we will not continue to pursue symmetry between rows and columns. To identify uniquely the parameters a_i, b_j, and c, the following restrictions are imposed:

$$\operatorname*{med}_i(a_i) = \operatorname*{med}_j(b_j) = 0 \quad \text{and} \quad \operatorname*{med}_i(f_{il}) = \operatorname*{med}_j(\lambda_{jl}) = 0, \tag{3}$$

for $l = 1, \ldots, k$. Furthermore, the scores are standardized by imposing $f_{1l}^2 + \cdots + f_{nl}^2 = 1$ for $l = 1, \ldots, k$. Note that there is no orthogonality condition for the factors, implying that the vectors of loadings $\lambda_{j\cdot}$ and scores $f_{i\cdot}$ are not uniquely determined, as is common in factor models.

By taking $k = 2$, and representing in the same two-dimensional plot the rows by (f_{i1}, f_{i2}) and the columns by $(\lambda_{j1}, \lambda_{j2})$, a biplot is obtained. The biplot allows us to investigate the row and column interaction by visual inspection of a two-dimensional graphical display.

Among others, Gabriel (1978) considered models like (2) and estimated the unknown parameters using a least squares fit. It is however well known that an LS-based method is very vulnerable in the presence of outliers. In this paper, we will propose a robust approach to fit model (2), show by a simulation experiment its merits and illustrate it with an example.

2 A robust fit

A first suggestion is to use the L_1-criterion to fit the model. If we denote by θ the vector of all unknown parameters in the model, and by $\hat{y}_{ij}(\theta) = \hat{\mu}_{ij}(\theta)$ the corresponding fit, then this procedure minimizes the objective function

$$\sum_{i=1}^{n} \sum_{j=1}^{p} |y_{ij} - \hat{y}_{ij}(\theta)|. \tag{4}$$

For the computation of the estimator we use an iterative procedure known as *alternating regressions*, which was originally proposed by Wold (1966) and used in the context of generalized bilinear models by de Falguerolles & Francis (1992). The idea is very simple: if we take the row index i in the model equation (2) fixed and consider the parameters b_j and $\lambda_{j\cdot}$ as known for all j, then we see that a regression with intercept of the ith row of the two-way table on the k vectors of loadings yields estimates for a_i and the vector of scores $f_{i\cdot}$. Reversely, if we take j fixed and suppose that a_i and $f_{i\cdot}$ are known for all i, and regress the jth column of the data matrix on the k vectors of scores, then we can update the estimates for b_j and $\lambda_{j\cdot}$. To make things robust, we will of course use a robust regression method, as was already proposed by Ukkelberg & Borgen (1993). Minimizing the criterion (4) results in performing alternating L_1 regressions.

Unfortunately, L_1-regression is sensitive to leverage points. Therefore we propose a weighted L_1-regression, corresponding to minimizing

$$\sum_{i=1}^{n} \sum_{j=1}^{p} |y_{ij} - \hat{y}_{ij}(\theta)| w_i(\theta) w_j(\theta). \tag{5}$$

These weights will downweight outlying vectors of loadings or scores. The row weights are defined by

$$w_i = \min(1, \chi^2_{k,0.95}/\mathrm{RD}_i^2) \quad \text{for } i = 1, \ldots, n,$$

where $\mathrm{RD}_1, \ldots, \mathrm{RD}_n$ are robust Mahalanobis distances computed from the collection of score vectors $\{f_{i\cdot} | 1 \le i \le n\}$ and based on the Minimum Volume Ellipsoid (Rousseeuw & van Zomeren, 1990). Analogously, we define the set of column weights w_j using the vectors of loadings. Since the true loadings and scores are unobserved, w_i and w_j depend on the unknown parameters, and will be updated at each iteration step in the alternating regression procedure. To start the iterative procedure one can take initial values obtained by robust principal component analysis (Croux & Ruiz-Gazen, 1996). It is recommended to orthogonalize the vectors of scores at the end of the iteration procedure.

It was shown by many simulations and experiments, that the above method works well, is highly robust and converges. As a byproduct of the algorithm, robust biplots can be produced. An S-plus program of the proposed algorithm is available at *http://www.statistik.tuwien.ac.at/public/filz/research.html.*

3 Simulation experiment

In this section we study the performance of the proposed method by a modest simulation study. We generated data sets with $n = 25$ rows and $p = 15$ columns according to a model with two factors:

$$Y_{ij} = c + a_i + b_j + \sum_{l=1}^{2} \lambda_{jl} f_{il} + \varepsilon_{ij}$$

$(i = 1, \ldots, n;\ j = 1, \ldots, p)$. Values for c, a_i, b_j, f_{il}, and λ_{jl} were randomly generated and fulfilled the restrictions discussed in Section 1. The noise term ε_{ij} was quite small (distributed according to a $N(0, 0.05)$) for $n \times p - n_{out}$ of the entries in the data matrix. However, for n_{out} entries, randomly placed in the data matrix, the noise term followed a $N(0, 10)$, which induced n_{out} outlying cells.

Fitting the model gave estimated parameters $\hat{c}^s$, $\hat{a}_i^s$, $\hat{b}_j^s$, $\hat{f}_{il}^s$, and $\hat{\lambda}_{jl}^s$, for $s = 1, \ldots, nsim = 150$ simulated samples. As a measure of deviation of the estimated parameters from the true ones we took the mean squared error (MSE):

$$\text{MSE}(c) = \frac{1}{nsim} \sum_{s=1}^{nsim} ||\hat{c}^s - c||^2, \quad \text{MSE}(a) = \frac{1}{nsim} \sum_{s=1}^{nsim} ||\hat{a}^s - a||^2,$$

where a^s is a vector of length n with components a_i^s and $||\cdot||$ is the Euclidean norm. (The expression for MSE(b) is obtained analogously.) It is also possible to compute proximity indices between the sets of estimated and true vectors of loadings, resp. scores, using e.g. angles between subspaces. We preferred, however, to compute an overall measure of the quality of the estimation procedure :

$$\frac{1}{nsim} \sum_{s=1}^{nsim} \sum_{i=1}^{n} \sum_{j=1}^{p} (\hat{\mu}_{ij}^s - \mu_{ij})^2, \tag{6}$$

with μ and $\hat{\mu}^s$ defined according to (2).

This simulation experiment was repeated for a percentage of outliers in the data set varying from 1 to 27. Figure 1 displays the summary measures as a function of the percentage of outliers when using the algorithm based on LS, L_1 and weighted L_1 regression. We clearly see that the approach based on LS is highly non-robust: even for a small percentage of outliers, we observe huge MSEs and a bad quality of the fit. For the estimation of the overall median, row and column effects, L_1 and weighted L_1 behave similarly. But the overall fit is much better for weighted L_1 than for L_1, since the latter approach is not capable of extracting the factor structure in the interaction terms when outliers are present.

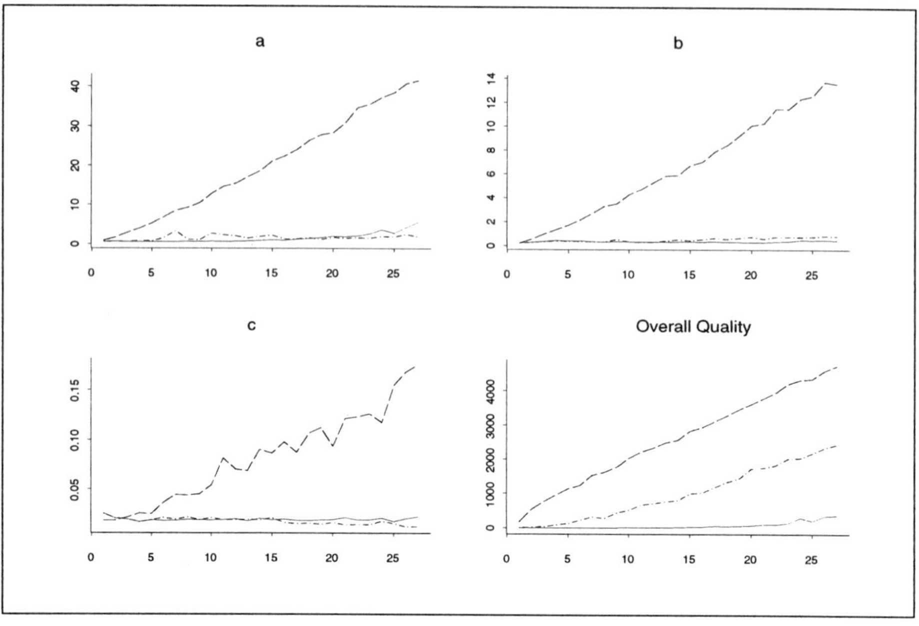

Fig. 1. MSE of the estimates for the row effects, column effects and for the overall median, and a general measure for the quality of the fit using the Least Squares (−−), the L_1 ($- \cdot -$) and the weighted L_1 (solid line) estimators, as a function of the percentage of outliers

4 Example

We measured $p = 13$ variables for the 17 Styrian political districts (Styria is part of Austria). One district is the capital Graz (G). The typical rural districts are Feldbach (FB), Hartberg (HB), Murau (MU), Radkersburg (RA), and Weiz (WZ), while typical industrial regions are Bruck/Mur (BM), Judenburg (JU), Knittelfeld (KN), and Mürzzuschlag (MZ). Graz-Umgebung (GU) is the surroundings of Graz. Liezen (LI) is a touristic region with beautiful nature. The remaining districts are Deutschlandsberg (DL), Fürstenfeld (FF), Leibnitz (LB), Leoben (LE), and Voitsberg (VO). As variables were considered: the proportion of children (< 15 years) (`chi`) and old people (> 60 years) (`old`) in each district. Furthermore, the proportion of people employed in industry (`ind`), trade (`tra`), tourism (`tou`), service (`ser`), and agriculture (`agr`), and the total proportion of unemployed people (`une`) were measured. Other variables are the proportion of mountain farms (`mou`), of people with university education (`uni`), of people who just attended primary school (`pri`), of employed people not commuting daily (`cnd`), and the proportion of employed people commuting to another district (`cdi`). The origin of these measurements is the Austrian census of 1991, and the data are available at the above mentioned web page.

We fitted the model, using weighted L_1 regression, with $k = 2$ to the raw data, although it may be more appropriate to apply the logit transformation first. In Table 1, we display the estimated row effect $\hat{a}_i$ and column effect $\hat{b}_j$, together with the residual matrix $y_{ij} - \hat{\mu}_{ij}$. We see that Graz (G) appears as an outlier for a lot of variables, indicating that it is clearly distinct from most other districts. The district GU has a high residual for commuting to another district (namely to Graz), which is also true for VO, and for people employed in industry (it is a quiet and refined district). District RA has an atypical row effect, and an outlying residual for the cell corresponding to people employed in agriculture.

The biplot (Figure 2) displays the estimates $(\hat{f}_{i1}, \hat{f}_{i2})$ and $(\hat{\lambda}_{j1}, \hat{\lambda}_{j2})$. The typical agricultural districts (FB, HB, MU, RA, WZ) have high loadings on the variable representing the people employed in agriculture, but they also have high values for commuting to another district (the latter is also true for GU, the surroundings of Graz). Additionally, the districts FB, HB, RA, and MU have high loadings for the variable "commuting not daily" (`cnd`). The industrial regions (BM, JU, KN, MZ) have high values at the vector "industry" (`ind`), but also GU and LE have high values there. LE additionally has a high value for people employed in service.

Graz appears as an outlier again. Fortunately the biplot is robust, implying that Graz will not influence the estimates of loadings and scores too much. A classical biplot would also reveal Graz as an outlier, but then the estimated loadings and scores would be heavily influenced by this outlier, making their interpretation subject to a lot of doubt.

Acknowledgment
We wish to thank A. de Falguerolles for bringing the technique of alternating regressions under our attention and the referee for his very helpful remarks.

References

Croux, C. & Ruiz-Gazen, A. (1996). A fast algorithm for robust principal components based on projection pursuit. In: *Proceedings in Computational Statistics, COMPSTAT 1996* (ed. A. Prat), 211-216. Heidelberg: Physica-Verlag.

de Falguerolles, A. & Francis, B. (1992). Algorithmic approaches for fitting bilinear models. In: *Computational Statistics, COMPSTAT 1992* (ed Y. Dodge & J. Whittaker), Vol. 1, 77-82. Heidelberg: Physica-Verlag.

Gabriel, K.R. (1978). Least squares approximation of matrices by additive and multiplicative models. *Journal of the Royal Statistical Society B*, **40**(2), 186-196.

Gollob, H. F. (1968). A statistical model which combines features of factor analytic and analysis of variance techniques, *Psychometrika*, 33, 73-116.

Rousseeuw, P.J. & van Zomeren, B.C. (1990). Unmasking multivariate outliers and leverage points. *Journal of the American Statistical Association*, **85**, 633-639.

Ukkelberg, Å. & Borgen, O. (1993). Outlier detection by robust alternating regressions. *Analytica Chimica Acta*, **277**, 489-494.

Wold, H. (1966). Nonlinear estimation by iterative least squares procedures. In: *A Festschrift for F. Neyman*, (ed. F.N. David), 411-444. New York: Wiley and Sons.

Table 1. Estimates for the row effects and column effects together with the residuals for the Styrian districts data set using the weighted L_1 approach (rounded values, in %)

WL_1	chi	old	ind	tra	tou	ser	agr	mou	une	uni	pri	cnd	cdi	$\hat{a}$
BM	-0.1	-0.1	0.2	1.1	0.7	-1.6	0.0	0.0	1.5	0.0	0.0	1.4	-2.8	0.1
DL	-0.7	0.4	0.1	-0.5	-0.1	0.0	0.0	0.0	0.1	0.2	-0.1	3.1	1.5	0.6
FB	-0.5	1.5	0.0	0.0	-0.2	0.0	4.1	-3.1	0.2	0.0	1.1	-0.9	-1.2	-0.3
FF	0.0	1.6	-2.2	4.3	-0.1	0.0	0.0	-2.0	0.0	1.0	-0.2	0.0	-0.7	0.0
G	2.0	0.0	0.0	12.8	0.0	8.1	0.0	-0.6	-0.2	12.8	-15.0	-13.5	-12.4	-4.8
GU	0.0	-1.4	-7.2	0.0	0.9	3.1	0.0	0.0	-0.9	0.8	-1.5	0.0	8.6	1.1
HB	0.4	-0.2	0.1	0.1	2.0	0.0	-0.2	0.3	0.0	0.0	0.0	6.3	-0.4	-0.1
JU	0.1	0.0	7.3	-0.2	-0.7	-2.5	-2.9	0.0	0.0	0.3	0.0	1.8	0.0	-0.3
KN	0.6	0.0	0.0	-1.4	-1.9	2.8	0.0	-0.1	-0.7	0.0	0.4	-0.4	0.0	0.4
LB	0.0	-1.3	-2.6	1.5	0.0	0.9	0.0	-2.5	0.3	-0.9	0.9	0.0	6.5	0.3
LE	-0.7	0.0	-3.8	0.0	-0.7	0.0	2.7	-1.0	1.1	-0.1	0.0	-0.5	4.8	-0.2
LI	1.2	-1.7	-3.3	1.3	4.1	0.0	-1.0	0.1	-0.1	0.0	-0.5	0.0	0.0	-0.2
MZ	0.0	1.2	5.9	-2.4	0.8	-5.1	0.0	0.2	-0.5	-1.0	0.8	0.0	-2.6	0.3
MU	0.0	-1.5	0.7	-2.6	0.6	0.0	0.0	0.1	-1.9	0.3	-3.8	6.2	7.5	0.3
RA	-1.7	3.5	-3.4	0.0	0.0	1.9	8.2	-3.4	0.0	0.8	0.0	-3.4	0.0	-1.1
VO	-0.4	0.3	0.0	-1.1	0.0	-3.6	0.0	1.2	2.0	-0.6	0.4	0.0	9.7	-0.2
WZ	0.5	0.0	3.2	-0.6	0.1	-2.9	-0.2	0.2	0.0	0.0	0.0	0.0	0.0	0.3
$\hat{b}$	5.7	8.8	27.9	-0.3	-7.1	16.7	0.0	-10.0	-6.1	-7.3	76.0	-1.1	40.7	$\hat{c} = 12.2$

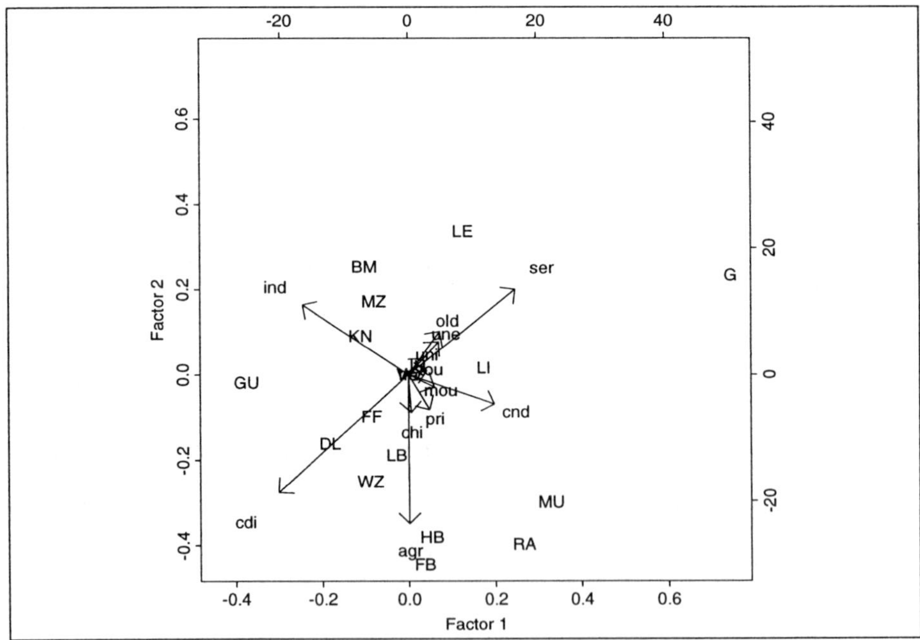

Fig. 2. Robust biplot representation of the interactions between rows and columns for the Styrian districts data set

Spatial Clustering Techniques: An Experimental Comparison

Mô Dang and Gérard Govaert

UMR CNRS 6599 Heudiasyc, Université de Technologie de Compiègne, B.P. 20529, 60205 Compiègne Cedex, France

Keywords. Clustering, spatial data, Markov random fields, mean filter, post-processing

1 Introduction

This study considers the problem of clustering spatially located observations, which arises in various fields like unsupervised image segmentation, quantitative biogeography, or mapping of soil properties. In those applications, it is often reasonable to assume that the partition changes slowly in the geographic space. This assumption is taken into account in a recently proposed fuzzy clustering method, the so-called Neighbourhood EM algorithm (Ambroise, 1996; Ambroise *et al.*, 1997): this method optimizes a criterion containing on the one hand the fuzzy sum of within-cluster inertia exhibited by Hathaway (1986), and on the other hand a spatial smoothing function of the classification. At each iteration of the resulting algorithm, the class memberships of the observations are updated based both on their fitness to the class parameters and on the class of the neighbours. This procedure is interpretable as an application of the Expectation-Maximization (EM) principle to a hidden Markov random field (Dang & Govaert, 1998). Alternatively, prior to applying traditional clustering techniques, the data may be preprocessed by filtering techniques in order to reduce the noise (Cocquerez *et al.*, 1995). Post-smoothing of the classification is also tested as an alternative approach to take into account the assumption of spatial regularity of the partition.

Those three approaches — the NEM algorithm, pre-smoothing and post-smoothing — are experimentally compared to non-spatial clustering on simulated spatial data, using various degrees of underlying spatial smoothness and noise levels. The mean and standard deviation of the error rate produced by the different clustering techniques are computed for each typical situation on a set of simulated images.

A non-spatial fuzzy clustering method based on EM and mixture models is presented in Section 2. It forms the basis of the NEM spatial fuzzy clustering algorithm, which is described in Section 3. The spatial mean filtering techniques to be compared with the NEM algorithm are presented in Section 4. The results of the experimental comparison are analyzed in Section 5. Concluding remarks and perspectives of work are outlined in Section 6.

2 Fuzzy clustering using EM and mixture models

The n observation vectors to be partitioned are noted by $\mathbf{x}_{(n \times d)} = (\mathbf{x}_1, \ldots, \mathbf{x}_n)'$ ($\mathbf{x}_i \in \mathbb{R}^d, 1 \leq i \leq n$). In the case of a grey-level image, $d = 1$ and x_i represents the grey-level intensity of pixel i. The number k of clusters is supposed to be known. A crisp partition of the n observations may be represented as a set of n binary indicator vectors, $\mathbf{z}_{(n\times k)} = (\mathbf{z}_1, \ldots, \mathbf{z}_n)'$, where $z_{ih} = 1$ if observation i is assigned to the class h, $z_{ih} = 0$ otherwise $(1 \le h \le k)$.

When clustering is based on a mixture model, the n observation vectors are supposed to arise independently from a mixture of k distributions $f_h(\cdot, \theta_h)$ with unknown parameters θ_h and in proportions π_h $(1 \le h \le k)$ (Symons, 1981). The unknown parameters Φ of the mixture may be estimated using only the unlabelled data by applying the EM algorithm (Dempster *et al.*, 1977). The iterative calculation of the EM algorithm for mixture identification is shown in Hathaway (1986) to be equivalent to the alternative maximization of the following fuzzy clustering criterion:

$$D(\mathbf{c}, \Phi) = \sum_{h=1}^{k}\sum_{i=1}^{n} c_{ih}\log(\pi_h f_h(\mathbf{x}_i;\theta_h)) - \sum_{h=1}^{k}\sum_{i=1}^{n} c_{ih}\log(c_{ih})$$

where $\mathbf{c}_{(n\times k)}$ is a fuzzy classification matrix, i.e. c_{ih} represents the grade of membership of $\mathbf{x}_i$ into class h $(0 \le c_{ih} \le 1,\ \sum_{h=1}^{k} c_{ih} = 1,\ \sum_{i=1}^{n} c_{ih} > 0,$ $1 \le i \le n, 1 \le h \le k)$. Indeed, starting from arbitrary parameters Φ^0 and maximizing the function $D(\mathbf{c}, \Phi)$ alternatively over the classification $\mathbf{c}$ and over the parameters Φ yields exactly the E-step and the M-step of the EM algorithm for a mixture model. The obtained fuzzy classification $\hat{\mathbf{c}}$ may then be interpreted as the posterior probabilities of membership of the observations to the classes. Assigning each observation to the class where it has the highest grade of membership, yields thus a crisp partition $\hat{\mathbf{z}}$ that maximizes the posterior probability given the data $\mathbf{x}$ and the estimated parameters $\hat{\Phi}$.

3 Spatial fuzzy clustering

In the case of a Gaussian mixture, the first double sum in $D(\mathbf{c}, \Phi)$ may be interpreted as the negative of a fuzzy sum of intra-class inertia (see e.g. Celeux & Govaert, 1995). The fuzzy clustering method described above aims thus to produce clusters that have low dispersions in the feature space $\mathbb{R}^d$. However, applying it directly to cluster spatially located data does not allow us to account for the hypothesis of geographic smoothness in a satisfactory way.

In order to use this spatial information, Ambroise *et al.* (1997) propose to add to $D(\mathbf{c}, \Phi)$ a function $G(\mathbf{c})$ which favours spatially smooth partitions :

$$G(\mathbf{c}) = \frac{1}{2}\sum_{i=1}^{n}\sum_{j=1}^{n}\sum_{h=1}^{k} v_{ij} c_{ih} c_{jh}.$$

The v_{ij} are the weights of the neighbourhood graph that links neighbouring observations. $G(\mathbf{c})$ may be interpreted as the number of pairs of observations having the same class. In this approach, the new criterion to be optimized is

$$U(\mathbf{c}, \Phi) = D(\mathbf{c}, \Phi) + \beta\ G(\mathbf{c})$$

where β is a scalar parameter which controls the degree of desired spatial smoothing.

Maximizing $U(\mathbf{c}, \Phi)$ alternatively over the classification and the parameters of the classes yields an iterative spatial clustering method, the so-called

Neighbourhood EM algorithm, which has the same structure as EM (Ambroise *et al.*, 1997). The parameters are given an arbitrary initial value Φ^0. At the E-step, the maximization of $U(\mathbf{c}, \Phi^m)$ over $\mathbf{c}$ yields a set of fixed-point equations, which may be used to compute the new classification matrix $\mathbf{c}^{m+1}$:

$$c_{ih} = g_{ih}(\mathbf{c}) = \frac{\pi_h f_h(\mathbf{x}_i|\theta_h^m) \cdot \exp\{\beta \sum_{j=1}^n c_{jh} v_{ij}\}}{\sum_{\ell=1}^k \pi_\ell f_\ell(\mathbf{x}_i|\theta_\ell^m) \cdot \exp\{\beta \sum_{j=1}^n c_{j\ell} v_{ij}\}}.$$

At the M-step, the parameters Φ of the classes are updated by the same calculations as in the M-step of the EM algorithm for a mixture model. This spatial fuzzy clustering procedure can be interpreted as an application of the EM principle in order to estimate the parameters of a particular hidden Markov random field (MRF) (Dang & Govaert, 1998). The underlying prior distribution of the unobserved classification $\mathbf{z}$ is the Potts MRF model, i.e. a Gibbs distribution with energy function $-\beta \sum_{\{i,j\}\ neighbours} v_{ij} \mathbf{z}_i' \cdot \mathbf{z}_j$. The fuzzy classification obtained can then be interpreted as the posterior probabilities of membership into the classes. A crisp partition can be computed from the fuzzy classification by the same principle as in Section 2.

4 Spatial linear filter

In low-level image segmentation tasks, the observed image $\mathbf{x}$ is often modelled as an ideal image $\mathbf{g}$ degraded by additive white noise:

$$\mathbf{x} = \mathbf{g} + \mathbf{y}$$

where $\mathbf{y} = (\mathbf{y}_1, \ldots, \mathbf{y}_n)'$, $\mathbf{y}_i \in \mathbb{R}^d$ and $\mathbf{y}_i \sim \mathcal{N}(0, \sigma^2 I)$. In simple segmentation tasks, the ideal image $\mathbf{g}$ is generally assumed to have constant value μ_h within regions having label h ($\mu_h \in \mathbb{R}^d, 1 \leq h \leq k$), i.e.

$$\mathbf{g} = \mathbf{z} \cdot \mathbf{M}$$

where $\mathbf{z}_{(n \times k)} = (\mathbf{z}_1, \ldots, \mathbf{z}_n)'$ and $\mathbf{M}_{(k \times d)} = (\mu_1, \ldots, \mu_k)'$.

4.1 Pre-smoothing

Filtering the image $\mathbf{x}$ prior to segmenting it aims to reduce the noise level and improve the final segmentation (Cocquerez *et al.*, 1995). It is then assumed that the spatial scale of the noise is narrow, compared to the size of the constant-valued regions in the noise-free image.

For the simple additive white noise model, the classical mean filter is known to minimize the noise level in the filtered image, denoted by $\tilde{\mathbf{x}}$ (see e.g. Cocquerez *et al.*, 1995, p. 77). Each observation of the filtered image is obtained by taking the average over a window of surrounding observations in the observed image:

$$\tilde{\mathbf{x}}_i = \sum_{j=1}^n w_{ij} \mathbf{x}_j \Bigg/ \sum_{j=1}^n w_{ij}$$

where w_{ij} are the weights of the observations j within the window surrounding the current observation i ($w_{ij} = 0$ for observations j outside of the window). For the simplest constant weighted mean filter, all the weights within the window have the same value, 1 for instance.

The smoothed image $\tilde{\mathbf{x}}$ may then be partitioned using non-spatial clustering techniques like the one described in Section 2.

4.2 Post-smoothing

Alternatively, the assumption of spatial regularity of the partition may also be taken in consideration by post-smoothing the classification (see e.g. Cressie, 1993, p.506). One simple and intuitive technique consists of firstly applying a non-spatial clustering technique on the original image $\mathbf{x}$. The obtained fuzzy classification $\mathbf{c}$ can then be smoothed using a mean filter, yielding a filtered fuzzy classification $\tilde{\mathbf{c}}$

$$\tilde{\mathbf{c}}_i = \sum_{j=1}^{n} w_{ij}\mathbf{c}_j \Big/ \sum_{j=1}^{n} w_{ij} \ .$$

A crisp partition may then be obtained from the fuzzy classification $\tilde{\mathbf{c}}$ by applying the rule described in Section 2.

5 Experimental comparison

This study aims to compare the spatial clustering techniques described above on spatial data with various degrees of spatial smoothness and noise levels. The artificial sets of observations consist of grey-level images ($d = 1$) of size 30×30. They have been generated in two steps. A segmented image $\mathbf{z}$, the "ground truth" is firstly generated according to a Potts MRF, with $k = 2$ classes and a specified spatial smoothness β. The observations $\mathbf{x}$ are then drawn according to the normal distribution of their class $\mathcal{N}(\mu_h, \sigma^2)$ — or equivalently, are generated by adding to $\mathbf{z} \cdot (\mu_1, \mu_2)'$ a white noise with variance σ^2. The two means μ_1 and μ_2 have been fixed to 0 and 1.

Twelve situations were simulated, combining three degrees of spatial smoothness ($\beta = 0.5, 1, 1.5$) and four noise levels ($\sigma^2 = 0.1, 0.2, 0.5, 1$). For each situation, 30 samples $\mathbf{x}$ have been generated, and the following clustering methods are applied on each sample:

1. $\mathbf{x} \stackrel{EM}{\longrightarrow} \mathbf{c} \rightarrow \hat{\mathbf{z}}_{EM}$
2. $\mathbf{x} \stackrel{NEM}{\longrightarrow} \mathbf{c} \rightarrow \hat{\mathbf{z}}_{NEM}$
3. $\mathbf{x} \stackrel{filter}{\longrightarrow} \tilde{\mathbf{x}} \stackrel{EM}{\longrightarrow} \mathbf{c} \rightarrow \hat{\mathbf{z}}_{pre}$
4. $\mathbf{x} \stackrel{EM}{\longrightarrow} \mathbf{c} \stackrel{filter}{\longrightarrow} \tilde{\mathbf{c}} \rightarrow \hat{\mathbf{z}}_{post}$.

The weights of the filters were adjusted in order to minimize the classification error. The EM (resp. NEM) algorithm is initialized 10 times from random parameter values (the initial means are picked randomly out of the observations) and the solution that maximizes the criterion $D(\mathbf{c}, \Phi)$ (resp. $U(\mathbf{c}, \Phi)$)is kept. The NEM algorithm was applied using the simulated value β.

The percentage of misclassified pixels is computed for each method. For each situation, the mean and standard deviation of the error over the 30 simulated samples are displayed in Table 1. The spatial clustering techniques considered appear to improve the classification in comparison to the non spatial clustering method. For low to medium noise levels (σ^2 from 0.1 to 0.5), NEM performs better than pre-smoothing and post-smoothing, especially for intermediate spatial smoothness. This is due to the fact that the filtering tends to smooth data in the same way over the whole image, even in more irregular areas which often occur when simulating images with $\beta = 1$. For a high noise level, NEM gives roughly the same performances as the filtering techniques, and even poorer segmentations for a high degree of spatial smoothing ($\beta = 1.5$). In this situation, the criterion optimized by NEM tends to select partitions in only one class. Pre-smoothing works generally better than post-smoothing on those simulated data, probably because it suits the simulated model better.

Table 1. Mean and standard deviation of the percent of misclassified observations over 30 samples

σ^2	β	EM	NEM	mean + EM	EM + mean
0.1	0.5	5.6 (0.8)	4.8 (0.8)	7.3 (0.8)	5.0 (0.7)
	1	7.2 (3.4)	2.2 (0.4)	2.8 (0.5)	3.6 (2.6)
	1.5	11.2 (8.8)	1.0 (0.4)	2.4 (4.4)	4.7 (6.6)
0.2	0.5	13.4 (1.1)	11.4 (1.0)	12.7 (1.1)	11.6 (1.1)
	1	15.9 (4.1)	4.9 (1.0)	7.9 (2.5)	8.7 (3.2)
	1.5	18.0 (6.9)	1.9 (0.6)	3.3 (0.9)	5.7 (6.0)
0.5	0.5	24.3 (1.1)	21.7 (1.2)	22.2 (1.2)	21.9 (1.2)
	1	25.9 (3.0)	10.4 (1.8)	12.5 (3.0)	13.5 (3.4)
	1.5	27.4 (4.4)	6.9 (7.3)	8.3 (8.1)	9.6 (5.8)
1	0.5	31.5 (3.7)	29.2 (2.1)	29.2 (2.1)	29.9 (4.0)
	1	32.2 (2.2)	16.9 (5.7)	15.8 (3.9)	16.4 (3.8)
	1.5	33.3 (3.5)	25.6 (16.1)	10.4 (6.6)	12.4 (7.5)

The behaviour of the four clustering techniques is illustrated on a biological image of muscular fibres (see Figure 1). Without spatial information, the segmentation is not very satisfying due to the rather high variance of grey-level intensities within the regions (see Figure 1.c). Using NEM with $\beta = 1.8$, most of the irregularities have been removed (Figure 1.d). Pre-smoothing tends to blur out the frontiers of the small dark cell at the bottom right of the image (Figure 1.b and Figure 1.e). Post-smoothing appears to give similar results to NEM.

6 Conclusion

Simple smoothing methods may work when all regions are large. The NEM algorithm gives consistent segmentations most of the time. However, in the most noisy situations with high spatial smoothness, using the simulated spatial smoothness coefficient β tends to select only one class partitions. Using data-driven techniques to estimate β may overcome this drawback. One may use for this purpose pseudo-likelihood based methods as described in Besag (1986), or the heuristic described in Dang & Govaert (1998) for the NEM algorithm. The comparison could also be extended by using different strategies of initialization of NEM as well as adaptive filters, and by considering other kinds of noise, such as impulsive or spatially correlated noise.

References

Ambroise, C. (1996). *Approche probabiliste en classification automatique et contraintes de voisinage.* PhD thesis, Université de Technologie de Compiègne.

Ambroise, C., Dang, M. & Govaert, G. (1997). Clustering of spatial data by the EM algorithm. In: *geoENV I - Geostatistics for Environmental*

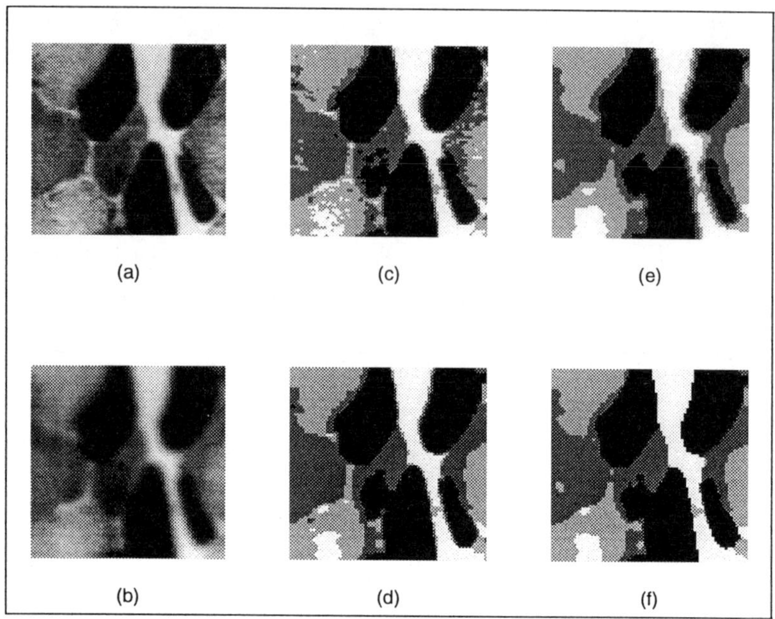

Fig. 1. Comparison on a biological image of muscular fibres: (a) Original image, (b) Pre-smoothed image, (c) Original image segmented using EM, (d) Original segmented using NEM, (e) Pre-smoothed image segmented using EM, (f) Post-smoothed image after segmentation by EM

Applications (ed. A. Soares, J. Gómez-Hernandez & R. Froidevaux), 493–504. Dordrecht: Kluwer Academic Publisher.

Besag, J. (1986). Spatial analysis of dirty pictures. *Journal of the Royal Statistical Society B*, **48**, 259–302.

Cressie, N. A. (1993). *Statistics for Spatial Data.* New York: Wiley.

Celeux, G. & Govaert, G. (1995). Gaussian parsimonious clustering models. *Pattern Recognition*, **28**, 781–793.

Cocquerez, J., Philipp, S. & coll. (1995). *Analyse d'images: filtrage et segmentation.* Paris: Masson.

Dang, M. & Govaert, G. (1998). Spatial fuzzy clustering by EM and Markov random fields. *International Journal of Systems Research and Information Science*, in press.

Dempster, A., Laird, N. & Rubin, D. (1977). Maximum likelihood from incomplete data via the EM algorithm. *Journal of the Royal Statistical Society B*, **39**, 1–38.

Hathaway, R. (1986). Another interpretation of the EM algorithm for mixture distributions. *Journal of Statistics & Probability Letters*, **4**, 53–56.

Symons, M.J. (1981). Clustering criteria and multivariate normal mixtures, *Biometrics.* **37**, 35–43.

A Visual Environment for Designing Experiments

Paul L. Darius[1], Wim J. Coucke[1] and Kenneth M. Portier[2]

[1] Lab Statistics & Exp. Design, KULeuven, Kard. Mercierlaan 92, B-3001 Leuven Belgium
[2] Dept Statistics, University of Florida, Gainesville, FL 32611-0339, USA

Abstract. A number of approaches to visually representing an experimental design are presented. Several of these representations are shown in a prototype software environment. Users can easily interact with each representation, and changes made in one view are immediately reflected in the others. The software provides a framework for easy access to the properties of a design, and for comparison of different candidate designs.

Keywords. Design of experiments, analysis of variance, Hasse diagram, graphics, interactive software

1 Introduction

Many statistical packages currently have some "design of experiments" facility. Their capability is often limited to providing a choice of one design from a list of standard designs. Each available design is represented by a name and the values for the parameters, and ultimately by a standard type dataset with missing information for the response variable. In this paper we discuss graphical representations for a design which might be better suited to exploring and comparing properties of alternative designs.

A Hasse diagram is one approach to graphically representing design information. It is primarily a general tool for displaying partially ordered sets. The elements of the set are shown as vertices, and links are added so that any couple of elements that obeys the partial ordering (and only those couples) is linked by a path following a chosen direction (e.g. downward).

Several authors have discussed the use of Hasse diagrams in relation to design and analysis of variance. In these, factors and/or effects are the elements of the set, and nesting is the partial ordering (Kempthorne *et al.*, 1961; Taylor & Hilton, 1981; Tjur, 1984; Speed & Bailey, 1987; Tjur, 1991; Nys *et al.*, 1994; Lohr, 1995). The first two references use the Hasse diagram as a representation with a one-to-one relationship to the design it represents. This limits its use to designs with balanced complete response structures. The later references use the Hasse diagram for a wider class of designs.

Our research has indicated that there are several relevant partial orderings in a design, hence several potentially useful Hasse diagrams. Moreover, the information in a Hasse diagram can be extended by adding one or more coordinate axes. Finally, we have developed some prototype software systems which facilitate the construction of these various Hasse diagrams and have used these systems to illustrate various designs.

2 Graphical representations for designs

2.1 The factor structure diagram

The simplest Hasse diagram displays the study factors and their nesting structure. The observational unit is explicitly considered as a factor nested in all the other factors. This is a departure from the usual notation found in textbooks and software package commands, but one which is very useful in dealing with practical design problems. In addition, a constant factor corresponding to an overall mean can be considered as nesting all other factors. The basic design diagram is then plotted with the constant factor at the top and the observational unit factor at the bottom.

A Hasse or factor structure diagram conveniently displays the most important features of a design and is useful when comparing different types of design. This is illustrated in Figure 1 where the diagrams for some well-known types of design are given.

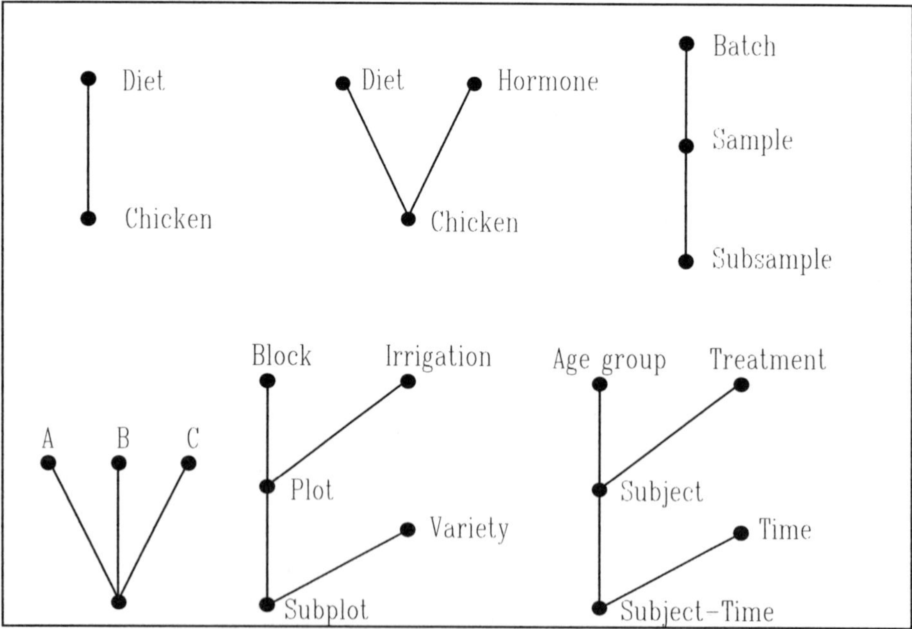

Fig 1. Hasse diagrams showing the factor structure for some well known designs. From top left to bottom right: one factor design, factorial design with 2 factors, subsampling design, factorial design with 3 factors (or a two-way design in blocks, or a Latin Square,..), split plot design, repeated measures study

The factor diagram can be extended by adding graphical clues to the plot to discriminate fixed from random factors, or to indicate which factors belong to the design structure and which to the treatment structure. One can also add a vertical axis and place the factors according to their number of levels.

2.2 The effects structure diagram

For each design, there are a number of models that can be used to analyse it. The effects structure diagram initially displays the maximal model containing all the factors plus all interactions consistent with the crossing-nesting structure. Specifically, the set consists of the partitions induced by each model term on the set of observational units. The partial ordering is the relation "finer partition than". Two model terms with partitions that refine each other are drawn on the same vertex.

Defined in this way, a new plot, the effects structure diagram can be drawn to include both the factor structure diagram information and interaction terms. This diagram can be further extended by adding other information, e.g. the number of degrees of freedom for each effect, or the number of levels of the induced partition.

2.3 The design layout view

An alternate view of the design is a graph showing which units receive a given treatment level combination, or which levels are assigned to a given unit. Each factor, including the observational unit, occupies one row, and on this row are icons for each level. The icons for the observational units are connected with a line to the icons for the other factors according to the assignments made. The links can also be drawn to account for any hierarchical relationships among factors.

The design layout view is the least abstract, and one particularly suited to discussions of the practical problems associated with realisation of a design. It is also useful in visualization of the result of the randomisation, or exploring the impact of missing values. Because this display can easily become too crowded, especially when the number of observational units is large, specific plotting conventions have to be implemented.

2.4 The model structure diagram and the GRANOVA plot

The set of possible models for a given design also presents a partial ordering. In this case, two models are related when one consists of a subset of terms from the other. The model structure diagram displays this partial ordering, plotting either the set of all possible models, or only the set of well-formulated models. This diagram can be further augmented by adding a vertical axis to which the error d.f. of the model are plotted. Now, the vertices are not labelled (i.e. the models), but the lines connecting two vertices are labelled with the effect that appears in the model of one vertex, and not in the other. The diagram then graphically depicts the effects that appear in the ANOVA table, ordinated using the degrees of freedom available for each.

After the experiment has been performed and the response values entered, the model structure diagram can be enhanced by adding a second, horizontal, axis to display the error s.s of the model. The Mean Square for each effect corresponds to the slope of the line labelled with the effect. The result is a graphical analogue of the ANOVA table, which we refer to as a GRANOVA plot. This display makes it possible to view the model for a design, then display analysis results within the same notational and graphical environment.

3 A prototype software environment

A JAVA applet implementing the graphical techniques described in the previous sections has been constructed. The environment consists of 5 linked views of the same underlying design. One or more views can be displayed on the screen at the same time. The views are linked, in the sense that changes made in one view are immediately reflected in all views. Each view has its own user interface, allowing user interaction in a point and click style. The user can, for example, build up a design in the factor structure view, pool effects in the effects structure view, check or change assignments in the design layout view and then examine the results in the GRANOVA plot view.

The fifth view is a tabular display in the classical dataset format. It can be used to enter data manually or to enter the response values for a design previously constructed. It can also be used to import designs or completed experiments from external sources. A problem can arise when designs are imported in which the nesting structure is not obvious from the factor level labels. In this case, a method adapted from Lorenzen & Anderson (1993) is used to help in reconstructing the correct nesting structure.

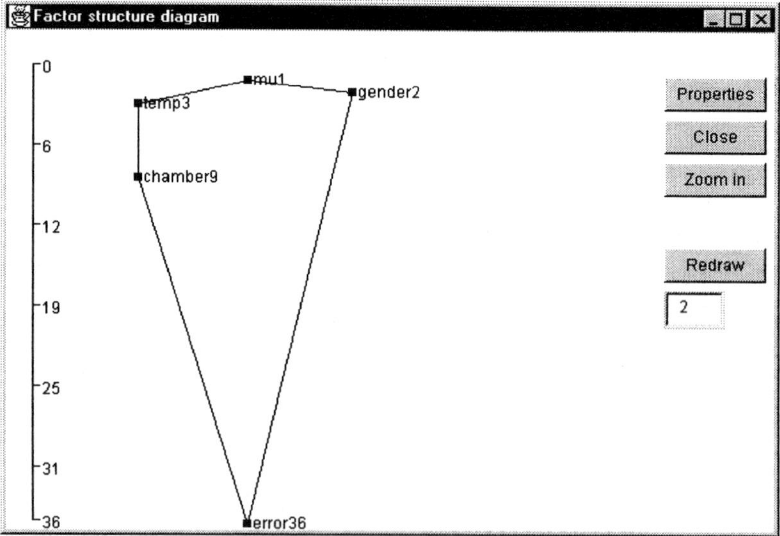

Fig 2. A factor structure diagram

Figures 2 to 5 show screen displays from the four graphical views, with an example adapted from Milliken & Johnson (1984) where male and female persons are placed in environment chambers and different temperatures are applied to the chambers.

The main aim of the prototype is to make it easier to study or discuss (with students or with clients) the properties of designs, and to provide a framework to compare different candidate designs.

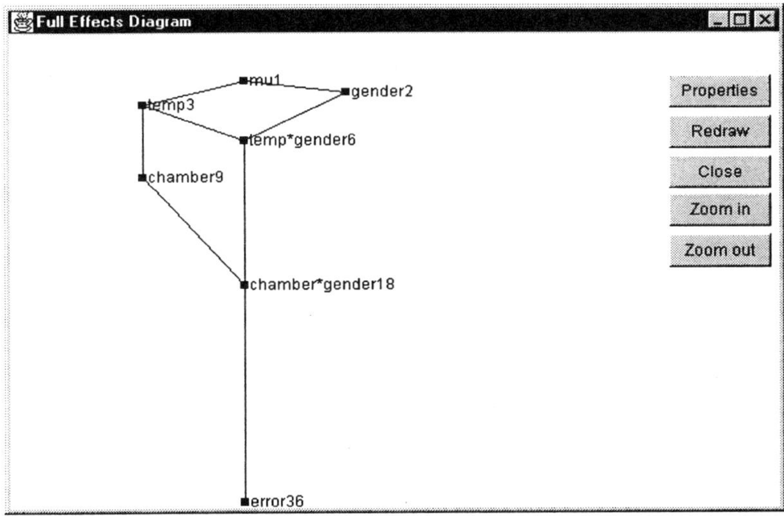

Fig 3. An effects structure diagram

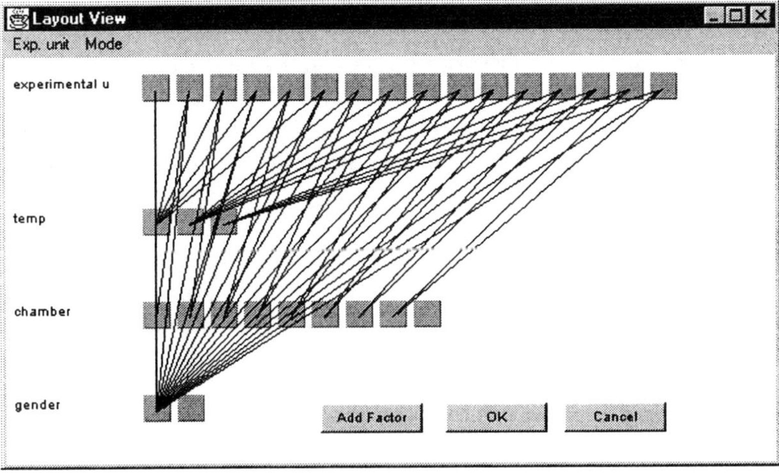

Fig 4. A design layout view

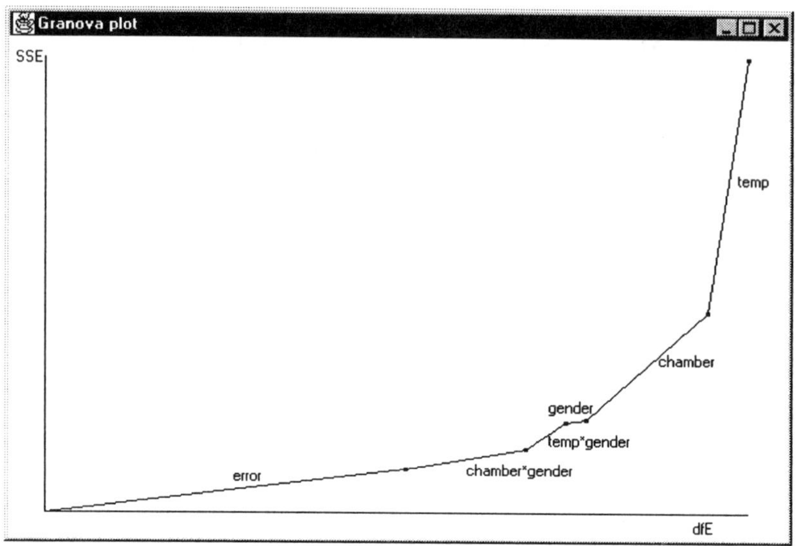

Fig 5. A GRANOVA plot

References

Kempthorne, O., Zyskind, G., Addelman, S., Throckmorton, T. N & White, R.F. (1961). Analysis of Variance Procedures. *Aeronautical Research Laboratory Technical Report 149.* Office of Technical Services, United States Department of Commerce.

Lohr, S.L. (1995). Hasse Diagrams in Statistical Consulting and Teaching. *The American Statistician,* **49**, 376-381.

Lorenzen, T.J. & Anderson, V.L. (1993). *Design of Experiments – A No-Name approach.* New York: Marcel Dekker Inc.

Milliken, G.A. & Johnson, D.E. (1984). *Analysis of Messy Data, Volume 1: Designed Experiments.* Belmont: Lifetime Learning Publications.

Nys, M., Darius, P. & Marasinghe, M. (1994). An Interactive Window-Based Environment to Explore the Design of an Experiment. In: *SoftStat '93 Advances in Statistical Software 4* (ed. F. Faulbaum). Stuttgart-Jena-New York: Gustav Fischer.

Speed, T.P. & Bailey, R.A. (1987). Factorial Dispersion Models. *International Statistical Review,* **55**, 261-277.

Taylor, W.H. & Hilton, G.H. (1981). A Structure Diagram Symbolization for Analysis of Variance. *The American Statistician,* **35**, 85-93.

Tjur, T. (1984). Analysis of Variance Models in Orthogonal Designs. *International Statistical Review,* **52**, 33-81.

Tjur, T. (1991). Analysis of Variance and Design of Experiments. *Scandinavian Journal of Statistics,* **18**, 273-322.

Essay of a Dynamic Regression by Principal Components Model for Correlated Time Series

M.J. Del Moral and M.J. Valderrama

Department of Statistics and O. R., University of Granada. 18071 Granada. Spain.

Abstract. In this paper we analyze the use of dynamic regression by principal components models for correlated time series forecasting. The choice of an appropriate cutting point on input and output series allows us to study their principal component analysis and the selection of a forecasting model. Two basic issues are discussed on studies with simulated and real data: parsimony and principal components selection in the forecasting model.

Keywords. Principal components, input/output, regression, forecasting, parsimony

1 Introduction

In this paper we are concerned with dynamic regression models in which one output and one input are involved. There are various model forms to capture the response of the output to the input in a parsimonious way. One of these comes out of the principal components analysis (PCA) for input and output.

The principal components of a stochastic process are the random variables of its Karhunen-Loéve orthogonal decomposition (Fukunaga, 1990; Del Moral & Valderrama, 1994).

Deville (1978) introduces a linear formulation to forecast a continuous stochastic process at a point future value in terms of the principal components associated with its evolution in a past interval. In this paper we consider a dynamic regression model for correlated time series obtained as an extension of Deville's formulation to those problems in which two stochastic processes are involved. The basic tools to set up this model are the orthogonal decomposition of a discrete stochastic process by means of its PCA, and linear regression performed on the principal components of input and output processes.

2 PCA and dynamic regression models on discrete stochastic processes

Let $\{X_t, t \in T\}$ be a discrete stochastic process defined on $T = \{1, \ldots, n\}$, $n \in \mathbb{Z}^+$. The covariance matrix of the process is $R_X = E[(X_t - \mu_X)(X_t - \mu_X)^T]$ where $\mu_X = E[X_t]$. We assume $E[X_t] = 0$, $t = 1, \ldots, n$.

The process X_t possesses an orthogonal decomposition

$$X_t = \sum_{j=1}^{n} \alpha_j \phi_j \tag{1}$$

where ϕ_j are eigenvectors associated with the eigenvalues λ_j of the covariance matrix, solutions of the matrix equation $R_X\phi_j = \lambda_j\phi_j$, $j = 1,\ldots,n$ and coefficients α_j are random variables determined by $\alpha_j = \phi_j^T X_t$ with $E[\alpha_j] = 0$ and $E[\alpha_j\alpha_i] = \lambda_j\delta_{ji}$.

Deville (1974) introduced the PCA of continuous processes as a natural extension of the multivariate technique. In discrete case, the combinations α_j in (1) are the principal components (PC) of the process.

The PCA of a stochastic process allows us to explain its variability in terms of the eigenvalues λ_j introduced above. The total variance of the process is given by $V(X_t) = E\|X_t\|^2 = \sum_{j=1}^n \lambda_j$ and the proportion of the variance accumulated by the first i principal components by $V_i^{acc}(X_t) = \sum_{j=1}^i \lambda_j/V(X_t)$.

The orthogonal decomposition of a discrete process by its principal components is optimum since the expression (1), when the eigenvalues λ_j are set up in decreasing order, is such that the mean square error of an approximation by dimensionality reduction,

$$\hat{X}_t = \sum_{j=1}^{m} \alpha_j\phi_j$$

minimizes the mean square error $\varepsilon^2(\phi_j, m) = E\|X_t - \hat{X}_t\|^2$ for every $m < n$ (Watanabe, 1965). This error is expressed through the eigenvalues λ_j as $\varepsilon^2(\phi_j, m) = \sum_{j=m+1}^n \lambda_j$.

The PCA of a discrete stochastic process leads us to introduce a dynamic regression model that states how an output is related to an input, allowing forecasting of future values. Let us assume the parallel evolution of two processes, $\{X_t, t \in T\}$ and $\{Y_t, t \in T\}$, $T = \{1, 2, \ldots, n\}$, $n \in \mathbb{Z}^+$, to be known until a given instant of time. We want to forecast the output process, $\{Y_t\}$, by using the additional information of the input process, $\{X_t\}$. The forecasts can be obtained by means of the PCA of both processes through the random variables in (1) for each process.

In this way, a dynamic regression by principal components model (DRPC model) expresses the output process as a function of the input process by the equation

$$Y_t = \mu_Y + \sum_{k=1}^{n} \rho_k\, \varphi_k$$

where $\mu_Y = E[Y_t]$, φ_k is the kth principal vector of Y_t and ρ_k is the kth coefficient obtained through the principal components of X_t, α_k, $k = 1,\ldots,n$, by the expression $\rho_k = r_0^k + r_1^k\alpha_1 + \cdots + r_n^k\alpha_n + \varepsilon^k$, $k = 1, 2, \ldots, n$.

3 DRPC models forecasting on correlated time series

Although the use of DRPC models is focused on the treatment of independent sample functions (Del Moral & Valderrama, 1997), it can be extended to treat problems in which two correlated time series are involved. By splitting the original series into several subseries through the choice of an appropriate cutting point, we can perform a study of their PCA and set up the forecasting model. In this paper we consider the seasonal patterns of the series as criteria for splitting them according to the seasonality period. So, let us assume that $m + d$ subseries (n observations each one) of X_t and m for Y_t are available. By considering the first m subseries of input and output we obtain the sets of principal components, $\hat{\alpha}_i$ and $\hat{\beta}_i$ respectively. The principal vectors are $\hat{\phi}_j$, $j = 1, \ldots, n$, for the input and $\hat{\varphi}_j$, $j = 1, \ldots, n$, for the output.

In order to forecast the output we can operate by different criteria: choose the whole set of principal components, choose a number of components such that the accumulated proportion of variance for every process is approximately the same, or choose the components for the output which show highly correlation with the ones of the input. We denote N for the input and M for the output.

Linear regressions between the principal components are performed

$$\hat{\beta}_k = \hat{c}_0^k + \hat{c}_1^k \hat{\alpha}_1 + \cdots + \hat{c}_N^k \hat{\alpha}_N, \qquad k = 1, 2, \ldots, M. \tag{2}$$

Let $X^\star$ be a new subseries of X_t. Its principal components are calculated by

$$\hat{\alpha}_k^\star = \hat{\phi}_k^T (X^\star - \mu_X), \qquad k = 1, 2, \ldots, N \tag{3}$$

The components for Y_t are obtained as

$$\hat{\beta}_k^\star = \hat{c}_0^k + \hat{c}_1^k \hat{\alpha}_1^\star + \cdots + \hat{c}_N^k \hat{\alpha}_N^\star, \qquad k = 1, 2, \ldots, M \tag{4}$$

The forecasts for Y_t are obtained through the use of the mean of the output, its principal vectors and the regression by

$$Y^\star = \mu_Y + \sum_{k=1}^{M} \hat{\beta}_k^\star \, \hat{\varphi}_k$$

4 Two DRPC models on simulated and real data

In this section we perform two studies that illustrate the use of DRPC models in forecasting correlated time series with seasonal patterns. We study their forecasting behaviour in selecting the principal components to be included in the model and the parsimony of their selection.

4.1 A model on simulated time series

We consider the correlated time series shown in Tables 1 and 2. The output is obtained from the input by the model

$$Y_t = -0.78 + 1.53\, X_t + 0.25 \,\log\, t + \varepsilon_t$$

where $\varepsilon_t \rightsquigarrow \mathcal{N}(0, 0.5)$. These series consist of 120 observations. To build a DRPC model we arrange the observations as several subseries. The cutting of the observations is performed by following the seasonal pattern of the series, according to the seasonal period $n = 8$. We build a DRPC model with the first 80 observations and use the last 50 to obtain and check the forecasts.

Table 1. Input series

3.45	3.51	4.22	4.38	4.55	4.47	4.09	3.86	2.89	3.64	3.99	4.28	4.17	3.90	3.65
3.22	3.04	3.11	3.32	3.75	4.01	3.82	3.80	3.19	3.91	4.08	4.57	4.61	4.88	4.50
4.16	3.78	3.26	3.10	3.69	3.85	3.97	3.52	3.33	3.23	2.99	3.11	3.54	4.05	3.82
3.79	3.60	3.40	3.01	3.40	3.70	3.90	3.97	3.87	3.67	3.40	3.50	3.60	3.90	4.00
4.07	4.00	3.77	3.42	2.71	3.00	4.00	4.15	4.20	4.17	3.87	3.41	3.80	4.10	4.40
4.60	4.62	4.58	4.22	3.70	3.00	3.50	3.80	3.98	4.13	4.10	3.99	3.57	3.37	3.95
4.18	4.28	4.32	4.25	4.15	3.60	3.10	3.35	3.60	3.70	3.78	3.69	3.50	3.43	2.93
3.65	4.06	4.50	4.65	4.13	3.78	3.25	3.37	3.57	3.98	4.17	4.25	4.18	4.05	3.79

Table 2. Output series

3.60099	4.90129	6.59257	6.32110	5.96909	6.18371	7.74998	6.32061
2.91485	4.39249	7.04326	5.82378	5.71290	5.23216	4.04824	4.22897
3.96694	3.20413	4.99870	5.10330	4.40241	5.50793	5.31003	3.64140
4.79818	5.86436	6.15023	6.45811	6.92391	7.96575	5.86083	6.77344
3.78969	3.34380	5.12518	4.86730	6.24862	3.60223	4.16137	4.96005
4.37449	3.48385	5.65002	4.99332	5.43798	4.98170	4.53992	5.42883
4.23558	4.95753	4.64967	5.62479	5.13518	4.94040	6.14075	6.07374
5.53444	6.41509	5.64941	6.43239	5.42476	5.89767	6.67696	4.28413
2.94207	3.94827	5.11712	6.35129	6.79835	6.07943	4.68950	4.92773
4.77315	4.70567	6.80041	7.56620	6.67405	6.85489	5.93872	5.85913
4.57375	3.73058	6.19733	5.17299	6.36657	4.98438	6.11180	3.57898
4.14985	5.45530	5.23376	7.28616	5.55920	6.99313	4.40988	4.86983
3.25445	4.25776	5.15624	4.18262	4.78883	3.73831	4.48744	5.26399
2.90058	5.85982	4.92401	6.36107	7.99738	6.41589	4.20001	5.08863
3.07741	4.83178	6.81171	5.74022	6.25781	6.61540	5.19555	4.99898

We calculate the PCA for input and output series. Table 3 shows eigenvalues along with variance explained for both series. It can be observed that for the input the first three principal components explain more than 95% of the variance. For the output, this percentage is obtained with the first six principal components.

Table 3. Eigenvalues and variance explained by PCA

X_t			Y_t		
Principal component	Eigenvalue	Cumulative variance (%)	Principal component	Eigenvalue	Cumulative variance (%)
P_1	0.768058	81.76	P_1	4.024600	52.08
P_2	0.082855	90.58	P_2	1.339200	69.41
P_3	0.042843	95.14	P_3	0.801440	79.78
P_4	0.025272	97.83	P_4	0.559151	87.02
P_5	0.011523	99.06	P_5	0.421026	92.46
P_6	0.005328	99.63	P_6	0.296328	96.30
P_7	0.003412	99.99	P_7	0.216133	99.10
P_8	0.000092	100.00	P_8	0.069928	100.00

We perform linear regression from the principal components of the output over those of the input by following (2). Then we calculate the principal components of the new five subseries of the input as indicated in (3) and the coefficients for the output forecasts through (4).

Table 4. DRPC forecasts

	t_1	t_2	t_3	t_4	t_5	t_6	t_7	t_8	Error
Original	4.57375	3.73058	6.19733	5.17299	6.36657	4.98438	6.11180	3.57898	
Mod 1	3.82560	4.24597	5.56607	5.90064	5.78314	5.79068	5.68392	5.25362	0.84641
Mod 2	4.07620	4.46644	5.73678	5.89671	5.82686	5.64219	5.46641	5.19621	0.81247
Mod 3	4.06241	4.45754	5.74961	5.90403	5.84604	5.65183	5.44696	5.20282	0.81548
Original	4.14985	5.45530	5.23376	7.28616	5.55920	6.99313	4.40988	4.86983	
Mod 1	4.23756	5.02688	6.30015	6.46371	6.34164	6.34111	5.69168	5.64322	0.81266
Mod 2	4.15493	4.91335	6.16846	6.44778	6.33600	6.42816	5.76934	5.70911	0.81742
Mod 3	4.33596	5.03009	6.00009	6.35170	6.08435	6.30160	6.02455	5.62241	0.83691
Original	3.25445	4.25776	5.15624	4.18262	4.78883	3.73831	4.48744	5.26399	
Mod 1	4.40337	4.17229	5.18582	5.15543	5.14553	4.59433	5.04299	4.61076	0.69551
Mod 2	4.09921	4.03060	5.14098	5.21003	5.09115	4.66931	5.28046	4.55880	0.69869
Mod 3	3.75119	3.80617	5.46465	5.39474	5.57493	4.91262	4.78985	4.72547	0.74106
Original	2.90058	5.85982	4.92401	6.36107	7.99738	6.41589	4.20001	5.08863	
Mod 1	3.17418	4.39723	6.67463	6.47390	6.93356	5.80012	4.37513	5.13043	0.92430
Mod 2	3.32877	3.96417	6.40607	6.26307	6.83420	6.12171	4.40292	5.52600	0.97819
Mod 3	4.04841	4.42825	5.73680	5.88113	5.83385	5.61859	5.41741	5.18136	1.17607
Original	3.07741	4.83178	6.81171	5.74022	6.25781	6.61540	5.19555	4.99898	
Mod 1	4.32624	4.73771	5.64466	6.14078	5.75316	6.32723	6.34572	5.66408	0.80574
Mod 2	4.45626	5.05671	5.82328	6.21160	5.83757	6.10933	6.16944	5.49398	0.77275
Mod 3	4.30078	4.95645	5.96788	6.29412	6.05370	6.21803	5.95026	5.56845	0.67309

We consider three forecasting models for this study. Firstly we set up 95 as the percentage of variance explained for input and output. In this way, the first three principal components of the input are selected and the first six of the output. If we take into account the correlation between both series, we obtain new models by choosing those components for the output that show higher correlation with those of the input. The second model is obtained by selecting P_1 and P_2 for the output, those components with higher correlations with the first three principal components of the input, $r^2_{\hat{\beta}_1} = 0.857$ and $r^2_{\hat{\beta}_2} = 0.582$. In order to analyze parsimony, a last model is obtained by selecting only P_1 for the output and the first three components for the input.

Table 4 shows the future observed values that have not been included in the model estimation along with forecasts for the three models and the global errors for every subseries (square root of the mean square errors).

4.2 A model on real time series

Let us now consider the input series *Monthly gasoline demand in Ontario from January 1960 to December 1974* and the output series *Monthly traffic fatalities in Ontario from January 1960 to December 1974*. These series have been studied by Abraham & Ledolter (1983). They each consist of 180 observations and show an annual seasonal pattern. We build a DRPC model with the first 168 observations and use the last 12 observations to obtain and check the forecasts.

Table 5 shows eigenvalues and variance explained for both series. It can be observed that the first two principal components of the input accumulate 99% of the variance. The accumulation process for the output is slower, the first nine principal components are needed to achieve 99%.

Table 5. Eigenvalues and variance explained by PCA

X_t			Y_t		
Principal component	Eigenvalue	Cumulative variance (%)	Principal component	Eigenvalue	Cumulative variance (%)
P_1	12975.40000	98.17	P_1	5132.10000	69.05
P_2	116.17000	99.05	P_2	699.19200	78.15
P_3	57.90550	99.49	P_3	605.24700	86.60
P_4	26.52620	99.69	P_4	342.95400	91.21
P_5	15.36140	99.80	P_5	210.56500	94.04
P_6	9.98027	99.88	P_6	177.17600	96.43
P_7	6.65339	99.93	P_7	84.31960	97.56
P_8	4.64660	99.96	P_8	71.82170	98.53
P_9	2.57941	99.98	P_9	48.13670	99.17
P_{10}	1.04926	99.99	P_{10}	36.95310	99.67
P_{11}	0.77791	100.00	P_{11}	18.15410	99.92
P_{12}	0.28683	100.00	P_{12}	6.28208	100.00

Table 6. DRPC forecasts

	Original	Model 1		Model 2		Model 3		Model 4	
1974		Forecast	Error	Forecast	Error	Forecast	Error	Forecast	Error
Jan	94	149.52	55.52	117.45	23.45	116.98	22.98	121.70	27.70
Feb	89	46.61	42.38	93.72	4.72	96.07	7.07	93.84	4.84
Mar	118	96.27	21.73	122.63	4.63	121.46	3.46	117.40	0.60
Apr	101	65.10	35.90	126.98	25.98	128.96	27.96	125.22	24.22
May	150	161.55	11.55	159.28	9.28	158.90	8.90	153.94	3.94
Jun	150	227.54	77.54	191.19	41.19	191.88	41.88	196.59	46.59
Jul	191	187.74	3.26	203.11	12.11	203.31	12.31	202.23	11.23
Aug	214	271.39	57.39	232.64	18.64	228.11	14.11	227.94	13.94
Sep	173	171.44	1.55	207.84	34.84	210.18	37.18	211.35	38.35
Oct	170	193.55	23.55	200.50	30.50	200.74	30.74	200.17	30.17
Nov	175	180.89	5.89	176.39	1.39	177.12	2.12	174.00	1.00
Dec	123	169.10	46.10	159.70	36.70	161.18	38.18	163.95	40.95
Global error			39.589		24.21		24.73		25.728

In this study we consider four models. In the first one we consider all the principal components for the output. By setting up 99 as the percentage of variance explained, the first two principal components are selected for the input and the first nine for the output. According to the correlation between both series, a third model is built by selecting P_1, P_3, P_5 and P_7 for the output. The correlations are $r^2_{\hat{\beta}_1} = 0.913$, $r^2_{\hat{\beta}_3} = 0.247$, $r^2_{\hat{\beta}_5} = 0.244$ and $r^2_{\hat{\beta}_7} = 0.171$. The last model consists of selecting only the first principal component for the output, the one with highest correlation. Table 6 shows output forecasts for every model.

5 Concluding remarks

The PCA of a discrete stochastic process allows us to set up DRPC models for seasonal correlated time series in a simple way.

It has been pointed out in this paper that parsimony is essential in building these models. Model 1 on real data shows larger error than the rest of the models. Nevertheless, the models with only one principal component for the output produce high forecasting errors.

As far as correlation goes, the selection of those components for the output that show higher correlation with those of the input does not improve the forecasting results significantly. The forecasting results do not exceed those obtained by selecting the principal components that accumulate a fixed percentage of variance explained. The last seems to be a better way to select the principal components to be consider in the forecasting model.

References

Abraham, B. & Ledolter, J. (1981). Parsimony and Its Importance in Time Series Forecasting. *Technometrics*, **23**, 4, 411-414.

Abraham, B. & Ledolter, J. (1983). *Statistical Methods for Forecasting.* New york: Wiley.

Box, G.E.P. & Jenkins, G.M. (1976). *Time Series Analysis. Forecasting and Control.* San Francisco: Holden-Day.

Del Moral, M.J. & Valderrama, M.J. (1994). A smoothing algorithm for observations corrupted by additive noises. In: *Selected Topics on Stochastic Modelling*, 121-128. Singapore: World Scientific.

Del Moral, M.J. & Valderrama, M.J. (1997). A Principal Component Approach to Dynamic Regression Models. *International Journal of Forecasting*, **13**, 237-244.

Del Moral, M.J., Valderrama, M.J. & Aguilera, A.M., (1994). Reducción de dimensión en un conjunto de señales aleatorias. *Estadística Española*, **36**, 135, 75-97.

Deville, J.C. (1974). Méthodes statistiques et numériques de l'analyse harmonique. *Annales de L'INSEE*, **15**, 3-101.

Deville, J.C. (1978). Analyse et Prevision des Series Chronologiques Multiples Non Stationaires. *Statistique et Analyse des Donnes*, **3**, 19-29.

Fukunaga, K. (1990). *Introduction to Statistical Pattern Recognition: Second Edition.* San Diego: Academic Press.

Pankratz, A. (1991). *Forecasting with Dynamic Regression Models.* New York: Wiley Interscience.

Watanabe, S. (1965). Karhunen-Loéve expansion and factor analysis. In: *Transactions on 4th Prague Conference on Information Theory*, 635-660. ACADEMIA, Publishing House of the Czechoslovak Academy of Sciences.

Traffic Models for Telecommunication

Dee Denteneer and Verus Pronk
Philips Research Laboratories Eindhoven, Prof. Holstlaan 4,
5656 AA Eindhoven, the Netherlands.

1 Introduction

Models that describe the traffic on the current (broadband-)integrated-services digital networks are a *hot topic* in telecommunication. They are relevant for at least the following two reasons:

- traffic description: It is assumed (at least for some types of networks) that potential users will have to give a traffic description. This will enable the network operator to decide whether the new connection can be admitted to the network without violating the quality of service guarantees of existing connections, i.e. without overloading the network;
- network simulation: with the aim to properly dimension future networks.

To develop such models is a statistical challenge, both mathematically and computationally. The mathematical challenge is brought about by the current insight that network traffic is long-range dependent, i.e. that the autocorrelation functions of such traffic approach zero very slowly in comparison with the exponential decay characterising short-range dependent (e.g. ARMA-type) processes. Since the discovery of this phenomenon for Ethernet traffic in Leland *et al.* (1994), this insight has been corroborated for other types of traffic such as variable bit rate video (see Beran *et al.*, 1995) and wide area traffic (see Paxson & Floyd, 1995).

We are hence confronted with a need for traffic models that are very different from the familiar ARMA and Poissonian ones. Moreover, performance models must be reevaluated in the light of this new class of models (see e.g. Jelenkovic *et al.*, 1997). The computational challenge is due to the enormous size of the data sets that have been gathered on network traffic.

In this paper we will briefly review the data used in traffic analysis and modelling (Section 2) and some of the mathematical models used in the field (Section 3). The current trend is towards models that exhibit long-range dependence. However, we will point out the shortcomings of standard models for long-range dependent processes, such as fractional Brownian motion or fARIMA, as traffic models. Our central tenet will be that these standard models concentrate exclusively on long-range aspects of the data to the exclusion of equally relevant short-range features. We will illustrate this statement with aggregate wide-area-network (WAN) traffic (Section 4.1) and with bit rate variations in variable-bit-rate (VBR) video (Section 4.2). In Section 5 we will comment on some of the computational difficulties in working with these huge data sets and put forth some proposals for simple functionality for statistical software that would greatly improve the ease with which the masses of traffic data can be analysed.

2 Data

In analysing telecommunication data, a broad understanding of the following characteristics is required.

Network hierarchy Networks are structured according to the OSI-protocol layer in which each layer offers increasingly sophisticated communication capabilities, extending the features offered by the previous layer, see e.g. Tanenbaum (1981). Passing data from a higher layer to a lower layer, on the sending side, involves fragmenting the data into packets and adding protocol-dependent information. The receiving side executes the inverse process.
As a consequence of this, measurements can be carried out at one of these various layers. They can be carried at the application layer (relating to what the user actually does), at the TCP/IP layer (the set of communication protocols used in the internet), and e.g. at the ATM layer ('close to' the physical layer).

Locality Networks are broadly classified as 'local-area' or 'wide-area', with implications for roundtrip delay.

Data type The data to be analysed are typically time series or (marked) point processes. So at e.g. the video application layer the data to be transmitted consist of image frames of variable sizes which must be transmitted at a constant rate of 25, say, per second. At the TCP/IP layer, the data consist of arrival times of variable-length packets and their sizes. At the ATM layer, the data consist of arrival times of fixed-length cells. In practice, one frequently reduces the point process type to the time series type by counting the number of arrivals (or bytes) in fixed-length intervals. In addition, the data may contain source-destination information so that aggregate traffic can be disaggregated into a number of bidirectional time series that record all the information passed between two communicating applications.

Data size The computational challenge is brought about by the enormous size of the data sets that have been gathered on network traffic. One hour of ATM traffic measurement records about 25 million cell arrival times, a trace of thirty days wide-area connections at LBL spawns about 700000 connections, the recordings of URL requests of UCB students working at home amount to about 250 Mbytes/5 days.

See the Internet traffic archive (`http://ita.ee.lbl.gov/index.html`) for some of the data sets used in this paper and for pointers to further information.

3 Models

In this section we will not provide detailed accounts of the numerous models for describing applications; see e.g. Paxson (1994) for applications such as http, ftp, or telnet that run on top of TCP/IP. Instead, we will indicate some broad characteristics that apply to most models.

Probably the most basic traffic model in use is the distribution function of the size of the units being transferred. If this size is discrete and limited (and small), as in the case of TCP/IP packets, this distribution can be estimated by means of a histogram. Even such simple models find important applications, e.g. in assessing the efficiency of various packetization schemes. In the case of distributions that have unbounded support we frequently encounter the first topic that dominates the current headlines: heavy tails. Heavy tails mean that either the full distribution or its tail is well approximated by a Pareto distribution:

$$\Pr\{X > x\} = (x/x_0)^{-\alpha}, \text{ for } x \geq x_0, \tag{1}$$

with $0 < \alpha \leq 2$. If the parameter α is smaller than 2, then the variance of the distribution is infinite, and α's smaller than 1, implying an infinite mean, have also found application. Such distributional models are of course also important in modelling interarrival times. It is now an established fact that numerous size distribution and interarrival times of relevance in telecommunication exhibit heavy tails; see Willinger *et al.* (1997) for a survey.

The next step in modelling is to take the serial structure into account and this leads to the second topic dominating the headlines: long-range dependence or even self-similarity or fractal behaviour. To introduce long-range dependence, let $X_t, t = 1, 2, \ldots$ be a stochastic process with mean m, variance σ^2, and autocorrelation function $r(k)$. Following e.g. Beran *et al.* (1995), we say that X_t exhibits long-range dependence if

$$r(k) \sim k^{1-2H} L_1(k), \quad \text{as } k \to \infty, \tag{2}$$

where $1/2 < H < 1$ is the so-called Hurst parameter, and L_1 is slowly varying for $k \to \infty$:

$$\lim_{k\to\infty} \frac{L_1(kx)}{L_1(k)} = 1 \tag{3}$$

for all $x > 0$. Long-range dependent processes are hence characterised by a slowly decaying correlation function, whereas exponential decay characterises short-range dependent processes.

The topics are interconnected, as is apparent from theorems in e.g. Willinger *et al.* (1997). They consider on/off-processes: during an on-period a source generates packets at a constant rate and during an off-period the source is silent. The source alternates between on and off periods and sojourn time in both on and off periods is heavy tailed. It can then be shown that the aggregate of such traffic converges to fractional Brownian motion, which is long-range dependent. For another such connection see Section 4.2.

4 Criticism

The presence of both heavy tails and long-range dependence is now a well-established fact. However, analysing and modelling the data exclusively on the basis of these viewpoints ignores a lot of relevant information. We will illustrate this statement for aggregate wide-area traffic and for variable-bit-rate video data.

4.1 WAN traffic

For want of models for aggregate wide-area traffic, we compared the distribution of packet interarrival times of wide-area traffic with the interarrival times predicted with the aggregation of on/off sources described in Section 3 that was originally developed for Ethernet LAN traffic. In comparing the tails of the histograms of observed interarrival times and interarrival times predicted on the basis of the model (see Figure 1(b)) one might very well conclude that the model fits the data nicely. However, the heads of the distributions are completely different: the fractal Brownian motion fails to reflect a very distinct bimodality which shows up in the head of the observed distribution. We have termed this effect 'the interaction effect' and have verified it in other sources (such as the LBL data set in the internet archive).

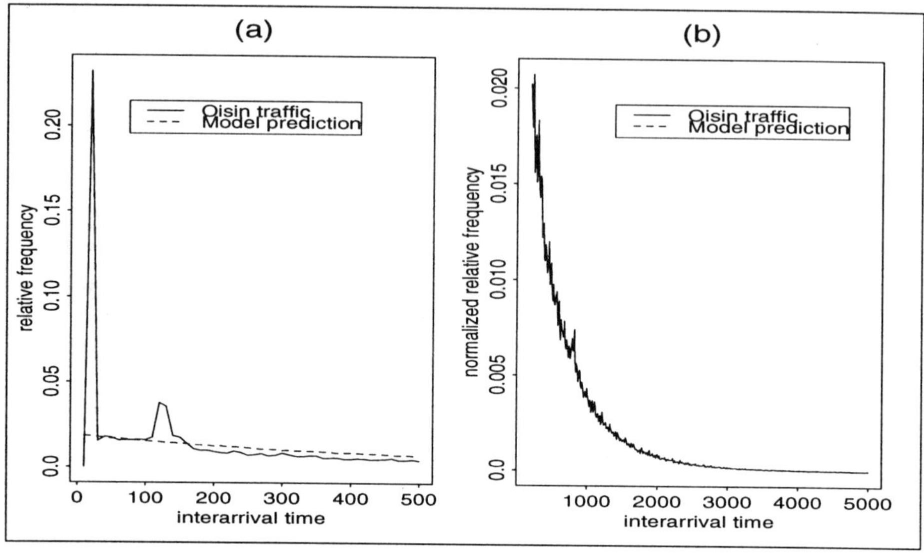

Fig. 1. Histograms of observed and predicted interarrival times (a) short interarrival times (b) long interarrival times

4.2 VBR video

An in-depth study of the serial structure of VBR video was undertaken in Beran *et al.* (1995). On the basis of 20 VBR videos of various origins and coded with various codecs (see their Table 1) and several estimates for the Hurst coefficient, they convincingly argue that long-range dependence is present. The aim of Beran *et al.* was to arrive at models that abstract from the specific video scenes and codecs and also capture the intrinsic properties of the observed video clips. They demonstrate that standard models for long-range dependent processes provide a useful family of processes that accurately reflect the 'long-range' serial structure of the data. However, they ignore equally relevant 'short-range' features, as we will now illustrate.

In the first place such models ignore the fact that consecutive image frames in an MPEG video consist of a regular arrangement of three different types and are compressed with distinct efficiency. So instead of modelling one process, it is better to model three distinct processes corresponding to the three frame types and take their crosscorrelation into account.

Secondly, more structure is present in the video data than the regular arrangement of I-, B-, and P-frames: see e.g. Figure 2, which shows the I-frame sizes of a short sequence from the movie 'Patriot Games'. The frame size evidently remains approximately constant for a while and then jumps to another level. This immediately suggests that these 'jumps' are related to the scene changes always present in videos. To further investigate the relevance of scenes for the frame-size process, we manually segmented the data from 'Four Weddings and a Funeral' into 290 scenes using the original analog data. We then computed an analysis of variance to assess the amount of variation due to scenes relative to the rest of the variation for each of the three frame types. The analysis reveals that the variation in scene means is highly significant. More importantly, the variation in scene means explains a relevant

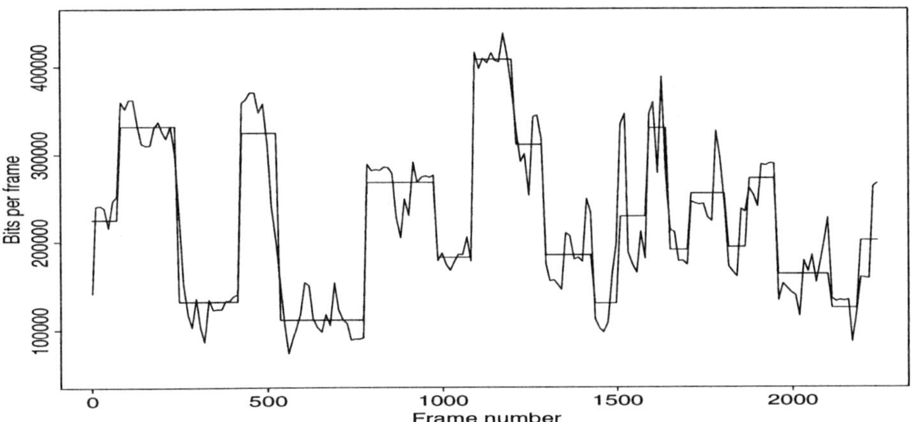

Fig. 2. Frame size variations of I-frames in 'Patriot Games'

portion of the overall variation: in the case of the I-frames, approximately 80% of the overall variation in frame size is due to scenes and in the case of the B- and P-frames this figure is approximately 65%.

In developing a scene-based model, assume that the series consists of scenes, $s = 1, \ldots, S$, and that each scene consists of l_s frames. In a scene-based model the number of bits per frame, X_k, is composed of a scene-level m_s plus a deviation, $\epsilon_{s,i}$, due to the current frame:

$$X_k = m_{s(k)} + \epsilon_{s(k),i(k)}, k = 1, \ldots, N. \tag{4}$$

Here $s(k)$ denotes the sequence number of the scene to which frame k belongs and $i(k)$ denotes a within-scene sequence number. In this model m_s, $\epsilon_{s,i}$, and l_s are the stochastic variables of interest, which can be serially correlated. The autocorrelation function of (4) is then given by

$$E\{X_k X_{k+l}\} - \mu^2 = (E\{m_s^2\} - \mu^2) \sum_{j=0}^{\infty} \rho_j \Pr\{s(k) + j = s(k+l)\}, \tag{5}$$

with μ being the overall mean scene level and ρ_j the correlation between successive scene levels. It can be shown that Pareto scene lengths imply long-range dependence, i.e. that (5) behaves as (2). However, it can also be shown that assuming a Pareto distribution for the scene lengths does not agree with the data. Moreover, it is true that lognormal scene lengths and ARMA-type dependence between successive scene levels does not imply long-range dependence. We have as yet no systematic answer to the key question: to what extent do short-range dependent processes give rise to long-range dependent-like correlations.

5 Computational issues

The masses of telecommunication data do not pose a problem per se: there is little difficulty in writing pieces of dedicated code to compute some predetermined quantities of interest. However, the amount of data is an obstacle in exploratory analysis and it very much limits the freedom to produce simple graphical representations of the data at hand and the possibility of fitting simple models.

Yet simple tools could lead to substantial improvement. Most importantly, procedures should be available for reducing the amounts of data without the need to have the data in core memory. The simplest way of doing this, by selection of a consecutive subset, is usually catered for. However, clustering, aggregating, and sampling are equally useful, but not always available. Clustering means the reduction of a (marked) point process to a marked point process by a user-defined separation criterion. Such a reduction is a natural operation on data that are at a lower level of the network hierarchy. Aggregation counts events or marks in fixed time intervals, where time intervals could be user-defined or adaptively defined by the maximum object size supported. Performing this operation with respect to marks rather than time computes a histogram. As to sampling: even random sampling would help, but the sampling of time series objects (identified by keys) from a large aggregate time series would be very useful.

The proposed procedures construct an object that can be computationally managed out of a (binary) large object, possibly using adaption. Such reduction schemes of course have consequences for the statistical efficiency with which operations can be performed. However, this is relatively unimportant in an exploratory setting and is part of a much broader field that surpasses these simple tools and includes estimation and simulation.

Acknowledgement
The authors gratefully acknowledge partial support by CEC ACTS project 014 CanCan.

References

Beran, J., Sherman, R., Taqqu, M., & Willinger, W. (1995). Long-range dependence in Variable-bit-rate video traffic. *IEEE Transactions on Communication*, **43**, 1566-1579.

Jelenkovic, P., Lazar, A., & Semret, N. (1997). The effect of multiple time scales and subexponentiality in MPEG video streams on queueing behaviour. *IEEE Journal of Selected Areas in Communications*, **15**, 1052-1071.

Leland, W., Taqqu, M., Willinger, W., & Wilson, D. (1994). On the self-similar nature of Ethernet traffic (Extended version). *IEEE/ACM Transactions on Networking*, **2**, 1-15.

Paxson, V. (1994). Empirically derived analytic models for Wide-Area TCP connections. *IEEE/ACM Transactions on Networking*, **2**, 316-336.

Paxson, V., & Floyd, S. (1995). Wide area traffic: the failure of Poisson modeling. *IEEE/ACM Transactions on Networking*, **3**, 226-244.

Tanenbaum, A. (1981). *Computer networks*. Englewood Cliffs, New Jersey: Prentice-Hall.

Willinger, W., Taqqu, M., Sherman, R., & Wilson, D. (1997). Self-similarity through high-variability: statistical analysis of Ethernet LAN traffic at the source level. *IEEE/ACM Transactions on Networking*, **5**, 71-96.

Construction of Non-Standard Row-Column Designs

A. N. Donev

Dept. of Medical Statistics, De Montfort University, Leicester, LE1 9BH, UK

Abstract. An algorithmic approach to the construction of row-column designs which can have a non-rectangular shape is presented. The rows and the columns of a row-column design correspond to the levels of two blocking variables. If information about correlation between the observations taken at each level of one of the blocking variables is available, it is taken into account in the construction of the design.

Keywords. A-optimality, row-column designs, correlated observations

1 Introduction

The sensitivity of tests for comparison of treatments can be increased by controlling the sources of variation that may inflate the experimental error. Small variances of the estimated treatment differences are required. This is usually achieved by dividing the observations into blocks of homogeneous units and adjusting the estimates of the treatment differences for the block effects. The precision of the estimated treatment differences depends on the total number of observations and the way they are split between the treatments to be compared.

Row-column designs have been used successfully in situations where there are just two blocking factors, i.e. where two sources of variation are controlled. Usually the data are easily summarized in a table of a rectangular shape. Its columns correspond to the levels of one of the blocking factors while the rows correspond to the levels of the other. Different treatments may be used in the resulting blocks, i.e. groups of homogeneous units.

Customarily balanced designs are used. Such designs replicate the treatments equal numbers of times at each level of each blocking variable (Latin Square Designs). When this is not practical, the balance is preserved as much as possible by ensuring that the treatments occur the same number of times together at each level of the blocking variable (Balanced Incomplete Block Designs). Many researchers have studied these types of designs and shown their good properties for a large class of practical situations where they can be used (for example, see Kiefer, 1959). The estimated treatment differences obtained using data from such designs have minimum possible variance. However, there are also situations where the use of balanced designs is not possible.

For example, five methods (treatments) of measuring the resistance to fatigue (response, Y) of an alloy have to be compared. The investigator would like to take into account that the variation in the results can come from the laboratory and the order in which the measurements are taken by the analysts. Four laboratories are available. Four measurements can be carried out by each of the analysts in two of the laboratories and three by those in the other two. Due to staff availability five analysts from each of the first two laboratories and three analysts from each of the other two laboratories will be used.

Apparently a standard block design for this experiment is not available. The following features make the required design irregular, or non-standard:

- a row-column incomplete block design is required: the rows correspond to the laboratories while the columns correspond to the order in which measurements are taken;
- the two blocks corresponding to the first two laboratories can contain four measurements, while the other two can have only three, i.e. the design will not have a rectangular shape;
- the number of the replicates in the blocks will be different as the number of analysts used in the different laboratories is not the same;
- as each analyst takes several measurements it could be expected that there may be correlation between them.

One way to solve such a design problem is to make compromises, for example to use 3 analysts from each laboratory and take 3 observations in each. However, using such an approach will not take into account the needs of the investigator. This paper proposes a better way of tackling such problems. It describes an algorithm which can be used to generate tailor-made row-columns designs according to the exact requirements of the investigator.

2 Background

Methods for construction of Incomplete Block Designs have been studied by many researchers. Some of them are based on combinatorial results and are limited to situations where such results are available. Other methods lead to a computer search for an optimum design using a specified criterion of optimality. Such algorithms include those proposed by Jones (1976) and Nguyen (1994). There has been a considerable interest in the situations where the observations within each block are dependent. For example, see Kiefer & Wynn (1981), Cheng (1983) and Martin & Eccleston (1991). Donev (1997) describes an algorithm for the construction of a special case of row-column designs, known as cross-over designs. Donev (1998) shows that the optimality of designs depends on the correlation structure of the observations so that a design which is optimum (according to a specified criterion of optimality) for one correlation structure may not be optimum for another. None of the available algorithms can be used to construct designs appropriate for the type of problems described earlier.

3 Model and criterion of optimality

The model that is assumed is

$$y_{klm}=\mu+\tau_k+\alpha_l+\beta_m+\varepsilon_{klm},$$

where μ is an overall mean, τ_k, k = 1, 2, ..., T, is the effect of the kth treatment, α_l, l= 1, 2, ..., R, is the effect corresponding to the lth level of the row blocking factor, β_m, m=1, 2, ..., C, is the effect corresponding to the mth level of the column factor, and ε_{klm} is the experimental error with zero mean. The model can easily be reparametrized and rewritten as a general linear model

$$\mathbf{y}=\mathbf{F}\beta+\varepsilon,$$

where $\mathbf{y}$ and ε are vectors of the observations and their errors, while β is a vector of estimable effects. An example of reparametrization is given in Donev (1998). Suppose that the treatments are not replicated in the blocks. The covariance matrix for the observations is assumed to be

$$\mathrm{var}(\varepsilon)=\sigma^2\Sigma=\sigma^2\mathbf{I}_{RC}\otimes\mathrm{W}(\rho),$$

where I_{RC} is an identity matrix and $\mathrm{W}(\rho)$ is the common correlation matrix for the observations at each level of one of the blocking factor. For example, if first order autocorrelation dependence between the observations exists, the ijth element of $\mathrm{W}(\rho)$ is

$$w_{ij}=\frac{\rho^{|i-j|}}{1-\rho^2}. \tag{1}$$

In this case Σ is block diagonal. Problems where more complicated correlation structures have to be assumed can be tackled in a similar way. For simplicity, only the case when $\mathrm{W}(\rho)$ depends on one parameter will be discussed.

Generalized least squares estimates of the model parameters are

$$\hat{\beta}=\left(\mathbf{F}^{\mathbf{T}}\Sigma^{-1}\mathbf{F}\right)^{-1}\mathbf{F}^{\mathbf{T}}\Sigma^{-1}\mathbf{y}=\mathbf{G}^{-1}\mathbf{F}^{\mathbf{T}}\Sigma^{-1}\mathbf{y}=\mathbf{M}\mathbf{F}^{\mathbf{T}}\Sigma^{-1}\mathbf{y}.$$

Note that the information matrix of the design $\mathbf{G}$ depends on the unknown parameter ρ. If ρ=0, i.e. the observations are independent, $\Sigma=\mathrm{I}_{RC}$.

If the elements of β are ordered so that the parameters corresponding to the treatment differences of interest are put first, it can be shown that the sum of the variances of all pairwise treatment comparisons can be calculated as

$$A=\sum_{i=1}^{T-1}\sum_{j=i+1}^{T}\mathrm{var}(\hat{\tau}_i-\hat{\tau}_j)=2T\sum_{i=1}^{T-1}\sum_{j=i}^{T-1}m_{ij}, \tag{2}$$

where m_{ij} is the ijth element of $\mathbf{M}$. Clearly, a design with as small as possible value of A is required. This criterion of optimality is often referred to as A-optimality. It can be used not just to compare but also to construct designs that are optimum with respect to this criterion, provided that the model, the shape and the size of the design are known.

The size of the matrices increases if the observations in the blocks are replicated, thus making the problem more complicated. However, an equivalent presentation is obtained if the matrix **F** is premultiplied with a diagonal matrix, say **R**, whose diagonal elements are equal to the square root of the numbers of replicates in the blocks, i.e. $\mathbf{M} = \left((\mathbf{RF})^{\mathbf{T}}\Sigma^{-1}\mathbf{RF}\right)^{-1}$. For instance, for the example given in the Introduction of the paper, the 14×14 matrix $\mathbf{R}=\text{diag}\left\{4\left(\sqrt{5}\right),\ 4\left(\sqrt{5}\right),\ 3\left(\sqrt{3}\right),\ 3\left(\sqrt{3}\right)\right\}$.

If the parameter of the correlation matrix between the observations is not known, the design problem becomes non-linear in the parameters. One way to solve such a problem is to start by assuming a parametric form for the correlation matrix. For example it can be defined by equation (1). Designs that are optimum with respect to criterion (2) can be easily found for various specified values of ρ. A design obtained in this way is guaranteed to be optimum only in the neighbourhood of the assumed value of ρ. However, if the value of ρ is not known or an accurate point estimate of it cannot be obtained, the design may not be optimum for the situation where it will be used. In most practical situations there is uncertainty about the value of ρ. In such a case a pseudo-Bayesian approach can be taken. Donev (1998) generalizes criterion (2) for this situation. The generalized criterion of optimality requires minimization of

$$A_I = E_\rho\left[A(\rho)\right] = \int_l^u A(\rho)f(\rho)d\rho\ , \tag{3}$$

where $A(\rho)$ is the sum of variances of the estimated treatment differences defined by (2) which depends on ρ. In (3) $f(\rho)$ is the probability density function of ρ and the integration is done over a plausible range of values of ρ, (l, u). In contrast to the standard approach to optimum Bayesian design (for a review see Chapter 19, Atkinson & Donev, 1992), prior information about the other parameters in the model is not required. If $f(\rho)$ is difficult to estimate, a user-defined function can be used. In the latter case, the empirical experience of the author suggests that assuming $f(\rho)$ to take the form of the probability density function of a uniformly distributed random variable can be very useful even if a relatively wide range (l, u) of values for ρ is specified.

4 An algorithm

Once a criterion of optimality is specified, an assumption about the model has been made and the required shape is known, a numerical search for the required design can be carried out. Equation (3) defines the criterion in the search. The algorithm which will be used is based on the iterative exchange technique introduced by Fedorov (1972). It takes into account the special features of the problem of interest. It consists of the following steps:

1. A starting design with the required shape and **R=I** is generated. The

procedure is similar to that explained by Donev (1997).
2. If the information matrix is singular, return to step 1. Otherwise go to step 3.
3. Consider for exchange each row of the current design with treatment sequences of the same length as the one considered for exchange. The trapezoidal method for numerical integration is used for calculation of (3) using 6 equally spaced values for ρ.
4. If beneficial exchanges exist, carry out the one for which the reduction in A is largest and go to step 3. Otherwise go to step 5.
5. If optimum group sizes are to be found, search for **R** that minimizes (3).
6. Repeat steps 1 to 5 a specified number of times (called tries). The best design which is found is used.

The efficiency of this algorithm depends on the complexity of the problem. Like virtually all existing computer algorithms for construction of various optimum designs, the algorithm described in this paper cannot guarantee that the best possible design is found. The local optimality problem increases with the size of the problem, i.e. with the increase of the number of rows and columns of the design. However, increasing the number of tries, i.e. the number of searches from different starting designs, increases the probability of finding the best possible design. The computational time required for each try is approximately the same. Therefore the total computational time increases approximately linearly with the number of tries. Due to the constantly increasing speed of the modern computers, the computational time is not an important issue for most of the practical situations where the algorithm is likely to be used.

Modifications of this algorithm are possible. For example, optimization for the number of replicates, i.e. for **R,** can be carried out in Steps 3 and 4.

The usefulness of the algorithm will be illustrated by using it to find a solution to the non-standard problem explained in the introduction of the paper.

5 Example

Optimum designs were found for three cases: (a) one analyst is used from each laboratory and no correlation is assumed to exist between the observations made by the same analyst (Design A); (b) as (a) but subsequent observations made by each analyst are correlated, following first order autocorrelation dependence (Design B); and (c) as (b) but five analysts are used from each of the first two laboratories and three from each of the other two. Design A would also be appropriate if four analysts are used in laboratories 1 and 2, and three analysts in laboratories 3 and 4. Design B and Design C are constructed under the assumption that the correlation coefficient ρ between the observations made by each analyst is equally likely to take any value in an interval $\{l,u\}$, i.e.

$$f(\rho)=\frac{1}{u-l}.$$

The optimum designs for l=0.3 and u=0.6 are given in the Table 1. Fifty tries were used to obtain each of the three designs.

The algorithm described in this paper can be easily modified to be used in situations where the correlation structure of the observations is different from that investigated in this paper.

Table 1 A-optimum designs for the comparison of 5 methods of analysis. The columns correspond to the order in which the methods of analysis are used by each analyst, while the rows correspond to the 4 laboratories.

Design A (ρ=0)	Design B (0.3≥ρ≥0.6)	Design C (0.3≥ρ≥0.6)
(1)* 1 3 4 5	(1)* 1 4 5 3	(5)* 1 3 5 1
(1) 2 1 5 4	(1) 2 1 3 5	(5) 2 1 4 5
(1) 3 5 2 -	(1) 3 2 4 -	(3) 3 4 2 -
(1) 4 2 3 -	(1) 4 5 1 -	(3) 4 2 3 -
$A = 5.398$	$A=5.154$	$A=1.269$

* - number of analysts

Acknowledgment

The author is grateful to J. Denne for his valuable suggestions during the preparation of the paper.

References

Atkinson, A.C. & Donev, A.N. (1992). *Optimum Experimental Designs*. Oxford: Oxford University Press.

Cheng, C. S. (1983). Construction of optimal balanced incomplete block designs for correlated observations. *Ann. Statist.*, **11**, 240-246

Donev, A.N. (1997) . An algorithm for the construction of cross-over trials. *Appl. Statist.*, **46**, 288-298.

Donev, A.N. (1998). Crossover designs with correlated observations. *J. Biopharmaceutical Statistics*, **8**, 249-262.

Fedorov, V.V. (1972). *Theory of Optimal Experiments*. New York: Academic Press.

Kiefer, J. (1959). Optimum experimental designs (with discussion). *J. R. Statist. Soc.* B, **21**, 272-319.

Kiefer, J. & Wynn, H.P. (1981). Optimum balanced block designs for correlated observations. *Ann. Statist.*, **9**, 737-757.

Jones, B. (1976). An algorithm for deriving optimal block designs. *Technometrics*, **18**, 451-458.

Martin, R.J. & Eccleston, J.A. (1991). Optimal incomplete block designs for general dependence structures. *J. Statist. Plann. Infer.*, **35**, 77-91.

Nguyen, N.-K. (1994). Construction of optimal block designs by computer. *Technometrics*, **36**, 300-307.

Computational Statistics for Pharmacokinetic Data Analysis

Lutz Edler

Biostatistics Unit-R0700, German Cancer Research Center,
Im Neuenheimer Feld 280, D-69120 Heidelberg

Abstract. Aims of pharmacokinetic computations are identified for an evaluation strategy useful to define criteria for software selection and software development.

Keywords. Pharmacokinetic, non-linear regression, model fit, software

1 Introduction

Pharmacokinetic (PK) and pharmacodynamic (PD) information are the basis of modern pharmacotherapy. PK investigations study the time-dependent fate of a drug and its breakdown products after administration to the body in terms of absorption, resorption, distribution, metabolism and elimination. PD is the study of biological effects induced by the drug's effective amount or concentration. PK and PD are linked in the so-called PK/PD modelling of combined dose-effective concentration relationships (Meibohm & Derendorf, 1997). PK and PD are extremely important in pre-clinical and early clinical trials in terms of efficacy and safety. Their impact is determined by the availability, appropriateness and applicability of computational methods and their implementations. The role and the use of computational statistics in PK analysis, will be presented below.

2 Pharmacokinetic Data Analysis

The statistical analysis of pharmacokinetic data addresses time-dependent repeated measurements of *drug of concentrations* in various organs (P) of the body with the goal to describe the time course $C_P(t)$ and to determine clinically relevant parameters by modelling the organism through compartments and flow rates. The mathematical solution is a system of differential equations with an explicit solution for most of the one- or two compartment models. Otherwise, numerical solutions have to be used. For basic methodology see e.g., Gibaldi & Perrier (1982) or Edler (1998). Intrinsic pharmacokinetic parameters are e.g., the area under the curve (AUC), clearance, distribution volume, half time, elimination rates, minimum inhibitory concentrations etc.

Numerous computer programs for linear and simple non-linear regression methods have been reported, see e.g. Gex-Fabry & Balant (1994) and Jackson (1996). Easy to use fitting procedures have been programmed on spreadsheet platforms, software packages have been maintained requiring special training, and macros have been suggested using standard statistical systems. Numerical integration and minimization is handled quite differently and often pose limitations. In any case, computational aspects and determination of statistical variability of parameters are important. Consideration of the subsequent stages of a PK analysis is relevant for the development, use, and assessment of computational systems in PK.

3 Stages of PK Analysis

3.1 Data Entry

Spread sheet *interfaces* (e.g., EXCEL or LOTUS) connected with databases (e.g., ACCESS, dBASE) facilitate a straightforward usage and are extremely helpful in solitary evaluations. Some programs use direct (matrix-type) data input or 'cut-and-paste' technique. For routine applications and large drug development programs, however, interfaces to the bigger pre-clinical or clinical databases are more efficient. It can be recommended to separate the data and the models in any case for avoiding uncontrolled modifications of the data.

3.2 Modelling

Mathematical modelling can improve the success in pre-clinical and clinical drug development and, especially, PK models aimed to predict time-concentration curves have been used to optimize the dosage regime in individual patients. Computational procedures should allow both, the prediction and the assessment of the prediction error.

3.2.1 Individual Model

In general, the concentration y_{ij} of individual i at time points t_{ij}, j=1, ...n_i is modelled by a nonlinear function f(x_{ij}, θ_i) with an individual parameter vector θ_i and covariates x_{ij} which include the time points t_{ij} as special case. One of the simplest types of PK 'modelling' is the estimation of θ_i by *curve fitting* of time-concentration curves of individual persons or patients. However, non-parametric curve fitting is as good for estimating $C_P(t)$ and a parameter like the AUC. Compartment models have served for a long time as paradigm of PK modelling. Program systems have been designed either as compartment *model library* or as compartment *model generator* where the user defines the model. Dosing schedules are translated to input functions depending on the locus of application (e.g., p.o., i.p., i.v.), the mode of administration (e.g., bolus, continuous infusion, pulse), or delays in time. Predefined model libraries tend to lack such options. Model generators would be preferred if they can comply with complex bioavailability of the drug.

3.2.2 Population Model

Methods of population pharmacokinetics are indicated when kinetic data of different individuals have to be combined and when the average behaviour of a population is used to predict an individual kinetic. The naive approach of calculating (weighted) means has been abandoned in favour of *mixed/random effects* models and *Bayesian* methods. Linearized maximum likelihood estimation in non-linear mixed effects regression was investigated by Beal & Sheiner (1992), see also Yuh *et al.* (1994). A three-stage hierachical Bayesian approach uses the individual time-concentration relationship, the distributional form for the pharmacokinetic parameters, optionally covariate information, and a prior distribution of the parameters of the second stage and of intra-individual variability (Wakefield & Racine-Poon, 1994).

3.2.3 Physiologically-Based PK Model (PBPK)

PBPK models define drug kinetics in terms of the physiology, anatomy and biochemistry of the organism and are composed of compartments which represent body organs and tissues. Further assumptions concern drug uptake, clearance and allometric scaling. The body compartments are linked together by a flow network. A PBPK model is defined by a system of deterministic kinetic equations (mass balance equations) of the amount or the concentration of the drug in the compartments as a function of time and initial dose. PBPK models are more complex than compartment models and they involve usually a large number of parameters.

3.3 Error Specification

Observed concentrations are subject to error. The individual model described in Section 3.2.1 is then written as a nonlinear regression model

$$y_{ij} = f(x_{ij}, \theta_i) + \varepsilon_{ij}$$

one has to account for a structure of the error variance and additional modelling of the error e.g. by

$$Var[\varepsilon_{ij}] = s_{ij}^2 = a + bf(x_{ij}, \theta_i)^c$$

where c models *heteroscedasticity*. The precision of the kinetic parameter estimates is improved when weights w_i are chosen proportional to the inverse of the error variance.

3.4 Parameter Estimation

3.4.1 The Objective Function and the Estimation Procedure

Weighted Least Squares (WLS) uses as objective function the sum of the squared terms

$$[y_{ij} - f(x_{ij}, \theta_i)]^2 \bullet w_i$$

whereas Extended Least Squares or penalized WLS add a penalty term g

$$\left([y_{ij} - f(x_{ij}, \theta_i)]^2 \bullet w_i\right) + g(s_{ij}, \phi_i)$$

Transform-both-sides (TBS) Models have been proposed for adjustment for a symmetric error distributions and heteroscedasticity (Carrol & Ruppert, 1988).

3.4.2 Numerical Solution

Among the numerical approaches to the minimization of the objective function approximate solutions have been used. The Downhill Simplex or *Nelder&Mead* algorithm is based on the geometry of the parameter space and the *Powell* method on directed one-dimensional minimization. Gradient methods require the first derivatives as e.g., the *Gauss-Newton* or the *Marquardt-Levenberg* method. Second derivatives are used by quasi Newton type methods. In more complex pharmacokinetic systems, the kinetic function is given implicitly by a system of differential equations, written formally in terms of an operator equation where the operator H defines the kinetic function f. Predicted values of f are calculated by solving the system H numerically. Differential equation solvers as the *Runge-Kutta* and extensions by adaptive step-sizes as the *Adams-Bashforth-Moulton* methods are used. Kinetic equations can be stiff because of rapid and slow reactions occurring simultaneously. Avoidance of local minima is attempted by the use of varying initial values, parameter perturbation (genetic algorithms) or simulated annealing. The stripping method or the peeling of exponential function is useful to define a sub-space of initial values.

Artificial Neural Networks (ANN) have been applied recently to clinical pharmacology for so-called ‘intelligent drug dosing’ defining relationships between patient characteristics, laboratory data and clinical PK, see e.g. Brier & Aronoff (1996). A comparison with NONMEM of Beal & Sheiner (1992) showed greater precision and less bias. Further work is needed to rule out data or model dependency. It is conceivable that the ANN copes better with covariates because of its flexibility but the handicap of lacking model explanation remains.

3.4.3 Model Fit and Model Validation

For obtaining parsimonious nonlinear kinetic models one has used the Akaike criterion which penalizes the likelihood by the number of parameters and the Schwartz criterion which penalizes the likelihood by the number of parameters multiplied by the square root of the number of observations. Influence of single measurements and single individuals can be investigated by importance sampling and a sensitivity analysis using MCMC. Model validation by using training and test sample and cross-validation has been prohibited in PK analyses so far perhaps because of the costs of obtaining large amounts of data.

3.5 Parameter Evaluation

Standard errors of the parameters and the correlation between parameter estimates are obtained from the estimated covariance matrix derived from the second partial derivatives of the objective function, as long as no re-sampling techniques are applied which appear to be rare in most of the software packages.

3.6 Presentation and Dissemination of the Result

Plots of observed values and model predictions on the original PK scale are standard, but too often restricted to the mean concentration curves. These plots should be supplemented by both, raw and weighted residual plots. An '*Evaluation Protocol Output*' should show besides the parameter estimates and their standard errors the raw data, all steps of the model definition, all steps of fitting iterations and initial value settings, confidence limits and model diagnostics.

3.7 User Interface and Program Documentation

Interfaces and handbooks should perhaps distinguish between the user interested in the methodology and the pharmacological user. Extended and user-specific help function technique together with executed and explained examples are of inestimable value.

3.8 Acceptability and Dissemination

Besides the hardware (speed and space), the software (languages, operating systems) and the periphery (plotter, new interactive media) an honest declaration of what is minimum in training courses is needed if programs want to be successfully used. NONMEM is using a programming language (Fortran77), NLIN MIXED in SAS a command language, and nlin in Splus is object oriented. A final issue concerns questions of license and installation.

4 Tools of Computational Statistics for PK Analysis

Naturally, PK computing was developed in parallel with general statistical computing: Starting with batch programs on the mainframes and, when PCs became available with BASIC programs. Batch programs were either command or menu driven. An early program for PK analysis was NONLIN (Metzler, 1969) now translated into PCNONLIN. Gex-Fabry & Balant (1994) reviewed 42 software packages and Jackson (1996), partly overlapping, 35 systems, providing also the purchase addresses. A provisional list of 72 PK-PD software packages can be found at http://dkfz-heidelberg.de/biostatistics/pkpd.

5 Discussion

The logical stages of a PK analysis described below should be helpful to select and evaluate PK software as described in recent reviews. Open problems are questions of *experimental design* where the timely location of concentration measurements and the number of measurements have to be determined, the areas of *uncertainty and sensitivity* analyses (Edler, 1994) and the inclusion of *covariates* as predictors of individual drug response.

References

Beal, S.L. & Sheiner, L.B. (1992). *NONMEM User's Guide*. NONMEM Project Group, UCSF San Francisco.

Brier, M.E. & Aronoff, G.R. (1996). Application of artificial neural networks to clinical pharmacology. *Int. J. Clinical Pharmacology and Therapeutics,* **34**, 510-514.

Carrol, R.J. & Ruppert, D. (1988). *Transformation and Weighting in Regression.* New York: Chapman & Hall.

Edler, L. & Berger, J. (1984). FITTEN - An APL Workspace for nonlinear regression. *APL Quote Quad*, **13**, 96-104.

Edler (1994). Computational aspects in uncertainty analyses of physiologically-based pharmacokinetic models. In: *COMPSTAT94 Proceedings in Computational Statistics*, 539-544. Heidelberg: Physica-Verlag.

Edler, L. (1998). 6/12/5000 Pharmakokinetik. In: *Verfahrensbibliothek. Versuchsplanung und -auswertung* (ed. Rasch, D., Herrendörfer, G., Bock, J., Victor, N. & Guiard,V), 629 - 638. Muenchen: R. Oldenburg.

Gex-Fabry, M. & Balant, L.P. (1994). Considerations on data analysis using computer methods and currently available software for personal computers. In: *Handbook of Experimental Pharmacology* (ed. Welling P. and H. Balant). **Vol 110**, Pharmacokinetics of Drugs, 507-527. New York: Springer.

Gibaldi, M. & Perrier, D. (1982): *Pharmacokinetics.* (2nd Ed and expanded). Marcel Dekker: New York.

Jackson, R. C. (1996). *Computer Techniques in Preclinical and Clinical Drug Development*. CRC Press: Boca Raton.

Meibohm, D. & Derendorf, H. (1997). Basic concepts of pharmacokinetic / pharmacodynamik (PK/PD) modelling. *Int. J. of Clinical Pharmacology and Therapeutics*, **35**, 401-413.

Metzler, C.M. (1969). *A user's manual for NONLIN.* The Upjohn Co. Techn. Rep. 7292/69/7292/005. Kalama 700. Mich.

Wakefield, J. & Racine-Poon, A. (1994): An application of Bayesian population pharmacokinetic/pharmacodynamic models to dose recommendation. *Statistics in Medicine*, **14**, 971-986.

Yuh, L., Beal, S., Davidian, M., Harrison, F., Hester, A., Kowalski, K., Vonesh, E. & Wolfinger, R. (1994): Population pharmacokinetic/pharmacodynamic methodology and applications: A bibliography. *Biometrics*, **50**, 566-575.

Frailty Factors and Time-dependent Hazards in Modelling Ear Infections in Children Using BASSIST

Mervi Eerola[1], Heikki Mannila[2] and Marko Salmenkivi[3]

[1] Rolf Nevanlinna Institute, University of Helsinki, P.O. Box 4, FIN-00014 University of Helsinki, Finland, Mervi.Eerola@rni.helsinki.fi
[2] Department of Computer Science, University of Helsinki, P.O. Box 26, FIN-00014 University of Helsinki, Finland, Heikki.Mannila@cs.helsinki.fi
[3] Department of Computer Science, University of Helsinki, P.O. Box 26, FIN-00014 University of Helsinki, Finland, Marko.Salmenkivi@cs.helsinki.fi

1 Introduction

The BASSIST system is a general purpose tool for MCMC sampling for intensity models. The system allows the user to specify an intensity model in a high-level language. The model is used to generate a simulation program that uses the Metropolis-Hastings algorithm to obtain the desired samples. In contrast to BUGS (Spiegelhalter *et al.*, 1996), BASSIST contains several primitives that are suited for modelling event data, including piecewise constant functions etc.

In this paper we describe the use of the BASSIST system when modelling the occurrences of middle ear infection (acute otitis media, AOM). Previous modelling on AOM by nonparametric Bayesian methods has been done in Andreev & Arjas (1998).

Typically, the infections occur several times to some particularly infection prone children. This recurrence causes dependence among the observations which can partly be modelled by observed information on the individuals but the risk is also related to the immunological development of a child. This process can of course not be fully observed. Moreover, there are anatomic and unobserved genetic differences which are presumably important risk determinants. Common to all these factors is that they are time and age dependent, perhaps in a very complex way. This suggests that there are several sources of unobserved heterogeneity which should be accounted for in the modelling.

In this paper we will show some basic models to illustrate the use of BASSIST when modelling the common baseline intensity nonparametrically and then adding more structure by observed time-dependent risk factors and individual specific unobserved random variables, commonly called frailties in survival analysis. The analysis is fully Bayesian and uses the flexible ideas of variable dimension MCMC methods suggested by Green (1995).

2 The data

In the pilot phase of a large vaccination trial on ear infections caused by Streptococcus pneumoniae, 329 children in the Tampere area in Finland were followed from the age of 2 to 24 months. The aim was to study the natural course of the disease in small children without vaccination.

In addition to the random occurrences of acute otitis media episodes (AOM's), changes in several risk factors were recorded repeatedly in health visits at the ages of 2, 3, 4, 5, 6, 9, 12, 15, 18, and 24 months. For illustration purposes, we will only consider here the effects of two well known risk factors: weaning and day care attendance.

3 Steps in the modelling

3.1 Homogeneous model

We started by modelling the age-dependent intensity $\lambda(t)$ by a piecewise constant function $f(t) = \sum_{j=1}^{N} 1_{\{j<t\leq j+1\}}\lambda_j$. The levels of the function, the jump times and the number of pieces were random variables with the following priors (hyper constants could easily be replaced by an additional hierarchy level):

- number of pieces $N \sim$ Poisson(10)
- levels of functions $\lambda_j \sim$ Uniform(0.0000001, 0.15)
- jump times $l_j \sim$ Uniform(1.5,24.9)

3.2 Observed history

We proceeded by adding observed history in terms of time-dependent indicator functions of weaning and day care multiplicatively into the model. In order to distinguish between risk factors for first and recurrent infections, we used the parameterisation in Eerola (1989) to model the infection intensity for individual i as

$$\lambda_i(t) = f(t)Y_i(t)\beta_i(t)\gamma_i(t)\alpha_i(t)\beta_i^*(t)\gamma_i^*(t)$$

where $Y_i(t)$ is an indicator function of being at risk at time t.

The time-dependent functions of weaning at T_W and daycare entrance at T_D, $\beta_i(t) = 1_{\{T_{W,i}\geq t\}}\beta$ and $\gamma_i(t) = 1_{\{T_{D,i}\leq t\}}\gamma$, indicate the relative risk of these factors compared to the baseline $f(t)$ in that the values $\beta, \gamma < 1$ correspond to protective effect and the values > 1 increased risk of AOM. More commonly, $\beta = exp(\tilde{\beta})$, as in the Cox model, but since we have shown the posterior results for β we write the intensity model in terms of it.

The indicator function $\alpha_i(t) = 1_{\{T_{1,i}+30\leq t\}}\alpha$ is used to distinguish between those who have had at least one infection already. The parameters for recurrent episodes are estimated among these individuals.

The functions $\beta_i^*(t) = 1_{\{T_{1,i}+30\leq t\}}1_{\{T_{W,i}\geq t\}}\beta^*$ (for $\gamma_i^*(t)$ respectively) are interactions between the covariates and the onset at $T_{1,i}$ where $T_{1,i}$ is the time of first infection for i. They correspond to the additional effect of the covariates on recurrent infections compared to the effect on the first infection. The total relative risk for recurrences is therefore $\beta \cdot \beta^*$. The lag time of 30 days is a convention to define distinct ear infection episodes.

In this model the baseline intensity for the first and subsequent episodes is the same, which seems to be a plausible assumption for recurrent events of the same nature. The prior distributions used for $\alpha, \beta, \beta^*, \gamma$ and γ^* were Gamma(2,2) in which case the prior mean of the relative risk is 1.

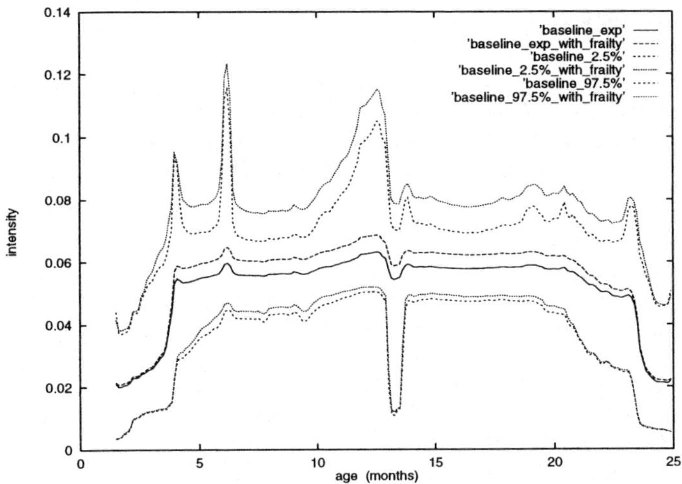

Fig. 1. Posterior baseline intensity curves $f(t)$ for the two models with no observed history: crude hazard only and unobserved frailties added (mean and 95 %-credibility-intervals)

3.3 Unobserved frailties

Finally, we added an individual constant frailty factor to account for the additional dependence due to possible genetic and anatomic variations among the children. The effect of the frailty is assumed to be multiplicative. For the case of no observed history, we get the model

$$\lambda_i(t) = f(t)\zeta_i,$$

where ζ_i is an unobserved frailty factor. When taking the observed history into account we get a model of the form

$$\lambda_i(t) = f(t)\zeta_i Y_i(t)\beta_i(t)\gamma_i(t)\alpha_i(t)\beta_i^*(t)\gamma_i^*(t).$$

The frailties are assumed to be a random sample from a common distribution Gamma(ν, η). We also make the usual assumption that the prior mean $E(\zeta|\nu, \eta) = 1$ which in the Gamma distribution means that $\nu = \eta$. It is of interest to let the hyperparameter ν also be random which allows us to consider the variability in the data in the light of the accumulating observations. The prior of ν was Uniform(0.00001,100).

The joint posterior distribution of the paramcters $\theta = (\lambda, k, l, \alpha, \beta, \gamma, \zeta, \nu)$ is then (by assuming independence of the parameters)

$$p(\lambda, k, l, \alpha, \beta, \gamma, \zeta, \nu|data) = p(\lambda)p(k)p(l)p(\alpha)p(\beta)p(\gamma)p(\zeta|\nu)p(\nu)L(data|\theta)$$

where the Poisson likelihood for the data is

$$L(data|\theta) = \prod_{i=1}^{329}\prod_{k=1}^{n_i} \lambda_i(T_{i,k})exp(-\int_0^\infty \lambda_i(s)ds).$$

4 The BASSIST system

Seminal work in software tools for Bayesian analysis was done by the BUGS group (Spiegelhalter *et al.*, 1996), who have created a very interesting tool for describing hierarchical models and performing MCMC simulations for obtaining estimates of posterior distributions. However, BUGS as such is not suitable for time-dependent data.

Thus we designed a new system, BASSIST, which contains several primitives suited for modelling event data, including piecewise constant functions etc. The system allows the user to specify an intensity model in a high-level language. For reasons of brevity, we omit most of the details in this paper; see Arjas *et al.* (1996) for a description of the first version of the program and Toivonen *et al.* (1998) for the new version.

Briefly, BASSIST is given a high-level description of a full probability model. The model is used to generate a simulation program that uses the Metropolis-Hastings algorithm to obtain the desired samples.

Below is a simple example of how the BASSIST description language is used to describe a homogenous model with unobserved frailties. We omit the details of the description language.

```
child(  var bdate;
        var sex;
        var frailty ~ dgamma(3,3);
 )


event_sequence infections(
    child;
    var time ~ Poisson process (child.frailty * f(time));
)

var piecewise constant function f(
        var start; var end;
        var pieces ~ dpoisson(10);
        piece(
           var start ~ dunif(1.5, 24.9);
           var level ~ dunif(0.0000001, 0.15); );
);
```

The BASSIST system was used to perform MCMC simulations of the models described above. Convergence was monitored by using the CODA package (Best *et al.*, 1995). Typically, the simulations took about 50 000 iterations (we picked up parameter values from every tenth sweep) before they passed the tests. For the case of the models with frailties and a hyperparameter for them, we needed about 500 000 iterations. On a 166 MHz Pentium, such a run took 15 hours.

5 Empirical results

The piecewise constant posterior baseline curves show that the risk of AOM's increases rapidly around the age of 5-6 months and another peak arises around the age of 12 months (Figure 1). We expect to be able to 'explain' this peak by adding information on day care because the maternal leave in Finland is 12 months long.

In Figure 1 baseline intensity is higher for the model with frailty factors. The reason is that the children with relatively many infections - and thus higher frailty factors - are absent from the risk set more frequently than those with few infections. Consequently, the average of frailty factors in the risk set is smaller than 1 during almost the whole time period and causes the baseline intensity to rise in proportion.

The recurrent infections tend to occur at the end of the follow-up which suggests that different factors might be responsible for their occurrence. Figure 2 indicates that day care loses significance for recurrent infections because the interaction parameter γ^* is clearly below 1. Since weaning was modelled as having the value 1 as long as breast feeding continues, the posterior mode of β^* around 2 indicates that the risk for recurrent AOM's increases considerably when weaning has occurred (Figure 2).

It is obvious that part of the proneness to ear infections can be captured by modelling the previous infections; when including the indicator of the first infection $1_{\{T_{1,i} \leq t\}}$ in the model the distribution of the frailty hyperparameter η is very skewed with a long tail (Figure 3). For the sake of comparison, Figure 4 shows the clearer differences between the frailty factors in the model with no observed history.

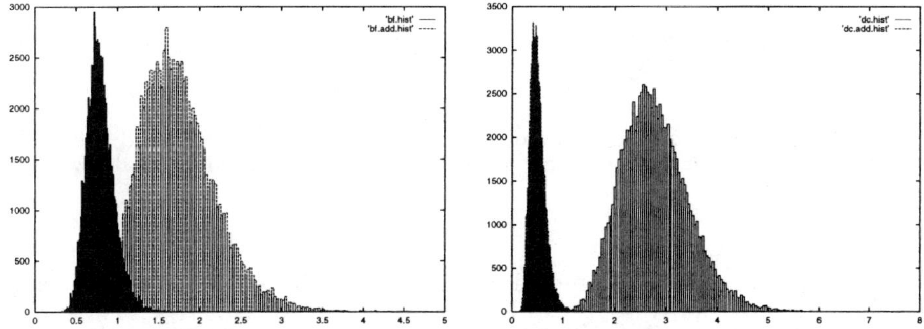

Fig. 2. The posterior distributions of the parameters β (dark), and β^* on the left side, γ and γ^* (dark) on the right side

Acknowledgements
We thank the Pnc study group from the National Public Health Institute (KTL), Finland, for letting us use the acute otitis media data.

References

Andreev, A. & Arjas, E. (1998). Acute Middle Ear Infection in Small Children: A Bayesian Analysis. In *Longitudinal Data Analysis (LIDA)*, in press. Boston: Kluwer Academic Publisher.

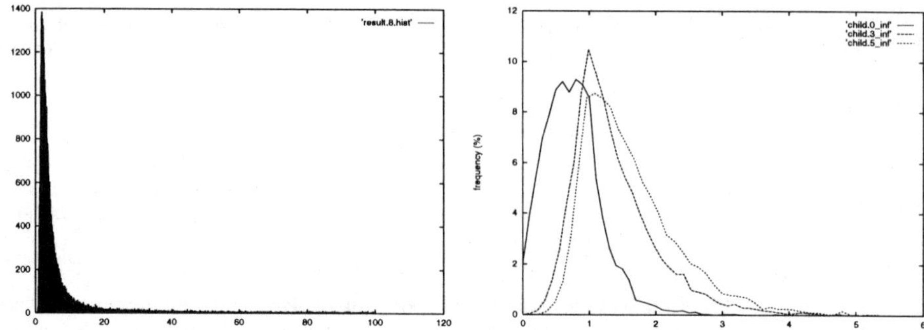

Fig. 3. The posterior distributions of the frailty hyperparameter η and the frailty factors ζ_i for children with 0,3 and 5 infections, respectively. The model with observed history

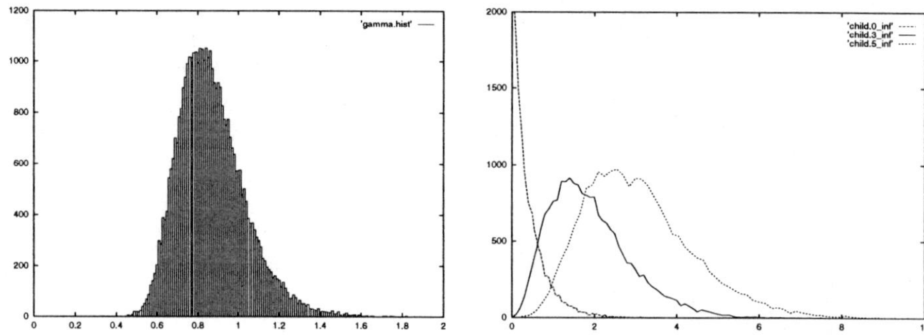

Fig. 4. The posterior distributions of the frailty hyperparameter η and the frailty factors ζ_i for children with 0,3 and 5 infections, respectively. The model with no observed history

Arjas, E., Mannila, H., Salmenkivi, M., Suramo, R. & Toivonen, H.(1996). BASS : Bayesian Analyzer of Event Sequences. In: *COMPSTAT'96 Proceedings in Computational Statistics* (ed. A. Prat), 199-204. Heidelberg: Physica-Verlag.

Best, N., Cowles, M. & Vines, K. (1995). *CODA. Convergence Diagnosis and Output Analysis Software for Gibbs sampling output. Version 0.30.* MRC Biostatistics Unit, Institute of Public Health, Cambridge.

Eerola, M. (1989). Repeatable events in event-history analysis; the effect of childhood separation on future mental health. *Bull. Int. Statist. Inst.*, **LIII**(1), 213-225.

Green, P (1995). Reversible jump Markov chain Monte Carlo computation and Bayesian model determination. *Biometrica,* **82**, 4, 711-732.

Spiegelhalter, D., Thomas, A., Best, N. & Gilks, W. (1996). *BUGS, Bayesian inference Using Gibbs Sampling Manual, Version 0.50.* MRC Biostatistics Unit, Institute of Public Health, Cambridge.

Toivonen, H., Mannila, H., Salmenkivi, M. & Laakso, K. (1998). BASSIST - a tool for MCMC simulation of statistical models. In: *EUROSIM'98 Proceedings of 3rd International Congress of the Federation of EUROpean SIMulation Societies*, in press.

Algorithms for Robustified Error-in-Variables Problems

Håkan Ekblom and Ove Edlund

Department of Mathematics, Luleå University of Technology, Sweden

Keywords. Error-in-variables, robust estimation, M-estimate, non-linear model, orthogonal regression

1 Introduction

We consider the problem of fitting a model of the form $y = f(\boldsymbol{x}, \boldsymbol{\beta})$ to a set of points $(\boldsymbol{x}_i, y_i)$, $i = 1, \ldots, n$. If there are measurement or observation errors in $\boldsymbol{x}$ as well as in $\boldsymbol{y}$, we have the so called errors-in-variables-problem with model equation

$$y_i = f(\boldsymbol{x}_i + \boldsymbol{\delta}_i, \boldsymbol{\beta}) + \varepsilon_i, \quad (i = 1, \ldots, n) \tag{1}$$

where $\boldsymbol{\delta_i} \in \mathbb{R}^m$, $i = 1, \ldots, n$ are the errors in $\boldsymbol{x}_i \in \mathbb{R}^m$. Then the problem is to find a vector of parameters $\boldsymbol{\beta} \in \mathbb{R}^p$ that minimizes the errors ε_i and $\boldsymbol{\delta}_i$ in some loss function subject to (1). We will present algorithms using more robust alternatives to the least squares criterion. Figure 1 gives examples where the least squares (L2), the least absolute deviation (L1) and the Huber criteria are used.

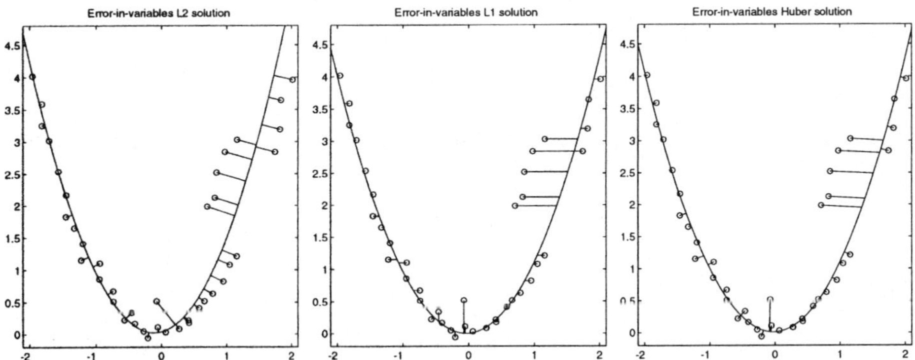

Fig. 1. A parabola fitted to data points using an error-in-variables model and 3 different criteria (L2, L1, Huber); notice the least squares (L2) fit gives orthogonal regression, while the L1 residuals are parallel to an axis

We will further discuss, from an algorithmic point of view, cases when the model is linear. Also another way to make the robustification will be pointed out.

2 Notation

Let $\boldsymbol{\delta}$ denote a vector containing all the $\boldsymbol{\delta}_i$, $i = 1,\ldots,n$, i.e. $\boldsymbol{\delta}^T = (\boldsymbol{\delta}_1^T,\ldots,\boldsymbol{\delta}_n^T) \in \mathbb{R}^{nm}$. Further assemble the unknowns $\boldsymbol{\beta}$ and $\boldsymbol{\delta}$ into one variable $\boldsymbol{\eta} \in \mathbb{R}^{p+nm}$ by setting $\boldsymbol{\eta}^T = (\boldsymbol{\beta}^T, \boldsymbol{\delta}^T)$. Now we let $\boldsymbol{g}(\boldsymbol{\eta}) : \mathbb{R}^{p+nm} \mapsto \mathbb{R}^{n+nm}$ be a function where the first n components are $y_i - f(\boldsymbol{x}_i+\boldsymbol{\delta}_i, \boldsymbol{\beta})$ $(= \varepsilon_i)$ and the last nm components are those of $\boldsymbol{\delta}$. Using the least squares criterion, the problem can be formulated as

$$\min_{\boldsymbol{\eta}} \boldsymbol{g}(\boldsymbol{\eta})^T \boldsymbol{g}(\boldsymbol{\eta})$$

i.e. we minimize

$$\sum_{i=1}^{n} \left[(y_i - f(\boldsymbol{x}_i + \boldsymbol{\delta}_i, \boldsymbol{\beta}))^2 + \boldsymbol{\delta}_i^T \boldsymbol{\delta}_i \right] \tag{2}$$

with respect to $\boldsymbol{\beta}$ and $\boldsymbol{\delta}$.

For a linear model we have

$$f(\boldsymbol{x}, \boldsymbol{\beta}) = x_1\beta_1 + x_2\beta_2 + \ldots + x_p\beta_p, \tag{3}$$

so the model can be formulated as

$$\boldsymbol{y} = [\boldsymbol{X} + \boldsymbol{D}]\boldsymbol{\beta} + \boldsymbol{\varepsilon}, \tag{4}$$

where $\boldsymbol{D} = [\boldsymbol{\delta}_1,\ldots,\boldsymbol{\delta}_n]$. Applying the least-squares criterion is equivalent to minimizing $\|[\boldsymbol{D}|\varepsilon]\|_F$, where $\|\cdot\|_F$ is the Frobenius norm. Algorithms for (4), based on singular value decomposition, are found in Golub & Van Loan (1989) and Van Huffel & Vandewalle (1991), while Boggs, Byrd & Schnabel (1987) and Schwetlick & Tiller (1985) give algorithms for the non-linear model (2). It should be noted that scaling matrices can also be introduced, but they will not essentially change the algorithms discussed in this paper.

3 An algorithm for the non-linear model

Starting from (2) consider a generalization to other criteria than least-squares, namely the following problem

$$\min_{\boldsymbol{\eta}} F(\boldsymbol{\eta}) \tag{5}$$

where

$$F(\boldsymbol{\eta}) = \sum_{i=1}^{n+nm} \varrho\left(\frac{g_i(\boldsymbol{\eta})}{\sigma}\right).$$

Hermey (1996) has used the ideas for Huber-estimation in Ekblom & Madsen (1989) to solve (5) using trust-region technique. Here we point to the possibility to use a similar approach to solve (5) for arbitrary ϱ-functions.

We solve a sequence of linearized model problems, like in Edlund, Ekblom & Madsen (1997). Thus $g_i(\boldsymbol{\eta} + \boldsymbol{h})$ is simplified to

$$l_i(\boldsymbol{h};\boldsymbol{\eta}) = g_i(\boldsymbol{\eta}) + \nabla g_i(\boldsymbol{\eta})^T \boldsymbol{h}, \quad i = 1, 2, \ldots, n + nm.$$

Letting

$$L(\boldsymbol{h};\boldsymbol{\eta}) = \sum_{i=1}^{n+nm} \varrho\left(\frac{l_i(\boldsymbol{h})}{\sigma}\right),$$

the linear subproblem is

$$\begin{array}{c} \min\limits_{\boldsymbol{\eta}} L(\boldsymbol{h};\boldsymbol{\eta}) \\ \text{subject to } \boldsymbol{h}^T\boldsymbol{h} \leq R^2. \end{array} \tag{6}$$

The trust region radius R is updated according to the usual updating procedure. It is based on the ratio between the decrease in the non-linear function and the decrease in the local model

$$r_k = \max\left\{0, \frac{F(\boldsymbol{\eta}_k) - F(\boldsymbol{\eta}_k + \boldsymbol{h}_k)}{L(\boldsymbol{0};\boldsymbol{\eta}_k) - L(\boldsymbol{h}_k;\boldsymbol{\eta}_k)}\right\}.$$

The trust region algorithm is (ϵ_1 and ϵ_2 are tolerance parameters)

Let $0 < s_1 \ll 0.25$ and $0.25 \leq s_2 < 1 < s_3$.
given $\boldsymbol{\eta}_0$ and R_0; $k := 0$;
while ($\|F'(\boldsymbol{\eta}_k)\| > \epsilon_1$ and $\|\boldsymbol{h}_k\| > \epsilon_2 \|\boldsymbol{\eta}_k\|$) **do begin**
  find $\boldsymbol{h}_k$ by solving the linear subproblem (6) with $\boldsymbol{\eta} = \boldsymbol{\eta}_k$;
  if $r_k > s_1$ **then** $\boldsymbol{\eta}_{k+1} := \boldsymbol{\eta}_k + \boldsymbol{h}_k$
    else $\boldsymbol{\eta}_{k+1} := \boldsymbol{\eta}_k$;
  if $r_k < 0.25$ **then** $R_{k+1} := R_k \cdot s_2$
  else if $r_k > 0.75$ **then** $R_{k+1} := R_k \cdot s_3$
    else $R_{k+1} := R_k$;
  $k := k + 1$
end

Here ϵ_1 and ϵ_2 are suitably chosen tolerance parameters.

When solving for the linearized model, the algorithm of Edlund (1997) can be used.

The matrices involved will be very sparse, a fact which can be utilized so that the large $(p+nm) \times (p+nm)$ system of equation to be solved reduces to a small $p \times p$ system (Hermey, 1996). This technique is very similar to what is presented in the next section.

4 Total linear M-estimators

Analogous to the total least squares problem (Huffel & Vandewalle, 1991), we can formulate the problem of finding total linear M-estimators: Given $\boldsymbol{X} \in \mathbb{R}^{n \times p}$, and $\boldsymbol{y} \in \mathbb{R}^n$, where $n > p$, we solve the problem

$$\begin{array}{ll} \text{minimize} & \sum\limits_{i=1}^{n} \varrho_1(\varepsilon_i/\sigma_1) + \sum\limits_{i=1}^{n}\sum\limits_{j=1}^{p} \varrho_2(\delta_{ij}/\sigma_2), \\ \text{subject to} & \boldsymbol{y} + \boldsymbol{\varepsilon} = (\boldsymbol{X} + \boldsymbol{D})\boldsymbol{\beta} \end{array} \tag{7}$$

for $\boldsymbol{\beta} \in \mathbb{R}^p$ and $[\delta_{ij}] = \boldsymbol{D} \in \mathbb{R}^{n \times p}$. In the above expression $\boldsymbol{\varepsilon} \in \mathbb{R}^n$, and ϱ_1 and ϱ_2 are M-estimator functions for errors in observations and errors in variables respectively, and σ_1 and σ_2 are scale factors.

4.1 Finding a solution

We will use Newton's method to find a local minimum to the problem (7). From (7) we get the objective function

$$G(\boldsymbol{\beta}, \boldsymbol{D}) = \sum_{i=1}^{n} \varrho_1(\varepsilon_i/\sigma_1) + \sum_{i=1}^{n}\sum_{j=1}^{p} \varrho_2(\delta_{ij}/\sigma_2),$$

where $\boldsymbol{\varepsilon} = \boldsymbol{\varepsilon}(\boldsymbol{\beta}, \boldsymbol{D}) = (\boldsymbol{X} + \boldsymbol{D})\boldsymbol{\beta} - \boldsymbol{y}$. To calculate a Newton step, first order and second order derivatives are needed. The following definitions will prove helpful in formulating those: The matrix form of the unknown $\boldsymbol{D}$ is troublesome, so let us introduce the vector $\boldsymbol{\delta}^* \in \mathbb{R}^{pn}$ containing the rows of $\boldsymbol{D}$ stacked on top of each other. Let us also define the matrix $\boldsymbol{B} \in \mathbb{R}^{n \times pn}$ by putting n copies of the vector $\boldsymbol{\beta}^T$ in a "staircase diagonal form" reaching from the upper left corner to the lower right corner of the matrix, i.e.

$$\boldsymbol{B} = \boldsymbol{I}_n \otimes \boldsymbol{\beta}^T = \begin{bmatrix} \boldsymbol{\beta}^T & & & \\ & \boldsymbol{\beta}^T & & \\ & & \ddots & \\ & & & \boldsymbol{\beta}^T \end{bmatrix}.$$

Now the first order derivatives are

$$\frac{\partial}{\partial \boldsymbol{\beta}} G = (\boldsymbol{X} + \boldsymbol{D})^T \boldsymbol{v}, \quad \text{where } v_i = \frac{1}{\sigma_1} \varrho_1'(\varepsilon_i/\sigma_1),$$

$$\frac{\partial}{\partial \boldsymbol{\delta}^*} G = \boldsymbol{B}^T \boldsymbol{v} + \boldsymbol{w}, \quad \text{where } w_i = \frac{1}{\sigma_2} \varrho_2'(\delta_i^*/\sigma_2).$$

Newton's method also requires second order derivatives. To write them in a convenient form we need an additional matrix $\boldsymbol{V} \in \mathbb{R}^{p \times pn}$. Now the second order derivatives are

$$\frac{\partial^2}{\partial \boldsymbol{\beta}^2} G = (\boldsymbol{X} + \boldsymbol{D})^T \boldsymbol{Q} (\boldsymbol{X} + \boldsymbol{D}), \quad \text{where } \boldsymbol{Q} = \operatorname*{diag}_{i=1\ldots n} \left((1/\sigma_1^2) \varrho_1''(\varepsilon_i/\sigma_1)\right),$$

$$\frac{\partial^2}{\partial \boldsymbol{\beta} \partial \boldsymbol{\delta}^*} G = (\boldsymbol{X} + \boldsymbol{D})^T \boldsymbol{Q} \boldsymbol{B} + \boldsymbol{V}, \quad \text{where } \boldsymbol{V} = \left[v_1 \boldsymbol{I}_p, v_2 \boldsymbol{I}_p, \ldots, v_n \boldsymbol{I}_p\right],$$

$$\frac{\partial^2}{\partial \boldsymbol{\delta}^{*2}} G = \boldsymbol{B}^T \boldsymbol{Q} \boldsymbol{B} + \boldsymbol{S}, \quad \text{where } \boldsymbol{S} = \operatorname*{diag}_{i=1\ldots pn} \left((1/\sigma_2^2) \varrho_2''(\delta_i^*/\sigma_2)\right).$$

Thus the Newton step $\left[\Delta\boldsymbol{\beta}^T \ \Delta\boldsymbol{\delta}^{*T}\right]^T$ is found by solving the system

$$\begin{bmatrix} (\boldsymbol{X} + \boldsymbol{D})^T \boldsymbol{Q} (\boldsymbol{X} + \boldsymbol{D}) & (\boldsymbol{X} + \boldsymbol{D})^T \boldsymbol{Q} \boldsymbol{B} + \boldsymbol{V} \\ \boldsymbol{B}^T \boldsymbol{Q} (\boldsymbol{X} + \boldsymbol{D}) + \boldsymbol{V}^T & \boldsymbol{B}^T \boldsymbol{Q} \boldsymbol{B} + \boldsymbol{S} \end{bmatrix} \begin{bmatrix} \Delta\boldsymbol{\beta} \\ \Delta\boldsymbol{\delta}^* \end{bmatrix} = - \begin{bmatrix} (\boldsymbol{X} + \boldsymbol{D})^T \boldsymbol{v} \\ \boldsymbol{B}^T \boldsymbol{v} + \boldsymbol{w} \end{bmatrix}. \tag{8}$$

This is an $(pn + p) \times (pn + p)$ system. Certainly we need to reduce the size of it, if possible. Let $\boldsymbol{P} = \boldsymbol{B}^T \boldsymbol{Q} \boldsymbol{B} + \boldsymbol{S}$, then the second row gives

$$\Delta\boldsymbol{\delta}^* = \boldsymbol{P}^{-1}\left[-\boldsymbol{B}^T \boldsymbol{v} - \boldsymbol{w} - (\boldsymbol{B}^T \boldsymbol{Q}(\boldsymbol{X} + \boldsymbol{D}) + \boldsymbol{V}^T)\Delta\boldsymbol{\beta}\right].$$

Putting this into the first row of (8) we get

$$\left[(\boldsymbol{X}+\boldsymbol{D})^T\boldsymbol{Q}(\boldsymbol{X}+\boldsymbol{D})-\left((\boldsymbol{X}+\boldsymbol{D})^T\boldsymbol{Q}\boldsymbol{B}+\boldsymbol{V}\right)\boldsymbol{P}^{-1}(\boldsymbol{B}^T\boldsymbol{Q}(\boldsymbol{X}+\boldsymbol{D})+\boldsymbol{V}^T)\right]\Delta\boldsymbol{\beta}$$
$$= -(\boldsymbol{X}+\boldsymbol{D})^T\boldsymbol{v} + \left((\boldsymbol{X}+\boldsymbol{D})^T\boldsymbol{Q}\boldsymbol{B}+\boldsymbol{V}\right)\boldsymbol{P}^{-1}(\boldsymbol{B}^T\boldsymbol{v}+\boldsymbol{w}). \tag{9}$$

This is a $p \times p$ system, what is left to figure out is how to take care of $\boldsymbol{P}^{-1}$.

4.2 Solving the block diagonal system

We are going to make use of the special structure of $\boldsymbol{P}$ to efficiently solve the system

$$\boldsymbol{P}\boldsymbol{u} = \boldsymbol{z}. \tag{10}$$

As a first step, let us divide the diagonal matrix $\boldsymbol{S}$ into blocks such that $\boldsymbol{S} = \text{diag}(\boldsymbol{S}_1, \ldots, \boldsymbol{S}_n)$, where $\boldsymbol{S}_k \in \mathbb{R}^{p\times p}$, $k = 1\ldots n$. Taking a closer look at the matrix $\boldsymbol{P}$ we see that it is block diagonal, i.e. $\boldsymbol{P} = \text{diag}(\boldsymbol{P}_1, \ldots, \boldsymbol{P}_n)$, where $\boldsymbol{P}_k \in \mathbb{R}^{p\times p}$, $k = 1\ldots n$. By the definition of $\boldsymbol{P}$, the diagonal blocks are $\boldsymbol{P}_k = q_{kk}\boldsymbol{\beta}\boldsymbol{\beta}^T + \boldsymbol{S}_k$. The system (10) can be solved by considering one diagonal block at a time, thus for $k = 1\ldots n$ we solve the system

$$(\boldsymbol{S}_k + q_{kk}\boldsymbol{\beta}\boldsymbol{\beta}^T)\boldsymbol{u}_k = \boldsymbol{z}_k. \tag{11}$$

Let $\boldsymbol{u}_k^*$ solve $\boldsymbol{S}_k\boldsymbol{u}_k^* = \boldsymbol{z}_k$, and $\boldsymbol{u}_k^{**}$ solve $\boldsymbol{S}_k\boldsymbol{u}_k^{**} = \boldsymbol{\beta}$. These calculations are particularly simple since $\boldsymbol{S}_k$ is a diagonal matrix. Then it is easy to see that the solution to (11) is

$$\boldsymbol{u}_k = \boldsymbol{u}_k^* - \frac{q_{kk}\boldsymbol{\beta}^T\boldsymbol{u}_k^*}{1 + q_{kk}\boldsymbol{\beta}^T\boldsymbol{u}_k^{**}}\boldsymbol{u}_k^{**}.$$

This result is utilized to solve (9). The complexity of the resulting algorithm turns out be $O(p^2n) + O(p^3)$. This is much better than solving (8) with Gaussian elimination, $O(p^3n^3)$. It is also better than solving (8) with general sparse matrix techniques, $O(p^3n)$.

4.3 The Newton iteration

Now we know that it is feasible to perform Newton iterations. The standard procedure for this is

given $\boldsymbol{\beta}_0$ and $\boldsymbol{\delta}_0^*$; $k := 0$;
while not $STOP$ **do begin**
  find $\Delta\boldsymbol{\beta}_k$ and $\Delta\boldsymbol{\delta}_k^*$ by solving (8);
  perform a linesearch to determine the steplength parameter α;
  $\boldsymbol{\beta}_{k+1} := \boldsymbol{\beta}_k + \alpha\Delta\boldsymbol{\beta}_k$; $\boldsymbol{\delta}_{k+1}^* := \boldsymbol{\delta}_k^* + \alpha\Delta\boldsymbol{\delta}_k^*$;
  $k := k + 1$
end

where $STOP$ can be a condition based on the norm of the first order derivatives and/or the steplength. In Newton's method, a linesearch is required to guarantee convergence, but this is no disadvantage since the computational cost for doing linesearch is modest, especially close to the solution.

5 A different robustification approach

It should be pointed out, that M-estimation for EIV problems can be applied in a different way. Jefferys (1990) considered fitting a straight line $f(x, \boldsymbol{\beta}) = \beta_1 + \beta_2 x$ in such a way that

$$\sum_{i=1}^{n} \varrho \left(\frac{\sqrt{(y_i - f(x_i + \delta_i, \boldsymbol{\beta}))^2 + \delta_i^2}}{\sigma} \right)$$

is minimized with respect to $\boldsymbol{\beta}$ and $\boldsymbol{\delta}$. This is a special case of "orthogonal regression M-estimates" proposed by Zamar (1989) for a general linear model. The algorithms used are of type steepest descent (Zamar) and Newton or IRLS (Jefferys). However, it seems reasonable to investigate how other computational approaches, similar to those presented in this paper, can be applied. This is a subject for future research.

References

Boggs, P.T., Byrd, R.H. & Schnabel, R.B. (1987). A stable and efficient algorithm for nonlinear orthogonal distance regression. *SIAM J. Sci. Statist. Comput.*, **8**, 1052–1078.

Edlund, O. (1997). Linear M-estimation with bounded variables. *BIT*, **37**(1), 13–23.

Edlund, O., Ekblom, H. & Madsen, K. (1997). Algorithms for non-linear M-estimation. *Computational Statistics*, **12**, 373–383.

Ekblom, H. & Madsen, K. (1989). Algorithms for non-linear Huber estimation. *BIT*, **29**, 60–76.

Golub, G.H. & Van Loan C.F. (1989). *Matrix Computations.* The Johns Hopkins University Press, second edition.

Hermey, D. (1996). *Numerical Methods for Some Problems in Robust Nonlinear Data Fitting.* PhD thesis, Department of Mathematics and Computer Science, University of Dundee, Scotland.

Van Huffel, S. & Vandewalle, J. (1991). *The Total Least Squares Problem: Computational Aspects and Analysis.* SIAM Publications.

Jefferys, W.H. (1991). Robust estimation when more than one variable per equation has errors. *Biometrika*, **77**(3), 597-607.

Schwetlick, H. & Tiller, V. (1985). Numerical Methods for Estimating Parameters in Non-linear Models With Errors in the Variables. *Technometrics*, **27**(1), 17–24.

Zamar, R.H. (1989). Robust estimation in the errors-in-variables model. *Biometrica*, **76**(1), 149–160.

Idaresa - a Tool for Construction, Description and Use of Harmonised Datasets from National Surveys

Joan Fairgrieve and Karen Brannen

Centre for Educational Sociology, University of Edinburgh, 7 Buccleuch Place, Edinburgh, EH8 9LW, Scotland

Keywords. Metadata, harmonisation, distributed databases, comparative analysis

1 Background

The Centre for Educational Sociology (CES) is a research centre at the University of Edinburgh. Research projects carried out at the centre are often based around secondary analysis of large and complex datasets and can require construction of harmonised datasets for comparative analysis. VTLMT is a two-year project which is funded under the Leonardo da Vinci Programme of the European Union.[1] It is co-ordinated by the Economic and Social Research Institute (ESRI) in Dublin with partners in the Netherlands, France and Scotland. The analysis is based around a subset of school leavers and requires the construction of a dataset combining data from school leaver surveys carried out in the four countries. 'Home Internationals' is a two-year project which is funded by the Economic and Social Research Centre of the UK.[2] It is based at CES and involves analysis of a dataset which has combined data from the different parts of the UK.

This paper will use examples from these substantive projects to illustrate some of the problems which can occur when trying to build a harmonised dataset for analysis. It will go on to describe the Idaresa project and how the concepts being developed in the project could be used to aid this process.

2 The process of harmonisation

In the projects mentioned above, the method for harmonisation seems to proceed in the same way: first, to look at what datasets are available, then to decide on the questions to be asked in the analysis and lastly, to define the variables which will make up the combined dataset. Using this method often means that the definition of the variables is arrived at after the analysis questions have been defined. This can result in difficulties either later on in the project or when the data come to be reused. These difficulties will be discussed later in the paper.

---

[1] "Vocational Training and Labour Market Transitions: A Comparative Perspective", EU, Leonardo, Dec 1996 - Nov 1998.

[2] "A 'Home International' Comparison of 14-19 Education and Training Systems in the UK", ESRC, Jan 1997 - Dec 1998.

Several different types of problem have become apparent when putting harmonisation into practice and these problems are repeated over the different projects. The following examples show some of the problems which actually occurred when constructing combined datasets.

2.1 Non-direct mappings

This example is taken from the 'Home Internationals' project whilst trying to construct a variable describing current activity.

The definition of the desired 'target' variable was agreed after taking a cursory look at the way the 'source' variables were asked in the questionnaire. The source variable from one of the datasets was easily mapped onto this 'target' variable with a simple recode. However, the second source dataset was a bit more problematic. One of the categories in the second source variable combined two categories from the target variable. Therefore, two further source variables were required in order to create the target variable. A diagrammatic representation of the harmonisation of this one variable is shown in Figure 1.

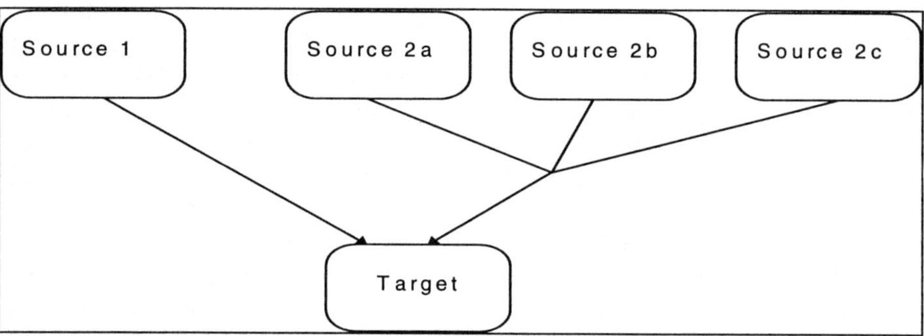

Fig. 1. A non-direct mapping

2.2 Collapsing values

This second example was also taken from the Home Internationals project. Using the two different source datasets available, there were two variables which appeared almost identical in their composition. However, the wording of some of the values was slightly different in each case. The example is of the variable asking about frequency of having played truant. Table 1 shows the differences in the wordings.

As can be seen from the table, there are two values which are exactly the same in the two source variables. However, there is a section of three values which are worded slightly differently. This was not seen as a problem for this particular project. The problem would only become apparent if the combined dataset were to be used by some other researcher who would wish to see a different categorisation. Since the finest level of granularity has not been retained, any different method of categorisation would be impossible.

Table 1. Collapsing values

Source variable 1	Source variable 2	Target variable
1 never	1 for weeks at a time	1 never
2 a lesson here and there	2 for several days at a time	2 occasionally
3 a day here and there	3 for particular days or lessons	3 days at a time
4 several days at a time	4 for the odd day or lesson	4 weeks at a time
5 weeks at a time	5 never	99 not answered
99 not answered	9 not answered	

What has happened in this case is that the definition of the target variable has been driven by the needs of the particular analysis which is being carried out in this specific project. Therefore, the variable which is produced may not be reusable for any other particular analysis.

2.3 Other issues

There are other types of problem which can occur when trying to carry out harmonisation or when trying to use data from different sources (especially different countries). These problems relate to the background to the data and therefore the information is usually much more difficult to obtain. For instance, sampling, population and weighting information should be available if the survey has been documented to a high standard - this is not always the case. Another set of problems occur because questions are asked in the context of a particular, in this case, educational system. If the background information about these country-specific educational systems is not made known then the data cannot be analysed in the correct context. For instance, the qualifications which can be obtained in England (A-levels) and in Scotland (Highers) are quite different and some kind of mapping must be defined before comparative analysis can be carried out.

3 Metadata research at CES

The process of creating a dataset of harmonised variables from data collected for, perhaps, different purposes at different times is, as shown in the earlier part of this paper, an extremely difficult process. It is also a costly one in terms of the time taken and the expertise required to carry out the exercise. It is therefore essential that the result of the work is not lost and that the result is more than a set of variables used for one research project and then stored with a minimum of documentation (or worse still, no documentation at all). At CES we have been committed for a number of years to the development of methods for capturing and using metadata.

Initially this work began with our own survey datasets from the Scottish Young Peoples' Surveys. Over the years, this has expanded into the field of investigation of international standards for metadata and software for the capture and use of metadata (see for example - Lamb, 1997). The work described in this paper, the Idaresa project, has been an opportunity to draw together CES's substantive interest in cross-national comparison of data, our experience in creating integrated datasets and our commitment to quality metadata.

4 Idaresa (Integrated Document and Retrieval Environment for Statistical Aggregates)

There are two stages which occur during the practical task of data integration. First is the 'analysis' stage; where decisions are made on the aim of the harmonisation, the datasets to be used and how the datasets differ. Secondly, the 'synthesis' stage where a definite data model is defined. (Froeschl, 1997). While not removing the need for the 'analysis' stage, where problems such as those discussed earlier have to be solved, the Idaresa system does provide a framework through which the resulting data model can be preserved and used.

Idaresa is a research and development project (no. 20478) of EUROSTAT's DOSIS (Development of Statistical Information Systems) initiative within the 4th EU Framework Programme for Research and Technological Development. It is co-ordinated by the University of Vienna and has partners at the University of Ulster, the University of Athens and DESAN Market Research of Amsterdam as well as the CES team. The goal of the project is to "design and implement a metadata-based statistical information and data processing system targeted at the practical needs of statistical agencies and offices in charge of supplying high-quality statistical information. Emphasis is laid on the harmonisation of statistical data originating from different sources and contexts".

Conceptually, Idaresa can be described as a structured 'space' into which a 'population' of data and metadata is introduced by three types of information *owners*: the *master administrator*, the *domain administrator* and the *data supplier*. Within the system there are different *scopes*, similar to the directory structure of computer operating systems, into which the different classes of owner can add metadata objects. This provides "... a platform for defining substantive data models" and a "language" to describe the models. A brief summary of the metadata categories and the role of each type of owner is given below.

5 Idaresa metadata categories

5.1 General information and terminology

This is entered into the *root* scope of the system by the master administrator and includes all system-wide definitions, classifications, legal regulations, statistical subjects and basic units of measurement and population. This metadata is available for reference from any other scope. The master administrator is also responsible for *registering* the *domains* which are the platforms for the harmonised variables.

5.2 Conceptual model and domain description

This type of metadata is restricted to the *domain scope* of the domain administrator who is responsible for a particular substantive area (for example R&D statistics or Wages & Earnings). It is here that the domain administrator provides descriptions of the *quality* variables which are the targets for harmonisation (arrived at by the process described in the earlier part of this paper, the 'analysis' stage). Details are also provided here about populations of study, definitions which differ from the ones given by the master administrator and the links to any relevant root metadata.

5.3 Source description

This is the metadata which accompanies each dataset, is entered by the data supplier and is contained within the *local* scope of that data supplier. It is at this stage that the original data is mapped to the *quality* variables defined by the domain administrator. The data supplier must decide on the closeness of fit and may have to re-work the original data to fit the quality. If this has to be done there is an opportunity to record this fact in the data-source metadata. The data supplier is also required to give details of data collection and data processing thus providing the context and background.

Although in practice, it may not be possible to get full information for older datasets with less complete documentation, the use of the Idaresa system should encourage full documentation by providing a template for dataset description. The aim for a later stage of the project is to produce a 'user friendly' interface to capture the information required.

5.4 Physical implementation

Physically, the system is being implemented as a distributed database using object oriented methods. The conceptual and the physical data models are the responsibility of the University of Vienna. Harmonised data will be held locally at a data supplier's site and the metadata objects along with the *InfoNet* linking them will be stored centrally and made available to all sites linked to the system. Potential data users will be able to use a *Browser* (currently being implemented by the University of Athens) to search the metadata and obtain a full description of the data available. The metadata connections built through the *InfoNet* will allow users to browse details of how, when, why and by whom each dataset was collected; what populations have been sampled and what processing methods have been used. They will also have access to information on data quality and weighting procedures, and the classifications and definitions used. Through an SQL server (under development at the University of Ulster) users can specify their data requests and obtain tables. The SQL server also uses information on weighting, bias and sample size for table production where appropriate.

At CES we are responsible for developing the Import Manager, through which the harmonised datasets are added to the system, and the Info Manager, through which the metadata is introduced. We are also, along with DESAN Market Research, providing test data in the domain of school leavers.

This has been a brief introduction to a complex and wide-ranging system. For a complete description of the system and updates on the work, which is due to end in December 1998, see the Idaresa web page on http://idaresa.univie.ac.at/.

6 Discussion

In our experience, harmonised datasets are often created with one research question in mind whereas the Idaresa aim is that each *quality* will describe (as the name suggests) the 'best' variable to describe a concept within a substantive area. Once a dataset is mapped to the *qualities* and documented then it is ready to be re-shaped for any research question and its relationship to the 'same' variable in another dataset is unambiguous. If a user feels that he or she would rather go back to the original data, relevant contact details should be available within the system.

It is important that, during the harmonisation process, a distinction is made between the theoretical framework and the conceptual data model. The theoretical framework is concerned with the research questions, the phenomena that are observed, the factors influencing those phenomena and the hypotheses which link the factors to the phenomena. The conceptual data model concerns the identification of data, actual and potential, which describes a particular domain. Stewart & Kamins (1993) discuss the importance of evaluating data sources thoroughly and suggest questions that a potential data user should ask about the data. These include the purpose of the study, what information was collected (including definitions and measures), when the data was collected, what was the data collection methodology, how consistent is the information with that from other sources. A well designed metadata system should capture and store this information so that it is available with the data. This is what the Idaresa *source description* attempts to capture, so that users of datasets accessed through the Idaresa *domains* can benefit from this information.

References

Froeschl, K.A. (1997). *Metadata Management in Statistical Information Processing*. Wien-New York: Springer-Verlag.

Grossman, W., Froeschl, K. & Walk, M. (1998). *The Idaresa Data Model Final Version*. Deliverable 3.4.2, DOSIS Project 20478.

Lamb, J.M., (1997). *Access to Distributed Databases for Statistical Information and Analysis*. Paper presented to the ISI - 5th Session of the International Statistical Institute (18-26 August), Istanbul.

Stewart, D.W. & Kamins M.A., (1993). *Secondary Research: Information Sources and Methods (Second Edition)*. Newbury Park: Sage Publications.

LLAMA: an Object-Oriented System for Log Multiplicative Models

David Firth

Nuffield College, Oxford OX1 1NF, United Kingdom

http://www.stats.ox.ac.uk/~firth/llama/

Abstract. *Llama* is an interactive environment, implemented in XLISP-STAT, for the specification and fitting of log linear and log multiplicative models to contingency tables. This paper describes the model syntax and object-oriented design of *Llama*, and some statistical and algorithmic features including the use of over-parameterized representations of models.

Keywords. Generalized inverse, iterative scaling, Lisp, social mobility

1 Introduction

Log multiplicative models are used quite commonly, in the social sciences especially, to simplify complex interaction terms in log linear models and thus to aid interpretation. A simple example is the 'uniform difference' model of social mobility studies (e.g., Xie, 1992; Erikson & Goldthorpe, 1993), in which

$$\log \mu_{ijt} = \alpha_{it} + \beta_{jt} + \delta_t \gamma_{ij}. \tag{1}$$

Here perhaps i and j represent origin and destination classes, and t indexes time periods: the term $\delta_t \gamma_{ij}$ simplifies the origin-destination-time interaction to an interpretable form in which γ_{ij} is a stable pattern of origin-destination association and δ_t measures the strength of that association at time t. There are many other standard instances of log multiplicative models, including the well known Goodman RC and $RC(m)$ association models (Goodman, 1985).

Much ingenuity has been exercised to fit such models in standard modelling packages, e.g. for two-way tables Falguerolles & Francis (1995) give a general GLIM macro and van Eeuwijk (1995) describes a similar approach which has been implemented in Genstat. However, the inherent limitations of the programming environment in a system such as GLIM have meant that model generalizations (e.g., to include two or more bilinear terms $\delta_t\gamma_{ij} + \epsilon_t\phi_{jk} + \ldots$, or perhaps even to include a three-way product such as $\delta_t\gamma_i\phi_j$) typically demand nontrivial programming effort and as a result are beyond the reach of most social science researchers.

This paper describes some elements of the development of the *Llama* (*Log linear and multiplicative analysis*) system, which runs within the XLISP-STAT environment (Tierney, 1990) on a variety of hardware and operating-system platforms. Section 2 introduces the model specification syntax of *Llama* and sets out some advantages of keeping models in their natural, over-parameterized form; Section 3 briefly discusses maximum likelihood algorithms in the context of over-parameterized models; while Section 4 outlines the object-oriented design and menu/dialog user interface of *Llama*.

2 Log multiplicative models in LLAMA

2.1 Example: Goodman RC models for a two-way table

For the cell means $\{\mu_{rc}\}$ in a table with rows indexed by r and columns by c, the log multiplicative form of the RC association model is

$$\log \mu_{rc} = \alpha_r + \beta_c + \gamma_r \delta_c. \tag{2}$$

The $RC(2)$ model is

$$\log \mu_{rc} = \alpha_r + \beta_c + \gamma_r \delta_c + \theta_r \phi_c, \tag{3}$$

and so on. Falguerolles & Francis (1995) describe GLIM macros for such models, and give an example in which the row variable is `CLASS`, the respondent's social class (5 categories), and the column variable is `RATE`, the frequency of attending meetings outside working hours (5 categories). In *Llama* the models (2) and (3) are specified by a straightforward translation from their algebraic representations as above; with row and column variables `CLASS` and `RATE`, model (2) would be expressed as

```
(CLASS) (RATE) ((CLASS)(RATE)),
```

while model (3) becomes

```
(CLASS) (RATE) ((CLASS)(RATE)) ((CLASS)(RATE)).
```

The independence model $\log \mu_{rc} = \alpha_r + \beta_c$, which is of course log linear, is expressed in *Llama* simply as

```
(CLASS) (RATE).
```

2.2 Example: 'uniform-difference' form of three-way interaction

The primary motivation for *Llama* was the increasing use in social science of models such as (1), involving a structured 3-factor (or higher) interaction: a need was felt for software which is flexible in terms of the models and data structures it can handle, and which has an intuitive user interface which encourages interactive modelling in the style of GLIM, for example. A fairly typical illustration is a 4-way table from political sociology, kindly provided by Bruno Cautrès. The table is constructed from post-election studies carried out in France, the four variables of interest being

- `C`: respondent's social class (6 categories)
- `R`: respondent's religion/religiosity (4 categories)
- `E`: which of 4 election studies the respondent took part in (each study relates to a different election)
- `V`: the respondent's reported vote (2 categories, 'left' and 'right').

Table 1 lists some candidate models for the log cell means $\{\log \mu_{crev}\}$. Of interest is whether either or both of the 3-factor interactions `(E C V)` and `(E R V)` can be simplified into the uniform-difference form for purposes of interpretation. The usual analysis of deviance applied to Table 1 suggests that changes in the religion-vote association across elections can indeed be simplified in this way, but that changes in the class-vote association cannot.

Table 1. Models for the change, across elections, in class-vote and religion-vote association patterns

Algebraic formula	*Llama* formula	d.f.	deviance
$\beta_{ecr} + \delta_{erv} + \phi_{ecv}$	`(E C R)(E R V)(E C V)`	60	79.5
$\beta_{ecr} + \gamma_e\delta_{rv} + \phi_{ecv}$	`(E C R)((E)(R V))(E C V)`	66	88.1
$\beta_{ecr} + \delta_{erv} + \theta_e\phi_{cv}$	`(E C R)(E R V)((E)(C V))`	72	125.4
$\alpha_{ev} + \beta_{ecr} + \gamma_e\delta_{rv} + \theta_e\phi_{cv}$	`(E V)(E C R)((E)(R V))((E)(C V))`	78	133.0

2.3 Over-parameterization

The model specifications exemplified above do not involve the constraints needed to turn quantities such as α_{ev}, β_{ecr}, etc., into estimable parameters. Such identifiability constraints are, of course, merely a matter of parameterization: different choices of constraints just yield different representations of the same model[1]. In *Llama*, identifiability constraints are not used: models are kept in their over-parameterized form throughout, from specification through fitting procedure to reporting of effects and standard errors.

This rather unusual approach not only reflects the conceptual separation of model and parameterization, but in some respects positively emphasizes it. The default behaviour of *Llama* is to use a randomly-determined parameterization, derived from pseudo-random starting values supplied to the algorithm used to maximize the likelihood (see Section 3 below). Thus if the same model is fitted twice in *Llama*, the results reported will differ in the parameterization used. As a concrete example, consider the second model of Table 1 above, in which substantive interest focuses on the uniform-difference term $\gamma_e\delta_{rv}$. Two particular *Llama* fits gave the following representations of the maximum likelihood solution:

$$\hat{\gamma}^{(1)} = \begin{pmatrix} 1.18 \\ 1.09 \\ .897 \\ .841 \end{pmatrix}, \quad \hat{\delta}^{(1)} = \begin{pmatrix} 1.55 & .127 \\ 1.19 & .564 \\ .893 & 1.01 \\ .166 & 1.41 \end{pmatrix};$$

$$\text{and} \quad \hat{\gamma}^{(2)} = \begin{pmatrix} 1.30 \\ 1.19 \\ .984 \\ .922 \end{pmatrix}, \quad \hat{\delta}^{(2)} = \begin{pmatrix} 1.35 & .268 \\ 1.22 & .865 \\ .895 & 1.22 \\ .126 & 1.47 \end{pmatrix}.$$

The two solutions reflect the parameter redundancy: there are only 6 non-redundant parameters here, not 12. Estimable combinations of the $\{\gamma_e, \delta_{rv}\}$ have the same estimated values in both solutions. For instance, γ_e/γ_f measures the relative strength, in elections e and f, of the religion-vote association; e.g., $\hat{\gamma}_3^{(i)}/\hat{\gamma}_4^{(i)} = 1.07$ regardless of the particular representation i that is used. Similarly, each of the quantities $\gamma_e(\delta_{rv} - \delta_{rw} - \delta_{sv} + \delta_{sw})$ is an estimable odds ratio specific to election e; for example, $\hat{\gamma}_1(\hat{\delta}_{22} - \hat{\delta}_{21} - \hat{\delta}_{12} + \hat{\delta}_{11}) = 0.94$ in either parameterization. Thus the estimability of any parameter combination is readily checked numerically, by comparing the estimates obtained

[1] Falguerolles & Francis (1995, p11) appear to miss this point in their formula for degrees of freedom and the supporting statement that 'the degrees of freedom...depends on the identification constraints chosen'. Happily, the default constraints used in their GLIM macros cause the correct d.f. to be reported; other choices would make the reported d.f. too large.

in two or more randomly-generated parameterizations. This is a useful feature, especially in connection with models where estimability is difficult to assess algebraically, such as some of the 'topological' or 'levels' models of Xie (1992) and Erikson & Goldthorpe (1993). The present version of *Llama* readily allows the user to make such calculations; it is likely that assessment of estimability in this way will be automated in a future version.

While *Llama* itself does not impose any particular standardization on estimated coefficients, the user is of course free to standardize the fitted model for purposes of presentation or comparison. Two or more randomly-generated representations of the maximum likelihood fit can again be used to check the validity of any proposed standardization.

From the computational standpoint, the use of an over-parameterized model representation has two main implications. First, it removes the need for the fitting program to 'know' what constraints are needed for identifiability or to determine constraints numerically. Secondly, it requires that the fitting algorithm used should avoid matrix inversion, since the matrices involved are rank-deficient. The former implication is a considerable advantage: the use of pre-programmed 'rules' for removing redundant parameters would restrict the class of models that could be handled, while the numerical detection of aliasing in nonlinear models is difficult and yields an arbitrary parameterization which may not be very useful for purposes of interpretation and presentation. The latter implication is not a serious restriction: two possibilities that avoid matrix inversion, discussed briefly in Section 3 below, are to use a generalized iterative scaling procedure or to use generalized inverses.

The problem of providing meaningful standard errors for coefficients in an over-parameterized model is solved by extending the ideas of Ridout (1989) and Easton, Peto & Babiker (1991). The resultant standard errors are more useful than those typically presented with the 'usual' type of identifiability constraints, in that they allow for covariance between parameter estimates. For further details see Menezes & Firth (1998).

3 Algorithms

The likelihood is maximized for the model in its natural, over-parameterized form. For concreteness, consider model (2) above for observed counts $\{y_{rc}\}$, for which the maximum likelihood equations are

$$y_{r.} - \hat{\mu}_{r.} = 0; \qquad \sum_c (y_{rc} - \hat{\mu}_{rc})\hat{\delta}_c = 0 \quad (\text{all } r);$$

$$y_{.c} - \hat{\mu}_{.c} = 0; \qquad \sum_r (y_{rc} - \hat{\mu}_{rc})\hat{\gamma}_r = 0 \quad (\text{all } c).$$

The iterative scaling procedure often used with log linear models is easily generalized to equations such as these. An appropriate generalization is a Gauss-Seidel iteration with each step a one-parameter Newton update, as described in Goodman (1979), Becker (1990) or Vermunt (1997)[2]. For a multiplicative parameter such as γ_r, for example, the update is

$$\hat{\gamma}_r^{(t+1)} = \hat{\gamma}_r^{(t)} + \left[\sum_c \left(y_{rc} - \hat{\mu}_{rc}^{(t)}\right)\hat{\delta}_c^{(t)}\right]\left[\sum_c \hat{\mu}_{rc}^{(t)}\left(\hat{\delta}_c^{(t)}\right)^2\right]^{-1}.$$

[2] Jeroen Vermunt's *LEM* program includes facilities for fitting log multiplicative models as part of a more general package for categorical data.

Starting values $\hat{\boldsymbol{\alpha}}^{(0)}, \hat{\boldsymbol{\beta}}^{(0)}, \hat{\boldsymbol{\gamma}}^{(0)}$ and $\hat{\boldsymbol{\delta}}^{(0)}$ are randomly generated; reasons for this are (i) the likelihood may be multimodal, (ii) certain models such as the *RC* model are degenerate when some parameters are equal, and (iii) the use of a random parameterization permits the kind of estimability and standardization checks described above in Section 2.3.

An alternative update is a variant of iterative weighted least squares as in Green (1984), applied first to the log linear model that has $\boldsymbol{\gamma}$ fixed at the current values (to update $\hat{\boldsymbol{\alpha}}, \hat{\boldsymbol{\beta}}, \hat{\boldsymbol{\delta}}$) and then to the complementary log linear model with $\boldsymbol{\delta}$ fixed. The variant uses a generalized inverse in place of the usual inverse information, to avoid the need for identifiability constraints. Iterative weighted least squares cannot be used with a generalized inverse to update all parameters simultaneously in the full log-multiplicative model, but a generalized inverse information matrix for all parameters together is useful after the last iteration as the basis for calculation of standard errors.

4 Program design and user interface

An early and highly influential example of object-oriented design in a statistical package is the notion in GLIM (Baker & Nelder, 1978) of the 'current model' which can be inspected or revised interactively. In *Llama* the basic philosophy is the same, and in particular the system is designed to encourage interactive elaboration and criticism of models. *Llama* defines two basic object types—table objects and model objects—with no restriction on the number of objects of each type that may coexist. A named table object holds the data and structure of the contingency table to be analysed, the names of the classifying factors, etc.; while a named model object contains the specification of a log linear or log multiplicative model for a particular table, along with estimates, fitted values, residuals, etc., once the model has been fitted. A table can be used simultaneously by many model objects, and every table object 'knows' which models apply to it; similarly, a model object 'knows' about the table to which it refers. Table objects have associated methods for producing marginal tables, for the addition and deletion of factors, etc.; and model objects are equipped with methods for fitting and display. For example, when displaying residuals, a model object is able to use the structure of the original table and to show specified marginal views rather than present a 'flat' list. Parameter estimates also are displayed in a tabular form, rather than as an unstructured list, for ease of interpretation.

Model specification, inspection and revision can be done either by typing commands directly to the Listener (the Lisp 'terminal'), or more conveniently through a simple menu and dialog system. As an example of the former method, the Lisp command needed to specify the second model in Table 1 using data from a table object called `election-study-data` would be

```
(make-model undiff-E-RV
  :data-table election-study-data
  :terms "(E V)(E C R)((E)(R V))(E C V)")
```

where `unidiff-E-RV` is the mnemonic name given to the newly-created model object. The menu/dialog interface eliminates the need ever to type such commands to the Listener in routine use of *Llama*. The Listener would, however, still be used if other XLISP-STAT facilities were needed, such as graphical displays or further calculations on model output.

Finally, it might be noted that the author is not a skilled programmer, and the code for *Llama* is undoubtedly clumsy and inefficient. But it was remarkably easy to write both the modelling routines and the simple graphical user interface, which works without modification or re-configuration on several widely-available platforms (Mac, Unix/X11, Windows95/NT/etc.). In this regard XLISP-STAT has been invaluable: it is an excellent environment for the rapid prototyping and development of a small, interactive statistical package with a convenient user interface.

Acknowledgement
The very helpful comments of the referee are gratefully acknowledged. The referee has pointed out that there is practical interest in generalized bilinear models with links other than logarithmic and data types other than counts; see van Eeuwijk (1995) for examples. For canonical links at least, such a generalization requires only fairly minor changes and is planned for a future version of *Llama*.

References

Baker, R.J. & Nelder, J.A. (1978). *The GLIM System, Release 3.* Oxford: Numerical Algorithms Group.

Becker, M.P. (1990). Maximum likelihood estimation of the RC(M) association model. *Applied Statistics*, **39**, 152–167.

Easton, D.F., Peto, J. & Babiker, A.G.A.G. (1991). Floating absolute risk: an alternative to relative risk in survival and case-control analysis avoiding an arbitrary reference group. *Statistics in Medicine*, **10**, 1025–1035.

Eeuwijk, F. van (1995). Multiplicative interaction in generalized linear models. *Biometrics*, **51**, 1017–1032.

Erikson, R. & Goldthorpe, J.H. (1993). *The Constant Flux: A Study of Class Mobility in Industrial Societies.* Oxford: Clarendon Press.

Falguerolles, A. de & Francis, B. (1995). Fitting bilinear models in GLIM. *GLIM Newsletter*, **25**, 9–20.

Goodman, L.A. (1979). Simple models for the analysis of association in cross-classifications having ordered categories. *J. Amer. Statist. Assoc.*, **74**, 537–552.

Goodman, L.A. (1985). The analysis of cross-classified data having ordered and/or unordered categories: Association models, correlation models, and asymmetry models for contingency tables with or without missing entries. *Ann. Statist.*, **13**, 10–69.

Green, P.J. (1984). Iteratively reweighted least squares for maximum likelihood estimation, and some robust and resistant alternatives (with discussion). *J. Roy. Statist. Soc.* B, **46**, 149–192.

Menezes, R. & Firth, D. (1998). More useful standard errors for group and factor effects. In: *COMPSTAT98: Proceedings in Computational Statistics, Short Communications and Posters*, in press.

Ridout, M.S. (1989). Summarizing the results of fitting generalized linear models to data from designed experiments. In: *Statistical Modelling: Proceedings of GLIM89 and the 4th International Workshop on Statistical Modelling held in Trento, Italy, July 17-21, 1989*, 262–269. New York: Springer.

Tierney, L. (1990). *LISP-STAT: An Object-Oriented Environment for Statistical Computing and Dynamic Graphics.* New York: Wiley.

Vermunt, J.K. (1997). *Log-Linear Models for Event Histories.* London: Sage.

Xie, Y. (1992). The log-multiplicative layer effect model for comparing mobility tables. *American Sociological Review*, **57**, 380–395.

Using Threshold Accepting to Improve the Computation of Censored Quantile Regression

Bernd Fitzenberger[1] and Peter Winker[2]

[1] Department of Economics, University of Konstanz and University of Mannheim, D–68131 Mannheim, F.R.G.
[2] Department of Economics, University of Mannheim, D–68131 Mannheim, F.R.G.

Abstract. Due to an interpolation property the computation of censored quantile regression estimates corresponds to the solution of a large scale discrete optimization problem. The global optimization heuristic threshold accepting is used in comparison to other algorithms. It can improve the results considerably though it uses more computing time.

Keywords. Censored quantile regression, interpolation property, BRCENS, threshold accepting

1 Introduction

Censored quantile regressions introduced by Powell (1984) have some appeal in econometric applications with fixed known censoring points (see Fitzenberger, 1997a, for a survey). A few applications exist, e.g. for modelling the conditional wage distribution. They have attracted more interest in the recent past as micro datasets are used more intensively for research purposes.

Unfortunately, to compute a censored quantile regression (CQR) estimate a non–differentiable and non–convex distance function has to be minimized. Various optimization routines suggested in the past fail to give satisfactory results for many instances. The algorithm BRCENS introduced by the first author is tailored to the CQR problem by taking into account special features of potential solutions – the interpolation property discussed in Section 2 and in the literature – improves the results, and hence the applicability, considerably. However, as pointed out in the simulation studies by Fitzenberger (1997a,b) it still fails to provide consistently high quality results, in particular if the degree of censoring is high.

In addition to the methods already discussed for the CQR problem (see Pinkse, 1993; Fitzenberger, 1997a,b, for some algorithms) we introduce an implementation of the global optimization heuristic threshold accepting (TA) to this problem. Pinkse (1993) also notes that optimization heuristics such as simulated annealing might improve the results, however, without giving an implementation. TA was introduced by Dueck & Scheuer (1990) for the travelling salesman problem. Afterwards it has been applied successfully to many problems in operational research but also to discrete optimization problems in statistics and econometrics (see Winker & Fang, 1997, for some references). As the interpolation property of the CQR problem provides a discrete optimization framework, TA seems to be also an appropriate choice for this problem. A suitable implementation will finally, i.e. with a huge number of iterations, converge to the true global optimum almost surely (cf. Althöfer

& Koschnick, 1991). Furthermore, and this seems to be the more convincing argument, it gives better results for many problem instances as compared to the algorithms used for this purpose so far. To preview the simulation results, TA can improve the estimation of CQR's considerably though it uses more computing time than BRCENS.

In this short contribution we cannot give an exhaustive description of the problem and its approximate solution by TA. Instead, we present a short introduction to the CQR problem in Section 2, describe two conventional and the threshold accepting algorithm in Section 3 and present some simulation results in Section 4.

2 Interpolation property

Introducing some notation, for a sample of size N, let the values of the dependent variable be $y = (y_1, \ldots, y_N)$, the design matrix be the $N \times k$ matrix $X = (x_1, \ldots, x_N)'$, with $x_i = (x_{i,1}, \ldots, x_{i,k})$, and the fixed known observation specific censoring values be $yc = (yc_1, \ldots, yc_N)$. Here, we consider a censored regression model with censoring from above, i.e. $y_i \leq yc_i$. For a given quantile $\theta \in (0,1)$, the CQR estimation problem is to minimize the piecewise linear distance function

$$\sum_{i=1}^{N} \{\theta I[y_i > g(x_i'\beta, yc_i)] + (1-\theta) I[y_i < g(x_i'\beta, yc_i)]\} \cdot \mid y_i - g(x_i'\beta, yc_i) \mid \qquad (1)$$

with regard to the $k \times 1$ parameter vector β, where the nonlinear response function is $g(x_i'\beta, yc_i) = \min[x_i'\beta, yc_i]$ and I[.] denotes the indicator function. Let $\widehat{\beta_\theta}$ denote a solution to the minimization problem. Then, the expression $x_i'\widehat{\beta_\theta}$ captures the estimated θ–quantile of the underlying uncensored dependent variable conditional on x_i.

Since the CQR distance function (1) is piecewise linear, the CQR minimization problem does not necessarily have a unique solution. Analogous to standard quantile regressions, cf. Koenker & Bassett (1978), the following interpolation property can also be established for CQR's, cf. Fitzenberger (1997a):

Interpolation Property: If the design matrix X has full rank k, then there exists a global minimizer $\widehat{\beta_\theta}$ of the CQR distance function such that $\widehat{\beta_\theta}$ interpolates at least k data points, i.e. there are k observations $\{(y_{i_1}, x_{i_1}), \ldots, (y_{i_k}, x_{i_k})\}$ with

$$\text{(IP)} \quad y_{i_l} = x_{i_l}'\widehat{\beta_\theta} \text{ for } l = 1, \ldots, k \text{ and the rank of } (x_{i_1}, \ldots, x_{i_k})' \text{ equals } k\,.$$

When evaluating the IP, the following two points deserve attention. First, if the CQR distance function exhibits a unique minimizer, it must satisfy IP. And second, the CQR can interpolate a censoring point where an observation is censored.

3 Algorithms

3.1 IPOL

The interpolation property IP suggests an enumeration algorithm to determine the CQR estimate exactly, i.e. an element of the set of global minimizers, cf. Pinkse (1993) and Fitzenberger (1997b). This algorithm, which we denote by IPOL, consists of an enumeration of the set of all k–tuples of data points with linearly independent regressor vectors. Among the corresponding interpolating regression lines, IPOL takes the one minimizing the CQR distance function. The algorithm involves the evaluation of at most $\binom{N}{k}$ k–tuples. For N large and k small, the computational effort grows approximately with $N^k/k!$ and thus, IPOL is already impractical for moderately sized problems. Nevertheless, it is much faster than grid search.

3.2 BRCENS

The algorithm BRCENS is introduced in Fitzenberger (1997a,b) as an adaptation of the standard **B**arrodale–**R**oberts–Algorithm (BRA) for standard Quantile Regressions – introduced by Koenker & Bassett (1978) – to the **Cens**ored Quantile Regression case. A standard quantile regression exhibits a linear programming structure. Barrodale & Roberts (1973) notice that the IP allows for a more efficient, condensed simplex approach. Only kinks of the distance function need to be considered for which k (design matrix exhibits full rank) observations are interpolated and for which the rank of the matrix formed by the regressor vectors is equal to k. Hence, the basic structure of BRA can be adapted to the CQR problem. The major modification involves the calculation of the directional derivative taking account of the changes at the censoring values. However, given the non–convex objective in (1) BRCENS can only guarantee convergence to a local minimum, which in cases involving a lot of censoring, often does not correspond to a global minimum, cf. Fitzenberger (1997a,b) and the results in Section 4.

3.3 Threshold Accepting algorithm

In contrast to the deterministic algorithms mentioned above, threshold accepting is a non deterministic global optimization heuristic similar to the more common simulated annealing.[1] It can be characterised as refined local search algorithm on a discrete search space.

Using the interpolation property of potential solutions, the search space for the CQR problem is given by all k–tuples of data points with linearly independent regressor vectors. A topology or local structure is introduced on this finite set by the following definition. Two k–tuples are neighbours if the second one can be obtained from the first by a move in one of $2k$ possible search directions. These search directions are implicitly defined by the $(k-1)$–tuples obtained when leaving out one point from the original k–tuple. In the dual coefficient space this defines a 1–dimensional subspace comprising two search directions as the orthogonal complement to the space spanned by the $(k-1)$ fixed data points. Calculating the residuals $(y_i - x_i'\beta)$ for the coefficient vector β interpolating the initial k–tuple, the corresponding data points are those for which the absolute size of the residual is strictly reduced when moving into one of the search directions. Given this local structure, the algorithm

[1] Dueck & Scheuer (1990) also present a completely deterministic version of threshold accepting which seems to perform only slightly worse than the non deterministic version.

proceeds as follows. It is initialized with some arbitrary k–tuple of linearly independent regressor vectors. Then, in each iteration one neighbour of the current solution is selected at random. The value of the objective function (1) is calculated for the new k–tuple. If it is better than the value of the current solution the new k–tuple becomes the current solution. The same holds true if it is not worse than the value of the current solution plus a positive threshold. In our implementation these thresholds are defined as a factor of the value of the current solution. This acceptance of a temporary worsening of the objective function is a central feature of the algorithm and enables it to escape bad local minima. A suitable threshold sequence (cf. Winker, 1995, for details), which decreases to zero as the algorithm proceeds, is responsible for the almost certain asymptotic convergence of threshold accepting. For our application, the threshold sequence is generated from the data of the problem based on an empirical jump function as described in Winker & Fang (1997). For the simulation study the values of the lower 5 percentile of this empirical jump function are used as threshold sequence. The algorithm is run with different numbers of iterations ranging from 5000 to 100000.

4 Simulation results

This section analyzes the performance of the algorithms described in Section 3 by means of a simulation study whose design is similar to Fitzenberger (1997a,b). Table 1 describes the data generating processes (DGP's), (A)–(D). The estimation problem is a censored LAD regression with two regressors and an intercept. It is also checked if the exact CQR estimate (determined by the enumeration algorithm IPOL) is unique.

Table 1. Data Generating Processes (DGP) (A) – (D) used in Simulation Study for the model $y_i = min(yc_i, \beta_1 + \beta_2 \cdot x_{i,2} + \beta_3 \cdot x_{i,3} + \epsilon_i)$ [a]

DGP	Censoring Values	True Coefficients	Regressor Values
(A)	$yc_i = Const$	$(\beta_1, \beta_2, \beta_3) = (0, 0, 0)$	$x_{i,2}, x_{i,3} \sim N(0,1)$
(B)	$yc_i = Const$	$(\beta_1, \beta_2, \beta_3) = (0, 0, 0)$	$x_{i,2} = -9.9 + 0.2 \cdot i,\ x_{i,3} = x_{i,2}^2$
(C)	$yc_i = Const + 0.5$	$(\beta_1, \beta_2, \beta_3) = (0.5, 0.5, -0.5)$	$x_{i,2}, x_{i,3} \sim N(0,1)$
(D)	$yc_i = Const + 0.5$	$(\beta_1, \beta_2, \beta_3) = (0.5, 0.5, -0.5)$	$x_{i,2} = -9.9 + 0.2 \cdot i,\ x_{i,3} = x_{i,2}^2$

a) *Const* denotes some constant taking various values and I(.) denotes the indicator function. The random variables ϵ_i are distributed as i.i.d. $N(0,1)$ and $i = 1, \ldots, N$

The DGP's differ in three dimensions. First, by whether the coefficients to generate the data are all 0 (A,B) or 0.5 (C,D). Since BRCENS starts with all coefficients at 0, it could make a difference whether the starting values are close to the truth. Second, the DGP's differ by whether the regressor is a random variable (A,C) or a fixed sequence of numbers (B,D). And third, the DGP's differ by the degree of censoring depending on $Const = 1, 0.5, 0$. $Const = 0$ represents a situation where on average 50% of the observations are censored, i.e. the exact CQR ($\theta = 0.5$) typically reaches the censored region.

For all twelve different situations (given by the DGP's (A)–(D) and three different values for Const) 1000 random samples of size 100 were generated. The enumeration algorithm IPOL described in Subsection 3.1 was used to obtain the global optimum out of at most $\binom{100}{3} = 161700$ 3–tuples. In no

Table 2. Absolute Frequencies among 1000 Random Samples that Algorithms Achieved Optimum

DGP	CONST	Algorithms BRCENS	TA–5000	TA–20000	TA–50000	TA–100000
(A)	1.0	993	443	842	986	998
(A)	0.5	847	845	974	971	984
(A)	0.0	121	411	431	428	442
(B)	1.0	993	479	859	978	998
(B)	0.5	747	781	890	930	938
(B)	0.0	17	172	193	173	193
(C)	1.0	955	371	774	951	993
(C)	0.5	857	495	865	967	992
(C)	0.0	649	827	956	964	977
(D)	1.0	745	942	984	986	989
(D)	0.5	749	973	981	984	987
(D)	0.0	735	975	983	984	991

situation did the global optimum prove to be non–unique. Table 2 displays the absolute frequencies that the two algorithms achieve the true optimum. The different values for the TA algorithm differ in the number of iterations performed ranging from only 5000 up to 100000.

Table 3. Absolute Frequencies among 1000 Random Samples that one Algorithm Achieves Lower Value of Distance Function Compared to Other Algorithm

		BRCENS vs. TA–5000		BRCENS vs. TA–20000		BRCENS vs. TA–50000		BRCENS vs. TA–100000	
DGP	CONST	BRCENS	TA	BRCENS	TA	BRCENS	TA	BRCENS	TA
(A)	1.0	553	7	157	4	14	5	2	4
(A)	0.5	128	142	10	159	9	162	4	159
(A)	0.0	128	653	148	638	134	627	122	640
(B)	1.0	517	7	141	5	21	8	2	2
(B)	0.5	122	206	20	210	8	196	8	226
(B)	0.0	105	817	118	809	120	823	112	827
(C)	1.0	608	34	213	45	42	42	4	38
(C)	0.5	454	117	112	157	16	165	2	154
(C)	0.0	115	326	19	359	7	173	5	364
(D)	1.0	35	244	7	252	8	244	1	257
(D)	0.5	13	243	11	257	5	254	7	229
(D)	0.0	16	261	3	263	4	255	4	243

For a low number of iterations the TA implementation is often not competitive with BRCENS for instances with a low degree of censoring ($Const = 1, 0.5$), while it already outperforms BRCENS for the instances with a high degree of censoring ($Const = 0$). Furthermore, as the number of iterations increases the probability to achieve the true optimum tends to one if there is not much censoring and to much higher probabilities than for BRCENS otherwise.

While Table 2 permits a comparison of the algorithms to the benchmark given by IPOL, Table 3 is based on a direct comparison. In each column the

Table 4. Average Relative Computation Times[a]

Const	IPOL	BRCENS	TA–5000	TA–20000	TA–50000	TA–100000
1.0	9000	1.0	800	1300	2400	4100
0.5	8900	1.7	800	1300	2300	4000
0.0	8800	18.0	780	1300	2300	4000

a) The reported numbers are the ratios of average computation times across DGP's (A)–(D) relative to BRCENS, *Const* = 1. The time results are obtained with the UNIX time command ("user time") on an IBM RS 6000 workstation (type 3AT), based on a Fortran implementation of the various algorithms. The numbers are rounded to two valid digits. In actual time, TA–100000 required on average 15 seconds per sample.

left (right) entries indicate the absolute frequency that the value of the objective obtained by BRCENS was strictly lower (higher) than for TA. Again, for a small number of iterations BRCENS is superior to the TA implementation for instances with a low degree of censoring, whereas TA outperforms BRCENS consistently if the number of iterations is increased.

Table 4 gives an overview on the average relative computation times required by the different algorithms. It becomes obvious that the higher efficiency of TA, in particular for large numbers of iterations, requires much more computation time. However, the results presented in this contribution refer only to a very small instance (3 regressors, 100 observations). In the future, we plan to investigate the application of TA in larger problems.

References

Althöfer, I. & K.-U. Koschnick (1991). On the Convergence of 'Threshold Accepting'. *Applied Mathematics and Optimization*, **24**, 183-195.

Barrodale, I. & F.D.K. Roberts (1973). An Improved Algorithm for Discrete l_1 Linear Approximation. *SIAM Journal of Numerical Analysis*, **10**, 839-848.

Dueck, G. & T. Scheuer (1990). Threshold Accepting: A General Purpose Algorithm Appearing Superior to Simulated Annealing. *Journal of Computational Physics*, **90**, 161-175.

Fitzenberger, B. (1997a). A Guide to Censored Quantile Regressions. In: *Handbook of Statistics, Volume 15: Robust Inference* (ed. G.S. Maddala & C.R. Rao), 405-437. Amsterdam: North–Holland.

Fitzenberger, B. (1997b). Computational Aspects of Censored Quantile Regression. In: *Proceedings of The 3rd International Conference on Statistical Data Analysis based on the L_1–Norm and Related Methods*, (ed. Y. Dodge), 171-186. Hayword, California: IMS Lecture Notes Series, **31**.

Koenker, R. & G. Bassett (1978). Regression Quantiles *Econometrica*, **46**, 33-50.

Pinkse, C.A.P. (1993). On the Computation of Semiparametric Estimates in Limited Dependent Variable Models. *Journal of Econometrics*, **58**, 185-205.

Powell, J.L. (1984). Least Absolute Deviations Estimation for the Censored Regression Model. *Journal of Econometrics*, **25**, 303-325.

Winker, P. (1995). Identification of Multivariate AR–Models by Threshold Accepting. *Computational Statistics and Data Analysis*, **20**, 295-307.

Winker, P. & K.–T. Fang (1997). Application of Threshold Accepting to the Evaluation of the Discrepancy of a Set of Points. *SIAM Journal on Numerical Analysis*, **34**, 2028-2042.

On the Convergence of Iterated Random Maps with Applications to the MCEM Algorithm

G. Fort[1], E. Moulines[1] and P. Soulier[2]

[1] Ecole Nationale Supérieure des Télécommunications, CNRS URA 820, 46, rue Barrault, 75634 Paris Cedex 13, France
[2] Université d'Evry Val d'Essonne, Département de Mathématiques, Boulevard des Coquibus, 91025 Evry Cedex, France

Abstract. Optimization via simulation is a promising technique to solve maximum likelihood problems in incomplete data models. Among the techniques proposed to date to solve this problem, the MCEM algorithm proposed by Wei & Tanner (1991) plays a preeminent role. Perhaps surprisingly, very little is known on the convergence of this algorithm and on the strategies to monitor this convergence. A particular emphasis is given on the stability issue (which is not guaranteed in the original proposal by Wei & Tanner, 1991). A random truncation strategy, inspired by Chen's truncation method for stochastic approximation algorithms, is proposed and analysed.

Keywords. Iterated random maps, Monte-Carlo Markov Chain, optimization

1 Introduction

In a number of situations, the objective function $g(\theta)$ that we have to maximize can be written as

$$g(\theta) = \int_{\mathcal{D}} f(z;\theta)\mu(dz), \quad \theta \in \Theta \tag{1}$$

where θ is the unknown parameter vector, Θ is the feasible set, $f(z;\theta)$ is, for all $\theta \in \Theta$, a positive borelian function, and μ is a given σ-finite positive measure on $\mathcal{B}(\mathcal{D})$, the Borel σ-field of $\mathcal{D}$. Eq. (1) includes as a particular example incomplete data problems: in such a case, $g(\theta)$ is the incomplete data likelihood (the dependence of this function on the observations is implicit) and $f(z;\theta)$ is the complete data-likelihood, z playing the role of the missing data. It is well-known that the EM algorithm (see Dempster *et al.*, 1977) finds the stationary points of $g(\theta)$ (the points such that $\nabla_\theta g(\theta) = 0$), by looking for the fixed point of an iterated map. This algorithm proceeds as follows. Define

$$p(z;\theta) \triangleq \begin{cases} \frac{f(z;\theta)}{g(\theta)} & \text{if } g(\theta) \neq 0 \\ 0 & \text{otherwise} \end{cases}$$

a real-valued function on $\mathcal{D} \times \Theta$.
The $n+1$-th iteration of the EM algorithm consists of **(1) E-step**: compute the *conditional* expectation given the current fit of the parameter θ_n, i.e., $Q(\theta;\theta_n) \triangleq \int_{\mathcal{D}} \log f(z;\theta)\; p(z;\theta_n)\; \mu(dz)$, **(2) M-step**: update the parameter so that $\theta_{n+1} \in \{\bar\theta \in \Theta; Q(\bar\theta;\theta_n) \;\geq\; Q(\theta;\theta_n)\ \forall \theta \in \Theta\}$.

There is a substantial number of models where the E-step cannot be performed in closed form. A simple idea, pushed forward by Wei & Tanner (1991), consists of replacing the conditional expectation $Q(\theta;\theta_n)$ by its Monte-Carlo approximation

$$Q_n(\theta,\theta_n) \triangleq \frac{1}{m_{n+1}} \sum_{j=1}^{m_{n+1}} \log f(Z_{j,n};\theta)$$

where $\{Z_{j,n}\}$, $j \leq m_{n+1}$ is a sequence of random variables identically distributed with distribution $p(z;\theta_n)\mu(dz)$ (or a Markov process with a probability transition kernel $P(z,\bullet;\theta_n)$ admitting $p(z;\theta_n)\mu(dz)$ as its unique stationary distribution), and $\{m_{n+1}\}$ is a non-decreasing sequence of positive integers. Provided that $\theta \longmapsto Q_n(\theta,\theta_n)$ has a single maximum, the MCEM thus defines an iterated random map (IRM) F_n as follows.

$$\theta_{n+1} = \mathrm{argmax}_{\theta\in\Theta} Q_n(\theta,\theta_n) \triangleq F_n(\theta_n).$$

The main purpose of this contribution is to study the stability and the convergence of such IRM, i.e. to find *verifiable* conditions upon which the IRM stays bounded w.p. 1 and eventually converges to some set or to some point.

The paper is organized as follows. In Section 2, we first derive conditions upon which the IRM eventually converges and identify the possible convergence set, under the assumption that the recursion is bounded. In Section 3, we address the stability problem, and we propose a modification of the original procedure to impose a recurrence condition which in turn implies the boundedness. Finally, some illustrations of our results are presented in Section 4.

2 Convergence of iterated random maps

In this section, we present results on convergence of iterated random maps. To keep the discussion simple, we focus on point-to-point Markovian algorithms. Extensions to point-to-set non-Markovian algorithms (the next parameter value effectively depends upon the whole past and not only the current value of the parameter) are considered in an extended version of this work. The convergence of an iterated random map is deeply linked with the convergence of an underlying deterministic map, which plays the role of the ODE (Ordinary Differential Equation) in the analysis of stochastic approximation (SA) algorithms. In fact the study of iterated random maps closely parallels the study of SA algorithms, the ODE being replaced by a deterministic iteration (or semi-dynamical deterministic system). Consider the following deterministic recursion

$$\theta_{n+1} = T(\theta_n). \tag{2}$$

Convergence of such semi-dynamical systems has been studied under many - sometimes very weak- assumptions for many years (see Hale, 1987; Haraux, 1991; and the references therein). Since we are mainly interested in iterative maximum procedures, we restrict our attention to the case where there exists a *Lyapunov function* i.e. a real-valued function V which increases monotonically along the path $\{\theta_n\}$: $V(\theta_{n+1}) \geq V(\theta_n)$. Under weak assumptions on V and T (see below), it is known that the iteration (2) converges to the set of fixed (equilibrium) points of T. Recall that $\theta^\star$ is an *equilibrium point of* T when $\forall n \geq 0$ $T^n(\theta^\star) = T(\theta^\star)$, and a continuous real function V on Θ is said to be *a Lyapunov function relatively to* T *and* $\mathcal{L}$ when $V \circ T(\theta) \geq V(\theta)\ \forall \theta \in \Theta$ and $V \circ T(\theta) = V(\theta)$ iff $\theta \in \mathcal{L}$. The Proposition 1 extends the classical iterated maps convergence result in the perturbed case. For we have a Lyapunov function V relatively to a given procedure T and a sequence $\{\theta_n\}$ which *approximates* the deterministic map $\{T(\theta_n)\}$ according to the Lyapunov criterion, i.e.,

$$|V(\theta_{n+1}) - V \circ T(\theta_n)| \to 0 \text{ as } n \to \infty.$$

Proposition 1. *Let* $(\mathcal{Z}, d)$ *be a complete metric space and* $\{\theta_n\}$ *be a* $\mathcal{Z}$*-valued sequence. Suppose*

- **(A1)** *for all* $n \in \mathbb{N}$, θ_n *is in a compact set* $\mathcal{K} \subset \mathcal{Z}$.
- **(A2)** *there exist a real-valued function* V, *a* $\mathcal{Z}$*-valued mapping* T *on* $\mathcal{Z}$ *and a subset* $\mathcal{L} \subset \mathcal{Z}$ *such that*
  - **(L1)** *the set* $\mathcal{L}$ *is closed and* $\mathcal{L} \cap \mathcal{K}$ *is non-empty,*
  - **(L2)** $V \circ T - V \geq 0$,
  - **(L3)** *for any compact set* $\mathcal{C} \subset \mathcal{Z} \setminus \mathcal{L}$, $\inf_{\mathcal{C}}(V \circ T - V) > 0$,
  - **(L4)** V *is continuous.*
- **(A3)** $|V(\theta_{n+1}) - V(T(\theta_n))| \longrightarrow 0$ *as* $n \longrightarrow \infty$.

Then $\{V(\theta_n)\}$ *converges to a connected component of* $V(\mathcal{L} \cap \mathcal{K})$.

Note that similar results have been established by Shapiro & Wardi (1996) under the stronger condition on the sequence F_n that $\theta_{n+1} = F_n(\theta_n)$ converges to T uniformly on any compact set $\mathcal{K}$,

$$\max_{\theta \in \mathcal{K}} d(F_n(\theta), T(\theta)) \to 0. \tag{3}$$

In the applications considered in this paper, the above condition requires the checking of uniform law of large numbers, which most of the time does not hold when using Metropolis-Hastings type simulation techniques.

In the applications below, the set $\mathcal{L}$ is the equilibrium set of T, $\mathcal{L} = \{x \in \mathcal{Z} : T(x) = x\}$ and Proposition 1 yields:

Corollary 2. *If* $\mathcal{L} \triangleq \{x \in \mathcal{Z} : T(x) = x\}$, $\{\theta_n\}$ *converges to the equilibrium set of the map* T.

Corollary 3. *If, in addition, the connected component of* $V(\mathcal{L})$ *is reduced to a point, i.e.* $V(\mathcal{L})$ *has empty interior, then* $\{V(\theta_n)\}$ *converges .*

Proposition 4. *Suppose* $\{V(\theta_n)\}$ *converges to* $v^\star \in V(\mathcal{L} \cap \mathcal{K})$, *and (A2) and (A3) hold, then* $d(\theta_n, \mathcal{L}) \longrightarrow 0$ *as* $n \longrightarrow \infty$.

The Proposition 4 generalizes the well-known La Salle's invariance principle (see Theorem 2.1.3, Haraux, 1991) i.e. any limit points of $\{\theta_n\}$ have the same *energy*.

3 Stability methods

It is a common problem in stochastic optimization procedure to control the behaviour of the algorithm when the current value θ_n gets too large (see for example, Kushner & Yin, 1997). In some cases, there is a *natural* compactification (constraint set) defined by the problem itself and we may impose constraints on the sequence so that the sequence $\{\theta_n\}$ always lies in that set. This is typically done in projection techniques where $\hat{\theta}_{n+1} \triangleq F_n(\theta_n)$ is projected onto the constraint set after each iteration. This kind of technique may introduce spurious convergence points on the boundary of the constraint set (see Kushner & Yin, 1997).

Another solution, first investigated by Chen (1997) for the SA procedure, consists of using a random truncation and a restart procedure i.e. substituting the parameter by a point generated at random in some suitable compact subset of the feasible set Θ. The key property behind this technique is the following proposition, which shows that the sequence $\{\theta_n\}$ is in some compact set $\mathcal{K}$ under the assumptions **(i)** there exists a recurrent set $\mathcal{G}^0$ ($\{\theta_n\}$ is infinitely often in $\mathcal{G}^0$) and **(ii)** there exists a Lyapunov function relatively to a set $\mathcal{L}$ that, loosely speaking, keeps θ_n close enough to the attractive set $\mathcal{G}^0 \cup \mathcal{L}$.

Proposition 5. *Let $(\mathcal{Z}, d)$ be a σ-compact metric space and $\{\theta_n\}$ be a $\mathcal{Z}$-valued sequence. Suppose*

- **(B1)** *there exist a real-valued function W and a $\mathcal{Z}$-valued mapping T on $\mathcal{Z}$ such that*
  - **(S1)** $\mathcal{L}_w \triangleq \{x \in \mathcal{Z} : (W \circ T - W)(x) = 0\}$ *is compact,*
  - **(S2)** $W \circ T - W \geq 0$,
  - **(S3)** *for any compact set* $\mathcal{C} \subset \mathcal{Z} \setminus \mathcal{L}_w$, $\inf_{\mathcal{C}}(W \circ T - W) > 0$,
  - **(S4)** *for all* $M \in \mathbb{R}$, $\{x \in \mathcal{Z} : W(x) \geq M\}$ *is a compact subset of* $\mathcal{Z}$ *and* $\mathcal{Z} = \bigcup_{M \in \mathbb{Z}} \{x \in \mathcal{Z} : W(x) \geq M\}$.
- **(B2)** $|W(\theta_{n+1}) - W(T(\theta_n))| \, \mathbb{I}_{\theta_n \in \mathcal{C}} \longrightarrow 0$ *as* $n \longrightarrow \infty$ *where* $\mathcal{C}$ *is any compact subset of* $\mathcal{Z}$.
- **(B3)** $\{\theta_n\}$ *is infinitely often in a compact set* $\mathcal{G}^0$.

Then there exists a compact set $\mathcal{K}$ such that, for all $n \in \mathbb{N}$, $\theta_n \in \mathcal{K}$.

This result in some sense extends *ultimate bounded* conditions (see Theorem 11.2.1, Meyn & Tweedie, 1993) for the iterative scheme.

The next step is to define a practical procedure to create a recurrent set. Let $\mathcal{Z}$ be a σ-compact metric space and $\{\mathcal{G}^n\}$ be a countable nondecreasing collection of compact set of $\mathcal{Z}$ such that $\bigcup_{n \in \mathbb{N}} \mathcal{G}^n = \mathcal{Z}$. We define recursively a sequence $\theta_n \in \mathcal{Z}$ and $p_n \in \mathbb{N}$ as follows. Let $\tilde{\theta}_0 \in G^0$.

$$\begin{aligned} & p_1 = 0 \text{ and } \theta_0 = \tilde{\theta}_0. \\ & \hat{\theta}_n = F_{n-1}(\theta_{n-1}). \\ & (\theta_n; p_{n+1}) = \begin{cases} (\hat{\theta}_n; p_n) & \text{if } \hat{\theta}_n \in G^{p_n} \\ (\tilde{\theta}_0; p_n + 1) & \text{otherwise.} \end{cases} \end{aligned} \tag{4}$$

Each time the current estimate $\hat{\theta}_n$ is outside the current truncation set $\mathcal{G}^{p_n}$, the algorithm is re-initialized at some point $\tilde{\theta}_0 \in \mathcal{G}^0$ and the truncation set is increased. This is a random truncation approach (in the sense that the truncation set is selected in an increasing family of compact sets using a non-decreasing integer-valued random process), similar to what has been suggested by Chen (1988,1997) for the SA procedure. Note that the point $\tilde{\theta}_0$ can be chosen at random in $\mathcal{G}^0$, but this does not affect the behaviour of the algorithm.

Proposition 6. *Let $\{F_n\} : \mathcal{Z} \to \mathcal{Z}$ be a sequence of applications. Suppose*

- **(C1)** *there exist a continuous function $W : \mathcal{Z} \to \mathbb{R}$ and a function $T : \mathcal{Z} \to \mathcal{Z}$ such that (B1) holds,*
- **(C2-a)** $|W \circ F_n(\tilde{\theta}_0) - W \circ T(\tilde{\theta}_0)| \longrightarrow 0$ *as* $n \longrightarrow \infty$,
- **(C2-b)** *for any compact set $\mathcal{C} \subset \mathcal{Z}$*

$$\sup_p |W \circ F_m(F^p_{m-1}(\tilde{\theta}_0)) - W \circ T(F^p_{m-1}(\tilde{\theta}_0))| \mathbb{1}_{F^p_{m-1}(\tilde{\theta}_0) \in \mathcal{C}} \longrightarrow 0 \text{ as } m \longrightarrow \infty$$

where $F^n_{n+k}(\theta) \triangleq F_{n+k} \circ F_{n+k-1} \circ \cdots \circ F_n(\theta)$ *and* $F^n_{n-1-k}(\theta) = \theta$, $k \geq 0$.

Then $\{\theta_n\}$ defined by the algorithm (4) remains in a compact set $\mathcal{K}$.

4 The MCEM algorithm for an exponential family

We will illustrate the convergence results obtained above in situations where the complete data model is from a parametric curved exponential family. This includes many situations where the EM algorithm has been successfully applied (the E-step reduces to compute the expectation of a data-sufficient statistics). Recall that we say that $f(z;\theta)$, for any $\theta \in \Theta$, is from a curved exponential family when

$$\log f(z;\theta) = \phi(\theta) + < S(z), \psi(\theta) >$$

where $z \in \mathbb{R}^d$, ϕ (resp. ψ) is a real (resp. $\mathbb{R}^q$-valued) function on Θ, and does not depend on z, and S is a μ-integrable $\mathbb{R}^q$-valued function on $\mathbb{R}^d$ which does not depend on θ. As emphasized in the introduction, the EM algorithm is widely used for this type of model. Provided that, for all $s \in S(\mathbb{R}^d)$ the function $L(s;\theta) \triangleq \phi(\theta) + < s, \psi(\theta) >$ has a single maximum $\hat{\theta}(s)$ and that $\hat{\theta}(s)$ is differentiable on the interior of $S(\mathbb{R}^d)$ the EM mapping may be explicitly written as

$$\theta_{n+1} \triangleq T(\theta_n) = \hat{\theta}(\bar{S}(\theta_n)), \quad \bar{S}(\theta_n) \triangleq \int_{\mathbb{R}^d} S(z) p(z;\theta_n) \mu(dz).$$

Under appropriate regularity assumptions it holds that: **(i)** $\{\theta \in \Theta : \theta = T(\theta)\} = \{\theta \in \Theta : \nabla_\theta g(\theta) = 0\}$ **(ii)** g is a Lyapunov function for T.

Assume now that the E-step cannot be performed in closed-form, and that we use a Metropolis-Hastings method to simulate the missing data. For each $\theta \in \Theta$, let $\Omega_\theta \subset \mathbb{R}^d$ be a Borel measurable state space such that $p(z;\theta)$ is positive in Ω_θ. Assume that $p^0(z;\theta)$ (the density distribution of the initial state) and $q(z'|z;\theta)$ (the proposal distribution) are positive in Ω_θ or a subset of Ω_θ. To generate a sample from $p(z;\theta_i)$, we generate an initial state $Z_{0,i}$ from the density $p^0(z;\theta_i)$ (conditionally independently from the past), and, for $0 \le j \le m_{i+1} - 1$ we **(i)** generate an alternative state y from the density $q(y|Z_{j,i};\theta_i)$, **(ii)** calculate

$$\alpha_{\theta_i}(Z_{j,i}, y) = \min\{1, \frac{p(y;\theta_i)\; q(Z_{j,i}|y;\theta_i)}{p(Z_{j,i};\theta_i)\; q(y|Z_{j,i};\theta_i)}\}$$

and **(iii)** set $Z_{j+1,i} = y$ with probability $\alpha_{\theta_i}(Z_{j,i}, y)$ and $Z_{j+1,i} = Z_{j,i}$ otherwise. Under weak assumptions, such MH chain may be shown to be geometrically ergodic (see, for example, Roberts & Tweedie, 1996). Denote

$$\mathcal{F}_n \triangleq \sigma(Z_{j,i}; i \le n-1, 1 \le j \le m_{i+1}).$$

Then $\{\theta_n\}$ is $\mathcal{F}_n$-adapted. Define

$$M_{i+1} \triangleq \frac{1}{m_{i+1}} \sum_{j=1}^{m_{i+1}} \left(S(Z_{j,i}) - \int_{\mathbb{R}^d} S(z) p^j(z;\theta_i)\mu(dz) \right)$$

where $p^j(z;\theta_i)$ is the probability distribution of $Z_{j,i}$ (i.e. the image of the initial distribution $p^0(z;\theta_i)$ after j iterations of the chain). To use the convergence results developed above, we need to check that M_i and $\bar{S}(\theta_i) - m_{i+1}^{-1} \sum_{j=1}^{m_{i+1}} \int_{\mathbb{R}^d} S(z)p^j(z;\theta_i)\mu(dz)$ converge to zero w.p. 1. The proof of these two properties relies on **(i)** Rosenthal's inequality for martingales and on **(ii)** the geometric convergence (in some appropriate norm) of $p^j(z;\theta_i)$ to $p(z;\theta_i)$. These two steps of course deeply rely on recent results on the MH algorithms.

References

Chen, H.F., Guo, L. & Gao, A.J. (1988). Convergence and robustness of the Robbins-Monro algorithm truncated at randomly varying bounds. *Stoch. Proces. Applic.*, **27**, 217-231.

Dempster, A., Laird, N. & Rubin, D. (1977). Maximum-likelihood from incomplete data via the EM algorithm. *Journal of the Royal Statistical Society B*, **39**, 1-38.

Hale (1987). *Asymptotic behavior of dissipative systems.* Providence: American Mathematical Society.

Haraux, A., (1991).*Systemes dynamiques dissipatifs et applications.* Paris: Masson.

Kushner, H. J. & Yin, G. G. (1997). *Stochastic approximation algorithms and applications.* New York: Springer Verlag.

Meyn, S.P. & Tweedie, R.L. (1993). *Markov chains and stochastic stability.* New York: Springer Verlag.

Robert, C., (1996). *Methodes de Monte-Carlo par chaines de Markov.* Economica.

Roberts, G. O. & Tweedie, R. L. (1996). Geometric convergence and central limit theorems for multidimensional Metropolis-Hasting algorithms. *Biometrika*, **83**, 95-100.

Shapiro, A. & Wardi, Y. (1996). Convergence analysis of stochastic algorithms. *Mathematics of operations research*, **21**, 615-628.

Wei, G. C. G. & Tanner, M. A. (1991). A Monte Carlo implementation of the EM algorithm and the poor's man data augmentation algorithms. *J. Amer. Statist. Assoc.* , **85**, 699-704.

Parameter Estimators for Gaussian Models with Censored Time Series and Spatio-temporal Data

C.A. Glasbey[1], I.M. Nevison[1] & A.G.M. Hunter[2]

[1] Biomathematics and Statistics Scotland, JCMB, King's Buildings, Edinburgh EH9 3JZ, Scotland

[2] Environmental Division, SAC, Bush Estate, Penicuik EH26 0PH, Scotland

Abstract. Computationally-fast algorithms are considered for estimating parameters in Gaussian time series and spatio-temporal models from censored and/or missing data. The problem arises in fitting models involving Gaussian latent variables to environmental data. Spectral estimators and least-squares fits of auto- and cross-covariances are found to be of similar efficiency for fitting models to rainfall and solar radiation data.

Keywords. Fourier transform, latent variable, multivariate time series, rainfall, solar radiation

1 Introduction

Environmental variables such as temperature can be modelled as Gaussian processes, whereas others, such as rainfall and solar radiation, are far from Gaussian but can possibly be transformed to normality. See Jones & Phelps (1996) for a review of weather models. In particular, many models have been proposed for rainfall, based either on point processes (Rodriguez-Iturbe *et al.*, 1988), or constructed in two stages: first a binary rain/no-rain process and then a rainfall distribution applied to the wet periods (Katz & Parlange, 1995). However, such models are far more difficult than Gaussian ones to study analytically, to combine with models of other environmental variables, or to make use of in forecasting. Glasbey & Nevison (1997) developed an alternative approach: they applied a monotonic transformation to rainfall data to achieve marginal normality. This defines a latent Gaussian variable, with zero rainfall corresponding to censored values below a threshold. A similar approach has been taken with solar radiation (Graham *et al.*, 1996): if observed values are divided by the elevation of the sun at that position in space and time (Page, 1986), the resulting variable is approximately a stationary Gaussian process. However, data sets are incomplete because the latent variable is unobservable during the night.

Missing and/or censored data are problematic in both time series and spatio-temporal modelling. Kleiner *et al.* (1979) estimated spectra from time series containing outliers; Jones (1980) used Kalman filters to fit autoregressive-moving average models to data with missing values; Kedem (1980), for computational speed, considered parameter estimation from binary time series obtained by a hard-limiting transformation. Glasbey & Nevison (1997) and Graham *et al.* (1996) both used an ad hoc procedure to estimate parameters in the latent Gaussian process, by minimising the sum of squares

$$L_C = \sum_{j=1}^{m} \sum_{k=1}^{m} \sum_{t=0}^{n'/2} \left(\hat{C}_{jkt} - C_{jkt} \right)^2 .$$

Here C_{jkt} and $\hat{C}_{jkt}$ are, respectively, the expected and sample cross-covariances between series j and k at time lag t, and there are m series of length $n \geq n'$. In this paper, for $n \gg m$ we consider alternative, computationally-fast estimators and the optimal choice of n'. A spectral approach is developed for multivariate, stationary processes in Section 2, and applied to rainfall data in Section 3 and solar radiation data in Section 4. Finally, conclusions are drawn in Section 5.

2 Spectral likelihood

The negative log-likelihood of a multivariate, stationary, Gaussian time series can be approximated by its spectral representation as a set of independent complex Wishart distributions,

$$L_S = \frac{1}{2} \sum_{l=-n/2}^{n/2-1} \left\{ \log |S_l| + \text{trace}\left(S_l^{-1} \hat{S}_l \right) \right\}$$

(Brillinger, 1974, p 238). Here S_l and $\hat{S}_l$ are, respectively, the $m \times m$ complex matrices of cross-spectral and cross-periodogram coefficients at frequency l/n, so that

$$\hat{S}_{jkl} = \frac{1}{n} \left(\sum_{t=1}^{n} y_{j_t} e^{-2\pi i l t/n} \right) \left(\sum_{t=1}^{n} y_{k_t} e^{+2\pi i l t/n} \right) = \sum_{t=-n/2}^{n/2-1} \hat{C}_{jkt} e^{-2\pi i l t/n},$$

where y_j and y_k are the jth and kth time series, and n is assumed to be even. The approximation is exact if covariances are circulant, i.e. $C_{jkt} = C_{jk(t-n)}$, and otherwise applies asymptotically as $n \to \infty$. In particular, for a bivariate series, at each frequency

$$S = \begin{bmatrix} S_1 & S_c + iS_q \\ S_c - iS_q & S_2 \end{bmatrix},$$

where S_1 and S_2 are the spectra of the two series, S_c is the co-spectrum and S_q is the quotient spectrum, and $L_S =$

$$\frac{1}{2} \sum_{l=-n/2}^{n/2-1} \left\{ \log \left(S_{1l} S_{2l} - S_{cl}^2 - S_{ql}^2 \right) + \frac{S_{1l}\hat{S}_{2l} + S_{2l}\hat{S}_{1l} - 2S_{cl}\hat{S}_{cl} - 2S_{ql}\hat{S}_{ql}}{S_{1l}S_{2l} - S_{cl}^2 - S_{ql}^2} \right\}.$$

To illustrate the use of L_S and L_C, 100 independent series of length 1000 were simulated from AR(1) and ARMA(1,1) processes with $\phi = 0.8$ and $(\phi, \theta) = (0.8, 0.5)$, respectively. For estimation using L_S, a value of $n' < n$ will suffice, with $\hat{S}$ replaced by

$$\hat{S}'_{jkl} = \sum_{t=-n'/2}^{n'/2-1} \hat{C}_{jkt} e^{-2\pi i l t/n'},$$

and similarly for S. It is only necessary for n' to be large enough so that autocorrelation coefficients $C_t \approx 0$ for $t > n'/2$. Table 1 shows the root-mean-square errors of parameter estimators obtained by minimising L_S and L_C for a range of values of n', using NAG routine E04JAF (NAG, 1993), a quasi-Newton algorithm which permits bounds on the parameters. The smallest values in each column are displayed in bold. For both models, with these parameter values $n' \geq 100$ is sufficient for estimators based on L_S to be fully efficient, because $\phi^{50} \approx 10^{-5}$, and therefore several root-mean-square

Table 1. 1000× root-mean-square errors of parameter estimators

model =	AR(1)		ARMA(1,1)			
parameter =	ϕ		ϕ		θ	
criterion =	L_S	L_C	L_S	L_C	L_S	L_C
$n' = 2$	**20**	**20**				
4	29	20	72	72	115	115
6	23	21	50	47	74	72
10	22	23	42	**40**	60	**58**
20	21	26	41	44	58	66
50	20	29	39	52	56	87
100	**20**	30	**39**	55	**56**	96
200	**20**	30	**39**	56	**56**	98
500	**20**	30	**39**	56	**56**	98
1000	**20**	30	**39**	56	**56**	98

errors are displayed in bold. As is well known, $\hat{C}_1/\hat{C}_0$ is an efficient estimator of ϕ in an AR(1) process, so $n' = 2$ is the optimal choice in L_C. For the ARMA(1,1) process, no choice of n' leads to fully efficient estimator using L_C, but $n' = 10$ is almost efficient for these values of the parameters.

3 Rainfall application

The data analysed by Glasbey & Nevison (1997) were a univariate time series of ten years of hourly rainfall data ($n = 87600$) at Turnhouse, Edinburgh. A monotonic transformation converted them to zero mean, unit variance, Gaussian variables, except that zero rainfall corresponded to censored values below a threshold. It is, therefore, not possible to compute $\hat{C}$ directly. We have considered two alternatives.

1. The faster method is to compute the sample autocorrelations of the observed data using Fourier methods, then apply a transformation which relates expected correlations of the data to expected correlations of the latent variable.
2. Alternatively, for each time lag t, we use the EM algorithm to obtain a maximum likelihood estimate for $\hat{C}_t$, by alternating between computing the *expected* correlation, conditional on the censored data, using standard bivariate Gaussian distributional theory (Johnson & Kotz, 1972), and *maximising* the likelihood by equating the correlation coefficient with its sample value.

In both cases, $\hat{S}'$ is then obtained by Fourier transforming $\hat{C}$. Note, L_S is a pseudo-likelihood rather than a log-likelihood, because $\hat{C}$ is not a set of sample correlation coefficients. Also, because the variance is known, we are using correlations rather than covariances, but the methodology in Section 2 applies equally to this situation. A third option would have been to use Markov chain Monte Carlo methods (Gilks *et al.*, 1996), by alternating between using a Gibbs sampler to simulate censored values and sampling parameter values from L_S. However, this would have been very computationally intensive and in this paper we are restricting ourselves to fast methods.

The EM-algorithm, in conjunction with minimising L_C for $n' = 960$, was used to fit to the data an ARMA(2,1) model parametrised as

$$C_t = \alpha\lambda_1^{|t|} + (1-\alpha)\lambda_2^{|t|},$$

with $1 \geq \alpha, \lambda_1, \lambda_2 \geq 0$. Values obtained using NAG routine E04JAF were $\hat{\alpha} = 0.83$, $\hat{\lambda} = (0.787, 0.979)$. The efficiency of this and alternative estimators,

Table 2. 1000× root-mean-square errors of parameter estimators in rainfall model

parameter =	α				λ_1				λ_2			
$\hat{C}$ =	transform		EM		transform		EM		transform		EM	
criterion =	L_S	L_C	L_S	L_C	L_S	L_C	L_S	L_C	L_S	L_C	L_S	L_C
$n' = 6$	526	486	474	411	362	354	334	296	107	108	93	84
12	453	465	309	340	249	239	84	97	92	90	62	67
24	293	270	165	153	123	99	29	33	62	52	41	35
48	215	121	65	70	101	33	12	17	47	26	19	18
96	158	71	45	37	50	24	9	**12**	29	15	10	8
240	127	55	32	**29**	46	21	7	12	20	**11**	6	**5**
480	**126**	**54**	**31**	31	**46**	**21**	**7**	13	**20**	11	**6**	5
960	**126**	60	**31**	32	**46**	24	**7**	13	**20**	12	**6**	6

and of different values of n', were compared by simulating 100 independent series with these values of the parameters and the same level of censoring, and then re-estimating the parameters. Results are summarised in Table 2, again by root-mean-square errors and with the smallest values in each column displayed in bold. We see that it is better to obtain $\hat{C}$ using the EM-algorithm than by transformation, in which case there is little to choose between L_S and L_C as criteria, provided we know the appropriate value for n'. For L_C a value of n' around 240 appears to be best, while for L_S it is sufficient for $n' \geq 240$.

4 Solar radiation application

Graham *et al.* (1996) analysed solar radiation data which had been recorded every 30 seconds between 8am and 4pm for 27 months at pairs of sites in Edinburgh. The sites were changed each month, and 12 different sites were used in total. It was found that dividing each observed radiation value by the elevation of the sun at that position in space and time was effective in removing temporal trends in both the mean and variance of solar radiation, provided times were restricted to those for which the solar angle exceeded 0.05 radians. Covariances were found to be well modelled by

$$C_{jkt} = \sigma^2 \phi^{\sqrt{D_{jk}^2 + \delta^2 (t + \kappa E_{jk})^2}},$$

where D_{jk} is the distance between sites j and k and E_{jk} is the distance site j is to the east of site k. Therefore, correlations between observations decay exponentially with increasing temporal and/or spatial separation, and in addition there is a time delay with more easterly sites experiencing fluctuations in radiation later.

The model was fitted separately to each of the 27 months of data, by minimising L_C with $n' = 48$ and $\hat{C}$ obtained by computing the sample autocovariance separately for each day and then averaging over the month. For this problem, full maximum likelihood estimation would have been possible, for example by approximating the series by bivariate autoregressive processes of high order (Jones & Vecchia, 1993), but this would have been computationally expensive. Average values obtained for parameters were $\hat{\phi} = 0.95$, $\hat{\delta} = 0.36$, $\hat{\kappa} = 0.95$ and $\hat{\sigma}^2 = 0.37$. Again, efficiencies of alternative estimators, and of different values of n', were compared by simulating 100 independent series with these values of the parameters and then re-estimating the parameters, subject to the bounds:

$$1 \geq \phi \geq 0, \quad 10 \geq \delta \geq 0, \quad 10 \geq \kappa \geq -10, \quad 100 \geq \sigma^2 \geq 0.$$

Table 3. Root-mean-square errors of parameter estimators in solar radiation model

par. =	$\phi \times 10^4$				$\delta \times 10^3$				$\kappa \times 10^2$				$\sigma^2 \times 10^3$			
series =	complete		8 h/day		complete		8 h/day		complete		8 h/day		complete		8 h/day	
crit. =	L_S	L_C	L_S	L_C	L_S	L_C	L_S	L_C	L_S	L_C	L_S	L_C	L_S	L_C	L_S	L_C
$n' =$ 6	1229	**33**	1158	**49**	722	10	699	18	77	17	80	32	337	21	337	31
12	858	33	886	49	628	10	669	**18**	89	17	88	31	353	21	351	31
24	1082	33	1080	49	726	**9**	732	18	78	16	77	30	358	21	357	**31**
48	1060	34	1073	51	743	10	743	19	53	**15**	53	**28**	355	21	358	31
96	977	36	953	53	698	11	716	23	134	16	156	31	364	**21**	354	31
240	317	37	578	56	507	14	475	32	345	23	470	43	307	21	302	33
480	1579	41	2130	62	139	18	224	38	517	27	622	52	243	22	258	36
960	564	54	630	93	137	19	163	76	484	28	597	56	104	23	156	44
2000	211	59	257	140	92	19	105	1777	314	28	397	56	187	23	409	49
4000	**31**	59	**46**	140	**8**	19	**15**	1777	**14**	28	**26**	56	**21**	23	**31**	49
10000	**31**	59	**46**	140	**8**	19	**15**	1780	**14**	28	**26**	56	**21**	23	**31**	49

A single month was simulated ($n = 86400$), with D and E set to typical values of 6 km and 4 km respectively, using a high-order bivariate autoregressive approximation. For larger values of n' a problem was encountered in that little or no data were available for $\hat{C}_t$ when $2160 \geq t \geq 720$ or $5040 \geq t \geq 3600$ (in units of 30 seconds), so these terms were omitted from L_C. It is not so straightforward with L_S, and three approaches were tried:

1. Shorten $\hat{C}$ by omitting the missing terms before applying the Fourier transform to obtain $\hat{S}'$, and do the same to C before obtaining S';
2. Set the missing terms in $\hat{C}$ to zero before obtaining $\hat{S}'$, and do the same to C before obtaining S';
3. Set the missing terms in $\hat{C}$ to the corresponding terms in C for current values of the model parameters, and then apply the Fourier transform to obtain $\hat{S}'$.

The final approach produced by far the best results, and these are the ones given in Table 3. For comparison, results are also given for the hypothetical case where the complete time series is observed. In both cases, for L_C a value of $n' \leq 48$ was found to be satisfactory, but L_S was marginally better provided that $n' \geq 4000$. However, for smaller values of n', L_S performed very poorly and many instances occurred where parameter estimates were at the limits of their ranges.

5 Discussion

Computationally-fast algorithms have been considered for estimating parameters in Gaussian time series and spatio-temporal models from censored data. Spectral estimators and least-squares fits of auto- and cross-covariances have been found to be of similar efficiency for fitting models to rainfall and solar radiation data. The advantages of the spectral approach are that it is slightly more efficient, it has better theoretical properties, such as being known to be fully efficient if data are not missing, the variance of the process is automatically constrained to be positive definite, and there is no problem in choosing an appropriate value for n', it simply has to be large. On the other hand, the least-squares approach is less sensitive to choices to be made between alternative ways of obtaining $\hat{C}$, is computationally faster because small values of n' are usually adequate, and is easier to generalise to larger numbers of series and irregular sampling schemes. The spectral method only gains in computational efficiency if spatial data are collected on a rectangular grid. Finally, the least-squares criterion is possibly more robust to distributional assumptions, as with variograms (Cressie, 1991, pp 90-99), and its efficiency can be improved by extending to weighted least squares and generalised least squares criteria.

Acknowledgements
The work was supported by funds from the Scottish Office Agriculture, Environment and Fisheries Department. The solar radiation data were collected under CEC Contract No. JOU2-CT92-0018.

References

Brillinger, D.R. (1974). *Time series : Data Analysis and Theory.* Holt, Rinehart and Winston: New York.

Cressie, N.A.C. (1991). *Statistics for Spatial Data.* Wiley: New York.

(ed.) Gilks, W.R., Richardson, S. & Spiegelhalter, D.J. (1996). *Markov Chain Monte Carlo in Practice.* Chapman and Hall: London.

Glasbey, C.A. & Nevison, I.M. (1997). Rainfall modelling using a latent Gaussian variable. In *Modelling Longitudinal and Spatially Correlated Data: Methods, Applications, and Future Directions* (T.G. Gregoire *et al.*, eds.). Lecture Notes in Statistics **122**, Springer: New York, 233-242.

Graham, R., Glasbey, C.A. & Hunter, A.G.M. (1996). Consequences of decentralised PV on local network management. Final report: variation of solar energy across a region: spatio-temporal models. *SAC Report*, Bush Estate, Penicuik EH26 0PH, Scotland.

Johnson, N.L. & Kotz, S. (1972). *Distributions in Statistics : Continuous Multivariate Distributions.* Wiley: New York.

Jones, J.E. & Phelps, K. (1996). A review of meteorological data and weather generators for practical use in agricultural and horticultural modelling. *Aspects of Applied Biology*, **46**, 5-12.

Jones, R.H. (1980). Maximum likelihood fitting of ARMA models to time series with missing observations. *Technometrics*, **22**, 389-395.

Jones, R.H. & Vecchia, A.V. (1993). Fitting continuous ARMA models to unequally spaced spatial data. *Journal of the American Statistical Association*, **88**, 947-954.

Katz, R.W. & Parlange, M.B. (1995). Generalizations of chain-dependent processes: applications to hourly precipitation. *Water Resources Research*, **31**, 1331-1341.

Kedem, B. (1980). *Binary Time Series.* Dekker: New York.

Kleiner, B. Martin, R.D. & Thomson, D.J. (1979). Robust estimation of power spectra (with discussion). *Journal of the Royal Statistical Society, Series B*, **41**, 313-351.

Numerical Algorithms Group (1993). *Library Manual Mark 16.* NAG Central Office, 256 Banbury Road, Oxford OX2 7DE, UK.

(ed.) Page, J.K. (1986). *Prediction of Solar Radiation on Inclined Surfaces.* Solar Energy R & D in the European Community: Series F, Volume 3 – Solar Radiation Data. D. Reidel Publishing Company: Dordrecht.

Rodriguez-Iturbe, I., Cox, D.R. & Isham, V. (1988). A point process model for rainfall: further developments. *Proceedings of the Royal Society, London, Series A*, **417**, 283-298.

Assessing the Multimodality of a Multivariate Distribution Using Nonparametric Techniques

S. Hahn[1] and P. J. Foster[2]

[1] Division of Statistics and Operational Research, Department of Mathematical Sciences, University of Liverpool, Liverpool, L69 3BX, UK

[2] Statistical Laboratory, Department of Mathematics, University of Manchester, Manchester, M13 9PL, UK

Keywords. Cluster validation, dip test, modes, single linkage clustering, smoothed bootstrap

1 Testing unimodality

1.1 The dip test

In the univariate setting a distribution function (d.f.) F is unimodal with mode m if F is convex in $(-\infty, m]$ and concave in $[m, \infty)$. Hartigan & Hartigan (1985) proposed the DIP statistic for testing whether a distribution is unimodal against a general multimodal alternative. The dip of a d.f. F is defined to be the maximum difference between F and the unimodal distribution function that minimises that maximum difference. i.e. the dip of a distribution function F is:

$$DIP(F) = \inf_{G \in \Lambda} \sup_x \mid F(x) - G(x) \mid$$

where Λ is the class of all distributions with unimodal density functions.

In practice we do not know F but may estimate it from a sample of data $X_1, \ldots, X_n$ by the empirical distribution function F_n. Hartigan & Hartigan (1985) show that since $\sup_x \mid F_n(x) - F(x) \mid \stackrel{a.s.}{\to} 0$ $DIP(F_n) \stackrel{a.s.}{\to} DIP(F)$. Therefore, a test based on $DIP(F_n)$ will asymptotically distinguish between the distribution F being unimodal and F being multimodal. The authors proposed an algorithm for computing the test statistic (essentially based on finding the best fitting unimodal distribution to the data) which is implemented in a Fortran subroutine by Hartigan (1985). They show that the uniform distribution is the distribution which asymptotically maximises the value of the dip statistic and so they empirically find a range of critical values to give a conservative test by simulating from the uniform.

The test as it stands cannot be easily generalised to a higher ($p > 1$) dimensional setting where we have n observations on the random vector $\mathbf{X} = (X_1, \ldots, X_p)^T$. One approach though is to project the data onto linear subspaces defined by linear combinations of the original variables and calculate the dip for each of the projected univariate distributions. We could seek to find the greatest dip over all linear combinations of the original variables but this results in an intractable optimisation problem. Projecting the data onto the principal components (PC's) presents a viable alternative. The set of p PC's are a set of p orthogonal linear transformations $Y_j = \mathbf{a}_j^T \mathbf{X}$ $(j = 1, \ldots, p)$ which define lines in p-space such that when the data is projected onto them the resulting covariance matrix is diagonal with $Var(Y_1) \geq, \ldots, \geq Var(Y_p)$

i.e. the Y_j's are uncorrelated and explain decreasing proportions of the total variation in the data. The PC's also have the geometrical property that Y_1 is the best fitting line in that the sum of squared perpendicular distances from all the data points to it is a minimum, Y_2 is the line, orthogonal to Y_1, which has the second best fit and so on. Jolliffe (1986) gives a detailed discussion of the derivation of PC's and their ensuing properties.

When multivariate data is projected onto a line then any structure seen in the projection is a shadow of an actual (usually sharper) structure seen in the full dimensionality. Hence, the above described properties of PC's should mean that if we separately estimate the densities of the scores on each PC (or indeed in planes defined by pairs of PC's) then this should be useful in investigating multimodality in p dimensions. More formallly, we can apply the dip test in turn to each of the p sets of PC scores.

As an example we consider the Swiss banknote data originally presented in Flury & Riedwyl (1988). This is a bivariate data set of size $n = 200$ where X_1 is the width of the bottom margin of a bank note (mm) and X_2 is the image diagonal length (mm). The data set actually consists of 100 notes known to be real and 100 notes known to be forged. Projecting the data onto the two PC's (based on the covariance matrix) shows the distribution for Y_1 to be bimodal while that for Y_2 is predominantly unimodal. The dip test was applied to the two sets of PC scores and the results are presented in Table 1.

Table 1. Values of the DIP statistic for the PC scores of the Swiss banknote data.

PC	DIP	p-value
Y_1	0.0522	0.0015
Y_2	0.0284	0.3234

We can see that unimodality is clearly rejected for Y_1 which explains 82% of the total variation while the relatively large p-value of the test for Y_2 provides no significant evidence aganst unimodality. These results therefore lead us to reject unimodality for the bivariate distribution.

1.2 Testing on trial modes

The *minimum spanning tree (MST)* for a set of data points in p dimensions is a set of lines drawn between pairs of points satisfying the following conditions:

(i) Each data point is connected to every other point by a sequence of at least one line.
(ii) There are no closed loops.
(iii) The sum of the Euclidean distances between points is a minimum.

The MST is a convenient method for highlighting close neighbours in a sample and an algorithm by Gower & Ross (1969) can be used for computational purposes.

Before describing the test statistic we need the following definition. If we have a d.f. F with a mode at the point m then we define the *least concave majorant (l.c.m)* of F in $[m, \infty)$ to be the inf $L(x)$ for $x \geq m$ and where the inf is taken over all functions L that are concave in $[m, \infty)$ and nowhere less than F.

A test of unimodality in a multivariate setting (originally briefly suggested by Hartigan & Hartigan, 1985) can be formulated by choosing one of the data points $\mathbf{x}_1, \ldots, \mathbf{x}_n$ as a trial mode which we can call $\mathbf{z}_0$. The closest data point to $\mathbf{z}_0$, denoted $\mathbf{z}_1$, is then determined from the MST and we let d_1 denote this distance. The next stage is to determine the closest point to either $\mathbf{z}_0$ or $\mathbf{z}_1$, denoted $\mathbf{z}_2$, with distance d_2 and then so on to find $\mathbf{z}_3, \mathbf{z}_4, \ldots, \mathbf{z}_{n-1}$ with distances $d_3, d_4, \ldots, d_{n-1}$. Now, let $y_k = \sum_{i=1}^{k} d_i$ for $k = 1, \ldots, n-1$ and H_{n-1} be the empirical distribution function of the y_k's. Then a suitable test statistic is given by:

$$INFD(\mathbf{z}_0) = \inf_{\mathbf{z}_0} \sup_{y} \mid H_{n-1}(y) - L(y) \mid$$

where $L(y)$ is the l.c.m of H_{n-1}. i.e. for a given trial mode we find the largest difference between H_{n-1} and L and then we find the minimum of these values over all the possible choices of trial mode (the data points). If a particular $\mathbf{z}_0$ is the unique mode then the $d_j's$ will be roughly increasing and H_{n-1} will be concave and close to L for all values of y. Therefore, large values of the test statistic will indicate departures from unimodality. This statistic has been calculated for the Swiss banknote data where we have $INFD = 0.0837$ but it remains to decide whether this value is significantly large.

The p-value of the observed test statistic, under the null hypothesis of a unimodal distribution, was determined empirically by repeatedly simulating samples of size 200 from the fitted l.c.m. The l.c.m curve is an estimated distribution of the y_k's fitted in such a way as to force the curve to have as close an agreement as possible with the unimodal shape of the original data. In the case here the p-value was found to be 0.04 leading to the rejection of unimodality which qualitatively agrees with the conclusion based on the PC analysis.

2 Cluster validation

2.1 Introduction

If the null hypothesis of unimodality is rejected then the next problem of interest is to determine how many modes the underlying density function has. In the univariate setting, Silverman (1981) based a suitable test on counting the number of modes in a kernel estimate of the density and assessed significance using the ideas of 'critical smoothing' and bootstrap calculations. This work was further explored and modified by Mammen, Marron & Fisher (1992) and Fisher, Mammen & Marron (1994). However, this approach is not readily extended to the multivariate case, not least because of problems in determining the values of the critical smoothing parameters in each dimension and also in locating the sample modes in the density estinate. We therefore propose to look at this problem using methodology which is based on cluster analysis.

If the density is multimodal then a random sample of obervations from the distribution will consist of a number of clusters whose location will tend to correspond to the location of these modes. How distinct the clusters are from one another will depend on how well separated the modes of the distribution are. A large number of clustering algorithms have been proposed in the literature for dividing a sample into separate clusters without making any priori assumptions about the number, form or even existence of any such groups. The algorithm to be used is the hierarchical single-linkage procedure which is said to be set-consistent for high density clusters (Hartigan, 1977). This means that asymptotically, the sets of enlarging hierarchical clusters of data points it constructs from the sample data are groups of points lying within successively lower density contours in the underlying distribution. Another important property of a set-consistent procedure is that it does not impose any geometrical structure on the clusters it produces. Hence, the use of the single-linkage method should help in the identification of groups corresponding to underlying modes. In this section we look at methods for trying to determine statistically the value of k for the number of clusters which provides the 'best' representation of the data. This validation will be done by first defining an appropiate statistic which measures the strength of a particular partition into k clusters and then assessing the stability of the k-cluster solution using a smoothed bootstrap approach.

2.2 Single-linkage and cluster validation

The single-linkage algorithm is an agglomerative procedure where the distance between two separate clusters is defined to be the minimum of all the pairwise distances between each member of one cluster and all the members of the second. While it has the desirable set consistency property described above, with finite samples there tends to be a chaining effect resulting in straggly clusters when there is at least one intermediate point between otherwise distinct groups. This is illustrated with a simulated bivariate dataset of size $n = 158$ where there are three clusters but the bottom two have a number of data points lying between them. Application of single-linkage results in a two cluster solution with the bottom two clusters fused together and a three cluster solution where the bottom two are still fused together and the third cluster consists of a single data point. One approach to overcoming this problem is to trim the dataset by removing, say 10%, of the observations at which an estimate of the underlying density is lowest. The product kernel density estimate which was used is defined by:

$$\hat{f}(\mathbf{x}) = \frac{1}{h_1, \ldots, h_p} \sum_{i=1}^{n} \left[\prod_{j=1}^{p} K\left(\frac{x_j - x_{ij}}{h_j}\right) \right]$$

where $\mathbf{x} = (x_1, \ldots, x_p)^T$, the kernel function K is a $N(0,1)$ density and the smoothing parameters h_1 and h_2 were set to c times the optimal values for Normally distributed data with $c = 0.5$ for the simulated data. Note that the formula for the Normal optimal h-values is:

$$\hat{h}_j = \left(\frac{4}{n(p+2)}\right)^{\frac{1}{p+4}} \hat{\sigma}_j$$

The density estimates $\hat{f}(x_1), \ldots, \hat{f}(x_{158})$ were ranked from largest to smallest and the bottom 16 values then trimmed from the dataset. For a detailed discussion of such a ranking see Bowman & Foster (1993). When single-linkage is then applied to the reduced data the three cluster solution accurately identifies the three groups.

In order to estimate the optimal number, k^*, of clusters, and hence modes, in the data we propose to use a method of cluster validation. A comprehensive review of recent work in this area is given by Gordon (1998). To measure the adequacy of a partition into k groups we will use the ratio of the sum of within-group distances to the sum of between-group distances, denoted by $G(k)$. For a given value of k, G will be small for an accurate partition and when $k = k^*$ the variability in G based on repeated samples from an estimate of the underlying distribution will be a minimum. Thus a measure of stability of a k-cluster solution is given by $\Delta G(k)$, the standardised length of a 68% confidence interval derived from the values of $G(k)$ in b bootstrap samples. (eg. $b = 100$). Using a 68% level interval removes the effect of small and large values on $\Delta G(k)$ and corresponds to an interval plus and minus one standard deviation from the mean for a Normal distribution. This is illustrated on the simulated data described above. If we choose $c = 0.5$ for trimming and $c = 0.3$ for the smoothed bootstrap resampling then we obtain the results given in Table 2. These indicate that we should choose $\hat{k}^* = 3$, a solution we know in fact to be correct.

Table 2. Values of the statistic $\Delta G(k)$ for the simulated and Swiss banknote data.

#modes, k	Simulated data $\Delta G(k)$	Swiss banknote data $\Delta G(k)$
2	1.090	2.025
3	0.107	2.203
4	0.144	2.191
5	0.302	2.204
6	0.321	2.356

As a second example, we again consider the Swiss banknote data analysed earlier. If we choose $c = 0.5$ for 10% trimming and $c = 1.0$ for the bootstrap resampling then we obtain the results in Table 2. The optimal value of k is estimated to be 2 and the two cluster solution after 10% trimming shows that the two clusters are associated correctly with the real and forged notes.

References

Bowman, A.W. & Foster, P.J. (1993). Density based exploration of bivariate data. *Statistics and Computing*, **3**, 171-177.

Fisher, N.I., Mammen, E. & Marron, J.S. (1994). Testing for multimodality. *Computational Statistics and Data Analysis*, **18**, 499-512.

Flury, B. & Riedwyl, H. (1988). *Multivariate Statistics: a practical approach.* London: Chapman and Hall.

Gordon, A.D. (1998). Cluster validation. In *Data Science, Classification and Related Methods* (ed. C. Hayashi, N. Ohsumi, K. Yajima, Y. Tanaka, H.H. Bock, & Y. Baba), 22-39. Tokyo: Springer-Verlag.

Gower, J.C. & Ross, G.J.S. (1969). Minimum spanning trees and single-linkage cluster analysis. *Applied Statistics*, **18**, 54-64.

Hartigan, J.A. (1987). Clusters as modes. In: *First International Symposium on Data Analysis and Informatics, Vol. 2*, IRIA, Versailles.

Hartigan, J.A. & Hartigan, P.M. (1985). The dip test of unimodality. *Ann. Statist.*, **13**, 70-84.

Hartigan, P.M. (1985). Algorithm AS 217. Computation of the dip statistic to test for unimodality. *Applied Statistics*, **34**, 320-325.

Jolliffe, I.T. (1986). *Principal Component Analysis.* New York: Springer-Verlag.

Mammen, E., Marron, J.S. & Fisher, N.I. (1992). Some asymptotics for multimodality tests based on kernel density estimates. *Probab. Theory Relat. Fields*, **91**, 115-132.

Silverman, B.W. (1981). Using kernel density estimation to investigate multimodality. *J. R. Statist. Soc. B*, **13**, 97-99.

Bayesian Signal Restoration and Model Determination for Ion Channels

M. E. A. Hodgson

Department of Mathematics, University of Bristol, UK

Abstract. A fully Bayesian method of ion channel analysis is developed and applied to simulated data. Our first model of channel kinetics is the alternating renewal process, with gamma distributed sojourn times, of Hodgson (1997). Having modelled the noise process masking the channel signal, we draw inference by generating a sample from the joint posterior distribution of all unknowns. The unknown dimensionality of the signal necessitates the use of Green's (1995) reversible jump Markov chain Monte Carlo method. Next we modify the methodology to cover selection between four simple hidden Markov models of channel kinetics (Hodgson & Green, 1998). Our sampler now includes reversible jump moves between these candidate models.

Keywords. Alternating renewal process, hidden Markov models, ion channels, reversible jump MCMC computation, signal restoration, simulated tempering, step functions

1 Introduction

Ion channels are large proteins spanning cell membranes which, in certain physicochemical states, conduct current in the form of selected ions. Understanding of these fundamental units of the central nervous system is at present poor, although neurophysiologists can measure the picoampere currents flowing through a single channel by the technique of patch clamp recording (Sakmann & Neher, 1995). An improved understanding of ion channel behaviour would aid the design of new drugs to act on the central nervous system.

Ball & Rice (1992) give an overview of the considerable statistical literature concerning ion channels. Their kinetics are most frequently modelled by a continuous-time Markov chain in which the state space is partitioned into classes of states having the same conductance, and since it is the current through the ion channel which is measured, states having the same conductance are indistinguishable. In theory patch clamp recording reveals which class the channel state belongs to, allowing inference about the postulated 'hidden' Markov chain to be based on this so-called 'aggregated' process. In practice, however, degradation of patch clamp records by additive noise and filtering prior to digitisation further complicates the statistician's task. We restrict attention to the usual case of just two conductance classes, termed 'open' and 'closed'.

In Section 2 we describe alternating renewal and hidden Markov models of channel kinetics and an autoregressive noise model. Section 3 outlines simulation of the resultant posterior distributions by Markov chain Monte Carlo (MCMC) methods, and Section 4 presents some aspects of the performance of our approach.

2 Modelling ion channel data

2.1 An alternating renewal model of channel kinetics

Let the indicator function of channel openness on the observation interval $[0, T]$ be denoted henceforth by $\mathbf{x}$. Hodgson (1997) treats $\mathbf{x}$ as the sample path of a steady-state alternating renewal process, with gamma distributed sojourn times:

$$\begin{aligned} \text{closed times} &\sim \Gamma(s_0, \lambda_0) \\ \text{open times} &\sim \Gamma(s_1, \lambda_1). \end{aligned}$$

He assigns independent $\text{Exp}(\xi)$ and $\Gamma(\alpha, \beta)$ priors to the 'shapes' (s_0, s_1) and 'scales' (λ_0, λ_1) respectively. Such alternating renewal models (Milne *et al.*, 1988) are atypical of ion channel models in that the aggregated process is modelled explicitly rather than implicitly through the unobservable stochastic process underlying it.

2.2 Discrimination between hidden Markov models

Hodgson & Green (1998) extend Hodgson's (1997) methodology to the more physically realistic hidden Markov models of channel dynamics. Colquhoun & Hawkes (1982) give a thorough exposition of the theoretical properties of these models. Attempting structural inference about the underlying Markov chain or inference about the transition rates for a given chain is made problematic by the loss of information in observing only the aggregated process (Fredkin & Rice, 1986). Hodgson & Green (1998) offer a pointer towards the feasibility of general Markov model selection for ion channels by attempting discrimination between a set of four simple models:

Model *11* $\quad \mathcal{C}_1 \underset{\lambda_o}{\overset{\lambda_c}{\rightleftarrows}} \mathcal{O}_1$

Model *21* $\quad \mathcal{C}_2 \underset{\mu_c}{\overset{\nu_c}{\rightleftarrows}} \mathcal{C}_1 \underset{\lambda_o}{\overset{\lambda_c}{\rightleftarrows}} \mathcal{O}_1$

Model *12* $\quad \mathcal{C}_1 \underset{\lambda_o}{\overset{\lambda_c}{\rightleftarrows}} \mathcal{O}_1 \underset{\nu_o}{\overset{\mu_o}{\rightleftarrows}} \mathcal{O}_2$

Model *22* $\quad \mathcal{C}_2 \underset{\mu_c}{\overset{\nu_c}{\rightleftarrows}} \mathcal{C}_1 \underset{\lambda_o}{\overset{\lambda_c}{\rightleftarrows}} \mathcal{O}_1 \underset{\nu_o}{\overset{\mu_o}{\rightleftarrows}} \mathcal{O}_2$

Each model is assigned prior probability 0.25. If there are two closed states, the transition rates governing closed times have independent gamma priors:

$$\nu_c,\ \mu_c,\ \lambda_c \sim \Gamma(\alpha, \beta/2).$$

If there is a single closed state, λ_c has the $\Gamma(\alpha, \beta)$ prior. Transition rates governing open times are assigned priors in the same manner. Closed and open sojourn times under all candidate models have easily derived densities (see Colquhoun & Hawkes, 1981), and as there is a single gateway between the classes, successive sojourn times are independent. When a class is comprised of a single state, the sojourn time in that class is of course exponential, and when a class has two states, the sojourn time is a positive mixture of two exponentials. Hence the sample path of the hidden Markov chain may be integrated out analytically, and only the aggregated process $\mathbf{x}$ need be included in the model.

2.3 The data set and the degradation model

In order to stimulate the development of new signal processing techniques (Eisenberg, 1994), physiologists Eisenberg and Levis have generated a time series of 10^5 data points representing the recorded current through a single ion channel with two conductance levels. Let $\mathbf{c}$ be the binary step function on $[0, T]$ representing the true current. Then denoting the currents through the channel in the open and closed states by μ_1 and μ_0 respectively, we have:

$$\mathbf{c}(t) = \begin{cases} \mu_1 \text{ if } \mathbf{x}(t) = 1 \\ \mu_0 \text{ otherwise} \end{cases}, \forall t \in [0, T].$$

The channel current, subject to additive dependent noise $\mathbf{z}$, is recorded at times $t \in \{0, 1, \ldots, T\}$, and then passed through a Gaussian linear filter F to produce the data $\mathbf{y}$:

$$\mathbf{y} = F(\mathbf{c} + \mathbf{z}) = F\mathbf{c} + F\mathbf{z}.$$

A zero-mean multivariate normal distribution $\text{MVN}(\mathbf{0}, \Sigma)$ is assigned to the digitised noise process $\mathbf{z}$. As Eisenberg and Levis' data are simulated and $\mathbf{z}$ is available, we can fit autoregressive processes of various orders m to the noise to determine Σ. Long spells of inactivity are identifiable in real ion channel recordings, and so knowledge of the noise process may be as good as we have assumed here and we justify this procedure on these grounds. Fitting $\text{AR}(m)$ processes affords a simple banded Toeplitz form for Σ, convenient for computational purposes. Independent normal priors are given to the conductance levels μ_0 and μ_1:

$$\mu_0,\ \mu_1 \sim \text{N}(\theta, \kappa^2).$$

We analyse a segment of Eisenberg and Levis' data of length 4096, previously examined using wavelet methods by Johnstone & Silverman (1997).

3 Computing the posterior

3.1 The alternating renewal case

The joint posterior distribution $p(s_0, s_1, \lambda_0, \lambda_1, \mathbf{x}, \mu_0, \mu_1 \mid \mathbf{y})$ from which we require to simulate exhibits variable dimensionality, since the number of discontinuities ('switches') of the indicator function $\mathbf{x}$ is unknown. Green's (1995) 'reversible jump' algorithm, a generalisation of the standard Metropolis–Hastings algorithm (Metropolis *et al.*, 1953; Hastings, 1970), provides a means of sampling from such distributions. A countable set of move types is used to traverse the state space. Some of these attempt to 'jump' between parameter subspaces of differing dimension and all attain detailed balance within themselves; hence the name for the algorithm. Green gives a recipe for dimension-changing moves which we follow in the design of moves updating $\mathbf{x}$. 'Type 2' moves, so called because they change the number of switches of $\mathbf{x}$ by two if accepted, attempt to create or delete a pair of consecutive switches (representing a channel sojourn). 'Birth' and 'death' are used to describe proposals to increase or decrease respectively the number of switches. In a 'type 2 birth' move, t is drawn uniformly on $[0, T]$. With probability 1, t will lie in the open interval (t_{j-1}, t_j) for some $j \in \{1, \ldots, s+1\}$. Given j, t' is generated uniformly on $[t_{j-1}, t_j]$ and the proposal is to 'give birth' to a new channel sojourn on $[\min(t, t'), \max(t, t')]$ within the existing sojourn on $[t_{j-1}, t_j]$, so creating an extra pair of switches. In the inverse 'type 2 death' move, s' is drawn uniformly from the set $\{1, \ldots, s-1\}$ and it is proposed to delete the switch pair $\{t_{s'}, t_{s'+1}\}$, thereby 'killing' the sojourn on the interval between them.

Green (1995) derives the acceptance probability satisfying the requirement of detailed balance within each move type. In the familiar Metropolis–Hastings acceptance probability $\min\{1, R\}$, R may be written as:

$$R = \text{prior ratio} \times \text{likelihood ratio} \times \text{proposal ratio.}$$

For moves in which there is a change of variable, such as the 'type 2' move, R above must be multiplied by its Jacobian. In the 'type 2 birth', if we denote the vectors of switch locations in the current and proposed new indicator functions by $\mathbf{t}$ and $\mathbf{t}'$ respectively then the change of variable $(\mathbf{t}, t, t') \mapsto \mathbf{t}'$ is trivial, so the Jacobian is unity. There are two other move types updating $\mathbf{x}$. 'Type 1' moves attempt to add or remove a single switch at either end of the interval, while dimension-preserving 'shift' moves perturb the location of a randomly chosen switch.

A single iteration of the MCMC sampler consists of sequential updates of the sojourn time distribution parameters $(s_0, s_1, \lambda_0, \lambda_1)$, the indicator function $\mathbf{x}$ and the channel conductance levels (μ_0, μ_1). As they involve no change of dimension, standard Metropolis–Hastings moves are used for the sojourn parameters and conductance levels.

3.2 Adaptation to the Markov model discrimination case

Since the vector of transition rates, denoted by Λ, has variable dimension across the competing Markov models, there is a second source of variable dimensionality in the target posterior distribution. Hence the reversible jump technique (Green, 1995) is needed to update the model indicator k. Suppose the current model is the two-state one, $k = 11$, and a second closed state is proposed; proposals to add a second open state are generated in the same way. The components of the new transition rate vector Λ' are given by:

$$\begin{aligned}
\lambda_c' &= \lambda_c \exp(z_1) \\
\mu_c' &= \lambda_c \exp(z_1 + z_2) \\
\nu_c' &= \frac{\lambda_c \exp(z_1 + z_2)}{\exp(z_1) - 1} \\
\lambda_o' &= \lambda_o,
\end{aligned}$$

where $z_1 \sim \Gamma(\rho_1, \rho_2)$, $z_2 \sim N(0, \tau^2)$ independently and ρ_1, ρ_2 and τ are simulation parameters. The Jacobian of this transformation, which preserves the mean closed sojourn time, appears in the move's acceptance probability. The inverse transformation must be used in the reverse move jumping from model *21* to model *11*.

4 Summary of results

4.1 Sampler performance

Hodgson's (1997) original sampler, outlined in Section 3.1, mixes very poorly between parameter subspaces of different dimension. The cause is diagnosed as the high degree of association between the indicator function $\mathbf{x}$ and the sojourn time distribution parameters $(s_0, s_1, \lambda_0, \lambda_1)$. Since this dependence becomes weaker as the data segment become shorter, a novel form of simulated tempering is proposed as a solution. The original data set is subdivided into segments of equal length, each with its own sojourn time distribution parameters. Realisations from the suitably revised sampler are recorded conditional on these 'local' parameters being equal globally. Such complications are seemingly unnecessary in Hodgson & Green (1998), presumably because the extra flexibility afforded by the choice of Markov models has an accelerating effect on mixing.

4.2 Signal restoration

For each $t \in \{0, 1, \ldots, T\}$, we may estimate the posterior probability of the channel being open at time t, and hence obtain a pointwise estimate of the posterior mean of $\mathbf{x}$. Thresholding the mean function at the 50% level yields an estimate $\widehat{\mathbf{x}}$ of the true indicator function. This estimate is found to have a misclassification rate of around 2%, similar to that reported in Johnstone & Silverman (1997) for a special-purpose detection algorithm devised by Eisenberg and Levis. However, $\widehat{\mathbf{x}}$ underestimates seriously the true number of switches: the insensitivity of the 50% threshold results in only those switches having strong posterior support being detected. Hodgson (1997) describes an algorithm to construct a signal estimate with a given number of switches, thereby making fuller use of all relevant information in the MCMC sample.

4.3 Comparison of predictive distributions

After every sampler iteration, the realised values of $(s_0, s_1, \lambda_0, \lambda_1)$ (in the alternating renewal case) and of (k, Λ) (in the hidden Markov case) imply realised closed and open sojourn time distributions. Averaging these across the MCMC output yields estimates of their posterior expectations, the Bayesian

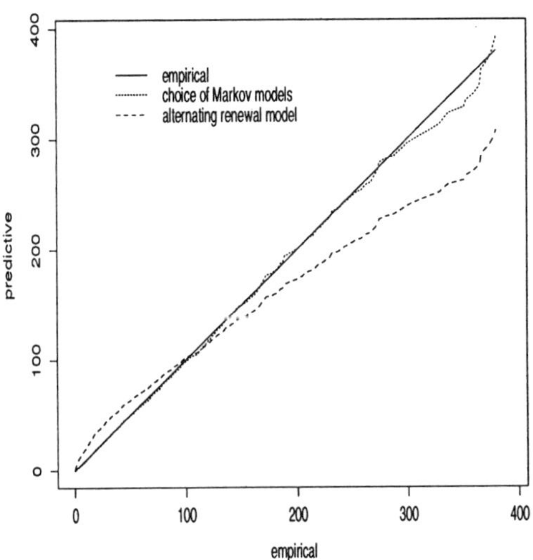

Fig. 1. Q–Q plot comparing predictive distribution function estimates

predictive sojourn time distributions. Figure 1 compares the closed time distribution function estimate in the hidden Markov case (unconditional on the model k) and that in the alternating renewal case with the empirical distribution function of the 811 completed closed times from Eisenberg and Levis' entire record. Clearly Hodgson & Green's (1998) choice of Markov models provides much the better fit.

Acknowledgements
I thank Peter Green for his guidance in the approach towards Bayesian inference for ion channels described in this paper, and also Bob Eisenberg and Rick Levis for permission to use their data.

References

Ball, F. G. & Rice, J. A. (1992). Stochastic Models for Ion Channels: Introduction and Bibliography. *Mathematical Biosciences*, **112**, 189-206.

Colquhoun, D. & Hawkes, A. G. (1981). On the stochastic properties of single ion channels. *Proceedings of the Royal Society of London* B, **211**, 205-235.

Colquhoun, D. & Hawkes, A. G. (1982). On the stochastic properties of bursts of single ion channel openings and of clusters of bursts. *Philosophical Transactions of the Royal Society of London* B, **300**, 1-59.

Eisenberg, R. (1994). Biological signals that need detection: Currents through single membrane channels. In: *Proceedings of the 16th Annual International Conference of the IEEE Engineering in Medicine and Biology Society*, 32a-33a.

Fredkin, D. R. & Rice, J. A. (1986). On aggregated Markov processes. *Journal of Applied Probability*, **23**, 208-214.

Green, P. J. (1995). Reversible jump Markov chain Monte Carlo computation and Bayesian model determination. *Biometrika*, **82**, 711-732.

Hastings, W. K. (1970). Monte Carlo Sampling Methods using Markov Chains and their Applications. *Biometrika*, **57**, 97-109.

Hodgson, M. E. A. (1997). A Bayesian restoration of an ion channel signal. Technical Report S-97-02, Department of Mathematics, University of Bristol. To appear in *Journal of the Royal Statistical Society* B.

Hodgson, M. E. A. & Green, P. J. (1998). Markov model discrimination for ion channels: a feasibility study. Technical Report S-98-01, Department of Mathematics, University of Bristol.

Johnstone, I. M. & Silverman, B. W. (1997). Wavelet threshold estimators for data with correlated noise. *Journal of the Royal Statistical Society* B, **59**, 319-351.

Metropolis, N., Rosenbluth, A. W., Rosenbluth, M. N., Teller, A. H. & Teller, E. (1953). Equations of State Calculations by Fast Computing Machines. *Journal of Chemical Physics*, **21**, 1087-1091.

Milne, R. K., Yeo, G. F., Edeson, R. O. & Madsen, B. W. (1988). Stochastic modelling of a single ion channel: an alternating renewal approach with application to limited time resolution. *Proceedings of the Royal Society of London* B, **233**, 247-292.

(ed.) Sakmann, B. & Neher, E. (1995). *Single-Channel Recording: Second Edition.* New York: Plenum Press.

ARGUS, Software Packages for Statistical Disclosure Control

Anco Hundepool and Leon Willenborg *

Statistics Netherlands, Department of Statistical Methods
P.O. Box 4000, 2270 JM Voorburg, Netherlands
Email: ARGUS@CBS.NL

Abstract. The paper describes two related software packages for producing safe data: μ-ARGUS for microdata and τ-ARGUS for tabular data.

Keywords. Statistical disclosure control, software, microdata, tables

1 Statistical Disclosure Control

Statistical offices collect information about persons, businesses, institutions, etc. through censuses and surveys. The data collected are ultimately released in a suitable form to policy makers, researchers and the general public for statistical purposes. The release of such information may have the undesirable side-effect that information on individual entities instead of on (sufficiently large) groups of individuals may be disclosed. The question then arises as to how the information available can be modified in such a way that the data released can be considered statistically useful and do not jeopardize the privacy of the entities concerned.

The aim of Statistical Disclosure Control (SDC) is to limit the risk that sensitive information about individual respondents can be disclosed from a data set. The data set can be either a microdata set or a table. A microdata set consists of a set of records containing information on individual respondents. A table contains aggregate information about individual entities.

In order to publish safe data one should first have criteria to check whether a particular data set is safe according to these criteria or not. If data are not safe according to these criteria they have to be modified in such a way that the resulting data meet these criteria. These modifications, while decreasing the risk of disclosure, also imply that the information content of the data is decreased, because certain variables are coded in a less detailed fashion or values are suppressed or replaced by other values. The idea is that the modifications should be applied in such a way that the resulting information loss is minimised. As a rule achieving this goal is quite complicated and requires the use of specialised software tools. Such tools are μ-ARGUS for microdata and τ-ARGUS for tabular data.

In the remainder of this paper two SDC packages are presented that can be used to produce safe microdata (μ-ARGUS) and safe tables (τ-ARGUS). Not only is the (main) functionality of both packages described, but also the background philosophy which tries to explain and motivate this functionality.

* The views expressed in this paper are those of the authors and do not necessarily reflect the policies of Statistics Netherlands.

2 μ-ARGUS

In the case of microdata, disclosure of sensitive information about an individual respondent can occur after this respondent has been re-identified. That is, after it has been deduced which record corresponds to this particular individual. So, disclosure control should hamper re-identification of individual respondents.

Re-identification can take place when several values of so-called identifying variables, such as 'Place of residence', 'Sex' and 'Occupation', are taken into consideration. The values of these identifying variables can be assumed known to friends and acquaintances of a respondent. When several values of these identifying variables are combined a respondent may be re-identified. Consider for example the following record obtained from an unknown respondent:

'Place of residence = Urk', 'Sex = Female' and 'Occupation = Statistician'.

Urk is a small fishing-village in the Netherlands, in which it is unlikely for many statisticians to live, let alone female statisticians. So, when we find a statistician in Urk, a female one moreover, in the microdata set, then she is probably the only one. When this is indeed the case, anybody who happens to know this rare female statistician in Urk is able to disclose sensitive information from her record if such information is contained in this record.

An important concept in the theory of re-identification is a key. A key is a combination of identifying variables. Keys can be applied to re-identify a respondent. Re-identification of a respondent can occur when this respondent is rare in the population with respect to a certain key value, i.e. a combination of values of identifying variables. Hence, rarity of respondents in the population with respect to certain key values should be avoided. When a respondent appears to be rare in the population with respect to a key value, then disclosure control measures should be taken to protect this respondent against re-identification. "Rare" means that a combination of characteristics occurs less than a certain threshold value D_k, where k is a key, implying that the threshold value depends on k. One can define the threshold value at the population level, and then use an equivalent threshold value for a sample, as usually is the case.

A key value that occurs less than D_k times in the population is considered unsafe, a key value that occurs at least D_k times in the population is considered safe. The unsafe combinations must be protected, while the safe ones may be published.

When the estimated frequency of a key value, i.e. a combination of scores, is at least equal to the threshold value D_k, then this combination is considered safe. When the estimated frequency of a key value is less than the threshold value D_k, then this combination is considered unsafe. An example of such a key is 'Place of residence' 'Sex' 'Occupation'.

μ-ARGUS has been developed to remove a set of unsafe combinations from a microdata set. The current version uses two techniques for this: global recoding and local suppression. In case of global recoding several categories of a variable are collapsed into a single one. In the above example, for instance, we can recode the variable 'Occupation'. For instance, the categories 'Statistician' and 'Mathematician' can be combined into a single category 'Statistician or Mathematician'. When the number of female statisticians in Urk plus the number of female mathematicians in Urk is sufficiently high, then the combination 'Place of residence = Urk', 'Sex = Female' and 'Occupation = Statistician or Mathematician' is considered safe for release.

The effect of local suppression is that one or more values in an unsafe combination are suppressed, i.e. replaced by a missing value. For instance, in the above example we can protect the unsafe combination 'Place of residence = Urk', 'Sex = Female' and 'Occupation = Statistician' by suppressing the value of 'Occupation' in the records in which the unsafe combination occurs. This only leads to a safe combination of scores if the number of females in Urk is sufficiently high. The resulting combination is then given by 'Place of residence = Urk', 'Sex = Female' and 'Occupation = missing'.

Both global recoding and local suppression lead to a loss of information, because either less detailed information is provided or some information is not given at all. A balance between global recoding and local suppression has to be found in order to make the information loss due to the application of SDC measures as low as possible.

μ-ARGUS has been designed to help the data-protector to efficiently find a set of global recodings. This selection is based on a set of tables of the identifying variables. These tables are generated from the original micro data file. As all the manipulations to inspect the results of certain global recodings are done at the level of these tables and therefore do not require lengthy runs through the micro data file, this can be done quite efficiently. This gives the data-protector the opportunity to experiment with the impacts of the different sets of global recodings. When a choice has been made μ-ARGUS will generate a safe file. In one run the selected set of global recodings are applied and the remaining unsafe combinations are removed by local suppressions.

As an alternative for the manual selection of the optimal set of global recodings, an automatic selection process has been developed. Finding an optimal balance between global recoding and local suppression and the selection of the set of global recodings leads to a big optimisation problem. The main problem is to find this optimal balance. Applying global recoding means that the codelist for that variable becomes less detailed and therefore some information will be lost for all records. When local suppression is applied all information for a specific variable in a selected set of records is removed. Both actions imply information loss. We use an entropy function to measure this information loss. It is the responsibility of the data-protector to give weights to the variables. The higher the weight the more important it is considered to keep the information of a variable in the data file. This problem has been solved by Sergey Tiourine and Cor Hurkens of the Technical University of Eindhoven, see Hurkens & Tiourine(1998). Their solution has been incorporated in μ-ARGUS.

In case the data-protector wants to apply microaggregation or controlled rounding (by specifying the necessary meta-information) to one or more numerical variables in the file, this is carried out in the final phase. In case of controlled rounding, this can be combined when the global recodings and local suppressions are executed. In case of microaggregation, a sorting of the data (of the variable in question) needs to be performed first, so that it is possible to form the respective groups and calculate the respective group means. This has to be repeated as many times as there are variables that are to be microaggregated. In case of a big microdata file the sorting may be rather time-consuming. In case of controlled rounding no preliminary sorting is needed but only a single pass through the data that can also be combined with a previous step in the process, that executed the global recodings and the local suppressions.

At the end of the μ-ARGUS process a report describing and documenting the actions performed is generated.

3 τ-ARGUS

τ-ARGUS is intended for producing safe tables. The current version of τ-ARGUS can only handle a single table (together with its marginals) of dimensions less than 5. τ-ARGUS can handle two kinds of tables, namely magnitude tables and frequency count tables. A magnitude table is a table where the cells are filled with the total of some numeric variable (like turnover, income etc.), while in frequency tables the cells display just the number of records pertaining to that cell. The difference is important for several reasons: first when employing a disclosure risk model to a (set of) table(s) in the way of defining sensitive cells, and second when protecting a table. In the latter case it makes a difference if the cell values can only take integer values or not.

The safety of a table is determined by the existence of sensitive cells and whether the cell values in these cells can be considered sufficiently protected. τ-ARGUS identifies sensitive cells in magnitude tables by employing a "dominance rule" (see e.g. Section 6.2 in Willenborg & De Waal, 1996). This rule states that a cell of a table is unsafe for publication if a few, n say, major contributors to a cell are responsible, when adding their contributions, for at least a certain percentage p of the total of that cell. A common choice is $n = 3$ and $p = 70\%$, but τ-ARGUS allows users to specify other parameter settings. Applying a dominance rule to a frequency count table implies a thresholding rule: nonempty cells with a frequency less than the threshold are considered unsafe, whereas those with a frequency above the threshold are considered safe (this is comparable to a thresholding rule for microdata). In some cases this approach makes sense, but in other cases it does not (see Section 6.3, Willenborg, 1996). When it does not, considerations motivated by group disclosure are taken into account, which go beyond those concerning individual disclosure considerations that are usually being applied.

It should be noted that for τ-ARGUS to identify the sensitive cells in a magnitude table, it needs, for a dominance rule with parameters n and p, apart from the cell totals, the sums of the top n contributors in each cell. If a user is permitted to lump rows or columns in the table together (in order to protect sensitive cells), then it is useful for each table cell to store the individual top n contributions of that cell instead of their sum. This allows τ-ARGUS to calculate the top n contributions for each cell that has been created by lumping two or more cells together. It simply requires that the top n contributions for all cells are merged and the top n contributions for this new cell is calculated.

Once τ-ARGUS has identified all sensitive cells in a table, it helps a user to protect them through the execution of certain SDC techniques, such as cell deletion, cell suppression, table redesign or rounding. Some of these operations have to be carried out interactively, using inputs provided by the data-protector, while others can be done automatically by τ-ARGUS itself.

A complication that typically exists in case of protecting tables — and what makes the exercise difficult — is the presence of additional constraints in the data, such as additivity constraints in case marginal tables are present or nonnegativity constraints of cell values. Due to the presence of these constraints cell suppression is usually not quite what it suggests: an interval of feasible values for a suppressed cell (in a pattern of such cells) can be calculated, rather than that the suppressed value is completely unknown. For rounding, the constraints that apply to the original table are imposed on the rounded table as well. Besides, the rounded table (and its rounded marginals) should be close to the original table (and its marginals) as well, assuming a suitable metric to measure distances.

Cell suppression and rounding require the solution of complex optimisation problems. These problems have been solved by Fischetti & Salazar (1998). Contrary to the use of μ-ARGUS the current version of τ-ARGUS requires the use of an externally called LP-solver package, to carry out local suppressions and controlled roundings [2].

τ-ARGUS generates the tables to be protected from scratch from a microdata file. Not only the actual table is constructed, but also the information required to apply the dominance rule is calculated. The first step will then be a possible redesign of the table. Since τ-ARGUS has stored all the necessary information to perform the table redesign without going back to the microdata file, these actions are performed quickly. This enables the data-protector easily to inspect the results of the different recoding schemes. After this table redesign a starting point has been created for the further protection of the table either by controlled rounding or cell suppression. The result will then be a safe table, which will be stored either as a text-file or a spread-sheet, suitable for further processing by e.g. the publication software. Not only the actual table will be stored but also a report describing the actions performed on the protected table.

Acknowledgement
This paper was written in the context of the SDC project, partly subsidized by the European Union through ESPRIT (contract no. 20462). Information on the SDC project can be found on the web page: http://www.cbs.nl/sdc/.

References

Fischetti, M. & Salazar J.J. (1998). *Modeling and Solving the Cell Suppression Problem for Linearly-Constrained Tabular Data.* Proceedings of the SDP98-conference Lisboa.

Hurkens, C.A.J. & Tiourine, S.R. (1998). *On Solving Huge Set-cover Models for the Microdata Protection Problem.* Proceedings of the SDP98-conference Lisboa.

Willenborg, L.C.R.J. & De Waal A.G. (1996). *Statistical Disclosure Control in Practice.* New York: Springer-Verlag.

[2] The LP-solver that τ-ARGUS requires should be independently purchased by the data-protector. We use the package XPress of Dash Associates in the UK

Building End-User Statistical Applications: An Example Using STABLE

Zivan Karaman

Limagrain Genetics Research, B.P. 115, 63203 Riom Cedex, France
E-mail: zivan.karaman@biocem.univ-bpclermont.fr

Abstract. In this paper we present an example of building a customised statistical application, suitable for use by people with no or minimal knowledge of statistics and computing. The application is being developed using the STABLE system, and its purpose is to provide plant breeders with customised, efficient, user-friendly software to perform all the data analysis tasks they need.

Keywords. Application building, data analysis, data visualisation, interactive graphics, statistical software, visual programming

1 Introduction

On many occasions (particularly in industrial applications), data must be analysed by people who are familiar with neither statistical methods, nor statistical terminology, and who often are not very proficient in use of computers, and particularly in programming. The majority of advanced statistical packages (like Genstat, SAS, or S-PLUS) nowadays offer a capability for performing statistical analyses (at least the most currently used ones) through a menu-driven interface. This is a very useful feature for statisticians who have not mastered the underlying command language perfectly, but it still requires understanding of statistical techniques and vocabulary.

On the other hand, modern Rapid Application Development (RAD) tools (like Borland's Delphi, Powersoft's PowerBuilder or Microsoft's Visual Basic), offer excellent possibilities for the development of customisable, user-friendly applications, but offer no built-in tools for statistics, and only very limited data plotting facilities ("business graphics"); this is a logical consequence of their data management orientation. Developing a statistical application using one of these tools therefore requires a lot of work, and produces very inflexible software, since any modification must be re-programmed in one of the general purpose programming languages like C or FORTRAN.

The STABLE system is currently being developed as a part of the European Union ESPRIT IV Project. Its goal is to bring together the application building framework and advanced data visualisation facilities from IRIS Explorer and the extensive range of reliable and widely-used statistical algorithms from Genstat (Payne *et al.*, 1993). The resulting system is a "Statistical Explorer", a visual programming environment where existing modules are easily combined together in order to perform specific data analysis tasks. The STABLE system allows advanced users (statistical applications programmers) easily and quickly to build flexible and powerful statistical applications with strong data visualisation capabilities. The components forming the application can be bundled into a single, fully customised user interface, that will use the terminology familiar to the user, and encompass all the statistical methods appropriate for the particular situation. Advanced data visualisation facilities are very important in this context, since statistical ideas are often more easily conveyed to the end-users by graphical representations than by numerical results only. The extensive description of the STABLE system is given by Morgan & Craig (1998).

2 Building an application

2.1 Motivation

Limagrain is the major European seed company, and its research division is in charge of the development of new varieties of major field crops. Limagrain scientists are conducting plant breeding programmes on crops such as barley, maize, rape seed, sunflower and wheat. The research programmes are conducted on a network of research stations dispersed over several countries. The staff of these distant sites is composed of plant breeders, agronomists, and field technicians, with no available local support in computing and statistics. On the other hand, to make the full use of winter nurseries (in order to advance two generations in a year), decisions must be taken very quickly (in extreme cases, a couple of hours after the last field plot has been harvested), with data being analysed and decision making done on-site, by people with limited knowledge of both computing and statistics. Therefore, customised, efficient, user-friendly software is needed to perform all the data analysis tasks.

The primary aim of many comparative agricultural field experiments is to obtain accurate and efficient estimates of treatment effects. The precision of the estimates can generally be improved by increasing the number of replicates of each treatment, but since the resources allocated to agricultural research tend to be decreasing, a more cost-effective solution is needed. The classical approach to the analysis of field experiments relies on the removal of within-site variability by blocking, using the appropriate experimental design. More recently there has been much research on the use of spatial models in the analysis of field experiments, in order to remove the trend due to uneven distribution of soil moisture, fertility, etc. Another important aspect of statistical methodology applied to plant breeding is the analysis of the genotype by environment interaction, where a mixture of general linear model theory and multivariate statistical methods is used, coupled with appropriate graphical displays such as biplots.

We propose to develop a tool that would enable the end-users (agronomists, plant breeders or field experimentation technicians - people who generally do not have very high proficiency with either computers or statistics) to analyse their data quickly and thoroughly in a user-friendly environment.

2.2 System description

The system we are building will allow the following operations to be performed in an easy and flexible way.

- Design and layout of field experiments: In order to obtain the maximum information from the field experiments, careful experimental design is necessary. This involves the choice of an appropriate experimental design, the choice of the parameters for the particular design and corresponding randomisation of the experimental units. Since many experiments are conducted simultaneously on the same field, they should be laid out within the field in a suitable way.
- Validation of experimental data: The validation procedures involve analysis of experimental data by appropriate statistical methods, visualisation of both raw data and the results of statistical calculations in an interactive graphical environment, choice of the model, and updating the external database with the model output.
- Decision making: The final step, after all the experimental data have been collected and validated, is to make conclusions and take the appropriate decisions. This involves fitting the models and doing hypothesis testing, viewing the data in variety of different ways, and presenting the data and the results of analyses both graphically and in a tabular form, allowing customised document layout.

The field experiment design module allows users to select from a variety of designs commonly used in agricultural experiments, such as randomised complete blocks and split-plot designs. It also provides the facilities for designing the experiments where a huge number of treatments - up to several hundreds - must be compared (the situation common in plant breeding), by the use of incomplete block designs such as lattice and alpha designs (Patterson & Williams, 1976). The appropriate randomisation procedures are also provided. The design specifications (treatment levels, number of replicates, design type) are retrieved from an external database. The field layout of experiments can be visualised in a raster-like display.

The data validation module provides facilities for the analysis of data from experiments conducted at one location. Data are first imported from the database, and can then be displayed and summarised by scatter plots, histograms, boxplots and other exploratory data analysis methods. The main statistical method used is analysis of variance. Both balanced and unbalanced data sets can be analysed, as well as models including both fixed and random factors. The usual diagnostic plots such as histograms of residuals, scatter plots of residuals vs. fitted values, half-normal plots, etc., are available, with interactive identification of points on the screen. This is particularly useful to spot outliers and potential errors in the data quickly, and thus significantly improve the speed of data validation. The alternative methods of data analysis based on spatial models (kriging, surface fitting) and neighbour models (Gleeson & Cullis, 1987) are also proposed, coupled with appropriate visualisation techniques including contour and surface plots and raster-like display. These are particularly useful for revealing the patterns and trends within a field that are otherwise difficult if not impossible to observe, but that can greatly influence the experimental results.

The tools for the final data analysis are grouped in the third module. Once again, analysis of variance is used as the basic statistical method. It is further enhanced by methods for assessing genotype adaptability and stability, and in particular by joint regression analysis (Digby, 1979). The genotype × environment interaction can be studied by the method known as the AMMI (Additive Main Mutliplicative Interaction) model (Gouch & Zobel, 1990), which performs principal components analysis on the two-way table of genotype × environment data with the main effects removed. The results of this model are displayed using biplots (Gabriel, 1971). It is also possible to classify the environments (or genotypes) according to their interaction patterns, using hierarchical clustering algorithms, and to plot the corresponding dendrograms.

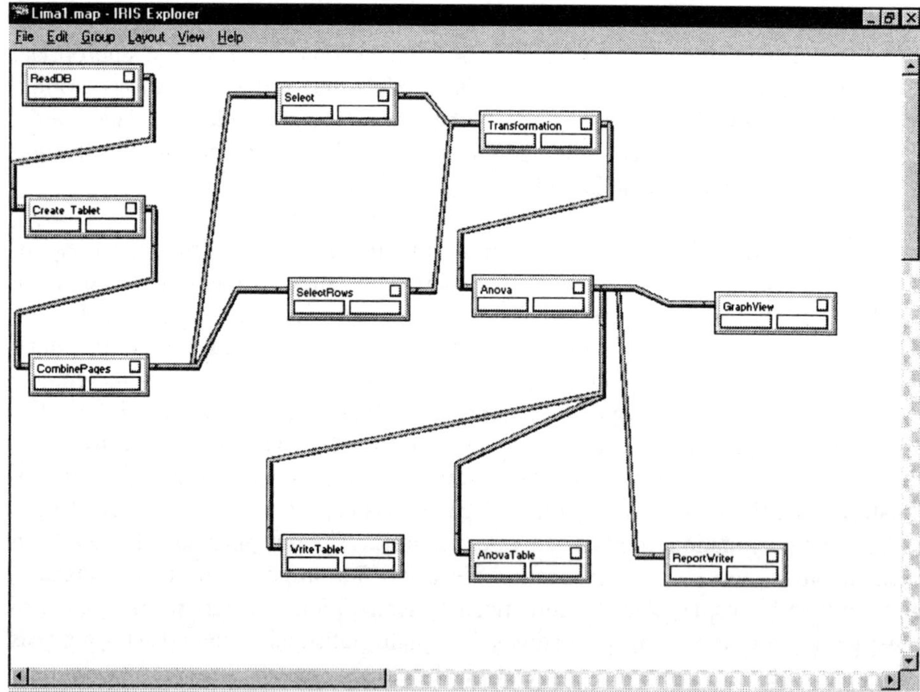

Fig. 1. STABLE map for analysis of variance

We expect two types of users of the STABLE system within our organisation:

- the applications programmers - who will build the maps like the one shown in Figure 1, based on the modules that come as a part of the standard release of the STABLE software, or, sometimes, develop their own modules for very specific tasks;

- the end-users, who will not necessarily see all the (relative) complexity of the map, but will just be requested to select a limited number of options from a user-friendly interface, as shown in Figure 2.

2.3 Example

Figure 1 illustrates the analysis of variance programme developed in the STABLE environment. Each grey box in the picture represents a *module*, which is the basic building block of the system. These modules are equivalents of commands or directives in a classical command-driven statistical system like Genstat. The lines linking the modules show the flow of the data through the program, which is called a *map* in the STABLE system. Each module can have one or more input ports and one or more output ports; the input and output ports of different modules are linked simply by a couple of mouse clicks.

In our example, data are first read from an external database, since the volume of the data dealt with requires the use of an RDBMS. After selection of the desired rows and columns of the data matrix, and possible transformations, the data are passed to the ANOVA module. From there, the results are passed to different modules that display data in both graphical (histogram of residuals, residuals vs. fitted values scatter plot, etc.) and tabular form (ANOVA table, table of means, etc.).

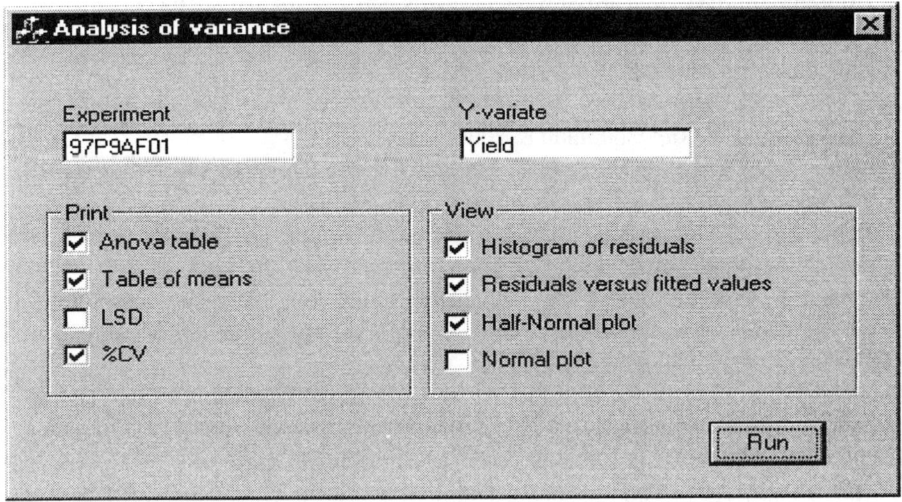

Fig. 2. STABLE dialogue for selecting options for analysis of variance

Once the map shown in the Figure 1 has been fully tested, it will be collapsed into a single application. The end-user will be presented with a unique dialogue window that will allow him to set all the parameters necessary for the analysis, and will mask the interface controls which have been pre-selected by the application programmer that should not be changed. Figure 2 shows an example of a dialogue window that would be presented to the end-user.

3 Conclusions

Our experience using the STABLE system shows the advantages of a visual programming environment for rapid development of statistical applications. We are able quickly to build powerful, flexible, user-friendly and fully customised statistical analysis applications. The users within our company will have access to a wide range of statistical techniques, without requiring knowledge of a command language. We shall be able in the future easily to enhance the developed applications by incorporating new statistical modules, or to make the application easier to use by combining modules in groups in a more intuitive manner.

Acknowledgements
STABLE is funded by European Union ESPRIT IV Project No. 22832.

More information about the STABLE project can be obtained from the project co-ordinators NAG Ltd, Wilkinson House, Jordan Hill Road, Oxford OX2 8DR, UK, or on the World Wide Web at the following address http://www.nag.co.uk/projects/STABLE.html.

References

Digby, P.G.N. (1979). Modified joint regression analysis for incomplete variety × environment data. *Journal of Agricultural Science, Cambridge*, **93**, 81-86.

Gabriel, K.R. (1971). The biplot graphic display of matrices with application to principal component analysis. *Biometrika*, **58**, 453.

(ed.) Kempton, R.A. & Fox, P.N. (1997). *Statistical Methods for Plant Variety Evaluation*. London: Chapman & Hall.

Gauch, H.G. & Zobel, R.W. (1988). Predictive and postdictive success of statistical analysis of yield trials. *Theoretical and Applied Genetics*, **76**, 1-10.

Gleeson, A.C. & Cullis, B.R. (1987). Residual maximum likelihood estimation of a neighbour model for field experiments. *Biometrics,* **43**, 277-288.

Morgan, G. & Craig, P. (1998). A visual future for statistical computing. In: *COMPSTAT 1998 Proceedings in Computational Statistics* (ed. R. Payne & P. Green), 389-394. Heidelberg: Physica-Verlag.

Patterson, H.D. & Williams E.R. (1976). A new class of resolvable incomplete block designs. *Biometrika*, **63**, 83-92.

Payne, R.W., Lane, P.W., Digby, P.G.N., Harding, S.A., Leech, P.K., Morgan, G.W., Todd, A.D., Thompson. R., Tunnicliffe Wilson, G., Welham, S.J. & White, R.P. (1993). *Genstat 5 Release 3 Reference Summary*. Oxford: Numerical Algorithms Group.

On Multiple Window Local Polynomial Approximation with Varying Adaptive Bandwidths

Vladimir Katkovnik

Department of Statistics, University of South Africa (UNISA),
P.O. Box 392, Pretoria 0001, South Africa
e-mail: katkov@risc6.unisa.ac.za

Keywords. Local bandwidth selection, nonparametric estimation, smoothing, threshold adjustment

1 Introduction

In this article we introduce an adaptive smoother that produces piecewise smooth curves with a small number of discontinuities in the function or its derivatives. This allows certain desirable features such as jumps or instantaneous slope changes to be present in the smooth curves.

Suppose that we are given noisy samples of a function $y(x)$ along the regular grid $x_s = s\Delta$, $z_s = y(x_s) + \varepsilon_s$, where ε_s i.i.d. Gaussian, $N(0,\sigma)$. It is assumed that $y(x)$ belongs to the nonparametric class of piecewise continuous $r-$differentiable functions $\mathcal{F}_r = \{\left|y^{(r)}(x)\right| \leq L_r\}$. Our goal is to estimate $y_s = y(x_s)$ and its derivatives $y^{(k)}(x_s)$, $k \leq r-1$, depending on observations $(z_s)_{s=1}^N$ with a pointwise mean squared error (MSE) risk which is as small as possible.

The following loss function is applied in the standard linear LPA (e.g. Fan & Gijbels, 1996):

$$J_h(x) = \frac{1}{N}\sum_{s=1}^{N} \rho_h(x_s - x)(z_s - C^T\phi(x_s - x))^2 \tag{1}$$

$$\phi(x) = (1, x, ..., x^{m-1}/(m-1)!)', \quad C = (C_0, C_1, ..., C_{m-1})',$$

where x is a "centre" and m is an order of the LPA. The window $\rho_h(x) = \rho(x/h)/h$ is a function satisfying conventional properties of the "kernel" estimates, in particular, $\rho(x) \geq 0$ and $\rho(0) = \max_x \rho(x)$. Here h is a window "size" or a bandwidth. Then the minimization of $J_h(x)$ with respect to C

$$\hat{C}(x,h) = \arg\min_{C\in R^m} J_h \tag{2}$$

gives $\hat{y}_k(x) \triangleq \hat{C}_k(x,h)$ as estimates of $y^{(k)}(x)$ if $y(x)$ is smooth enough, i.e. $0 \leq k \leq \tilde{m}-1$, $\tilde{m} = \min(m,r)$.

It is well known that bandwidth selection is a crucial point of the efficiency of the LPA estimators. The new bandwidth selection procedure, which we name the intersection of confidence intervals (ICI), is proposed by Goldenshluger & Nemirovski (1994) (see also Goldenshluger & Nemirovski, 1997, page 875). It is shown that the LPA equipped with the ICI statistic possesses simultaneously many attractive asymptotic properties, namely, 1) it is nearly ideal within $\ln N$ factor in the pointwise risk for estimating the function and its derivatives; 2) it is spatial adaptive over a wide range of the classes of $y(x)$ in the sense that its quality is close to that which one could achieve if smoothness of $y(x)$ was known in advance.

This paper presents a modification and development of the results obtained in Goldenshluger & Nemirovski (1994). This modification mainly concerns a choice of the threshold parameter of the ICI according to the mean squared error accuracy criterion. It is shown that this threshold is an important design parameter of the algorithm, which influences the accuracy in a crucial way and that the cross-validation proves to be a good criterion for selection of an adjusted data-driven threshold. The multiple window estimator which combines left, right and symmetric window LPA estimates, each with the adjusted threshold, is used in order to repair edge effects and discontinuities of the function and derivatives.

2 Algorithm

2.1 The background of the ICI

As $\Delta, h \to 0$ the mean squared risk

$$r_k(x,h) = Ee_k^2(x,h),\ e_k(x,h) \triangleq y^{(k)}(x) - \hat{y}^{(k)}(x,h),$$

of the estimate (2) can be represented in the form

$$\begin{aligned} r_k(x,h) = std_k^2(x,h) + \omega_k^2(x,h) = \\ \sigma^2 \Delta b_{k,m}/h^{2k+1} + a_{k,\tilde{m}}^2 (y^{(\tilde{m})}x))^2 h^{2(\tilde{m}-k)}, \end{aligned} \tag{3}$$

where $std_k(x,h)$ and $\omega_k(x,h)$ are the standard deviation and the bias of the estimate respectively, and the constants $b_{k,m}$ and $a_{k,\tilde{m}}$ depend only on k, m and $\tilde{m} = \min(m,r)$. Minimizing the risk gives the ideal bandwidth $h^*(x,k)$ and the ideal risk $r_k^*(x)$ as follows:

$$r_k^*(x) = \min_h r_k(x,h) = std_k^2(x,h^*(x,k))(1+\gamma_{k,\tilde{m}}^2), \tag{4}$$

$$h^*(x,k) = (\gamma_{k,\tilde{m}}^2 \sigma^2 \Delta b_{k,m}/a_{k,\tilde{m}}^2 (y^{(\tilde{m})}x))^2)^{1/(2\tilde{m}+1)},\ \ \gamma_{k,\tilde{m}}^2 = \frac{2k+1}{2(\tilde{m}-k)},$$

where $\gamma_{k,\tilde{m}} = \omega_k(x,h)/std_k(x,h)$ is a ratio of the bias and standard deviation at $h = h^*(x,k)$.

For the linear estimate (2) the estimation error can be presented in the form $|e_k(x,h)| \leq \omega_k(x,h) + |\zeta_k(x,h)|$, where $\zeta_k(x,h)$ is a Gaussian random error, $N(0,\ std_k(x,h))$. Then with probability $p = 1-\alpha$

$$|e_k(x,h)| \leq \omega_k(x,h) + \chi_{1-\alpha/2} std_k(x,h), \tag{5}$$

where $\chi_{1-\alpha/2}$ is the $(1-\alpha/2)-th$ quantile of the standard Gaussian distribution.

Let us introduce a finite set of bandwidth values $H = \{h_1 < h_2 < .... < h_L\}$, starting with quite small h_1, and define the optimal bandwidth $\hat{h}(x,k)$ as follows:

$$\hat{\imath} = \max_{1 \leq i \leq L} \{i : \omega_k(x,h_i) \leq X \cdot \chi_{1-\alpha/2} std_k(x,h_i)\},\ \ \hat{h}(x,k) = h_{\hat{\imath}}, \tag{6}$$

where $X > 0$ is a parameter which determines a desirable proportion between the bias and random error.

Now for $i \leq \hat{\imath}$ the inequality (5) can be strengthened to

$$|e_k(x, h_i)| \leq (1 + X)\chi_{1-\alpha/2} std_k(x, h_i). \tag{7}$$

According to (7) introduce a sequence of the confidence intervals $\mathcal{D}_k(i)$ of the biased estimate as follows:

$$\begin{aligned} \mathcal{D}_k(i) &= [L_i,\ U_i],\ \ U_i = \hat{y}_k(x, h_i) + \Gamma \cdot std_k(x, h_i), \\ L_i &= \hat{y}_k(x, h_i) - \Gamma \cdot std_k(x, h_i),\ \ \Gamma = (1 + X)\chi_{1-\alpha/2}, \end{aligned} \tag{8}$$

where Γ is a threshold of the confidence intervals. Then (7) is of the form $y^{(k)}(x) \in \mathcal{D}_k(i)$ and we can conclude from (6) and (5) that while $i \leq \hat{\imath}$ all the intervals $\mathcal{D}_k(i)$ have a point in common, namely, $y^{(k)}(x)$.

The following ICI statistic tests the very existence of this common point and gives an estimate of $\hat{\imath}$.

Consider the intersection of the intervals $\mathcal{D}_k(j)$, $1 \leq j \leq i$, with increasing i, and let i^+ be the largest of those i for which the intervals $\mathcal{D}_k(j)$, $1 \leq j \leq i$, have a point in common. This i^+ defines the adaptive bandwidth and the adaptive LPA estimate as follows:

$$\hat{y}_k^+(x) = \hat{y}_k(x, h^+(x, k)),\ \ h^+(x, k) = h_{i^+}. \tag{9}$$

The following choices are considered for the parameter X (or Γ) in (8):

(a) $X = 1$ assumes the equality of the random error and the bias. The corresponding $\Gamma = 2\chi_{1-\alpha/2}$ has been used in Goldenshluger & Nemirovski (1994).

(b) $X = \gamma_{k,\tilde{m}}$, $\Gamma = (1 + \gamma_{k,\tilde{m}})\chi_{1-\alpha/2}$, assumes a proportion between the bias and random error corresponding to that for the ideal bias and standard deviation (4).

(c) $X = \gamma_{k,\tilde{m}}^2$, $\Gamma = (1 + \gamma_{k,\tilde{m}}^2)\chi_{1-\alpha/2}$ is an optimal choice, minimizing the upper bound of the estimation error. It is shown later.

(d) X and Γ are data-driven adjusted to observations.

2.2 Threshold optimization

Let the bandwidth $\hat{h}(x, k)$ be a solution of the balance equation corresponding to (6)

$$\omega_k(x, h) = X \cdot \chi_{1-\alpha/2} std_k(x, h). \tag{10}$$

It can be verified that (3) gives the following formula for the standard deviation: $std_k(x, \hat{h}(x, k)) = std_k^*(x)(\gamma_{k,\tilde{m}}/X\chi_{1-\alpha/2})^{(2k+1)/(2\tilde{m}+1)}$.

Substituting $std_k(x, \hat{h}(x, k))$ into (7), we obtain

$$\left|e_k(x, \hat{h}(x, k))\right| \leq (1 + X)\chi_{1-\alpha/2} std_k^*(x) \left(\frac{\chi_{1-\alpha/2} X}{\gamma_{k,\tilde{m}}}\right)^{-(2k+1)/(2\tilde{m}+1)}. \tag{11}$$

It can be seen that the minimum of the right hand side of (11) is achieved at $X^* = \frac{2k+1}{2(\tilde{m}-k)} = \gamma_{k,\tilde{m}}^2$. This proves that this choice of X and the corresponding threshold $\Gamma^* = (1+\gamma^2_{k,\tilde{m}})\chi_{1-\alpha/2}$ minimize the upper bound of the estimation error for the optimal bandwidth $\hat{h}(x,k)$.

The following two statements provide insight into optimization properties imbedded in the ICI bandwidth selection rule.

Let us consider the asymptotic of the estimates provided Δ and $(h_{i+1} - h_i) \to 0$, $\Gamma = \Gamma^*$ and assume that $|\zeta_k(x,h)| \leq \chi_{1-\alpha/2} std_k(x,h)$, i.e. we do not have large random errors. Then: (a) For the optimal bandwidth $\hat{h}(x,k)$:

$$\left| e_k(x, \hat{h}(x,k)) \right| \leq \gamma_{k,\tilde{m}}^{-(2k+1)/(2\tilde{m}+1)} \sqrt{1 + \gamma^2_{k,\tilde{m}} \chi_{1-\alpha/2}^{2(\tilde{m}-k)/(2\tilde{m}+1)}} \cdot \sqrt{r_k^*(x)}. \quad (12)$$

(b) For the adaptive bandwidth h_k^+:

$$|e_k(x, h_k^+)| \leq 3\gamma_{k,\tilde{m}}^{-(2k+1)/(2\tilde{m}+1)} \sqrt{1 + \gamma^2_{k,\tilde{m}} \chi_{1-\alpha/2}^{2(\tilde{m}-k)/(2\tilde{m}+1)}} \cdot \sqrt{r_k^*(x)}. \quad (13)$$

The inequalities (12) and (13) show that the error of the estimate both with the optimal and the adaptive bandwidths within the constant product is bounded by the square root of the ideal risk.

2.3 Adjustment of the threshold Γ

In the analysis produced, the threshold Γ is a constant depending on the smoothness r of $y(x)$ and on the quantile $\chi_{1-\alpha/2}$. Both the value of r and the probability $p = 1-\alpha$, minimizing the MSE risk are unknown in advance. We considered Γ as a natural design parameter of the algorithm to be optimized and found that the cross-validation method proves to be efficient for this goal. Our attempts to use other quality-of-fit statistics, in particular C_P and Akaike criteria, instead of the cross-validation, have not shown an improvement in the accuracy.

The adaptive LPA estimation consists of the following basic steps:

1. Set $\Gamma = \Gamma_l$, $l = 1,2,...,G$ and $x = x_s$, $s = 1,2,..,N$.
2. For $h = h_i$, $i = 1,...,L$, calculate the estimates $\hat{y}_k(x_s,h)$ and

$$\bar{L}_{i+1} = \max[\bar{L}_i,\ L_{i+1}],\ \ \underline{U}_{i+1} = \min[\underline{U}_i,\ U_{i+1}], \quad (14)$$
$$i = 1,2,...,J,\ \ \bar{L}_1 = L_1,\ \ \underline{U}_1 = U_1.$$

3. The largest of those i for which $\bar{L}_i \leq \underline{U}_i$ gives i^+ and adaptive estimate $\hat{y}_k^+(x)$ (9).
4. Repeat Step 2 for all x_s, $s = 1,2,..,N$, and Γ_l, $l = 1,2,...,G$.
5. Find $\hat{\Gamma}$ minimizing the cross-validation criteria and determine the corresponding adjusted LPA estimates.

The standard deviation σ of the noise for $std_k(x,h)$ is estimated by the robust estimate: $\hat{\sigma} = \{median(|z_s - z_{s-1}| : s = 2,..,N)\}/(0.6745 \cdot \sqrt{2})$.

For the data given on the regular grid the implementation of the fast algorithm is done in MATLAB.

2.4 Multiple window estimate

Let ρ_L, ρ_R and ρ_S be the left, right and symmetric window functions, i.e. $\rho_L(u) = 0$ for $u > 0$, $\rho_R(u) = 0$ for $u < 0$ and $\rho_S(u) = \rho(-u)$, and $\hat{y}_L$, $\hat{y}_R$, and $\hat{y}_S$ be the corresponding estimates of $y(x)$. Then the combined LPA estimate $\hat{y}$ can be produced as a linear combination of the left, right and symmetric window estimates with the weight inversely proportional to the variances of the corresponding estimates. Similar fusing of the left, right and symmetric window estimates was used in McDonald & Owen (1986) and in Goldenshluger & Nemirovski (1994).

Thus we arrive at the concept of the LPA filter bank which consists of the elementary filters with weights obtained for the left, right and symmetric window functions with different LPA degrees m, reasonably restricted, say, to $m = 0, 1, 2$. These combined estimates produce sets of possible estimates. However, as these combinations do not always result in an improved estimate, the selection of the best estimate is a problem. We found that the cross-validation once more presents an effective choice.

3 Simulation

Experiments produced for some smooth test functions considered in Fan & Gijbels (1996) and Ruppert (1997) showed that the developed algorithm performs comparably and easily achieves the equivalent accuracy. However, for functions with discontinuities the advantage of the developed algorithm is quite clear. For this type of problem a comparison with wavelets was done. In particular, we used two combined LPA estimates (with $m = 2$) obtained by fusing: the left and right estimates and all three estimates, including the symmetric one. The cross-validation is applied for a selection of the best of these two combined estimates. The results are compared with those achieved by the wavelet with the adaptive thresholds on the test functions *Blocks* and *HeavySine*. These functions, noise, and the conditions of the Monte-Carlo statistical modelling exactly correspond to the results given in Table 2 in Donoho & Johnstone (1995, p. 1218).

The square root mean squared errors ($SRMSE$) are presented in Table 1. The figures in the first column are the numbers of observations N. The second and third columns give the $SRMSE$ for the LPA estimator respectively with the adjusted threshold parameter $\Gamma = var$ and fixed $\Gamma = 4.4$, as used in the simulation given in Goldenshluger & Nemirovski (1994). The fourth column presents the interval of the $SRMSE$ values obtained in Donoho & Johnstone (1995) for the different adaptive wavelets.

It is evident from the table that the algorithm developed with the adjusted threshold parameter in all cases achieves a better accuracy than the wavelet estimators, while this is not true for the algorithm with a fixed value of the threshold parameter Γ. The algorithm with the adaptive threshold parameter demonstrates an accuracy improvement of around 1.5-2 in comparison with the algorithm with the fixed value of Γ.

It is interesting to note that the optimal accuracy is achieved for the comparatively small values of Γ from the interval $(0.8 \div 1.2)$. As the quantile $\chi_{1-\alpha/2} < \Gamma$ this means that the accuracy optimization results in a high level of risk for the inequalities (5) and (7) to be violated.

Table 1. Root mean squared errors of estimation using the LPA and various wavelet methods

N	$\Gamma = adjusted$	$\Gamma = 4.4$	$Wavelets$
$Blocks$			
256	0.61	1.56	(0.68-1.20)
512	0.46	0.92	(0.59-1.12)
1024	0.33	0.74	(0.47-1.03)
2048	0.25	0.49	(0.41-0.85)
$Heavy$			
256	0.47	0.79	(0.49-0.62)
512	0.37	0.67	(0.40-0.55)
1024	0.31	0.49	(0.32-0.46)
2048	0.24	0.34	(0.38-0.61)

References

Donoho D.L. & Johnstone I.M. (1995). Adapting to unknown smoothness via wavelet shrinkage. *JASA*, **90** (432), 1200-1224.

Fan J. & Gijbels I. (1996). *Local polynomial modelling and its application.* London: Chapman and Hall.

Goldenshluger A. & Nemirovski A. (1994). On spatial adaptive estimation of nonparametric regression. *Research report,* 5/94, Technion, Israel.

Goldenshluger A. & Nemirovski A. (1997). Adaptive de-noising of signals satisfying differential inequalities. *IEEE Trans. Inf. Theory,* **43**, 873-889.

McDonald J.A. & Owen A.B. (1986). Smoothing with split linear fits. *Technometrics,* **28**(3), 195-208.

Ruppert D. (1997). Empirical-bias bandwidths for local polynomial and density estimation. *JASA*, **92** (439), 1049-1062.

Minimization of Computational Cost in Tree-Based Methods by a Proper Ordering of Splits

Jan Klaschka[1] and Francesco Mola[2]

[1] Academy of Sciences of the Czech Republic, Institute of Computer Science, Pod vodárenskou věží 2, CZ – 182 07 Prague 8, Czech Republic
[2] University of Naples *Federico II.*, Dept. of Mathematics and Statistics, Complesso Mt. St. Angelo, Via Cinthia, I – 80 126 Naples, Italy

1 Introduction

Tree-based methods such as CART (Breiman *et al.*, 1984), CHAID (commercialized by SPSS) or RECPAM (Ciampi, 1993) have a common typical feature: Analogous calculations are performed many times with one and the same portion of data. An unsophisticated implementation of such methods may result in a low computational effectiveness. Various computational enhancements are dealt with in a series of papers by Mola, Siciliano, Klaschka and Antoch – see, e.g., Mola & Siciliano (1992, 1997), Siciliano & Mola (1996), Klaschka & Antoch (1997), and Klaschka, Siciliano & Antoch (1998).

This contribution continues the above series. We propose a new algorithm which allows us to optimize calculations related to utilization of so called *auxiliary statistics.* In Section 2 we recapitulate some basics of tree-based methodologies. Section 3 explains in brief the concept and standard usage of the auxiliary statistics, a technique of restriction of handling raw data. The core of the paper is Section 4 where we present the new algorithm and show how it eliminates unnecessary calculations pertinent to the standard usage of the auxiliary statistics. A concluding remark can be found in Section 5.

2 Basics of the tree-based methodologies

Each of the tree based methods, when applied to a data set, recursively grows a tree that represents a tree-structured model (a decision tree, a piecewise constant regression function, a kind of piecewise defined multivariate model, etc.) Throughout the paper, we shall deal only with *binary* trees (which is the most typical case, though some methods may grow n-ary trees with $n > 2$).

One of the common building blocks of all these methods is the task of optimal partitioning of a set of cases $\mathcal{L}$, representing a node t of a tree, into two subsets $\mathcal{L}_L$ and $\mathcal{L}_R$ corresponding to new nodes t_L and t_R, i.e. the left and right "sons" of t, respectively. The optimal partition, usually called the *best split*, should maximize (or minimize) a kind of statistic $\phi(s, \mathcal{L})$ – the *splitting criterion* – over a given set $\mathcal{S}$ of *candidate splits s*.

In the standard setting, each case l in the data set has its values ${}^1X_l, \ldots, {}^mX_l$ of *predictors* ${}^1X, \ldots, {}^mX$, and each of the candidate splits for a node of a tree is based on one of these predictors.

For a *categorical predictor* X that takes (within a node) values from the set $\mathcal{X}$, respective splits are generated by the questions of the form "*Is the value of* X *in* B?", where B is any proper nonempty subset of $\mathcal{X}$.

For a *numerical* (ordered) *predictor* X, splits are generated by questions "*Is the value of* $X < c$?", where c is an element of a finite set of cutpoints (e.g., the set of the midpoints between the ordered values from $\mathcal{X}$).

For any split s, the positive and negative answers, respectively, to the corresponding question determine whether a case from $\mathcal{L}$ belongs to the *left subset* $\mathcal{L}_L(s)$ or *right subset* $\mathcal{L}_R(s)$, i.e., whether the case *is sent by* s *to the left* or *to the right.*

3 Auxiliary statistics

A straightforward way of maximizing the splitting criterion $\phi(s, \mathcal{L})$ within a node over the set of all splits based on all predictors is as follows.

(A) All predictors are treated as possible candidates to generate the optimal split.
(B) For each predictor X, the values $\phi(s, \mathcal{L})$ are calculated for all splits based on X.
(C) For each split s, the value $\phi(s, \mathcal{L})$ is calculated from the raw data.

Neither of (A), (B), and (C) is inevitable for all kinds of splitting criteria. As regards (A), a fast splitting algorithm developed by Mola & Siciliano (1997) enables us, when the splitting criterion satisfies some natural conditions, to omit calculation of $\phi(s, \mathcal{L})$ for the splits based on "weak" predictors without losing the global optimum. Concerning (B), it is well known that for some splitting criteria an exhaustive search within the set of *all* splits based on a categorical predictor can be reduced to a search over a smaller subset – see Breiman *et al.* (1984), Theorem 4.5 and Proposition 8.16. As regards (C), multiple handling the raw data may be restricted by applying so called *auxiliary statistics* – see Klaschka, Siciliano & Antoch (1998).

By the auxiliary statistic for a set of cases $\mathcal{L}$ we mean a structured set $\alpha(\mathcal{L})$ of numbers (a vector, a matrix, a list of vectors and/or matrices, etc.) possessing the following properties.

- The splitting criterion can be calculated from the auxiliary statistics for $\mathcal{L}_L(s)$ and $\mathcal{L}_R(s)$ (the sets of cases sent by s to the left and to the right) as $\phi(s, \mathcal{L}) = \Phi\big(\alpha(\mathcal{L}_L(s)), \alpha(\mathcal{L}_R(s))\big)$, where Φ is a computationally inexpensive procedure.
- There exist computationally cheap procedures `paste` and `cut` such that $\alpha(\mathcal{L}_1 \cup \mathcal{L}_2) = \texttt{paste}\big(\alpha(\mathcal{L}_1), \alpha(\mathcal{L}_2)\big)$, and $\alpha(\mathcal{L}_1) = \texttt{cut}\big(\alpha(\mathcal{L}_1 \cup \mathcal{L}_2), \alpha(\mathcal{L}_2)\big)$ for any pair of disjoint sets of cases $\mathcal{L}_1$ and $\mathcal{L}_2$. Procedures `paste` and `cut` will also be referred to as *recalculation* operations.

Standard usage of the auxiliary statistics (Klaschka, Siciliano & Antoch, 1998) depends on the type of predictor.

- For a *categorical* predictor X with values $\{x_1, \ldots, x_n\}$, the set $\mathcal{L}$ of cases that belong to a node of a tree is partitioned into the subsets $\mathcal{L}_{x_k} = \{l \in \mathcal{L}; X_l = x_k\}$, $k = 1, \ldots, n$. Notice that for every split s both sets $\mathcal{L}_L(s)$ and $\mathcal{L}_R(s)$ are finite unions of some of the sets $\mathcal{L}_{x_k}$. Thus, for each split s based on X, the auxiliary statistics $\alpha\big(\mathcal{L}_L(s)\big)$ and $\alpha\big(\mathcal{L}_R(s)\big)$, needed for calculation of the splitting criterion $\phi(s, \mathcal{L})$, are recalculated from $\alpha\big(\mathcal{L}_{x_k}\big)$, $k = 1, \ldots, n$ by repeated application of the operation `paste`.
- For a *numerical* predictor X, the splits are processed in the natural order, so that cutpoints increase. Let c and c' be adjacent cutpoints, $c < c'$. Auxiliary statistics for sets $\{l; X_l < c'\}$ and $\{l; X_l \geq c'\}$, needed for calculation of the splitting criterion for the split corresponding to cutpoint c', can be recalculated from the stored auxiliary statistics $\alpha\big(\{l; X_l < c\}\big)$

and $\alpha(\{l; X_l \geq c\})$, and from $\alpha(\{l; c \leq X_l < c'\})$, which is computed from the raw data.

In both cases calculations from the raw data are much reduced in comparison with the evaluation of each split "from scratch".

4 New algorithm minimizing the recalculation costs

The standard usage of the auxiliary statistics for categorical predictors, as described in Section 3, is far from being optimal. As we show in this section, the number of operations performed with the auxiliary statistics during the process of the search for the optimal split may be considerably reduced. We present a new algorithm, which has been inspired by the standard usage of the auxiliary statistics for *numerical* predictors. Though the set of all splits based on a categorical predictor lacks a *natural order*, we apply a proper *artificial ordering*, and then perform recalculation operations analogous to those that take place in the case of a numerical predictor.

For the rest of this section we shall use the following assumptions and notation. Let $\mathcal{L}$ denote the set of those cases that belong to a fixed node t. Let X be a fixed categorical predictor and $x_1, x_2, \ldots, x_n$ the values of X in t, $n > 2$.

4.1 Coding of splits

The $2^{n-1} - 1$ possible binary splits based on X can be coded by integers from 1 to $2^{n-1} - 1$ in the following way. At first, we assume that x_n always belongs to the right subset $\mathcal{L}_R(s)$ and $L(s) = \{k; s \text{ sends } x_k \text{ to the left}\}$. Then we assign an integer called the *split code* to any split s:

$$\texttt{splitcode}(s) = \sum_{k \in L(s)} 2^{k-1}.$$

Thus, the i-th rightmost digit of the binary representation of $\texttt{splitcode}(s)$ is 1 if and only if s sends x_i to the left. For example, for $n = 5$ the split sending x_1, x_4 to the left and x_2, x_3, x_5 to the right is coded by $\texttt{splitcode}(s) = 1001_{BIN} = 9$.

4.2 Ordering of split codes

The codes ranging from 1 to $2^{n-1} - 1$ of the $2^{n-1} - 1$ binary splits based on X can be ordered into the sequence $o_1, o_2, \ldots, o_{2^{n-1}-1}$ by the following rule.

1. We put formally $o_0 = 0$.
2. For $i \geq 1$,

$$o_i = 2^{\lfloor \log_2 i \rfloor} + o_h, \qquad h = 2^{\lfloor \log_2 i \rfloor + 1} - i - 1,$$

where $\lfloor x \rfloor$ denotes the integer part of x.

For example, $o_5 = 2^2 + o_2 = 4 + (2 + o_1) = 4 + 2 + (1 + o_0) = 4 + 2 + 1 + 0 = 7$.

Notice that the parameter n does not enter the definition. Thus, the sequences for different values of n are finite initial segments of one and the same infinite sequence $o_1, o_2, \ldots$.

The fundamental property of the sequence $o_1, o_2, \ldots$ is as follows. *For every $i \geq 1$, the binary representations of o_i and o_{i-1} differ in exactly one digit.* This enables, as shown in the next paragraph, economic recalculation of auxiliary statistics.

The above definition of $o_1, o_2, \ldots$ is not suitable for straightforward programming. Instead, we propose the following algorithm generating codes o_1, $o_2, \ldots$.

1. Let $i = 0$ and $o_0 = 0$.
2. Let `oldcode` $= o_i$.
3. Increment i by 1. If $i = 2^{n-1}$, then stop.
4. Find the rightmost digit equal to 1 in the binary representation of i and denote its position j.
5. Flip (i.e., change from 0 to 1, or from 1 to 0) the j-th rightmost digit in the binary representation of `oldcode`. Assign the result to o_i and go to step 2.

For example, let $i = 12 = 1100_{BIN}$. The rightmost binary digit 1 is at the 3-rd position from the right. Therefore from $o_{11} = 14 = 1110_{BIN}$ we get $o_{12} = 1010_{BIN} = 10$.

The algorithm is designed so that bit operations can be utilized. Unlike in the definition, code o_i is derived only from i and the immediate predecessor o_{i-1}.

Sequence $o_1, \ldots, o_{15}$ (for $n = 5$) together with corresponding splits is listed in Table 1.

Table 1. Sequence of splits and their codes for $n = 5$

i (split number)		o_i (split code)		split	
decadic	binary	decadic	binary	left subset	right subset
0	0000	0	0000	$(\emptyset)$	$(x_1, x_2, x_3, x_4, x_5)$
1	0001	1	0001	x_1	x_2, x_3, x_4, x_5
2	0010	3	0011	x_1, x_2	x_3, x_4, x_5
3	0011	2	0010	x_2	x_1, x_3, x_4, x_5
4	0100	6	0110	x_2, x_3	x_1, x_4, x_5
5	0101	7	0111	x_1, x_2, x_3	x_4, x_5
6	0110	5	0101	x_1, x_3	x_2, x_4, x_5
7	0111	4	0100	x_3	x_1, x_2, x_4, x_5
8	1000	12	1100	x_3, x_4	x_1, x_2, x_5
9	1001	13	1101	x_1, x_3, x_4	x_2, x_5
10	1010	15	1111	x_1, x_2, x_3, x_4	x_5
11	1011	14	1110	x_2, x_3, x_4	x_1, x_5
12	1100	10	1010	x_2, x_4	x_1, x_3, x_5
13	1101	11	1011	x_1, x_2, x_4	x_3, x_5
14	1110	9	1001	x_1, x_4	x_2, x_3, x_5
15	1111	8	1000	x_4	x_1, x_2, x_3, x_5

4.3 Evaluation of splits

In this paragraph we present a new algorithm which evaluates splits in sequence $s_1, \ldots, s_{2^{n-1}-1}$, where `splitcode`$(s_i) = o_i$ for $i = 1, \ldots, 2^{n-1} - 1$. We make use of the fact that for each $i > 0$ codes o_{i-1} and o_i differ in exactly one binary digit. The subsets of values from $\{x_1, ..., x_n\}$ sent to the left and to the right by s_i are thus obtained from analogous subsets related to s_{i-1} by "moving" just one element either from the left subset to the right one, or from the right subset to the left one.

Let $\mathcal{L}_{x_k}$, $k = 1, \ldots, n$ denote (as in Section 3) the sets of cases $\{l \in \mathcal{L}; X_l = x_k\}$. A slightly simplified version of the new algorithm reads as follows.

1. Compute $\alpha_k = \alpha(\mathcal{L}_{x_k})$, $k = 1, \ldots, n$, from the raw data.
2. Let $i = 1$, `code` $= 1$, $\alpha_L = \alpha_1$ and, using the `paste` operation iteratively, $\alpha_R = \alpha\left(\cup_{k=2}^{n} \mathcal{L}_{x_k}\right)$.
3. Calculate the splitting criterion for s_i as $\phi(s_i, \mathcal{L}) = \Phi(\alpha_L, \alpha_R)$.
4. Let `oldleft` $= \alpha_L$, `oldright` $= \alpha_R$ and `oldcode` $=$ `code`.
5. Increment i by 1. If $i = 2^{n-1}$, stop.
6. Let `code` $= o_i$.
7. Let j be the position (from the right) of the binary digit where `code` differs from `oldcode`.
   If `code` $>$ `oldcode`, then
   let $\alpha_L =$ `paste(oldleft,` α_j`)` and $\alpha_R =$ `cut(oldright,` α_j`)`,
   else
   let $\alpha_L =$ `cut(oldleft,` α_j`)` and $\alpha_R =$ `paste(oldright,` α_j`)`.
8. Go to step 3.

The "real-life" version of the algorithm differs from the above description in the following details.

- The splits coded by powers of 2 send only one of the x_k's to the left. If the simplified description of the algorithm was followed, then the auxiliary statistics for the left sets corresponding to these splits would be calculated by `cut` operations, despite the fact that they have already been calculated and stored (see step 1 of the algorithm). The algorithm can recognize the splits of the given kind and replace recalculations by recalling the saved auxiliary statistics. The problem concerning the split sending nothing but x_n to the *right* is resolved in an analogous way.
- During the calculation of $\alpha\left(\cup_{k=2}^{n} \mathcal{L}_{x_k}\right)$ (see step 2 of the algorithm), auxiliary statistics for the right sets $\mathcal{L}_R(s)$ of splits coded by $2^k - 1$, $k = 2, \ldots, n-2$ are obtained as intermediate results. The algorithm stores these auxiliary statistics during step 2, so that they can be recalled instead of recalculated at proper moments.

4.4 Recalculation cost savings

We shall concentrate here on the reduction of the number of recalculation operations `paste` and `cut`.

Without our new algorithm, the auxiliary statistics for sets $\mathcal{L}_L(s)$ and $\mathcal{L}_R(s)$, needed for the evaluation of any split s, are obtained by $n-2$ `paste` operations. Thus, for all $2^{n-1} - 1$ splits based on X, operation `paste` is executed $(2^{n-1} - 1)(n-2)$ times.

Let us take a look at the recalculation cost of the new algorithm. There are $2^n - 2 - n$ sets of cases, the auxiliary statistics of which must be obtained by recalculation. (These correspond to all the subsets of $\{x_1, \ldots, x_n\}$ of cardinality from 2 to $n-1$.) For each of them, the auxiliary statistic is obtained by only one execution of either `paste`, or `cut`. On assumption that operations `paste` and `cut` are of about the same computational costs, all the recalculations are approximately as costly as if operation `paste` was applied $(2^n - 2 - n)$ times.

The ratio of costs of the new and naive algorithms (on the above assumption of equal costs of the recalculation operations) is asymptotically equal to $2/(n-2)$. The number of operations for $n = 3, \ldots, 10$ is listed in Table 2.

Table 2. Number of recalculation operations without and with the new algorithm for splits based on a categorical predictor with n values

n	3	4	5	6	7	8	9	10
naive algorithm	3	14	45	124	315	762	1785	4088
new algorithm	3	10	25	56	119	246	501	1012
ratio new/naive	1	0.71	0.56	0.45	0.38	0.32	0.28	0.25

5 Conclusions

We have to admit that in spite of the impressive data of Table 2, the computational cost savings that can result from the new algorithm are rather minor *at present*. The reason is that the recalculation operations for all those kinds of auxiliary statistics that we have implemented up to now are quite simple; they consist in mere addition or subtraction of vectors or matrices, and they are cheap in comparison with other calculations involved in the process of the optimal split search. Thus, though we can demonstrate a considerable reduction *of the cost of recalculation of auxiliary statistics*, we cannot attain by our "trick" any large savings *of the total tree-growing cost.*

We are performing, nevertheless, analyses aimed at the development of more sophisticated auxiliary statistics for some splitting criteria. The results should, on one hand, enable to reduce the cost of calculation of the splitting criteria from the auxiliary statistics but, on the other hand, the respective recalculation rules may be non trivial and more costly. Therefore the practical importance of our present results is likely to grow in the future.

Acknowledgements
This research was partially supported by the following grants: Czech Ministry of Environment VaV/520/2/97, Grant Agency of the Czech Republic 102/96/0183 and CNR grant No. 94.4464-CT12 coordinated by Prof. Lauro.

References

Breiman, L., Friedman, J.H., Olshen, R.A. & Stone, C.J. (1984). *Classification and Regression Trees*. Belmont CA: Wadsworth.

Ciampi, A. (1993). Constructing prediction trees from data, the RECPAM approach. In: *Computational Aspects of Model Choice* (ed. J. Antoch), 105-156. Heidelberg: Physica-Verlag.

Klaschka, J. & Antoch, J. (1997). How to grow trees fast. In: *Robust'96* (ed. J. Antoch & G. Dohnal), 91-106. Prague: JČMF. (*In Czech).*

Klaschka, J., Siciliano, R. & Antoch, J. (1998). Computational enhancements in tree-growing methods. In: *Proceedings of IFCS'98* (ed. A. Rizzi & M. Vichi), in press.

Mola, F. & Siciliano, R. (1997). A fast splitting procedure for classification trees. *Statistics and Computing*, **7**, 209-216.

Mola, F. & Siciliano, R. (1992). A two-stage predictive splitting algorithm in binary segmentation. In: *COMPSTAT 92 Proceedings in Computational Statistics* (ed. Y. Dodge & J. Whittaker), Volume 1, 179-184. Heidelberg: Physica-Verlag.

Siciliano, R. & Mola, F. (1996). A fast regression trees procedure. In: *Statistical Modeling, Proceedings of the 11th International Workshop on Statistical Modeling*, (ed. A. Forcina *et al.*), 332-340. Perugia: Graphos.

Stable Multivariate Procedures - Strategies and Software Tools

Siegfried Kropf and Jürgen Läuter

Institute of Biometrics and Medical Informatics, Otto von Guericke University Magdeburg, Leipziger Str. 44, D - 39120 Magdeburg, Germany

Abstract. In 1996, Läuter proposed a new class of exact multivariate parametric tests for small samples of high-dimensional observations, which has been extended in subsequent papers. Here we give an overview on the two main strategies – the use of special linear scores in standard parametric tests and the construction of new tests based on random matrices with a uniform left-spherical matrix distribution. It is shown how these tests can be integrated into standard packages. In particular, SPSS macros are offered.

Keywords. Multivariate tests, stabilization, scores, correlation, software tools

1 Introduction

Traditional affine invariant multivariate methods require large samples to be effective with high-dimensional observations. In the framework of tests, for example, the question arises, whether it is more important to have a test statistic which is equally sensitive against all deviations from the null hypotheses, or to have a test which is powerful only in practically relevant situations, but then also for small samples.

Several proposals have been made for exploratory and confirmatory techniques which abandon the condition of affine invariance, for example ridge methods in regression analysis (Hoerl & Kennard, 1970) or discriminant analysis (Läuter, 1992) or the well-known tests of O'Brien (1984) for multiple endpoints. Such stabilized procedures offer alternative strategies to the methods that involve variable selection. Usually, however, these methods lead to intractable distributions of the statistics used for tests or prediction. Especially in the case of tests, this is a problem, since such tests do not keep the error of first kind exactly.

Recently, Läuter (1996) proposed a new class of exact tests for multivariate normal data. These tests are based on linear scores with data dependent weights. In subsequent papers (Läuter, Glimm & Kropf, 1996; Läuter, Glimm & Kropf, 1997), this approach was generalized, thus allowing for a surprising variety of adaptations to the data situation considered. A lot of heuristically motivated techniques can be included as long as some general strategy is maintained. Typically, such a test consists of two steps. In the first one, one-dimensional or q-dimensional scores are calculated from the p-dimensional observation vectors ($q << p$). Then these scores can be analyzed in standard tests of univariate or multivariate analysis. But it is also possible to use the scores or the original data in tests which are not known in the classical Gaussian set-up. The mathematical background is the theory of spherical matrix distributions (Fang & Zhang, 1990).

In Section 2, the basic mathematical ideas warranting the null distribution of the proposed tests are outlined. An example for a non-Gaussian test can be found there, too. Section 3 gives an overview on some possibilities for the derivation of effective scores for some common data structures, and it is shown, how these procedures can be included in standard software. In particular, macros for SPSS 7.5 are described.

2 Theoretical background

We consider an $n \times p$ -sample matrix $\boldsymbol{X} = \begin{pmatrix} \boldsymbol{x}_{(1)}' \\ \vdots \\ \boldsymbol{x}_{(n)}' \end{pmatrix}$ with independent p-dimensional normal data vectors $\boldsymbol{x}_{(j)} \sim N_p(\boldsymbol{\mu}_{(j)}, \Sigma)$, $j = 1,\ldots,n$, and a p-dimensional data dependent weight vector $\boldsymbol{d}$ or a $p \times q$ -weight matrix $\boldsymbol{D}$ which produces an n-dimensional score vector $z = \boldsymbol{Xd}$ or an $n \times q$ score matrix $\boldsymbol{Z} = \boldsymbol{XD}$, respectively. Then these n score values or score vectors are no longer independent nor normal. But under special restrictions on the weights, the theory of spherical distributions can be utilized to find the null distribution of the score based tests. The concept of spherical distributions replaces that of the multivariate normal distribution. We give here a short summary of definitions, theorems, and conclusions which are the background of the proposed procedures.

(i) An $n \times r$ matrix $\boldsymbol{Y}$ of random variables has a left-spherical distribution if for every fixed $n \times n$ orthogonal matrix $\boldsymbol{C}$ the distribution of $\boldsymbol{CY}$ is the same as that of $\boldsymbol{Y}$ (Fang & Zhang, 1990).

(ii) A left-spherically distributed matrix has uncorrelated, identically distributed rows. However, the rows are independent if and only if the rows have a p-dimensional normal distribution with zero expectation (Fang & Zhang, 1990).

(iii) With $\boldsymbol{U} = \boldsymbol{Y}(\boldsymbol{Y}'\boldsymbol{Y})^{-\frac{1}{2}}$ (the expression $\boldsymbol{B} = \boldsymbol{A}^{\frac{1}{2}}$ with a positive semidefinite symmetric matrix $\boldsymbol{A}$ denotes the symmetric square root, i.e. $\boldsymbol{B}^2 = \boldsymbol{A}$), a left-spherical $n \times q$ matrix $\boldsymbol{Y}$ is transformed into a matrix $\boldsymbol{U}$ that has a so-called $n \times q$ uniform matrix distribution (Fang & Zhang, 1990).

(iv) Consider a one-sample test for the null hypothesis $\mathbf{M} = \begin{pmatrix} \boldsymbol{\mu}_{(1)}' \\ \vdots \\ \boldsymbol{\mu}_{(n)}' \end{pmatrix} = \boldsymbol{0}$ with the above sample matrix $\boldsymbol{X}$. If the weights $\boldsymbol{d}$ or $\boldsymbol{D}$ are determined as arbitrary but uniquely defined functions of the matrix argument $\boldsymbol{X}'\boldsymbol{X}$, then the resulting scores have a left-spherical distribution under the null hypothesis, and the usual one-sample t test for the vector of scores $z = \boldsymbol{Xd}$ or Hotelling's T^2 test for the score matrix $\boldsymbol{Z} = \boldsymbol{XD}$ can be used. These tests with the scores are exact parametric tests, just as if they had been done with the original normally distributed variables (Läuter, 1996).

(v) In a one-way layout, a test of the hypothesis $\mathbf{M} = \begin{pmatrix} \mu_{(1)}' \\ \vdots \\ \mu_{(n)}' \end{pmatrix} = \begin{pmatrix} \mu' \\ \vdots \\ \mu' \end{pmatrix}$ can be based on scores with a weight vector or weight matrix derived from the sums of squares matrix $(\boldsymbol{X} - \overline{\boldsymbol{X}})'(\boldsymbol{X} - \overline{\boldsymbol{X}})$, where $\overline{\boldsymbol{X}} = \boldsymbol{1}_n \overline{\boldsymbol{x}}'$ is the matrix of total means with n identical rows of mean vectors $\overline{\boldsymbol{x}}' = \frac{1}{n} \boldsymbol{1}_n' \boldsymbol{X}$, and $\boldsymbol{1}_n$ is a vector consisting of n ones. The scores are then analyzed in a univariate or multivariate ANOVA with the scores (Läuter, 1996).

(vi) Suppose we have a test problem on q-dimensional multivariate normal data (matrix $\boldsymbol{Y}$) and a corresponding affine invariant statistic $F(\boldsymbol{Y})$, i.e. $F(\boldsymbol{YA} + \boldsymbol{B}) = F(\boldsymbol{Y})$ for every fixed positive definite matrix $\boldsymbol{A}$ and for every fixed shift matrix $\boldsymbol{B}$ which is irrelevant for the null hypothesis of the test problem. Then this test can also be applied on an $n \times q$ matrix of scores $\boldsymbol{Z}$ derived from $\boldsymbol{X}'\boldsymbol{X}$ as in (iv). The null distribution of the test statistic is the same for each covariance matrix Σ, for each special choice of weights and for each original dimension p.

(vii) If in (vi) the null hypothesis and the test statistic are invariant with respect to changes of the global mean, then we can also use scores derived from the term $(\boldsymbol{X} - \overline{\boldsymbol{X}})'(\boldsymbol{X} - \overline{\boldsymbol{X}})$. These tests should be more powerful than those in (vi).

(viii) New tests for normal data can be constructed by a suitable transformation into a random matrix $\boldsymbol{U}$ with a left-spherical uniform distribution under the null hypothesis of the test problem (utilizing (iii)). When a statistic can be defined on $\boldsymbol{U}$ which is sensitive against deviations from the null hypothesis, then the null distribution depends only on the two dimension parameters of $\boldsymbol{U}$. In general, this distribution will not be a known and tabulated one, such that critical values have to be derived.

Examples for tests corresponding to (iv) to (vii) are given in the next section. Here, we illustrate the above item (viii) by two proposals for a test of correlation between two sets of variables.

Suppose, two sets of variables, $x_1, x_2, \ldots, x_p$ and $y_1, y_2, \ldots, y_p$ are measured on n individuals. The common distribution is the $(2p)$-dimensional normal distribution, and we want to test the independence between the block of the x- variables and the block of the y-variables. Our background knowledge supports the idea, that correlation (if any) will occur between the variables with the same subscript. The tests should be exact for arbitrary correlation structure but should have a high power in case of the independence of the pairs of variables (x_i, y_i) $(i = 1, \ldots, p)$. For example, we might have two physiological parameters measured at several well-defined time points. We expect a correlation between the two measurement at the same time, but we do not believe in correlation across different points in time (without being sure of this).

Then we can use two new proposals for an exact test of independence between the x-variables and the y-variables. First we compute the usual matrix $\boldsymbol{R}$ of pairwise correlation coefficients between all $2p$ variables

$$\boldsymbol{R} = \begin{pmatrix} \boldsymbol{R}_{xx} & \boldsymbol{R}_{xy} \\ \boldsymbol{R}_{yx} & \boldsymbol{R}_{yy} \end{pmatrix}.$$

Then we calculate the matrix $\boldsymbol{C}_{xy} = \boldsymbol{R}_{xx}^{-\frac{1}{2}} \boldsymbol{R}_{xy} \boldsymbol{R}_{yy}^{-\frac{1}{2}}$. This matrix can be shown to be the upper $p \times p$-part of a $(n-1) \times p$ uniformly distributed matrix (cf. Kropf, Läuter & Klein, 1998). We define the trace $F_1 = \operatorname{tr}(\boldsymbol{C}_{xy})$ as the first proposal for a test statistic. When the data from different times are nearly independent, then F_1 is approximately equal to the sum of the pairwise correlation coefficients. The second proposal applies the transformation (iii) to the matrix $\boldsymbol{C}_{xy}$ to yield a $p \times p$ uniformly distributed matrix with the trace $F_2 = \operatorname{tr}(\boldsymbol{C}_{xy}(\boldsymbol{C}_{xy}{}' \boldsymbol{C}_{xy})^{-\frac{1}{2}})$. Critical values for the tests can be obtained from the authors by demand. Whereas the second statistic has the advantage that its distribution depends only on the parameter p, the first one seems to be more powerful in typical situations.

3 Use of stabilized methods with standard software

Stabilized procedures are a supplement to traditional univariate and multivariate procedures, not a substitute. Tests based on scores still utilize the classical tests. That is, why these methods should be used in the framework of a statistical standard package.

The spherical distributions are distributions of the whole sample matrix. As a consequence, tests utilizing this background are usually derived in matrix notation and can easily be implemented using the matrix language of the standard systems. Macros for these methods enable the effective interaction of special stabilized techniques with standard procedures. The above mentioned correlation tests could be the contents of such a macro as well as algorithms for the calculation of scores (see below). We prepared several SAS and SPSS macros. For a convenient use of these macros, the new version 7.5 of SPSS has the advantage of an easy-to-use script language which allows the incorporation of the new macros in the menu system, such that the new procedures can be started from the same graphical surface as the standard procedures of the system.

Here we want to describe an SPSS macro for the derivation of scores corresponding to the above items (v) / (vii) in a bit more detail. These scores are originally designed for the one-way layout, but can also be used in higher-factorial designs, in correlation analysis or for multiple testing problems (Läuter, Glimm & Kropf, 1996, 1997; Kropf, Hothorn & Läuter, 1997). The general strategy to derive weights for the scores from the term $(X - \overline{X})'(X - \overline{X})$ leaves a lot of possibilities for an adaptation to special properties of a given data set. At the present state the macro allows for four basic score versions (cf. Läuter, Glimm & Kropf, 1996):

- standardized sum (SS) score: useful when the p variables are expected to have equal directions of effects and similar effectiveness, 1 score;
- principal component (PC) score: useful when the p variables are expected to be controlled by one latent variable, 1 score;
- covariance sum (CS) score: intermediate properties of the above two versions, 1 score;
- q-dimensional principal component (PC_q) scores: useful if a multiple-factor structure is assumed, q scores; the number q is estimated from the data.

For the PC score and the CS score, one-sided and two-sided versions exist. All four score versions can be combined with a procedure for the selection of variables (Kropf, Läuter & Glimm, 1998).

Finally, one of three imputation methods for missing data can be included, which

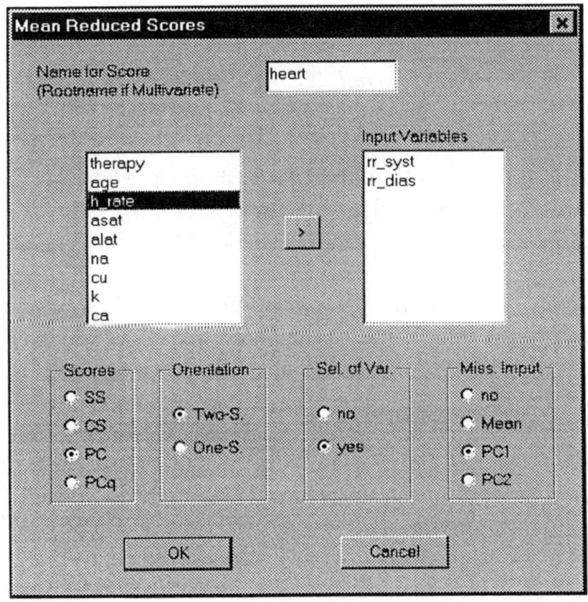

Fig 1. Dialog menu for calculation of scores

are approximately compatible to the theory of spherical distributions (Kropf, Läuter & Glimm, 1997):

- mean imputation,
- principal component based imputation,
- iterated principal component based imputation.

The macro can be invoked with a special dialog menu (Figure 1) written in a BASIC script. It appends the new score (or several new scores) to the data matrix, such that these scores can be used for graphics, standard tests from SPSS or for newly developed tests (as the correlation test above) just like original variables.

Other macros will treat scores based on $X'X$ or scores for repeated measurements and growth curves.

Acknowledgements. The authors thank the referee for his helpful criticism. This research was supported by German governmental grant BMBF 01ZZ9510.

References

Fang, K.-T. & Zhang, Y.-T. (1990). *Generalized Multivariate Analysis.* Berlin: Springer-Verlag.

Hoerl, A.E. & Kennard, R.W. (1970). Ridge Regression: Biased Estimation for Nonorthogonal Problems. *Technometrics*, **12**, 55-67.

Kropf, S., Hothorn, L.A. & Läuter, J. (1997). Multivariate Many-to-One Procedures with Applications to Preclinical Trials. *Drug Infromation Journal*, **31**, 433-447.

Kropf, S., Läuter, J. & Glimm, E. (1997). Stabilized Multivariate Tests - the Inclusion of Missing Values. *Biometrical Journal*, **39**, 149-169.

Kropf, S., Läuter, J. & Glimm, E. (1998). Stabilized Multivariate Procedures in Tests and Prediction. Submitted to *Statistics in Medicine*.

Kropf, S., Läuter, J. & Klein, S. (1998). Exact multivariate tests for correlation in a special data structure with pairs of variables. In: *Proceedings of the 43th Annual Meeting of the GMDS, Bremen, 1998*, MMV Medizin Verlag München, in press.

Läuter, J. (1992). *Stabile Multivariate Verfahren. Diskriminanzanalyse, Regressionsanalyse, Faktoranalyse.* Berlin: Akademie Verlag.

Läuter, J. (1996). Exact *t* and *F* Tests for Analyzing Studies with Multiple Endpoints. *Biometrics*, **52**, 964-970.

Läuter, J., Glimm, E. & Kropf, S. (1996). New Multivariate Tests for Data with an Inherent Structure. *Biometrical Journal*, **38**, 5-22.

Läuter, J., Glimm, E. & Kropf, S. (1997). Multivariate Tests Based on Left-Spherically Distributed Linear Scores. Preprint no. 6/1997 at the Faculty of Mathematics, University of Magdeburg, Germany; accepted by *Ann.Stat.*

O'Brien, P.C. (1984). Procedures for Comparing Samples with Multiple Endpoints. *Biometrics*, **40**, 1079-1087.

An Alternating Method to Optimally Transform Variables in Projection Pursuit Regression

Annalisa Laghi and Laura Lizzani

Department of Statistics, University of Bologna, Italy

Abstract. An alternating method to optimally transform both the response and the regressors in projection pursuit regression is proposed. It is based on alternating the model building stage and the transformation stage. Transformations are deemed optimal with respect to a goodness of fit measure. The main feature of the method is the possibility to deal with mixed data.

Keywords. ACE, distance-based regression, mixed data, MORALS, principal co-ordinate analysis, projection pursuit regression

1 Introduction

Projection pursuit regression (Friedman & Stuetzle, 1981; Friedman, 1984a) is a non-parametric regression method for modelling a response variable *Y*, given a *p*-dimensional random vector *X* of explanatory variables, on the basis of a sample of *n* matched observations $y_i, x_{i1}, x_{i2}, \ldots, x_{ip}$, $i = 1,2,\ldots,n$. The response variable is modelled as a linear combination of smooth functions of linear projections of the regressors. The model takes the form:

$$E(Y|x) = \mu_Y + \sum_{j=1}^{M} \beta_j f_j(\alpha_j' x) \qquad (1)$$

where $\mu_Y = E(Y)$, α_j are *p*-dimensional unit projection directions and f_j are univariate smooth functions, with zero mean and unit variance, of the projections $\alpha_j' x$, $j=1,2,\ldots,M$. Friedman's projection pursuit regression algorithm (Friedman, 1984a) estimates the coefficients β_j and α_j by least squares, and the smooth functions f_j, for each selected projection direction, using a variable span smoother, called the supersmoother (Friedman, 1984b). The algorithm proceeds by finding a model with $M_0 \geq M$ terms and then pruning the model back to a total of *M* terms, where *M* and M_0 are user-specified parameters.

The introduction of linear combinations of smooth functions and predictors allows projection pursuit regression to model non-linear regression surfaces and interactions between explanatory variables, respectively.

The projection pursuit regression method has been developed for continuous explanatory variables. In this work we propose to extend it to the case of mixed data through an alternating method which optimally transforms both the response and the regressors.

The paper is structured as follows: in Section 2 we present the proposed approach for the treatment of mixed variables in projection pursuit regression, together with two previously analysed procedures (Laghi & Lizzani, 1997). Examples and conclusions are discussed in Section 3.

2 Projection pursuit regression with mixed variables

The possibility to extend the projection pursuit regression method to the case of mixed predictors has been already analysed (Laghi & Lizzani, 1997) following two different approaches. The former consists of replacing, in model (1), each categorical predictor (with k categories) by k dummy variables (PPR-1). The latter consists of transforming the covariates by means of principal co-ordinate analysis (or metric multidimensional scaling) (Gower, 1966). A monotone transformation of Gower's similarity coefficient (Gower, 1971), purposely proposed to deal with mixed data, has been adopted to construct the Euclidean matrix of dissimilarities from which principal co-ordinate analysis moves. The p principal co-ordinates presenting the largest absolute correlation coefficient with the dependent variable are selected and inserted in model (1) (PPR-2). The idea to adopt projection pursuit in conjunction with metric multidimensional scaling has been suggested by Gower (1987).

Both procedures, PPR-1 and PPR-2, are based on preliminary transformations of the predictors.

A new method for optimally transforming the data is presented (PPR-3). It is based on alternating the model building stage and the transformation stage. Transformations are deemed optimal in the sense of minimising the fraction of unexplained variance (*FUV*) of the dependent variable.

The algorithm can be summarised as follows:

(0) Initialisations

$$\tilde{y} = (y - \bar{y}) / s_y$$

$$\tilde{x}_k = (x_k - \bar{x}_k) / s_{x_k} \quad k = 1, \ldots, p$$

(1) Model building and computation of *FUV*

$$\hat{y} = \sum_{j=1}^{M} \hat{\beta}_j \hat{f}_j (\tilde{X} \hat{\alpha}_j) = \sum_{j=1}^{M} \hat{\beta}_j \hat{f}_j (z_j)$$

where $\widetilde{X}$ is the $(n \times p)$ matrix containing the transformed regressors, $z_j = \widetilde{X}\hat{\alpha}_j$.

$$FUV = \sum_{i=1}^{n}\left[\widetilde{y}_i - \sum_{j=1}^{M}\hat{\beta}_j\hat{f}_j\left(\hat{\alpha}_j{}'\widetilde{x}_i\right)\right]^2$$

(2) Transformation of the dependent variable

$$y^* = s(\hat{y}|y)$$

where $s(\hat{y}|y)$ denotes the smoothing of $\hat{y}$ versus y.

$$\widetilde{y} = (y^* - \bar{y}^*)/s_{y^*}$$

(3) Transformations of the regressors

For *j=1,...,M*

updating of the partial residuals on the basis of current transformations:

$$r_j = \frac{\left(\widetilde{y} - \hat{y} + \hat{\beta}_j\hat{f}_j(z_j)\right)}{\hat{\beta}_j}$$

updating of the *j*-th linear combination: $\hat{z}_j = s(z_j|r_j)$

For *k=1,...,p*

$$\hat{x}_k = \frac{\left(\hat{z}_j - z_j + \hat{\alpha}_{kj}\widetilde{x}_k\right)}{\hat{\alpha}_{kj}}$$

$$x_k^* = s(\hat{x}_k|x_k)$$

$$\widetilde{x}_k = (x_k^* - \bar{x}_k^*)/s_{x_k^*}$$

$$z_j = \widetilde{X}\hat{\alpha}_j$$

end for *k*

evaluation of the regression function in correspondence of the current transformed values of the regressors: $\hat{y} = \sum_{j=1}^{M}\hat{\beta}_j\hat{f}_j(z_j)$

end for *j*

(4) Go to (1)

The procedure is iterated a fixed number of times (specified by the user) and the transformations yielding the minimum *FUV* value are retained as the solution. We could not specify a stopping rule because the algorithm fails to converge; in fact the sequence of the criterion of fit values does not show a monotonic behaviour. Supersmoother (Friedman, 1984b) is employed as the scatterplot smoother and linear interpolation to evaluate the smooth functions over the updated projections of the transformed carriers.

It is worthwhile to remark that PPR-3 inserts the variable transformations in the model building stage, while PPR-1 and PPR-2 use preliminarily transformed regressors.

Both PPR-3 and PPR-1 allow the interpretation of the transformations in terms of the original variables, while PPR-2 does not.

PPR-1 could become unreliable when the number of categorical predictors or categories increases. In such a situation it is advisable to resort to PPR-2 or PPR-3.

3 Examples

We test the performances of the three proposed procedures in handling the case of mixed predictors both on simulated and real data sets, and compare them with the results obtained with classical linear regression (CR), distance-based regression (DB) (Cuadras & Arenas, 1990), MORALS (Young *et al.*, 1976) and ACE (Breiman & Friedman, 1985).

The DB model assumes a linear relationship between the dependent variable and the p principal co-ordinates presenting the largest absolute correlation coefficient with the dependent variable. The MORALS method consists of maximising the multiple correlation coefficient by using an algorithm based on the alternating least squares and optimal scaling principles. The ACE method finds smooth non-linear transformations, both of the response and independent variables that produce the best fitting additive model.

PPR-3 algorithm structure is similar to MORALS, but transformations are obtained through the supersmoother as in ACE procedure.

The S-Plus functions *ppreg*( ) and *ace*( ) are used for PPR (PPR-1, PPR-2, PPR-3) and ACE method, respectively, while the SAS *proc transreg* is used for MORALS.

The performances of the different methods are evaluated on the basis of the fraction of the unexplained variance.

Simulated data

Data are generated according to the following models, characterised by a strongly non linear relationship between the response and the regressors, and by the presence of interaction terms:

$$C1:\ Y = 0.17X_1 \cdot 0.52X_3 + 0.26X_2 \cdot \left(0.2X_4 + 0.43X_5 + 0.64X_6\right) + \varepsilon$$

$$C2:\ Y = sin\left(0.17X_1 \cdot 0.26X_2\right)^2 + sin\left(0.52X_3 \cdot \left(0.2X_4 + 0.43X_5 + 0.64X_6\right)\right)^2 + \varepsilon$$

where X_1 and X_2 are normally distributed with zero mean and unit variance, X_3 is binary, X_4, X_5, X_6 are dummy variables representing a three-state categorical predictor and $\varepsilon \sim N(0,0.04)$.

In both cases (*C1* and *C2*) 50 samples, each of 200 observations, are generated (throughout all replications the *X* values are kept fixed).

Our three PPR based methods outperform the other ones in terms of the fraction of unexplained variances averaged over the 50 replications (*MFUV*). In particular PPR-3 shows the best performances in both situations (Table 1 and 2).

The poor performances of the DB model are due to its inability to model strongly non-linear regression surfaces. MORALS and ACE fail, as they are not able to capture interactions between predictors. CR performs the worst because it cannot deal with either non-linear relationships or interactions.

Table 1. Mean *FUV* and its standard deviation for *C1*

	MFUV	*SD(MFUV)*
PPR-3	0.1796	0.0051
PPR-2	0.1959	0.0051
PPR-1	0.2739	0.0069
MORALS	0.4907	0.0121
ACE	0.5483	0.0142
DB	0.6635	0.0075
CR	0.7472	0.0102

Table 2. Mean *FUV* and its standard deviation for *C2*

	MFUV	*SD(MFUV)*
PPR-3	0.2196	0.0048
PPR-2	0.2315	0.0067
PPR-1	0.3621	0.0079
MORALS	0.4839	0.0098
ACE	0.6676	0.0153
DB	0.7522	0.0063
CR	0.9192	0.0059

Real data

This example is taken from SAS/IML User's guide (1985, p. 67). The data come from an experiment in which nitrogen oxide emissions from a single cylinder engine were measured for various combinations of fuel, compression ratio and equivalence ratio. Only two kinds of fuel, ethanol and indolene, are considered, as in Cuadras & Arenas (1990) where the same data set is used to test the performances of the DB model. The data set consists of 110 observations. Two predictors, compression ratio and equivalence ratio, are continuous, while the remaining one, fuel, is a two-state categorical variable.

All methods, but classical linear regression, show good performances (see Table 3). Only one term ($M = 1$) is sufficient for PPR-3 to obtain the smallest fraction of unexplained variance.

Table 3. *FUV* for fuel data

		FUV		*FUV*
PPR-3	*M*=1	0.0116	*M*=2	0.0096
PPR-2	*M*=1	0.0643	*M*=2	0.0300
PPR-1	*M*=1	0.0407	*M*=2	0.0222
ACE		0.0359		
MORALS		0.0297		
DB		0.1126		
CR		0.7709		

References

Breiman, L. & Friedman, J. H. (1985). Estimating optimal transformations for multiple regression and correlations. *Journal of the American Statistical Association*, **80**, 580-619.

Cuadras, C. M. & Arenas, C. (1990). A distance-based regression model for prediction with mixed data. *Communications in Statistics - Theory and Methods*, **19**, 2261-2279.

Friedman, J. H. (1984a). SMART: User's Guide, Technical Report no. 1, Department of Statistics and Stanford Linear Accelerator Center, Stanford University.

Friedman, J. H. (1984b). A variable span smoother, Technical Report no. 5, Department of Statistics, Stanford University.

Friedman, J. H. & Stuetzle, W. (1981). Projection pursuit regression. *Journal of the American Statistical Association*, **76**, 817-823.

Gower, J. C. (1966). Some distance properties of latent root and vector methods used in multivariate analysis. *Biometrika*, 53, 325-338.

Gower, J. C. (1971). A general coefficient of similarity and some of its properties. *Biometrics*, 27, 857-872.

Gower, J. C. (1987). Contribution to the discussion of Jones, M. C. & Sibson, R. *Journal of the Royal Statistical Society, Series A*, **150**, 19-21.

Laghi, A. & Lizzani, L. (1997). Projection pursuit regression with mixed variables. *Book of Short Papers - Classification and Data Analysis,* Pescara (Italy), 3-4 July 1997, 181-184.

SAS/IML (1985). *User's guide*. Cary, North Carolina: SAS Institute Inc.

Young, F. W. & de Leeuw, J. & Takane, Y. (1976). Regression with qualitative and quantitative variables: an alternating least squares method with optimal scaling features. *Psychometrika*, **41**, 505-529.

Predicting from Unbalanced Linear or Generalized Linear Models

Peter W. Lane

Statistics Department, IACR-Rothamsted, Harpenden AL5 2JQ, England

Abstract. The formation of predictions to summarize the fit of a model is an important stage of practical analysis. In unbalanced linear models, and in generalized linear models, choices have to be made about the basis of standardization of subsidiary effects so that primary effects can be summarized. Two problems in the formation of predictive summaries are discussed and solutions are proposed. First, the large storage requirement of calculations for relatively modest models can be greatly reduced by modifying the algorithm to take account of the structure of the intermediate matrices. Second, the presence of non-estimable parameters can be dealt with satisfactorily by a further modification to keep track of 'unset' values throughout the calculations.

Keywords. Prediction, standardization, computer intensive methods, aliasing, estimability, unbalanced linear model, generalized linear model

1 Introduction

The goal of fitting models to data is often to allow the construction of predicted values on the basis of the established relationships. Such predictions can be intended as summaries of the fitted model, as in the 'predicted values' or 'fitted values' formed at each set of observed values of the explanatory variables used in multiple regression, and in the adjusted means from an analysis of covariance. Alternatively, they can be intended as forecasts of future values corresponding to chosen values of the explanatory variables. With models for time series, forecasting is recognized as an important final stage of an analysis, and software makes provision for it. But with regression models, and particularly with generalized linear models (McCullagh & Nelder, 1989), the final stage is rarely addressed, making it difficult to produce effective summaries or future predictions.

In a balanced linear model, the construction of predictions is straightforward. For example, in the analysis of a balanced experiment with several possibly interacting factors, it is usual to summarize the effects of a factor, or of a pair of interacting factors, with a table of means which ignores the other factors in the model. The effects of the other factors are orthogonal to those of the factors being summarized, so no adjustment is necessary; standard errors of means are constructed taking account of replication. In the presence of quantitative covariates, an analysis of covariance is summarized with means adjusted in addition to correspond to chosen values (often the means) of the covariates.

In an unbalanced model, the formation of summaries of this kind is not so easy. The effects of some factors in the model are partially aliased with the effects of others, and decisions have to be taken about the basis on which adjustments are made for the other factors. These choices were described in Lane & Nelder (1982), particularly the ideas of *marginal, equal-weights*, and *population-weights* standardization. Exactly the same problem arises in generalized linear models, where non-orthogonality between effects can be produced by the nonlinear link function as well as by unbalanced replication. The formation and interpretation of predictions from these models, particularly in log-linear and logistic regression, was investigated in Lane (1984).

This paper describes advances in the construction of predictions which address two problems identified in Lane (1984). Firstly, the computation involved in constructing measures of variability such as standard errors of the predictions can quickly become prohibitive as the number of combinations of factor levels increases. Secondly, difficulties arise when parameters in the model cannot be estimated because of absent factor combinations.

2 Prediction from models with many combinations of factor levels

The straightforward way of constructing predictions for one or more of the factors in a generalized linear model starts by forming a table of 'linear predictors' classified by all the factors in the model. Each value in the table is easily derived from the estimates of parameters representing the factor effects, together with the effect of each quantitative variable in the model corresponding to some standard value of the variable (commonly the mean). The table is then transformed by the inverse of the link function, before averaging over the levels of the subsidiary factors: that is, those not chosen for examination. The final predictions are then interpretable as summaries of some of the factors on the natural scale, without difficulties caused by taking means on the transformed scale; but see Ridout (1987) for a discussion of circumstances when it might be preferable to average on the transformed scale.

The predictions are more informative if accompanied with standard errors. These can be formed from the variance-covariance matrix of the parameters, and modified by the derivative of the inverse link function to give an approximate variance-covariance matrix of values in the full table on the natural scale, from which the standard errors of the averaged values can easily be derived. However, this requires storage of a covariance matrix with one row and column for each cell, before taking means to produce the final summary for one or two factors only. If any of the factors in the model have many levels, the matrix can be very large indeed, as illustrated by the following example.

2.1 Example: analysis of edge effects in *Biolog* plates

The microbiological properties of soil can be investigated by the use of *Biolog* plates, which consist of a series of 'wells' containing a range of different carbon substrates. A given amount of soil solution is added to each well, and microbial activity is measured after a fixed time in terms of the colour-change of a reagent. Various methods have been proposed for using these multivariate measurements of soil to characterize important properties; see Hackett & Griffiths (1997) for a comparison of some of these methods.

A study was recently carried out to investigate some of the properties of these measurements, which will be published in a microbiological journal. One aspect required the calculation of predictions. The study used plates with 96 wells, grouped into three replicate sets of 32 substrates. Five different soil samples were investigated, adding a solution made from each soil to 11 plates which were allowed to 'develop' over different lengths of time, from 0 to 140.5 hours, before taking colour measurements. In total, therefore, there were 5×11×96 = 5280 measurements. The measurements lie in a range from 0 to about 2.8 units, and show the typical variance pattern of such restricted observations, with declining variance at each end of the range. I therefore used a generalized linear model with a binomial-like relationship between mean and variance, which can be fitted by quasi-likelihood methods (Wedderburn, 1984).

One aspect of the study was to assess whether the unrandomized positioning of the substrates on these mass-produced plates could bias the results. The most likely effect of positioning is uneven colour development related to distance of each well from the edge of the plate. Plates are stacked in a cabinet during development, and air reaches the wells through small gaps between the stacked plates. I fitted a model to assess whether there was a difference between wells on the edge of a plate, and those not on the edge. This model required in addition the main effects of factors representing the 32 different sources, the 5 soil samples and the 11 times. On the logit scale, there appeared to be no interactions between these factors, and the effect of the 'edge' factor seemed small and not statistically significant, with a deviance ratio of about 1.5. On average, the wells on the edge tended to have developed more, with an estimated effect of 0.0450 (s.e. 0.0368) on the logit scale.

To understand what this effect means in scientific terms, it is necessary to transform it back onto the natural scale of the measurements. Using the Genstat system (Genstat Committee, 1993), which provides a PREDICT command to form predictions, I formed predictions for the edge factor, averaging on the natural scale over the other factors – substrate, soil and time – and scaling by the range of 2.8:

```
              Prediction          S.e.
      edge
        no         1.0466      0.00742
       yes         1.0658      0.01095
```

This shows that the size of the edge effect, as well as being statistically non-significant, is scientifically uninteresting, compared to the range of variation (0 to 2.8) of the measurements.

The calculation of the standard errors in this simple table required the formation of a variance-covariance matrix with 32×11×5×2 = 3520 rows and columns (one for each combination of levels of the factors in the model), giving 6,196,960 values to store (taking advantage of symmetricity). Though modern computers may have such amounts of space available, it seems wasteful for such a simple summary; moreover, another type of Biolog plate has 96 different substrates rather than three replicates of 32, so more than 55 million values would have to be stored.

2.2 An algorithm to form standard errors of prediction with reduced storage

The problem can be circumvented by an algorithm to accumulate the contributions to each final predicted value and measure of variability, without the need to store the full table or variance-covariance matrix.

The expected values of the response variable Y in a generalized linear model can be represented by the equation

$$E(\mathbf{Y}) = g^{-1}(\mathbf{X}\boldsymbol{\beta})$$

where $\mathbf{X}$ is the design matrix, consisting of values of explanatory variates and of dummy variates representing explanatory factors, and g( ) is the link function. The full set of predictions on the scale of the linear predictor is then

$$\mathbf{p} = \mathbf{X}_p\mathbf{b}$$

where $\mathbf{b}$ are maximum-likelihood estimates of the parameters and $\mathbf{X}_p$ is a modified design matrix for all possible combinations of levels of explanatory factors, with explanatory variates at their mean values. If $\mathbf{V}$ estimates the variance-covariance matrix of the parameters, then the variance-covariance matrix of $\mathbf{p}$ is

$$\mathbf{S} = \mathbf{X}_p\mathbf{V}\mathbf{X}_p^T$$

Transforming to the natural scale of the response, the predictions become

$$\mathbf{p}' = g^{-1}(\mathbf{p})$$

with approximate variance-covariance matrix (derived by a Taylor expansion)

$$\mathbf{S}' = \mathbf{D}\mathbf{S}\mathbf{D}^T$$

where $\mathbf{D}$ is a diagonal matrix, containing the derivative of the inverse link function g^{-1}( ) with respect to the linear predictor, at each combination of factor levels.

The full set of predictions can then be standardized over most of the factors in the model, to give a summary for one or two of them only. This can be represented by a diagonal matrix of weights, $\mathbf{W}$, calculated according to the chosen type of standardization, and an a×b block matrix $\mathbf{C}$ to combine the weighted values, where a is the final number of predictions and b is the number of predictions in the full set. The final predictions are then

$$\mathbf{p}'' = \mathbf{C}\mathbf{W}\mathbf{p}' = \mathbf{C}\mathbf{W}g^{-1}(\mathbf{X}_p\mathbf{b})$$

with approximate variance-covariance matrix

$$\mathbf{S}'' = \mathbf{C}\mathbf{W}\,\mathbf{S}'\,\mathbf{W}\mathbf{C}^T = \mathbf{C}\mathbf{W}\mathbf{D}\mathbf{X}_p\mathbf{V}\mathbf{X}_p^T\mathbf{D}\mathbf{W}\mathbf{C}^T$$

In this expression, it is the component $\mathbf{X}_p\mathbf{V}\mathbf{X}_p^T$ which poses the computational problem; it is symmetric with dimension b×b, and so requires $b(b+1)/2$ storage locations: 6,196,960 in the example above. The component $\mathbf{V}\mathbf{X}_p^T$, however, requires only mb locations, where m is the number of parameters. In the example above, this is 47×3520 = 165,440. An algorithm for calculating $\mathbf{S}''$ is as follows, and has been implemented in Genstat (from Release 3.1):

1) initialize $\mathbf{S}''_{pq} = 0$ for p=1...a, q=1...p

2) loop for i=1...b

2.1) for combination i, calculate and store $\mathbf{VX_p}^{\mathrm{T}}{}_{ki}$ for k=1...m

2.2) loop for j=1...i

2.2.1) calculate $\mathbf{S}_{ij}$ from $\mathbf{VX_p}^{\mathrm{T}}{}_{kj}$ k=1...m, j=1...i

2.2.2) multiply $\mathbf{S}_{ij}$ by $\mathbf{D}_{ii}\mathbf{W}_{ii}\mathbf{W}_{jj}\mathbf{D}_{jj}$

2.2.3) add to $\mathbf{S}''_{pq}$ where i is mapped to p by $\mathbf{C}$, and j to q

If the predictions are to be interpreted as estimates of new values of the response, rather than as adjusted fitted values for the observed data, an extra component of variation needs to be added to the diagonal of $\mathbf{S}''$, according to the assumed distribution of the response. In this case, though, it is unlikely that standardization would be carried out because averages of future values would be difficult to interpret.

3 Avoiding misleading predictions when parameters are aliased

Effects in a model can be *aliased*, and so non-estimable, either as a result of linear dependence between explanatory variables or because of non-representation of some combinations of factor levels in the collected data. In the latter case, it is essential to avoid producing misleading summaries which make unwarranted assumptions about the unknown effects. As an example, consider the effect of losing the data, perhaps by contamination, from one of the Biolog plates in the experiment above, say for soil sample 3 tested after 65 hours. With a model containing main effects only, there is no difficulty because all parameters could be estimated. But if the soil samples show different patterns of development over time, it would be necessary to fit an interaction between soil and time, and one component of the interaction could not be estimated.

When forming predictions to summarize a model in these circumstances, there are several possible approaches. For example, in Genstat the default approach is to avoid forming predictions and to display a diagnostic message. This can be overcome by setting options to specify how to deal with the missing parameter estimates. A computationally simple approach is to assume that the parameter is zero, which in the context above corresponds to an assumption that the colour development from the first time to the missing time is the same for the missing soil as for whichever has been chosen as the reference soil. Here is the resulting summary for the effect of soil sample, standardized for the effects of time and substrate:

```
Soil sample        1        2        3        4        5
 Prediction   0.8974   1.2048   0.9262   1.1220   1.0909
```

An alternative is to specify that standardization should be carried out only over the set of combinations that have actually been observed in the data:

```
Soil sample        1        2        3        4        5
 Prediction   0.8974   1.2048   0.9223   1.1220   1.0909
```

This method may be appropriate if the combinations are necessarily missing; but when they are accidentally missing, as here, the resulting prediction for sample 3 is based on a different standardization strategy than those for the other samples, and so can be misleading.

To avoid this problem in a general system for forming predictions, in the absence of specific weights for standardization, it is best to avoid forming any predictions that involve contributions from non-estimable parameters. This requires modification of the algorithm used to calculate the predictions, to allow any intermediate value in the calculations effectively to be 'unset' with the effect that consequent values formed from it are also 'unset'. This has been implemented in Release 4.1 of Genstat, leading to the following summary when this option is selected:

```
Soil sample        1        2        3        4        5
 Prediction   0.8974   1.2048        *   1.1220   1.0909
```

4 Conclusion

The formation of predictions from models involving factors with many levels can be made more efficient by taking advantage of the form of the matrices involved. The calculations are still computer-intensive, but the amount of storage space required is a small fraction of that needed otherwise. The handling of aliased parameters can also be improved to provide partial summaries of effects without assumptions about unestimated effects. This method of dealing with aliasing is likely to become the default in a future release of Genstat.

Acknowledgements. I am grateful to Dr Kirsten Lawlor (Soil Science Department, IACR-Rothamsted) for permission to use data from her experiments with Biolog plates. IACR receives grant-aided support from the Biotechnology and Biological Sciences Research Council of the United Kingdom.

References

Genstat 5 Committee (1993). *Genstat 5 Release 3 Reference Manual.* Oxford: Numerical Algorithms Group.

Hackett, C.A., & Griffiths, B.S. (1997). Statistical analysis of the time-course of Biolog substrate utilization. *Journal of Microbiological Methods*, **30**, 63-69.

Lane, P.W. (1984). Methods for summarizing the analysis of categorical data. In: *COMPSTAT 1984 Proceedings in Computational Statistics* (ed. T. Havranek, Z. Zidak & M. Novak), 366-371. Vienna: Physica-Verlag.

Lane, P.W, & Nelder, J.A. (1982). Analysis of covariance and standardization as instances of prediction. *Biometrics*, **38**, 613-621.

McCullagh, P. & Nelder, J.A. (1989). *Generalized linear models (second edition).* London: Chapman & Hall.

Ridout, M.S. (1987). Prediction on the scale of the linear predictor for generalized linear models. In: *Genstat Newsletter*, **20**, 20-30. Oxford: Numerical Algorithms Group.

Wedderburn, R.W.M. (1984). Quasi-likelihood functions, generalized linear models and the Gauss-Newton method. *Biometrika*, **61**, 439-447.

MCMC Solution to Circle Fitting in Analysis of RICH Detector Data

A. Linka[1], G. Ososkov[2], J. Picek[1], P. Volf[3]

[1] Department of Applied Mathematics, Technical University of Liberec, 461 17 Liberec, Czech Republic
[2] JINR Dubna, Russia
[3] UTIA AV CR, Prague, Czech Republic

Abstract. The present contribution deals with a data analysis problem which is of great importance for many experiments in high energy physics, namely the problem of recognition of circles in observed noisy planar data. The situation is formulated in terms of the Bayesian estimation problem, the solution is based on the Metropolis–Hastings algorithm.

Keywords. Markov chain Monte Carlo (MCMC), Metropolis-Hastings algorithm, Bayesian estimation, ring imaging Cherenkov (RICH) detector

1 Introduction

Markov chain Monte Carlo (MCMC) generates a Markov chain whose probability distribution converges to a given target one. When combined with simulated annealing, MCMC is used as a method of randomized optimization. In the present contribution the MCMC approach is applied to a problem of recognition of certain structures in planar data, namely the problem of fitting circles to data obtained from Cherenkov photons detector. As the data are contaminated with random noise, the problem is formulated as a problem of statistical estimation of a multivariate parameter in the framework of a probabilistic model of the physical phenomenon. The paper is organized as follows: First, properties of Metropolis-Hastings MCMC algorithm are recalled. Then the performance of Cherenkov photons detector RICH is described. Finally, the method for off-line analysis of Cherenkov rings is presented.

2 On hybrid MH algorithm

The idea of Metropolis-Hastings algorithm is described elsewhere (e.g. in Roberts & Smith, 1994). In the context of the Bayesian estimation problem, the limit (target) distribution is the posterior distribution of model parameters given the data. In standard cases, this distribution is either a discrete one or has a density w.r. to some fixed measure. Let $g(\boldsymbol{\theta}|\boldsymbol{x})$ be such a density, $\boldsymbol{\theta}$ the parameter of interest, $\boldsymbol{x}$ the data. The Metropolis–Hastings algorithm generates the chain of values of parameters $\{\boldsymbol{\theta}^{(m)}, m = 0, 1, \ldots\}$ in the following way: In state $\boldsymbol{\theta}^{(m)}$, it first proposes a new value $\boldsymbol{\theta}^*$, drawing it from a conditional distribution $P(\mathrm{d}\boldsymbol{\theta}^*|\boldsymbol{\theta}^{(m)})$. Then, $\boldsymbol{\theta}^{(m+1)}$ is set to $\boldsymbol{\theta}^*$ with probability

$$\min\left\{1, \frac{g(\boldsymbol{\theta}^*|\boldsymbol{x})\, P(\mathrm{d}\boldsymbol{\theta}^{(m)}|\boldsymbol{\theta}^*)}{g(\boldsymbol{\theta}^{(m)}|\boldsymbol{x})\, P(\mathrm{d}\boldsymbol{\theta}^*|\boldsymbol{\theta}^{(m)})}\right\},$$

otherwise $\boldsymbol{\theta}^{(m+1)} = \boldsymbol{\theta}^{(m)}$. If the proposals generate an irreducible and aperiodic sequence, the convergence of distribution of $\boldsymbol{\theta}^{(m)}$ to the distribution given by $g(\boldsymbol{\theta}|\boldsymbol{x})$ is guaranteed. If we denote by $p(\boldsymbol{x};\boldsymbol{\theta})$ the probability density of the data for a given parameter value $\boldsymbol{\theta}$ and by $g_0(\boldsymbol{\theta})$ the density of the prior distribution, then $g(\boldsymbol{\theta}|\boldsymbol{x}) \sim p(\boldsymbol{x};\boldsymbol{\theta}) \cdot g_0(\boldsymbol{\theta})$. In the special case, when the prior distribution is used as a proposal distribution, the acceptance probability reduces to $\min\{1, \frac{p(\boldsymbol{x};\boldsymbol{\theta}^*)}{p(\boldsymbol{x};\boldsymbol{\theta}^{(m)})}\}$.

In the case of a multidimensional parameter the standard version of the method updates one component of $\boldsymbol{\theta}$ after another (visiting them either randomly or systematically). However, in many cases of Bayesian parameter estimation, the parameter dimension is not known. It is then necessary to consider several different types of transitions, some of them changing the dimension of the parameter. Their combination then leads to what is called a hybrid algorithm. The problem is that now the probabilities $P(\mathrm{d}\boldsymbol{\theta}|\boldsymbol{\theta}^*)$ and $P(\mathrm{d}\boldsymbol{\theta}^*|\boldsymbol{\theta})$ may be defined in different spaces. Such a situation is discussed and cleared up for instance in Green (1995), for further explanation see also Tierney (1995). It is shown that two mutually reverse steps have to be defined with respect to a symmetrical joint measure (which actually is a product of measures corresponding to individual steps) and restricted to subspaces of items which can be reached one from the other. These conditions are more precisely characterized by the “dimension matching” assumption of Green (1995).

3 Application – the RICH detector

The Rich detector registers (in a finite two-dimensional grid) the incidences (hits) of Cherenkov photons emanated by particles passing through the detector. The hit points create a number of rings, with different centres and radii. From the radius the type of the original particle is to be recognized. However, the incidence points are not observed directly, the data consist of measurements of energy at cells of the grid of the detector. Typically, the energy released by a photon hit is not concentrated to one site but is (randomly) dissipated in several adjacent cells. Thus, we observe a number of points (cells) with positive energy (amplitude). The objective is to recognize the rings (and their number).

The method which is used up to now consists of two steps: in the first one the centres of photon hits are estimated with the aid of a clustering procedure, the second step fits rings to these centres (Agakichiev *et al.*, 1996). Both steps were combined to one robust procedure in Chernov *et al.* (1995). The simplest (and not too reliable) method uses weighted averaging, with the advantage that it can be employed also for on-line computations. As regards an off-line analysis, a number of different techniques (including methods of pattern recognition) can be considered. We have explored an application of the MCMC method, in the framework of Bayesian estimation of unknown parameters. The distribution of observed data is rather complicated, for instance it includes a distribution of points of hits and this distribution should be integrated out. Moreover, the presence of frequent additional ‘noisy’ background photons has to be taken into account. In a traditional approach the robust technique was used (e. g. a part describing the contamination was added to Gaussian distribution of hit points). We avoid this by introducing a Poisson model of occurrence of background photons.

4 Probabilistic model of RICH

Let us first consider one ring created by photons emitted by one particle. At least partial information about the particle trajectory is available, so that the position of centre of the ring is known to a certain extent. This information can be used for a choice of prior distribution. As regards the radius, a simplified case deals with only a few different radii, so that the prior distribution of the radius is selected either as a discrete one or as a mixture of normal distributions, each concentrated around one of expected radii values.

4.1 Notation

$\mathcal{N}$ is a $N \times N$ grid of cells of the detector screen. Each cell $s \in \mathcal{N}$ is given by its coordinates $s = (i, j)$, $i, j \in \{1, \ldots, N\}$.

$\boldsymbol{c} = (c_x, c_y) \in \mathcal{N}$ is the position of the centre of a Cherenkov ring, R is the radius of a ring.

K – random number of Cherenkov photons emitted by one particle collision, λ – the mean of K.

L – number of background photons, μ their intensity (mean number). In a more general case, μ can also be an unknown parameter.

$(x, y) \in \mathcal{N}$ is a point (cell) hit by a photon (either a Cherenkov or a noisy one).

A_{uv} is the energy observed at cell $(u, v) \in \mathcal{N}$, so that $\boldsymbol{A} = \{A_{uv}\}$ are the data registered by the detector.

$\overline{A}_{uv}$ is the energy expected at site (u, v). Naturally, it depends on the position of $\boldsymbol{c}$, on R, K, L, on locations of the hit points (x, y).

The energy produced by one photon is random. It can be modelled with the aid of the exponential distribution. In order to avoid analytical problems caused by the local dependence of $\overline{A}_{uv}$ we approximate the energy by its expected value $\overline{E}$ which is supposed to be known.

4.2 Description of distributions entering the model

$g_0(\boldsymbol{c})$ is the prior distribution of $\boldsymbol{c}$ (actually a two-dimensional density discretized to $\mathcal{N}$).

$f_0(R)$ – prior distribution of R (or its density).

$P_\lambda(K)$, $P_\mu(L)$ – Poisson distributions of numbers of photons, with parameters λ, μ respectively. It is assumed that background photons are distributed uniformly in the area $\mathcal{N}$.

As regards the hit points of Cherenkov photons, we assume that they are mutually independent (given K, $\boldsymbol{c}$, R), and that each is given (in radial coordinates around centre $\boldsymbol{c}$) by the angle φ and radius r. Angle φ is distributed uniformly in $(0, 2\pi)$, $r \sim \mathcal{N}(R, \sigma_R^2)$. Then, for $(\boldsymbol{\varphi}, \boldsymbol{r}) = \{(\varphi_k, r_k),\ k = 1, \ldots, K\}$,

$$f(\boldsymbol{\varphi}, \boldsymbol{r}|K, R) = \prod_{k=1}^{K} \left\{ f_1(r_k|R) \cdot \frac{1}{2\pi} \right\}, \quad f_1(r|R) = \frac{1}{\sqrt{2\pi}\,\sigma_R} \cdot \exp\left(-\frac{(r-R)^2}{2\sigma_R^2} \right) \tag{1}$$

and corresponding (x, y) coordinates are

$$x_k = c_x + r_k \cos\varphi_k, \qquad y_k = c_y + r_k \sin\varphi_k. \tag{2}$$

The expected contribution of a photon hitting the cell (x, y) to the energy observed at cell (u, v) is given by $\overline{E} \cdot \rho[(u,v),\,(x,y)]$, where

$$\rho[(u,v),\,(x,y)] = \frac{1}{2\pi\,\sigma_d^2} \cdot \exp\left(-\frac{(u-x)^2 + (v-y)^2}{2\sigma_d^2}\right).$$

Therefore, the total expected energy at (u, v) is the sum of expected contributions from Cherenkov photons and from background photons, namely

$$\overline{A}_{uv} = \overline{E} \cdot \lambda \cdot \int_x \int_y \rho[(u,v),\,(x,y)] \cdot f_1^*(x,y|\boldsymbol{c},R)\,\mathrm{d}x\mathrm{d}y + \frac{\overline{E}}{N^2} \cdot \mu, \qquad (3)$$

where f_1^* is derived from the distribution of r and φ, i.e. from (1) and (2).

Finally, the distribution of the energy A_{uv} actually observed at cell (u, v) is $\mathcal{N}(\overline{A}_{uv}, \sigma_a^2)$. We assume that variables A_{uv} are mutually conditionally independent, given the values of $\overline{A}_{uv}$. In other words, we model the dependence of energies at neighbouring cells through the dependence of expected energies $\overline{A}_{uv}$ in (3). Thus, for each given $\boldsymbol{c}$ and R we are able to compute the probability $p(\boldsymbol{A};\boldsymbol{c},R)$ of observed data $\boldsymbol{A} = \{A_{uv}, (u,v) \in \mathcal{N}\}$. This probability distribution depends naturally on a set of parameters. We assume that these parameters, namely $\overline{E}$, λ, μ, σ_R, σ_a, σ_b are known, from the physical background of the experiment. Parameters of priors g_0, f_0 are selected by an analyst, but they also may follow from the knowledge of the physical context.

Remark: All normal distributions should be (more realistically) taken as trimmed normal, either symmetrically (e.g. two-dimensional density ρ) or with a non-symmetrical threshold (e.g. energy A is observed only between some $A_{min} \geq 0$ and A_{max} given by the detector limitations).

The integral in expression (3) will be approximated in a Monte Carlo way: For sufficiently large n, we sample independently $\varphi_1, \ldots, \varphi_n$ from the uniform distribution $U(0, 2\pi)$ and, for a given R, we sample independently $r_1, \ldots, r_n$ from $f_1(r|R)$. Then, for a given $\boldsymbol{c}$, n pairs (x_i, y_i) are obtained from (2) and the integral is approximated by $\frac{1}{n}\sum_{i=1}^{n} \rho[(u,v),\,(x_i,y_i)]$.

4.3 MCMC algorithm

The objective is to derive optimal (in the Bayesian sense) values of $\boldsymbol{c}$ and R from observed data $\boldsymbol{A}$. Once the likelihood function $p(\boldsymbol{A};\boldsymbol{c},R)$ is available, we can construct a Metropolis–Hastings algorithm for approximate sampling from the posterior distribution of $\boldsymbol{c}$ and R. Details of practical implementation of the algorithm can be modified, from the choice of initial values, up to the method of updating. As the MCMC is a method of global random search, knowledge leading to a reduction of space of the search is very valuable. A basic variant of the procedure consists of the following steps:

1. Select initial $\boldsymbol{c}^{(m)}$, $R^{(m)}$ $(m = 0)$.
2. Propose new $\boldsymbol{c}^*$, R^* from their priors.
3. Compute $\pi = \frac{p(\boldsymbol{A};\boldsymbol{c}^*,R^*)}{p(\boldsymbol{A};\boldsymbol{c}^{(m)},R^{(m)})}$.
4. Set, with probability $\min(1, \pi)$, $\boldsymbol{c}^{(m+1)} = \boldsymbol{c}^*$, $R^{(m+1)} = R^*$, otherwise $\boldsymbol{c}^{(m+1)} = \boldsymbol{c}^{(m)}$, $R^{(m+1)} = R^{(m)}$.

Repeat loop 2.–4. J-times. Choose J_1 such that both J_1 and $J - J_1$ are sufficiently large. Take averages $\overline{\boldsymbol{c}}$, $\overline{R}$ from the last $J - J_1$ iterations as final estimates of $\boldsymbol{c}$, R.

In a simulated annealing variant the acceptance probability defined at step 4 is $\min(1, \pi^{s(m)})$, where the function $s(m)$ is selected in such a way that $s(m) \to \infty$ as $m \to \infty$ (e.g. $s(m) \sim \log(1+m)$).

4.4 Case of several rings

Let us now explore the case of the data created by M Cherenkov photon rings together with a set of background photons. We assume that the rings can have different centres and radii $\boldsymbol{c}_j$, R_j, $j = 1, \ldots, M$ and, eventually, $\overline{E}_j$, and common remaining parameters λ, σ_a, σ_d, σ_R. Naturally, M is not known to an analyst and is also the subject of estimation. We select a prior distribution $G_0(M)$ on $\{0, 1, 2, \ldots\}$. In Linka *et al.* (1996) it has been shown that a decreasing prior (e. g. $G_0(M) \sim \exp(-M)$) acts like a penalty and reduces an uncontrolled growth of M.

The MCMC procedure now generates a sequence $M^{(m)}, \boldsymbol{C}^{(m)}, \boldsymbol{R}^{(m)}$ and, therefore, some of its transitions change the dimension of the state space. The standard approach to such a situation updates one component of state after another – here one ring after another – in the following way: Let M be the current number of rings. At the next step, with probability $\frac{1}{3}$, a completely new $M+1$-st ring is proposed ($M^* = M+1$). Otherwise (i.e. with probability $\frac{2}{3}$) we select one (j-th) ring from the existing ones. Then, with equal (0.5) probability it is proposed either to update its $\boldsymbol{c}_j$ and R_j ($M^* = M$) or to discard the j-th ring ($M^* = M - 1$). Thus, the proposals of new M are symmetrical, $Q(M^*|M) = Q(M|M^*)$. Further, we have to choose the conditional priors of ring parameters, for given M, $g_0(\boldsymbol{C}, \boldsymbol{R}|M)$ and probabilities generating new ring parameters, $P(\mathrm{d}\boldsymbol{C}^*, \mathrm{d}\boldsymbol{R}^*|\boldsymbol{C}, \boldsymbol{R}, M, M^*)$, where $\boldsymbol{C}^*$, $\boldsymbol{R}^*$, M^* denote the updated configuration of rings and $\boldsymbol{C}$, $\boldsymbol{R}$, M the old one. In such a general setting, the acceptance probability is $\min(1, \pi)$, with

$$\pi = \frac{p(\boldsymbol{A}; \boldsymbol{C}^*, \boldsymbol{R}^*, M^*)}{p(\boldsymbol{A}; \boldsymbol{C}, \boldsymbol{R}, M)} \frac{g_0(\boldsymbol{C}^*, \boldsymbol{R}^*|M^*) G_0(M^*)}{g_0(\boldsymbol{C}, \boldsymbol{R}|M) G_0(M)} \frac{P(\mathrm{d}\boldsymbol{C}, \mathrm{d}\boldsymbol{R}|\boldsymbol{C}^*, \boldsymbol{R}^*, M, M^*)}{P(\mathrm{d}\boldsymbol{C}^*, \mathrm{d}\boldsymbol{R}^*|\boldsymbol{C}, \boldsymbol{R}, M^*, M)}.$$

Naturally, it can be simplified by a convenient choice of the components of functions g_0 and P.

Example. We simulated a simple example of two overlapping circles (with additional noisy photons). 'True' parameters of the rings were $R_1 = R_2 = 9.66, c_{x1} = 15.9, c_{y1} = 14.84, c_{x2} = 25.3, c_{y2} = 14.17$. At each hit point, the value of energy released by the photon was sampled from an exponential distribution with the mean $\overline{E}$. Parameters $\lambda, \mu, \sigma_R, \sigma_d, \overline{E}$ were known, parameter σ_a was used, instead of annealing function $1/s(m)$, as a parameter controlling the acceptance probability. The MCMC procedure ran for about 2000 iterations and gave $\hat{R}_1 = 9.59, \hat{c}_{x1} = 15.59, \hat{c}_{y1} = 14.61, \hat{R}_2 = 9.92, \hat{c}_{x2} = 25.06, \hat{c}_{y2} = 13.97$. The algorithm has been implemented in MATLAB.

We also tested the traditional method (and we optimized only the centres, radii were fixed to 9.7). The method yielded $\tilde{c}_{x1} = 16.17, \tilde{c}_{y1} = 14.16, \tilde{c}_{x2} = 25.20, \tilde{c}_{y2} = 15.00$.

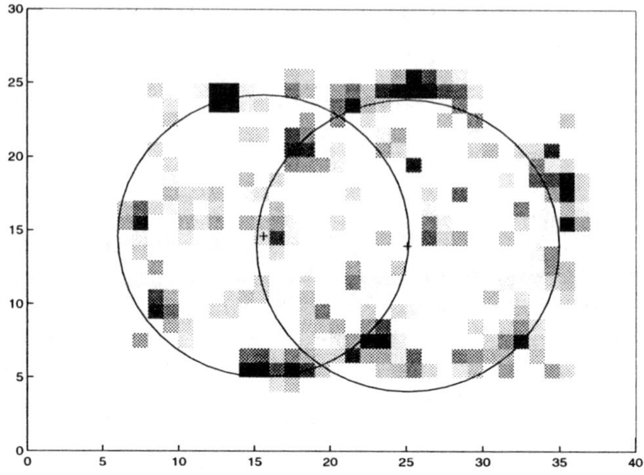

Fig. 1. The data and estimated rings; the darkness of cells is scaled according to observed amplitudes

5 Conclusion

The MCMC procedure has been proposed for the solution of an important physical problem of identification of the RICH rings parameters. The method processes the raw data directly, no preliminary clustering is necessary. Another advantage of the approach is its generality: unlike the conventional methods, the MCMC procedure can be applied without knowledge of the number and locations of the rings.

Acknowledgement
The research is partly supported by GA CR grant No 201/97/0354, by MSMT CR project No VS97084 and by RFBR grant No 97-01-01027 (Russia).

References

Agakichiev, G., Drees, A. & Glässel, P. (1996). Cherenkov ring fitting techniques for the CERES RICH detectors. *Nuclear Instruments and Methods in Physics Research*, **A 371**, 243–247.

Chernov, N., Kolganova, E. & Ososkov, G. (1995). Optimal weights for circle fitting with discrete granular data. *Communication of JINR*, E 10–95–468, Dubna.

Green, P. J. (1995). Reversible jump MCMC computation and Bayesian model determination. *Biometrika*, **82**, 711–732.

Linka, A., Picek, J. & Volf, P. (1996) Bayesian analysis for likelihood-based nonparametric regression. In: *Proceedings of Compstat'96*, Physica-Verlag, 343–348.

Roberts, G. O. & Smith, A. F. M. (1994). Simple conditions for convergence of the Gibbs sampler and Metropolis–Hastings algorithm. *Stoch. Processes Applic.*, **49**, 207–216.

Tierney, L. (1995). A Note on Metropolis–Hastings Kernels for General State Spaces. *Techn. Report. No. 606*, School of Statist., Univ. of Minnesota.

A Visual Future for Statistical Computing

Geoffrey Morgan and Patrick Craig

The Numerical Algorithms Group Ltd, Wilkinson House, Jordan Hill Road, Oxford OX2 8DR, UK

Abstract. A visualisation system is adapted to provide statisticians with an innovative statistical system that uses visual programming and provides access to visualisation and application building facilities.

Keywords. Visualisation, visual programming, statistical software, application building

1 Introduction

Statistical systems such as Genstat and S-PLUS have provided users with a flexible command language environment and a wide range of graphics. The problem with such systems is the effort required by the user to master the command language. Recognising this problem, virtually every statistical system now has a graphical user interface with a range of analyses available through menus. While the systems now have ease-of-use, they retain the original flexibility only by returning to the old command language.

Traditionally, statistical graphics have been static and two-dimensional, reflecting the practice when all that was available was graph paper. Some progress has been made with techniques such as brushing and spinning now being available in several systems. However, these are generally one- or two-dimensional objects in plain three-dimensional space.

It is true to say that 20 years ago statistical computing was at the forefront of interactive scientific computing and computer graphics. Since that time there have been major improvements in the environments used for scientific computing, in particular in visualisation. One such system is IRIS Explorer that not only provides facilities for visualising three-dimensional data, but also makes use of a visual programming environment that combines flexibility with ease of use.

The European Commission funded project STABLE is developing a statistical system that provides a visual programming environment along with flexible visualisation facilities and the ability to produce tailored end-user applications by combining the IRIS Explorer environment with the statistical algorithms available within Genstat. The resulting innovative statistical computing system represents a significant enhancement to the environment in which statistical analyses can be carried out. This paper describes the approach that has been taken in the project. The aim is to draw out the wider implications of the work for statistical software.

2 Visual programming

An IRIS Explorer visual program, called a map, consists of a number of modules linked by data flow; see Figure 1. The modules will perform the basic operations of a statistical analysis such as data input, data calculations and transformations, data displays and model fitting. Data are passed to a module (from the left) and then subsequent results are passed on to other modules (from the right) via ports. For example, data consisting of a response variable and explanatory variable may be passed into a generalized linear model module, and parameter estimates, residuals etc. may be passed to modules for display or further calculation. A map is constructed by selecting modules from a librarian and placing them on the map editor. Modules are then connected by clicking on the required input and output ports.

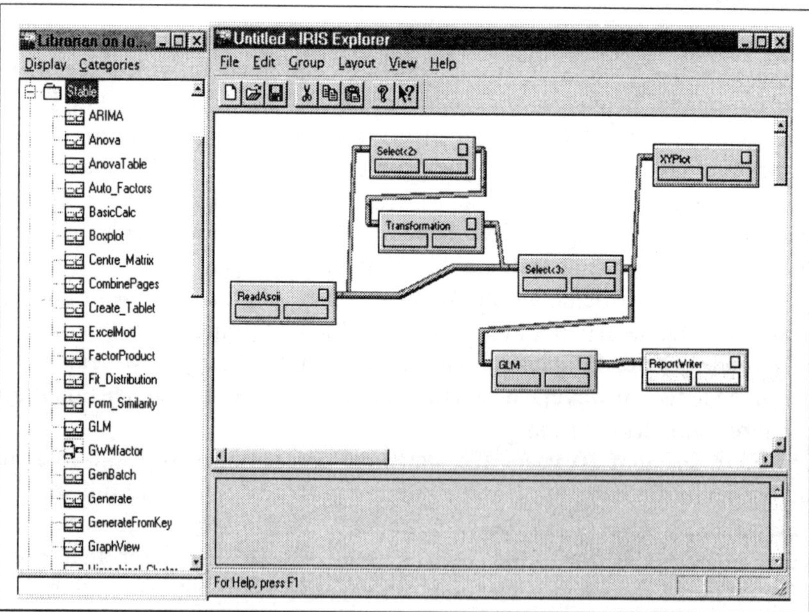

Fig 1. Librarian and Map Editor

Each module has its own graphical user interface with controls such as sliders and text boxes. The controls allow the user to modify the way in which a module operates, for example, selecting the response variable from the incoming data or setting the type of analysis to perform. Modules are normally set to operate (known as firing) when they receive new input data or when settings are changed in the user interface. This makes exploring changes in models or data very easy.

Several modules can be grouped together and given a single interface constructed from the interfaces of the component modules. Such a *group module* can then be treated like an ordinary module. This gives a way of constructing sub-programs using visual programming. Maps can also be saved for future use, providing a way of storing programs.

Modules can be on a different machine from the map, provided there is a copy of IRIS Explorer on the machine. Modules that require heavy computation or access to large databases can be located on suitable machines, while the users run the system from their local PCs. Further, recent developments should allow users on different machines to view the same map and displays, making it possible for collaborative working across different sites.

There are two fundamental differences between the manner in which IRIS Explorer is programmed and how many traditional statistical systems are used. First, there is no concept of a common data pool, either in the form of a central spreadsheet, a common data array or a directory containing data files. A module only knows about data that are passed to it and creates new data objects that have to be passed explicitly to other modules. In some ways this is similar to SAS data sets except that there is no concept of the default previous data set. While this approach is different, it does have several advantages. For example, having set up a map to perform an analysis on data from one file, this can be used on similar data from another data file simply by pointing the data input module at that file. Another advantage is that it reduces the likelihood of corrupting the data by inputting a false instruction.

The second major difference is the absence of the sequential nature of a typical statistical program. That is, a typical program may consist of a sequence of commands for Data-read, Data-calculate, Model-fit and Results-display. If in an interactive session the wrong calculation has been made, then at best the Data-calculate, Model-fit and Results-display commands will have to be re-entered, and at worst the data may have been corrupted and the program will have to start from scratch. Within the IRIS Explorer visual programming environment, the correction can be made to the appropriate module and then all the downstream modules will fire to produce the correct results. Further, visual programming provides a natural way of carrying out different analyses in parallel rather than sequentially. For example, if different types of model are fitted to a data set, the effect on all the models of deleting an observation can be viewed simultaneously.

3 Constructing a statistical system

Two major adaptations were needed to make the IRIS Explorer environment suitable for statistical analysis. First, the creation of suitable data types and second the provision of statistical algorithms.

The data types available within IRIS Explorer are limited to those required in scientific visualisation. A much richer class of data structures is needed in statistical work; including variate, classification factors, matrices, tables etc. These have been added to the system by what is known as *soft typing*. IRIS Explorer uses hard typing, that is, each module port is associated with one of the IRIS Explorer data types and can be connected only to another port of the same type. This would mean, for example, that if there were separate IRIS Explorer types for matrices and tables, there would have to be separate print modules for each type. To avoid the proliferation of modules for each data type a single IRIS Explorer data type has been developed, with different data types soft-typed within it. This means that there can be a single print module that recognises the type and prints appropriately. The different types are stored in an hierarchical way within a database, so users can create new data type simply by editing the database.

Statistical algorithms are available either as part of an existing system or as stand-alone routines in libraries or other collections. The main source of statistical algorithms for the STABLE system has been the Genstat statistical system (see Payne *et al.,* 1993). Genstat provides a collection of algorithms written in Fortran. These however are imbedded within the system code and some work is required to extract the algorithms so they can function as stand-alone modules. An additional source of algorithms has been the NAG libraries. Because the algorithms in the libraries already exist in a stand-alone form, it is relatively straightforward to construct modules based on library algorithms. The experience gained from including these algorithms will enable the provision of templates so that other users can easily add their own computational modules.

Statistical systems such as Genstat or S-PLUS have a range of techniques available that are not written in a base language of Fortran or C but are instead written in the system's own language. The complexity of such programs would often rule out the possibility of writing them using visual programming. So, there is the need to provide a command language interpreter. Modules written using this interpreter will appear like standard modules, the only difference being that the data and parameters will be passed along with the required program to the command language interpreter, running as a separate process.

Modules can also be used to provide links to other systems. For example, a module that links to Excel or another spreadsheet application will allow spreadsheet facilities to be available within the system, without the developers having to re-create what is already available, and the users having to learn a new system.

The challenge of adding statistical facilities is not just one of finding the algorithms but also of deciding on the functionality required and the level of modularity. As mentioned above the IRIS Explorer system allows modules to be grouped and a new interface constructed from the interfaces of the component modules. This has meant that it has been possible to construct modules at a low level of granularity knowing that they can easily be combined at a higher level. For example a principal component module would take all the incoming data and perform the analysis to produce loading, roots and scores. This module could then be combined with a selection module, a display module and a graphics module to produce a full analysis module.

4 Visualisation

Systems like IRIS Explorer have been developed to visualise data in subjects such as computational fluid dynamics, geology and chemistry. The data is usually three dimensional and often dynamic. In order to represent this type of data, the graphical facilities in these systems allow a wide range of three-dimensional objects rather than just the three-dimensional scatter diagrams, histograms or pie charts that are currently available in statistical systems.

The visualisation in three dimensions allows the objects to have properties of shape, size, colour, texture and position as well as being able to view them from different angles (i.e. allowing rotation) and under different lighting. These extra facilities allow several different properties of observations or objects to be visualised. What is more, the resulting displays look like real objects rather than the abstract displays generally used by statisticians, and should be more accessible, particularly to a generation brought up on computer games.

A system like IRIS Explorer can output its visualisations in VRML (virtual reality mark-up language). This will allow the results to be made available over the World Wide Web. The person wishing to view the results does not need a full visualisation system, only a VRML viewer. This will provide a new way of presenting the results of statistical analysis. In particular, it may be possible to develop ways of presenting complex results to non-specialists by using visualisation techniques.

5 Application building

Many statistical analyses are not carried out by statisticians but by professionals from other fields (hopefully) with advice from statisticians. We have already seen that two users at different sites (say the statistician and the experimenter) can share the same map and displays, and so explore the data together. However, when users need to perform the same basic type of analysis on many different data sets, there is a need to construct a program that is tailored to that particular analysis so that the user does not feel overwhelmed by the full statistical system.

The IRIS Explorer environment allows the construction of applications from the visual programming maps. This is similar to the way in which grouped modules are formed. Visual programming is used to tailor an analysis for a particular user or application. From this, an application is produced with the interface consisting of a selection of the interfaces of the modules making up the map. When such an application is run, for example from a short-cut icon, the user will see only the interface and need not know of the existence of the underlying software. The main advantage is that the statistician can build the application without having to learn an application building language like C++ and indeed without having to do conventional programming. Karaman (1998) gives an illustration of using the STABLE system in building an application.

6 Conclusions

This paper has sought to show how a modern visualisation system can be used to provide the basis of a new type of statistical system, which provides statisticians with a visual programming environment and access to modern visualisation techniques. The visual programming environment provides far greater flexibility than the standard windows menu environment, without the difficulties that inexperienced users encounter with command languages. The easy access to modern visualisation facilities should encourage statisticians to develop new ways of looking at data and results of analyses, and new ways of presenting the results to non-specialists.

Acknowledgements
The STABLE project (ESPRIT Project: 22832) is funded by the European Commission through Eurostat. The work presented in this paper represents the collaborative effort of those engaged on the project.

References

Craig, P. (1997). Implementing a Statistical Data Type in IRIS Explorer. *IRIS Explorer Technical Report TR3/97/10(NP3160)* Oxford: The Numerical Algorithms Group.

IRIS Explorer Users' Guide (1995). Oxford: The Numerical Algorithms Group.

Karaman, Z. (1998). Building end-user statistical applications: an example using STABLE. In: *COMPSTAT 1998 Proceedings in Computational Statistics* (ed. R. Payne & P. Green), 347-352. Heidelberg: Physica-Verlag.

NAG Fortran 77 Library Manual, Mark 18 (1997). Oxford: The Numerical Algorithms Group.

Payne, R.W., Lane, P.W., Digby, P.G.N., Harding, S.A., Leech, P.K., Morgan, G.W., Todd, A.D., Thompson. R., Tunnicliffe Wilson, G., Welham, S.J. & White, R.P. (1993). *Genstat 5 Release 3 Reference Summary*. Oxford: Numerical Algorithms Group.

Development of Statistical Software SAMMIF for Sensitivity Analysis in Multivariate Methods

Yuichi Mori[1], Shingo Watadani[2], Tomoyuki Tarumi[3] and Yutaka Tanaka[3]

[1] Department of Socio-Information, Okayama University of Science, 700-0005, Japan
[2] Department of Liberal Arts and Science, Kurashiki University of Science and the Arts, 712-8505, Japan
[3] Department of Environmental and Mathematical Science, Okayama University, 700-8530, Japan

Abstract. Statistical software SAMMIF is being developed for sensitivity analysis in multivariate methods where influence functions or their analogues are available. It can be used for detecting jointly as well as singly influential observations and also for obtaining information on influential directions from the aspect of Cook's local influence. A numerical example illustrates its performance in factor analysis.

Keywords. Influence function, local influence, multiple-case diagnostics

1 Introduction

There are two major tools in sensitivity analysis in statistical methods. One is Hampel's influence function (Hampel, 1974), and the other Cook's local influence (Cook, 1986). Methods of sensitivity analysis using either of these tools have been proposed by many authors including Radhakrishnan & Kshirsagar (1981), Critchley (1985), Tanaka (1988), Tanaka & Watadani (1992), and Wang & Lee (1996). We are now developing a statistical package SAMMIF (Sensitivity Analysis in Multivariate Methods based on Influence Functions) for sensitivity analysis in multivariate methods. So far we have developed some statistical packages such as SAM (Tarumi & Tanaka, 1986) and SACS (Watadani & Tanaka, 1994). Compared to them, SAMMIF has the following characteristics: (1) It is a unified package in the sense that it can be used for detecting not only singly but also jointly influential observations in any multivariate method where influence functions or their analogues are available; (2) Compared to the packages such as SAM and SACS, which are developed on MS-DOS (BASIC) platform, the so-called GUI is reinforced under the Windows environment and some options are provided for both beginners and specialists. Though it contains the expression "based on influence functions" in its name, it also provides information on influential directions in the sense of Cook's local influence, utilizing the relationship between the two approaches based on influence functions and Cook's local influence (see Tanaka, 1994). The present version can analyze confirmatory and exploratory factor analyses (FA), principal component analysis (PCA) and canonical correlation analysis (CCA).

2 Methodology

2.1 Influence function approach

Consider the case where we analyze a sample of n observations $\{\boldsymbol{x}_i; i=1,\mathrm{K}, n\}$ using a multivariate method which contains an m-dimensional parameter vector θ. In influence functions a perturbation is introduced to the cdf from $\hat{F}$ to $(1-\varepsilon_i)\hat{F}+\varepsilon_i\delta_{\boldsymbol{x}_i}$, where δ is the cdf with a unit point mass at $\boldsymbol{x}_i$, and the first derivative of $\hat{\theta}=\theta(\hat{F})$ with respect to ε at $\varepsilon=0$, which is simply denoted by $\hat{\theta}_i^{(1)}$ and is called the empirical influence function (*EIF*) of $\hat{\theta}$ at $\boldsymbol{x}_i$, is computed to evaluate the influence of observations. We usually summarize the *EIF* vector into some scalar measures to evaluate the influence of a single observation. Let $\hat{\theta}$ and $\hat{\theta}_{(A)}$ be the estimates based on the sample with/without a subset A of k observations. Then it is easily verified that the additive relation $\hat{\theta}_{(A)} \cong \hat{\theta}-(n-k)^{-1}\sum_{i\in A}\hat{\theta}_i^{(1)}$ holds. This relation suggests that, as a possible policy to detect influential subsets of observations, we should search for observations which have relatively large *EIF* vectors with similar directions from the origin. To do this by taking into account the correlations among the components of $\hat{\theta}$, we can use PCA with metric $[\widehat{\mathrm{acov}}(\hat{\theta})]^{-1}$ and search for the observations as discussed above by inspecting the plots of principal components (PCs) obtained by solving the eigenvalue problem (EVP)

$$\left(\frac{1}{n}\sum_{i=1}^{n}\hat{\theta}_i^{(1)}\hat{\theta}_i^{(1)T}-\lambda[\widehat{\mathrm{acov}}(\hat{\theta})]\right)\boldsymbol{u}=0. \qquad (1)$$

Instead of the influence function type perturbation we may consider case-weight or variance perturbation. It is obvious that the additive property as stated above holds and therefore PCA can be used for the first derivative vectors also in this case.

The problem of searching for influential subsets is similar to the problem of searching for simple structure of the coordinates. Based on this idea the varimax method can be applied to PC scores for detecting influential subsets in our program.

2.2 Cook's local influence

Denote the unperturbed weights for n observations by $w_0=(1,1,\ldots,1)^T$, and introduce a perturbation to the weight vector from w_0 to w. Let $\hat{\theta}$ and $\hat{\theta}_w$ be the estimates for the unperturbed and perturbed cases, respectively. In Cook's local influence the effect of the perturbation from w_0 to w is measured with the likelihood displacement defined as $LD(w)=2\left[L(\hat{\theta}|w_0)-L(\hat{\theta}_w|w_0)\right]$, where $L(.)$ is a log likelihood function, and the effect is represented by a graph called influence graph $(w, LD(w))$. Considering the change of $LD(w)$ along a straight line $\boldsymbol{w}=\boldsymbol{w}_0+a\boldsymbol{h}$, where $\|\boldsymbol{h}\|=1$, Cook searches for the direction which has the maximum normal curvature at w_0. The maximum curvature and the most influential direction are obtained as the largest eigenvalue $\lambda_{\max}$ and the associated eigenvector $\boldsymbol{h}_{\max}$ of an $n\times n$ EVP as

$$\left(2\left[\partial\hat{\theta}_w^T/\partial w\right]\left[\widehat{\mathrm{acov}}(\hat{\theta})\right]^{-1}\left[\partial\hat{\theta}_w/\partial w^T\right]-\lambda I\right)\boldsymbol{h}=0, \tag{2}$$

respectively, where $[\widehat{\mathrm{acov}}(\hat{\theta})]^{-1}=-[\partial^2L/\partial\theta\partial\theta^T]$, which is obtained from the theory of maximum likelihood (ML) estimation. Observations with large values of components in $\boldsymbol{h}_{\max}$ are regarded as an influential subset of observations.

Usually we may assume $m<n$ and $\partial\hat{\theta}_w/\partial w^T$ is of full rank. Then, after some algebraic calculation the $n\times n$ EVP (2) can be transformed to an $m\times m$ EVP

$$\left(2\left[\partial\hat{\theta}_w/\partial w^T\right]\left[\partial\hat{\theta}_w^T/\partial w\right]-\lambda[\widehat{\mathrm{acov}}(\hat{\theta})]\right)\boldsymbol{u}=0, \tag{3}$$

where $\boldsymbol{u}$ is defined as $\boldsymbol{h}=[\partial\hat{\theta}_w/\partial w^T]\boldsymbol{u}$. Eq. (3) is equivalent to eq.(1) except for the multiplying constant. In particular when we define the weights $w_\alpha^*=nw_\alpha/\sum_\beta w_\beta$, the relationship $n^{-1}\hat{\theta}_i^{(1)}=\partial\hat{\theta}_w/\partial w_i$ holds and the multiplying constants in eq. (1) and eq. (3) are $1/n$ and $2/n^2$, respectively.

3 Statistical software SAMMIF

SAMMIF is written in Microsoft Visual Basic. It has the following features other than Windows functions. A clickablemap-type flowchart, whose shapc is the same as Figure 1, is displayed to indicate where the user is in the flow; graphical displays such as index plot and scatter plot are available to visualize the results; the results including intermediate reports can be saved as text files; brief tutorial and suggestion windows are supplied for beginners; many options and some details outputs are available. SAMMIF consists of the following five parts (Figure 1 shows its flow).

1) ***Data entry*** (*Data*): SAMMIF reads a data file. Basic statistics are computed on demand.
2) ***Prior analysis*** (*Pre*): For a specified multivariate method, in which users want to evaluate the influence of observations, SAMMIF estimates parameters ordinarily using all observations. We can specify a method among PCA, CCA and FA in this version.
3) ***Diagnostics***: Users can do 3-1) and 3-2) in parallel.

3-1) ***Single-case diagnostics*** (*SD*): This part computes the *EIF* or its analogue of each

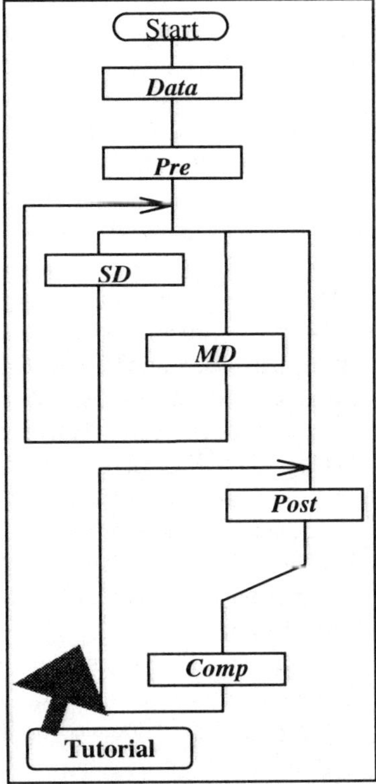

Fig.1 Flow of SAMMIF

observation for the estimated parameters and summarize them into influence measures such as generalized Cook's distance and *COVRATIO*-like measure. Users can display the measures in index plots, histograms and/or scatter plots to detect singly influential observations.

3-2) ***Multiple-case diagnostics*** (*MD*): PCA is applied to the *EIF* for all parameters or for a part of the parameters to detect candidates for influential subsets of observations. Here our program solves eq.(3) rather than eq.(1). Therefore the eigenvalues indicate the curvatures in the directions of eigenvectors. Users can display the PC scores in index plots, histograms and/or scatter plots. Influential directions in the sense of Cook's local influence can be displayed in index plots.

4) ***Posterior analysis*** (*Post*): For the sample without each candidate for singly or jointly influential observations, SAMMIF provides the results of the reanalysis using the multivariate method selected in 2). They are displayed in the same form as 2).

5) ***Comparison*** (*Comp*): Using the results of 2) and 4) SAMMIF outputs the comparison of the results for the sample with and without specified set of observations. Users can confirm whether they are really influential or not.

4 A numerical example

To illustrate our procedure we analyze a set of data taken from HATCO data sets (Hair *et al.*, 1984) using exploratory FA and its sensitivity analysis procedure. HATCO data sets consist of 100 artificial observations on 14 variables for business study. Among these variables we use 6 metric variables in the group of "benefits sought" variables.

We open the data file in the *Data* step and select "exploratory FA" in the 'Method' menu in the *Pre* step. Based on the specified conditions that the number of factors is two (based on the fact that eigenvalues of the correlation matrix are 2.5135 > 1.7395 > 0.5975 > … in order of magnitude) and that the estimation method is ML, SAMMIF estimates the parameters such as factor loadings and unique variances.

Next we proceed to the diagnostics step. In the *SD* step we try to find singly influential observations. SAMMIF computes the *EIF*s of the parameters and, at the users' choice, summarizes them into influence measures such as:

1) Generalized Cook's distance of the *EIF* to represent the influence on the estimate;
2) *COVRATIO*-like measure to represent the influence on the precision;
3) Approximate change of likelihood ratio statistic to represent the influence on the goodness-of-fit.

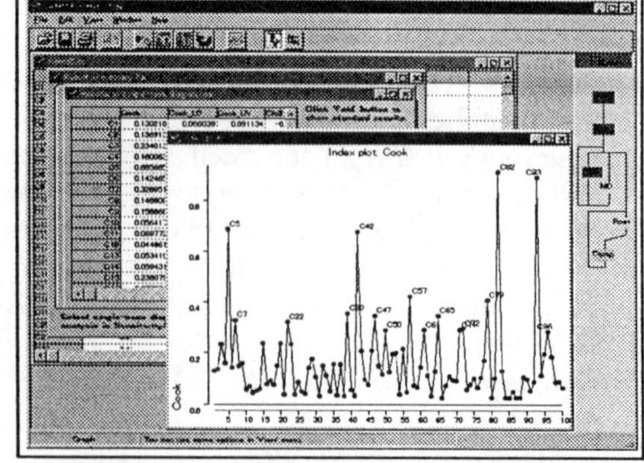

Fig. 2 Index plot of Cook's distance summarizing all *EIF*s

Figure 2 is the hardcopy of a SAMMIF window displaying the index plot of Cook's distances based on all *EIF* vectors. From this it can be stated that observations C5, C42, C82 and C93 are more influential than others.

In the *MD* step we try to find influential subsets. Here we apply PCA to all *EIF*s using $[\widehat{\mathrm{acov}}(\hat{\theta})]^{-1}$ as its metric. The eigenvalues are 1) 4.729, 2) 3.769, 3) 3.438, 4) 2.805, 5) 2.427, 6) 2.119, ... Figure 3 is a scatter plot of the 1st and 2nd PCs. From this plot two subsets {C82, C93} and {C65, C79} can be regarded as candidates for influential subsets since observations in each subset are located far from the origin and have directions very close to each other. We may search for more candidates by drawing scatter plots of other PCs. Instead we applied the varimax rotation to all dimensions of PCs to search for simple structures. Figures 4 and 5 show the index plots of the 1st and 2nd rotated scores. From these plots subsets {C82, C93} and {C65, C79} are formed as candidates for influential subsets. Similarly we can find another subset {C5, C42} in the index plot of the 3rd rotated scores. The curvatures corresponding to the varimax rotated axes are 1) 4.278, 2) 3.039, 3) 3.370, 4) 2.319, 5) 1.839, 6) 1.884,.... Then we found three candidates for influential subsets, i.e., {C82, C93}, {C65, C79} and {C5, C42}.

In the *Post* step exploratory FA is re-applied to the data set without a specified candidate, and then a table is given for convenience to compare between the *Pre* and *Post* results. When there is more than one candidate, we go back to the *Post* step and repeat the *Post* and *Comp* steps. In our numerical example it is found that the goodness-of-fit becomes better by omitting {C65, C79} while it becomes worse by omitting each of {C82, C93} and {C5, C42}.

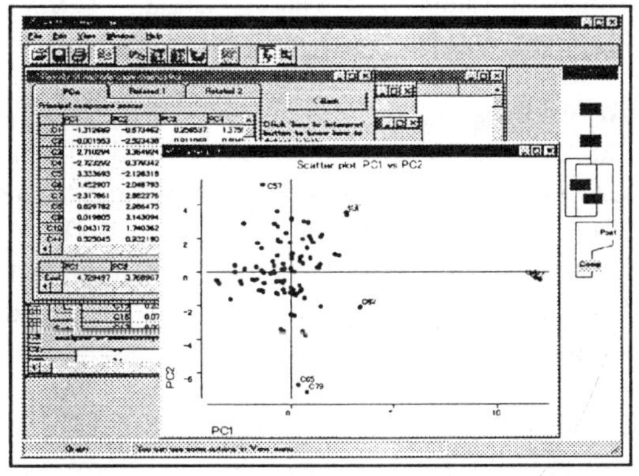

Fig. 3 Scatter plot of the 1st and 2nd PC scores using $[\widehat{\mathrm{acov}}(\hat{\theta})]^{-1}$ as a metric

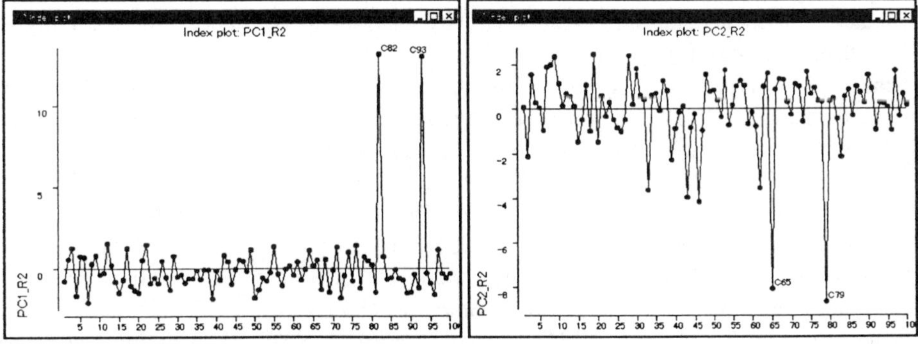

Fig.4 Index plot of the 1st rotated PCs

Fig.5 Index plot of the 2nd rotated PCs

5 Concluding remarks

The statistical software package SAMMIF is being developed for sensitivity analysis in multivariate methods. As illustrated in the numerical example, it can be used conveniently for detecting not only singly but also jointly influential observations. In particular, PCA of *EIF* along with its interpretation from the perspective of Cook's local influence and the varimax method applied to PC scores are very effective for detecting influential subsets of observations. In our experience our procedure, which we call the general procedure of sensitivity analysis based on influence functions, works well in the circumstances where the so-called masking effect is not severe. For the case where it is very severe, we are planning to implement the robust version of the general procedure (Tanaka & Watadani, 1994). In the numerical example we showed only the results of the case where the influence function type perturbation is introduced. However, similar results were obtained when the other type, the case-weight or variance perturbation, was introduced.

SAMMIF project: Yuichi Mori[1], Shingo Watadani[2], Yoshimasa Odaka[2], Yoshiro Yamamoto[3], Tomoyuki Tarumi[3] and Yutaka Tanaka[3]. ***URL:*** http://www.f7.ems.okayama-u.ac.jp/sammif/, http://www.soci.ous.ac.jp/~mori/sammif/, http://www.kusa.ac.jp/~wat/sammif/

References

Cook, R. D. (1986). Assessment of local influence. *J. R. Statist. Soc.*, **B48**, 133-169.

Critchley, F. (1985). Influence in principal component analysis. *Biometrika*,**72**,627-636.

Hair, Jr., J. F., Anderson, R. E., Tatham, R. L. & Black, W. C. (1984). *Multivariate data analysis*. New York: Macmillan Publishing Company.

Hampel. F. R. (1974). The influence curve and its role in robust estimation. *J. Amer. Statist. Assoc.*, **69**, 383-393.

Radhakrishnan, R. & Kshirsagar, A. M. (1981). Influence function for certain parameters in multivariate analysis. *Comm. Statist.*, **A10**, 515-529.

Tanaka, Y. (1988). Sensitivity analysis in principal component analysis: Influence on the subspace spanned by principal components. *Comm. Statist.*, **A17**, 3157-3175. (Corrections, **A18** (1989), 4305).

Tanaka, Y. (1994). Recent advance in sensitivity analysis in multivariate methods. *J. Jpn. Soc. Comp. Statist.*, **7**, 1-25.

Tanaka, Y. & Watadani, S. (1992). Sensitivity analysis in covariance structure analysis with equality constraints. *Comm. Statist.*, **A21**, 1501-1515.

Tanaka, Y. & Watadani, S. (1994). Unmasking influential observations in multivariate methods. In: *COMPSTAT94 Proceedings in Computational Statistics* (ed. R.Dutter & W.Grossman), 292-297. Heidelberg: Physica-Verlag.

Tarumi, T. & Tanaka, Y. (1986). Statistical software SAM: Sensitivity analysis in multivariate methods. In: *COMPSTAT86 Proceedings in Computational Statistics* (ed. F.De Antoni, N.Lauro & A.Rizzi), 351-356. Heidelberg: Physica-Verlag.

Wang, S.-J. & Lee, S.-Y. (1996). Sensitivity analysis of structural equation models with equality functional constraints. *Comp. Statist. & Data Analysis*, **23**, 239-256.

Watadani, S. & Tanaka, Y. (1994). Statistical software SACS: Sensitivity analysis in covariance structure analysis. *J. Jpn. Soc. Comp. Statist.*, **7**, 105-118.

Data Imputation and Nowcasting in the Environmental Sciences Using Clustering and Connectionist Modelling

F. Murtagh[1], G. Zheng[1], J. Campbell[1], A. Aussem[2], M. Ouberdous[3], E. Demirov[3], W. Eifler[3] and M. Crépon[4]

[1] Faculty of Informatics, University of Ulster, BT48 7JL L'Derry, Northern Ireland
[2] Université Blaise Pascal, Clermont-Ferrand II, I.S.I.M.A., Campus des Cezeaux, B.P. 125, 63173 Aubière Cedex, France
[3] Space Applications Institute, Marine Environment Unit, Joint Research Centre, 21020 Ispra (Va), Italy
[4] Laboratoire d'Océanographie Dynamique et de Climatologie, Université Pierre et Marie Curie, 75252 Paris Cedex 05, France

Abstract. We discuss data fusion in the context of measured sea surface temperature data, wind stress and radiation budget data, topographic feature information, and output of physical oceanographic models. Our immediate set of objectives are data imputation and feature selection. Our longer term goal is nowcasting and forecasting of oceanic upwelling.

Keywords. Cluster analysis, neural networks, clusterwise regression, connectionist modelling, oceanography, environmental sciences, knowledge discovery

1 Introduction

Data in the environmental sciences are often characterized by the following, and they hold for the work described in this paper.

1. Large sections of the data may be missing. In the case of this work, satellite sea surface temperature (SST) data had considerable areas missing due to cloud cover. Inferring missing data using interpolation in the spatial domain, or using time series prediction in the time domain, may be quite insufficient to infer very large quantities of missing data. For this reason, various hybrid strategies are investigated in this work.
2. Data are often quite uncertain. E.g. cloud cover in part of the area spanned by a pixel, or thin high-level cirrus cloud, may damagingly "pollute" a pixel's value. We would suggest that this makes a classical statistical problem-solving approach quite questionable. The data may appear to be real-valued and interval-scaled, but in practice they have severely heterogeneous certainties. This leads to the need for a *classificatory* approach to problem-solving: similar patterns in the solution space must be sought, in order to facilitate not only the finding of high-quality data analytic solutions (predictions, fits) but also – and very importantly – the quantifying of the relevance and practicality of those results.
3. The raw data on which we are working are very diverse (and a fortiori heterogeneous): SSTs, simulated SSTs, wind vectors, heat and energy fluxes, topographical measurements. It is clearly inadvisable to take many such

variables in a single formulation of the problem based on multivariate analysis. Instead the approach we adopt has been a bootstrapping one: if reasonably satisfactory results cannot be obtained with the most basic data we are dealing with (e.g. SSTs), then we seek additional information from other datasets available to us. Problem decomposition – analysis – is a traditional problem-solving approach (with implementation greatly aided by database management languages and systems). The framework for our work, however, has a large element of bottom-up problem *synthesis*.

We can stress then two innovations in this work, – problem synthesis being at times more important than problem analysis, and what we might describe as *classificatory reasoning*, or reasoning through pattern finding and recognition, as being an important part of the knowledge discovery process.

These two aspects of our work lead *grosso modo* in the direction of neural network (connectionist or self-organizing) modelling approaches.

2 The problem and the data

Our work relates to prediction of oceanic upwelling off the Mauretanian coast, using SST images, and real and model meteorological data for the year 1982. Upwelling (Tomczak, 1996) is the periodic replenishment of coastal surface waters with cold deep water, which has various attendant effects. Among these, in particular, are the movement of nutrients from the colder waters, with further effects on higher-level life in the food-chain, and with consequent economic effects for nearby human populations. One mechanism for producing upwelling is through wind forcing parallel to the coast, engendering ocean currents through a Coriolis mechanism.

The Mauretanian coast in the region of Cap Blanc is associated with upwelling having a strong and well-characterized signal. The data available to us consists of:

1. Daily advanced very high resolution radiometer (AVHRR) satellite SST data for 1982 using a geographic window covered by 82×70 $0.1'$ pixels off the north-west African coast (e.g. Nykjær & Van Camp, 1994).
2. Daily wind and surface heat flux data, from the European Centre for Medium-Range Weather Forecasting, Reading, interpolated on the same latitude-longitude grid.
3. Data on the topography of the region.
4. Output from the ISPRAMIX Ocean General Circulation Model (OGCM).

Forecasting of upwelling is essentially a function-mapping task. In a restricted region, the ocean state (temperature, salinity, currents, etc.) may be considered as the output of a complex transfer function forced by a specific input consisting of wind, solar radiation, thermohaline exchanges, and so on.

3 SST data imputation through clusterwise regression

Over 70% of the SST pixel values are missing. We briefly summarize the initial approach adopted to achieve high-quality imputation of these values, based primarily on spatial and temporal information, and meteorological information. Further description can be found in Murtagh *et al.* (1998).

We carried out the mapping between a vector of 10 successive daily values of a meteorological variable, and Δ SST,

$$\{w(v,i,j,d-10),\ldots,w(v,i,j,d-1)\} \longrightarrow \mathrm{SST}(i,j,d-1)-\mathrm{SST}(i,j,d)$$

independently for four available meteorological variables (wind stresses and heat fluxes), v, using a clusterwise regression method. Here i,j are the pixel coordinates and d is the day. Thus, for example, one wind stress component was taken, clusters of wind components found (16 were used as a good compromise between parsimony and compactness), and function fits to Δ SSTs (see above) were sought within each cluster.

Following this phase of missing data estimation, further phases using spatial and temporal information were carried out. This led to more than 97% of the missing SST values being imputed.

The results obtained using the clusterwise regression are locally Euclidean. We believe that results of better quality can be obtained by a connectionist modelling approach (see, e.g., Aussem *et al.*, 1995, 1996). The following points in favour of this can be noted.

1. Since inputs in a connectionist model are weighted on input, we have a mechanism for automated *feature selection*, which should be superior to the Euclidean framework used in the clusterwise regression.
2. Weights on the inputs would help with necessary automated rotation of the orthogonal wind components.
3. We require an integrated model for data imputation and nowcasting, with quality assessment of output, which is easy to apply.

The clusterwise regression has the hallmarks of exploratory data analysis. I.e. we seek patterns in the data, and then use these for our prediction or nowcasting or other objectives. In the next section, we will more directly address the issue of an appropriate system for tackling these problems.

4 Multiple task learning

A novel supervised feed-forward neural network was developed to address the problems of nowcasting (or forecasting) and data imputation. The principle is reminiscent of the perspective on the (unsupervised) Kohonen self-organizing map method which was advanced in Murtagh & Hernández-Pajares (1995), namely that it simultaneously attempts to optimize a clustering compactness criterion, and a criterion related to inter-cluster proximities. In our supervised work, we will seek to simultaneously impute the present day's SST, and to predict the next day's SST. We will compare this with the single-task objective.

As mentioned above, the neural network approach used here has the additional advantage of feature selection, and of being able to find automatically the most appropriate rotation of the all-important wind vectors, whose forcing plays a major role in the upwelling event.

The architecture used for multi-task learning is illustrated in Figure 1. We initially trained such an architecture on the following tasks:

$$\{w(i,j,d-10),w(i,j,d-9),\ldots,w(i,j,d-1)\} \longrightarrow \mathrm{SST}(i,j,d)\}$$

and the mapping from the same inputs to the previous day's SST value:

$$\{w(i,j,d-10),w(i,j,d-9),\ldots,w(i,j,d-1)\} \longrightarrow \mathrm{SST}(i,j,d-1)\}$$

The network architecture used for the single task learning of forecasting was 10–14–1 and for the multi-task learning 10–14–2. The single forecasting task learning and testing results are shown in Figure 2 and the training and testing results of the forecasting part of the multi-task learning are shown in Figure 3. The two feedforward networks were both trained for 5000 epochs. During the first 1000 training iterations, the multi-task learning is much faster than single task learning. After further training, though the learning performance of both learning methods are similar, the single forecasting task learning cannot hold reliable generalization ability. Its internal representations only satisfy the training set within its more limited learning ability. However in the multi-task learning method, the additional (similar) task helps to control generalization performance. Even if with 5000 training iterations, the learning performance in both cases is similar, the generalization ability is more reliable in the multi-task method.

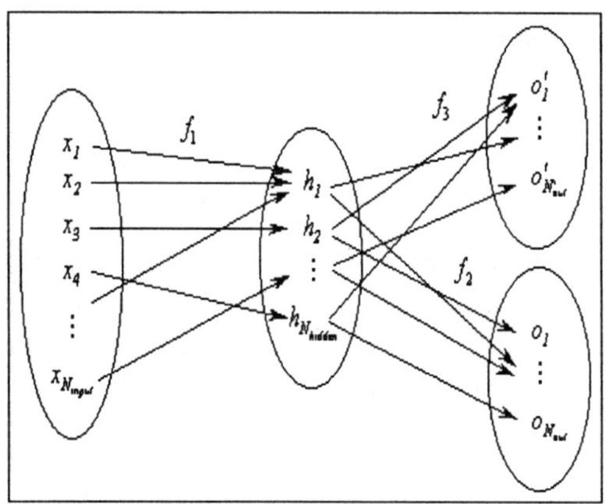

Fig. 1. Mapping in multi-task learning

Using both (orthogonal) wind directions, with the pixels in the general geographical area under investigation not being differentiated in any way, led to MSE (mean square error) results for training and testing of, respectively, 0.0696 and 0.3422. (Single-task MLP predictions yielded MSE results of 0.0715 and 0.7048, respectively, for training and testing.) This translates into an error in generalization of better than approximately 0.6^o.

In our experimentation we have consistently found that one wind component (the y-direction, which is approximately along the coast) was better. We tested the above multi-task learning with one versus both wind components and found the MSE for the y-direction alone to be, respectively for training and testing 0.1665 and 0.3783. This indicates that our results are not only better, but are in keeping with our initial hope to have the neural network do the tricky job for us of rotating wind components. What was less satisfactory was that the y-direction wind component required only 15000 training iterations, while 500000 were used with both wind components.

We believe improved efficiency can be obtained, to complement the improved effectiveness which we have found. Experiments currently being carried out are based on small geographical areas. This allows for approximate

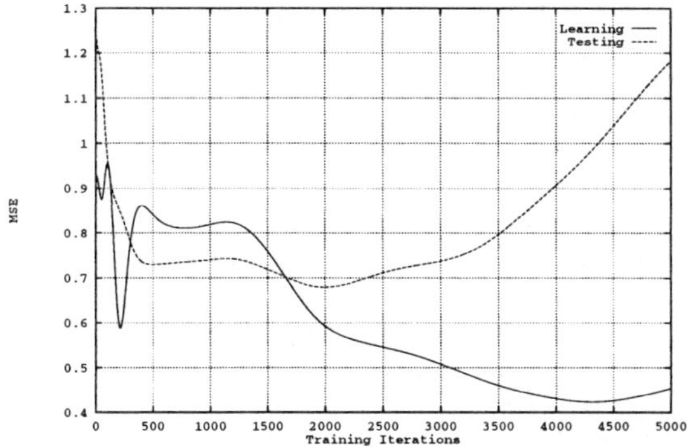

Fig. 2. Single forecasting task learning and testing

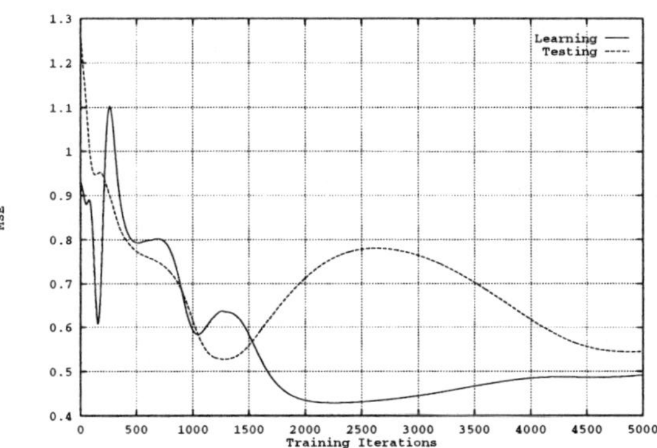

Fig. 3. Learning and testing of the forecasting part of the multi-task learning

constancy of angular direction in such geographical areas relative to the coastline. In addition, the small geographical areas will implicitly take the topography into account.

In this current work, we see again examples of the principles expressed in the Introduction: we are using locality (spatial clustering); and we are incorporating further facets of the problem into our solution in order to ratchet up our effectiveness and efficiency results.

5 Conclusion

"Data-driven" pattern recognition and neural network methodologies have an important role to play in the modelling of data. They may yet help us to provide solutions for some of our society's most pressing problems in the environmental and other arenas.

A Web area has been set up, which displays various datasets. Visualization is provided by animations of available SSTs, and completed (or imputed) SSTs. In addition various papers and reports are being made available at this site. The address is: `http://hawk.infm.ulst.ac.uk:1998/neurosat`

Acknowledgement
Support by the Environment and Climate Programme (EU, DG XII) project, "Neurosat – Processing of Environmental Observing Satellite Data with Neural Networks" (1997–1999), is acknowledged.

References

Aussem, A., Murtagh, F. & Sarazin, M. (1995). Dynamical recurrent neural networks – towards environmental time series prediction. *International Journal of Neural Networks*, **6**, 145–170.

Aussem, A., Murtagh, F. & Sarazin, M. (1996). Fuzzy astronomical seeing nowcasts with a dynamical and recurrent connectionist network. *Neurocomputing*, **12**, 359–373.

Murtagh, F. & Hernández-Pajares, M. (1995). The Kohonen self-organizing map method: an assessment, *Journal of Classification*, **12**, 165–190.

Murtagh, F., Campbell, J., Zheng, G., Aussem, A., Ouberdous, M., Demirov, E., Eifler, W. & Crépon, M. (1998). Data imputation and nowcasting using clustering and connectionist modelling. *Proc. IFCS-1998*, in press.

Nykjær, L. & Van Camp, L. (1994). Seasonal and interannual variability of coastal upwelling along northwest Africa and Portugal from 1981 to 1991. *Journal of Geophysical Research*, **99**, 14197–14207

Tomczak, M. (1996). Upwelling dynamics in deep and shallow water, Chapter 6. In: *Shelf and Estuarine Oceanography: an Introduction.* Flinders University of South Australia,
http://gaea.es.flinders.edu.au/ mattom/Shelf/chapter06.html

Graphical User Interface for Statistical Software Using Internet

Junji Nakano

Faculty of Economics, Hitotsubashi University, Kunitachi, Tokyo 186-8601, Japan
E-mail: nakanoj@stat.hit-u.ac.jp

Abstract. There exist many statistical packages, like the SHAZAM package, which have character user interfaces and have been widely used for a long time by professional users. As computing environments are rapidly changing, users require these packages to adopt new innovations such as graphical user interfaces (GUIs) and Internet abilities. For adding these technologies to SHAZAM, we have made client/server GUI programs by using the composite user interface approach and recently developed tools: Tcl/Tk, Tclet, WWW browsers and CGI.

Keywords. Client/server application, distributed computing, GUI, SHAZAM, Tcl/Tk, Tclet, WWW

1 Recent computing environments

Computers and their networks have become part of the infrastructure of modern society. This was brought about by cheap but powerful personal computers, easy to use graphical user interfaces (GUIs) and the Internet.

1.1 GUI

Formerly, software was operated by commands or programs, and results from computers were given in plain text. This interface is called a character user interface (CUI). CUIs are manageable for professional users, but are difficult to use by other people. So, user friendly GUIs are required, in which users operate computers mostly by clicking locations on the screen using a pointing device such as a mouse.

Because GUIs need more computer resources and higher programming skills than CUIs, making GUI programs was a difficult task for non-professional programmers. Recently, the situation has changed. We have inexpensive and powerful computers with GUI operating systems, and many convenient tools for GUI programming have been developed.

1.2 WWW and Internet

The World Wide Web (WWW) has become the most frequently used service that the Internet provides for two main reasons. The first is the easily operated and beautifully designed common GUIs provided by WWW browsers. The second is the simple and well designed communication protocols: HyperText Transfer Protocol (HTTP) is used to transfer files among computers, and HyperText Markup Language (HTML) is used to decorate WWW pages.

WWW provides various facilities to extend its abilities. The Common Gateway Interface (CGI) is available for executing software on HTTP daemon servers. Plugin libraries can add new functions to browsers. Java language is available as Applets on browsers.

These technologies make WWW browsers usable as common operating interfaces for many applications. For example, applications of the X Window System are operated on WWW browsers by the "Broadway" technology.

The success of WWW promotes other ways of using the Internet. An example was the project of decrypting an RC5 encrypted message on the Internet. This project tried to check all the possible keys using many voluntary computers. Distribution of checking programs and test key sets and assembling of results were done through the Internet. This project found the right answer after checking 34 quadrillion keys by more than 4000 active teams (each team consisted of many computers). This showed the power of distributed computing.

2 GUI and Internet usage of modern statistical packages

2.1 Composite user interface

For a long time, statistical packages had CUIs, and were used by professional researchers. As GUIs become popular and the number of naive users has increased, statistical packages also began to have GUIs. Command languages, however, still have their own merits: the ability to describe complex and long operations.

Therefore, many statistical packages employ composite user interfaces (Liu *et al.*, 1995), in which the user interface part of the software is as independent of the original computational portion of the software as possible. One merit of this approach is the stability of the program. If we isolate the user interface program, we can avoid inserting new bugs into the original computational codes. This approach is also appropriate for the modern "gluing" tools for combining existing software.

2.2 Recent network abilities

Statistical analyses were originally performed on one computer. Today, statistical procedures, data, CPU for GUI and CPU for calculation may be on different computers. For example, the amount of data sometimes becomes very huge because some computers automatically record mega bytes of data every hour. Such data are difficult to move to other computers.

For this situation, many WWW based GUIs have been developed. CGI was used at first, see for example, Schmelzer *et al.* (1996) and Nakano & White (1997). A recent trend is the use of Java Applets. Kötter (1997) has proposed a sophisticated GUI Applet. West (1997) is making Applets which include statistical procedures and GUI.

3 Outline of our GUIs

3.1 Design principles

SHAZAM is a statistical package mainly for econometricians (White, 1997). It runs on almost all platforms. Although its CUI supports interactive use, it is so simple that batch type use is appropriate. Windows and Macintosh versions have simple GUIs. Its site licenses allow unlimited distribution of the program.

We hope to make GUIs for SHAZAM which support distributed computing environments by the composite user interface approach. Our GUI should be able to edit programs and results, save programs and results to local computers, and load programs and data from local and remote computers. The GUI should also be able to distribute programs and/or data to remote computers on which SHAZAM is installed.

Our GUIs communicate with remote SHAZAM server computers. Therefore, if we prepare many SHAZAM servers, we can use much computing power from GUIs. When a SHAZAM calculation is executed on the server, all programs and data should be gathered there. Hence, it is convenient that the computer on which the huge amount of data are stored, should also be a SHAZAM server.

We use Tcl/Tk language by Ousterhout (1994) for realizing our GUIs. Tcl/Tk has many merits for our purpose. It has well designed graphics and network commands. Plugin libraries are available for executing restricted versions of Tcl/Tk programs inside two major WWW browsers: Netscape and Internet Explorer. Network transferred Tcl/Tk programs by HTTP are called Tclet. Restrictions of Tclet are set for security and are changeable by a security policy mechanism. Tcl/Tk and Tclet plugin are freely distributed. Differences between scripting languages such as Tcl/Tk and system programming languages such as Java are described in Ousterhout (1998).

We made three GUIs for SHAZAM: stand-alone GUI, Tclet based GUI on WWW browsers and CGI based GUI on WWW browsers.

3.2 Stand-alone GUI

When we start this GUI directly from the operating system, one window opens (Figure 1).

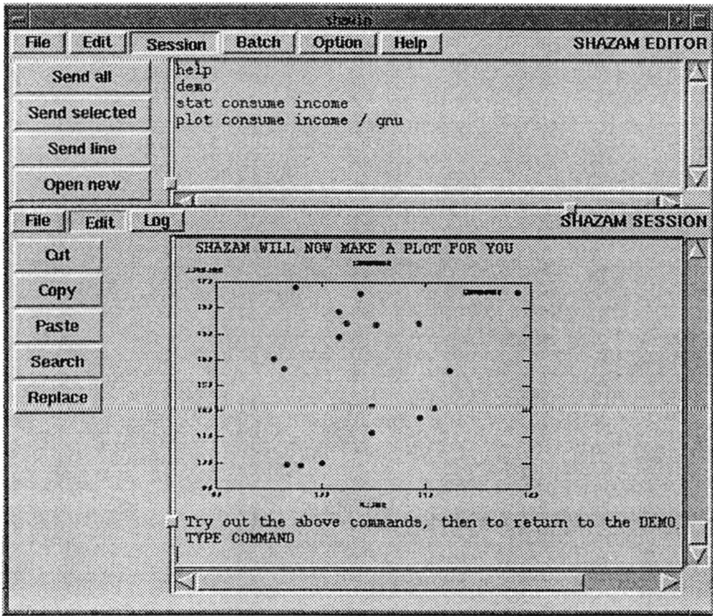

Fig. 1. Stand-alone GUI

This window consists of upper and lower windows. The upper window is for editing a SHAZAM program. The lower window is for interactive use and/or output display. Each window consists of an upper main menu, left and right windows. Left windows are for sub menus, and right windows are work areas. Borders between windows can be moved by dragging small buttons on them.

Two main menu buttons on the upper window, **File** and **Edit**, are used for the usual editor operations. By the **Load** sub menu button of **File**, we can read programs from local and remote computers. Throughout our GUIs, files on the local computer, on which the GUI program are executed, are specified by `file:` prefix, and files on remote computers are specified by `http://` prefix. These notations are same as the Universal Resource Locator (URL) of WWW. File names without these prefixes indicate files on the SHAZAM server computer.

Next two main menu buttons, **Session** and **Batch**, are for connecting with SHAZAM servers. **Session** enables interactive use of the SHAZAM program. Figure 1 shows sub menu of **Session** on the upper left window. **Open new** button starts communication with a SHAZAM server. If the name of the server is not specified, an available SHAZAM server is selected automatically. We can specify a SHAZAM server by the **Option** main menu button or in the program by the newly added `*run*` command. If the specified server denies the execution, the GUI gives up executing that program. When the GUI connects with a SHAZAM server, the lower window can be used like usual SHAZAM interactive sessions, with useful support such as back-scroll ability or history records. As SHAZAM can generate GNUPLOT programs and data for plotting graphs, our GUI can display such graphs. It is possible to send all or part of the program written in the upper window to servers. If interactive use is not required, the **Batch** main menu button is appropriate. The GUI then sends all programs in the editor window to servers, receives and shows results in the lower window and closes connections.

In statistical calculations, it sometimes happens that many similar calculations are required for different parameter values, for example, simulation experiments or non linear maximization by grid search. These calculations can be performed by many computers simultaneously, just like the above mentioned RC5 decryption project. Our GUI allows some additional commands in the editor window for this purpose. One of them is the `*run*` command, and full Tcl commands are available for generating many similar but slightly modified programs. For example, the program

```
set lambda {1.0 2.0 3.0}; set machine {a b c}
foreach lvar $lambda mvar $machine {
  write {
    *run*  $mvar.mydomain
        (SHAZAM programs including lvar)
  }
}
```

in the editor window will send the `(SHAZAM program including lvar)` part to three servers (a.mydomain, b.mydomain and c.mydomain) with variable `lvar` values of 1.0, 2.0 and 3.0 respectively. All results from the three servers are returned to the lower window.

3.3 Tclet based GUI on WWW browsers

The Tclet based GUI on WWW browsers is started by opening one HTML document, and can be used in almost the same way as the stand-alone GUI, except for security restrictions. For example, we can not specify the directory name in the `file:` notation. The directory that Tclet programs may use for reading and writing is decided by the Tclet plugin. As our security restrictions are looser than those of original browsers or Java Applets, this GUI may be most appropriate inside Intranet.

3.4 CGI based GUI on WWW browser

In some cases, the Tcl/Tk or Tclet plugin can not be installed for security or maintenance reasons. For this situation, we have made a primitive GUI on pure WWW browsers utilizing CGI of HTTP daemon programs. The editor is realized by the HTML input tag `textarea` and has very limited functions compared to the Tclet based GUI. However, some abilities for supporting distributed computing are still available.

4 Implementation

Our systems are client/server applications, and divided into roughly three parts: a GUI part, a client part and a server part.

In the stand-alone and Tclet based GUI, a GUI part and a client part are realized in one Tcl/Tk program. In the CGI based GUI, a GUI part is written in HTML and a client part is realized as a CGI program in Tcl/Tk.

4.1 Internet protocol for SHAZAM

For the client/server communication, we defined an Internet protocol for SHAZAM, called SIP (SHAZAM Internet Protocol). SIP is designed following HTTP and MIME (Multipurpose Internet Mail Extensions).

SIP uses only plain text. An SIP session begins when a client sends an SIP request (SESSION, BATCH, etc.) followed by some headers to a server. The server checks the permission of the client and its own state of loading. If both checks are OK, the server returns an ACCEPT response to the client. Otherwise, the server returns a REJECT response with the reason for rejection. When the client receives an ACCEPT response, it sends the body of SIP, in which the program and data from the local computer are packed. To include more than one file in the body, we use the MIME format. After the SHAZAM calculation, the server returns the result to the client as ASCII text, which may contain several files in MIME format. This connection is kept until the SHAZAM command "stop" is sent from the client to the server.

4.2 Tcl/Tk GUI program

The editor part has usual editor functions written in Tcl/Tk. Loading remote files is performed by HTTP.

The editor can expand the Tcl extended SHAZAM programs in the upper window, and can return to the original one. The expansion is always done when the editor passes programs to the client part, and at the same time, local files specified by `file:` prefix are included by MIME format.

If the result from the server includes GNUPLOT files in the MIME format, the GUI program interprets them and display graphs in the lower window together with the result as ASCII texts.

4.3 Client program

The client program has a list of available SHAZAM servers. If the name of the SHAZAM server is not specified, the client program tries to connect with the server at the top of the list. If the server denies the connection, the server second in the list is tried. If all listed servers deny or the server specified by the user denies the connection, the client gives up the execution. The client can add new available server names to the list by sending a SERVERS request to the servers already listed.

Newly added SHAZAM commands (`*run*`, etc.) are interpreted by the client to construct appropriate pure SHAZAM programs for each server. Then they are sent to servers as the body of the SIP.

When the CGI client program receives results which include files for GNUPLOT from the server, it generates a GIF image files of the graph for WWW browsers.

4.4 Server program

The server program uses some functions of Expect, an extension of Tcl/Tk by Libes (1995), for a slight technical reason.

It waits for connections at one TCP/IP port. When an SIP request from a client is detected, negotiation begins. Permission checks are performed by comparing the domain name of the client with the list of permitted domain names stored in the server. This list also contains other available server names for each permitted client domain name, and can be sent to the client by the SERVERS request.

The server checks its state of loading by the number of SHAZAM processes and load average values, whose limits can be determined by the server administrator. If the body of an SIP is received, the server program saves the MIME part into a temporary directory, and input files whose names are prefixed by `http://` are fetched by HTTP and saved to the temporary directory. Then the server rewrites these parts appropriately to make a pure SHAZAM program, and starts a SHAZAM process with it. The result of the execution is then returned to the client. If GNUPLOT related files are generated, they are packed with the result in MIME format.

References

Kötter, T. (1997). Implementation of networked applications for statistical computing. In: *Bulletin of the International Statistical Institute, 51st Session Proceedings Book 2*, 47-50. Turkey.

Libes, D. (1995). *Exploring Expect: A Tcl-based Toolkit for Automating Interactive Programs.* Sebastopol: O'Reilly.

Liu, L.-M., Chan, K.-K., Montgomery, A.L. & Muller, M.E. (1995). A system-independent graphical user interface for statistical software, *Computational Statistics & Data Analysis*, **19**, 23-44.

Nakano, J. & White, K. (1997). Using WWW abilities from SHAZAM statistical program, *Proceedings of the Institute of Statistical Mathematics*, **45**, 41-47. (in Japanese).

Ousterhout, J.K. (1994). *Tcl and the Tk Toolkit.* New York: Addison-Wesley.

Ousterhout, J.K. (1998). Scripting: Higher level programming for the 21st century. *IEEE Computer Magazine*, **31**, 23-30.

Schmelzer, S., Kötter, T., Klinke, S. & Härdle, W. (1996). A new generation of a statistical computing environment on the Net. In: *COMPSTAT96 Proceedings in Computational Statistics* (ed. A. Prat), 135-148. Heidelberg: Physica-Verlag.

West, R. W. (1997). Statistical applications for the World Wide Web. In: *Bulletin of the International Statistical Institute, 51st Session Proceedings Book 2*, 7-10. Turkey.

White, K. J. (1997). *SHAZAM User's Reference Manual Version 8.0.* New York: McGraw-Hill.

A Wavelet Approach to Functional Principal Component Analysis

Francisco A. Ocaña, Olga Valenzuela and Ana M. Aguilera

Department of Statistics & Operations Research, University of Granada, Campus de Cartuja, 18071-Granada, Spain

Abstract. The aim of this paper is to approximate the estimates in the principal component analysis of a continuous time stochastic process (functional PCA) by using wavelet methods. A short review of estimating in the functional PCA leads to the problem of solving the integral equation with the covariance function as kernel. An estimating procedure based on wavelet methods is then provided to obtain approximate estimates. Wavelet methods and multiresolution analysis (MRA) are jointly considered. Furthermore, MRA provides an approximating framework to estimate functional PCA when data are observed at discrete knots on a real interval. This wavelet approach is tested by simulating at discrete knots sample functions of Brownian motion. The PCA of this process is compared with those estimated by means of the wavelet approach.

Keywords. Principal components, functional data, Karhunen-Loeve expansion, multiresolution analysis, wavelets, Brownian motion

1 Introduction

Data involved in some real situations (Ramsay & Silverman, 1997) describe the evolution of a random magnitude with respect to another deterministic magnitude (e.g. the time) which is varying in a continuous way. The probabilistic tool to model such a random phenomenon used to be a second order continuous time stochastic process, where the data are thus the sample functions of the process.

In practice, expressing a phenomenon as a combination of other ones (harmonics) is widely used in several scientific fields. In statistics, the decomposition of a stochastic process into harmonic components is a very useful tool for studying its statistical properties. In fact, the harmonic analysis of a stochastic process (functional PCA) was defined by Deville (1974) through a generalization of the principal component analysis (PCA) of a finite set of real random variables.

Section 2 reviews the basic concepts on functional PCA (FPCA) required to introduce the problem of the estimation from a set of independent sample paths. The major difficulty is the fact that the estimators of the principal factors are the eigenfunctions of the Fredholm integral equation whose kernel is the sample covariance function.

Owing to the inherent difficulties of obtaining analytical solutions for such equations, in Section 3 we resort to approximating the estimates in the FPCA by those obtained from a wavelet approximation of the process.

Finally, in Section 4 the sample paths of Brownian motion are simulated at discrete knots, so the estimates approximated in the FPCA by wavelets can be compared to the exact ones of the process considered.

2 Overview of functional PCA

Let $X = \{X(t,w) : (t,w) \in T \times \Omega\}$ be a second order continuous time stochastic process mapping from a probability space $(\Omega, \mathcal{A}, P)$ into $L^2(T)$, where $L^2(T)$ denotes the separable Hilbert space of square-integrable functions on an interval $T \subseteq \mathbb{R}$. For the space $L^2(T)$, we denote its inner product by $< f, g >= \int_T f(t)\, g(t)\, dt$, for all $f, g \in L^2(T)$, and its norm by $\|*\|$. We assume throughout the text that $\mathrm{E}[\|X\|^2] < \infty$.

For such a process X, the covariance operator, denoted by $\mathcal{C}_X : L^2(T) \longmapsto L^2(T)$, is defined as follows

$$\mathcal{C}_X(f)(s) = \int_T R(s,t)\, f(t)\, dt, \qquad \forall f \in L^2(T), \tag{1}$$

where $R(s,t) = \mathrm{Cov}[X(s), X(t)]$ is the covariance function of X. As usual, the mean function of X is given by $\mathrm{E}[X](t) = \mathrm{E}[X(t)]$, for all $t \in T$.

Under this setting, it is well known (Ramsay & Silverman, 1997) that the spectral equation associated to $\mathcal{C}_X$, given by

$$\mathcal{C}_X(f_i) = \lambda_i\, f_i, \qquad \forall i \geq 1,$$

provides an orthogonal decomposition of X (the Karhunen-Loeve expansion) as follows

$$X(t,w) = \mathrm{E}[X](t) + \sum_{i \geq 1} b_i(w) f_i(t), \tag{2}$$

where $\{\lambda_i\}_{i\geq 1}$ is the non increasing sequence of non null eigenvalues of $\mathcal{C}_X$, which are called principal values (the eigenvalues are repeated in the sequence according to their multiplicity order), $\{f_i\}_{i\geq 1}$ is an orthonormal set of eigenfunctions of $\mathcal{C}_X$, which are called principal factors, and $\{b_i\}$ is a collection of uncorrelated and centred real random variables defined by

$$b_i(w) = < X(*,w) - \mathrm{E}[X], f_i >, \qquad \forall i \geq 1, \tag{3}$$

which are called principal components.

As in the PCA of a multivariate vector, the i-th principal component b_i associated with the process X is the normalized (generalized) linear combination of the process variables with maximum variance out of all the generalized linear combinations uncorrelated with $b_1, \ldots, b_{i-1}$, where $\lambda_i = \mathrm{Var}[b_i]$.

Deville (1974) and Dauxois *et al.* (1982), among others, have considered the estimation problem from a set of n independent sample paths, $X(*,1)$, ..., $X(*,n)$, drawn from X. This approach requires all the sample paths to be known in the entire interval T. Accordingly, the natural estimates (Deville, 1974) of the elements defining PCA are obtained by solving the sample spectral equation

$$\widehat{\mathcal{C}}_X(\widehat{f}_i) = \widehat{\lambda}_i\, \widehat{f}_i, \qquad \forall i \geq 1,$$

subject to $< \widehat{f}_i, \widehat{f}_j >= \delta_{i-j}$, where $\widehat{\mathcal{C}}_X$ is defined from (1) by the usual estimate of the covariance function on the given real interval. Thus the principal components are derived from (3) for each sample path.

3 Wavelet-based estimation

Multiresolution analysis (MRA) provide an ideal framework to introduce wavelet functions. Roughly speaking, MRA attempts to establish an approximating framework in terms of dilations and translations of a single function. In what follows, the translations and dilations to be consider for a function f are indexed by integers j and k as follows

$$f_{j,k}(t) = 2^{j/2} f\left(2^j t - k\right), \qquad t \in \mathbb{R}.$$

Let $\{0\} \subseteq \ldots \subseteq V_{-1} \subseteq V_0 \subseteq V_1 \subseteq \ldots \subseteq L^2(\mathbb{R})$ be a MRA of $L^2(\mathbb{R})$ such that there exists a compactly supported function $\phi \in V_0$, called the scaling function, such that $\{\phi_{j,k} : k \in \mathbf{Z}\}$ makes up an orthonormal basis of the vector subspace V_j, for each $j \in \mathbf{Z}$.

Assume another compactly supported function $\psi \in V_1$, called the wavelet function, such that the collection $\{\psi_{j,k} : j, k \in \mathbf{Z}\}$ makes up an orthonormal basis of $L^2(\mathbb{R})$ and the subspaces $W_j = \overline{\mathrm{Lin}\{\psi_{j,k} : k \in \mathbf{Z}\}}$ satisfy that $V_j \oplus W_j = V_{j+1}$, for each $j \in \mathbf{Z}$. It can then be shown that $L^2(\mathbb{R}) = V_{J_0} \oplus \bigoplus_{j \geq J_0} W_j$.

Throughout the paper the scaling and the wavelet functions are compactly supported to obtain computational efficiency (Alpert, 1992). A more detailed summary on wavelet theory can be found, for example, in Chui (1990), Daubechies (1988) and Morettin (1996). Furthermore, Daubechies (1988 & 1993) provide several examples of orthonormal and compactly supported wavelet (scaling) functions.

Let us now consider a second order stochastic process X, as in the previous section, with $\mathrm{E}[X] = 0$. And assume n sample functions drawn from X, denoted by $\{X(*, w) : w = 1, \ldots, n\}$. We now consider the process defined, without loss of generality, over the interval $[0,1]$.

Note that X is defined on interval $[0,1]$ and, on the other hand, the functions involved in the definition of the MRA are defined over $\mathbb{R}$. It is then usual to extend the sample paths of X on $\mathbb{R}$, which can be performed in several ways. The most obvious way is to set zero outside $[0,1]$, for each sample path. In our approach, we consider a MRA on $[0,1]$, $\{V_j^{[0,1]}\}$, by periodizing the basis functions (Sweldens & Piessens, 1993) as follows

$$\phi^*_{j,k} = \chi_{[0,1]} \sum_{l \in \mathbf{Z}} \phi_{j,k-l2^j}, \qquad \forall\, k = 0, \ldots, 2^j - 1 (j \geq 0).$$

Because of notational simplicity, we omit from now on the asterisk in the notation of the periodized scaling and wavelet functions on $[0,1]$.

By projecting each sample path of X onto the subspace $V_J^{[0,1]}$, we can define a wavelet approximation of the process X, for a given $J \geq 0$, as follows

$$\mathcal{P}_J X(*, w) = \sum_k c_{J,k}(w) \phi_{J,k}, \qquad \forall w = 1, \ldots, n, \tag{4}$$

where $\mathcal{P}_J$ denotes the orthogonal projection onto $V_J^{[0,1]}$. Since the periodized basis functions are orthonormal, the random coefficients are obtained by

$$c_{J,k}(w) = \int_0^1 X(t, w)\, \phi_{J,k}(t)\, dt, \qquad \forall\, k. \tag{5}$$

In this way, MRA provides a framework to approximate by different degrees the process X. In fact, Cambanis & Masry (1994) establish that the integrated mean square error satisfies $\mathrm{E}[\|X - \mathcal{P}_J X\|^2] = \int_0^1 \mathrm{Var}[X(t)]dt - \sum_k \mathrm{Var}[c_{J,k}] = \mathcal{O}(2^{-J})$, which can even be improved by assuming convenient regular conditions for the wavelet family and the covariance function of X.

From (4) we can obtain an approximation of X in terms of wavelets as follows

$$\mathcal{P}_J X(*,w) = \mathcal{P}_{J_0} X(*,w) + \sum_{j=J_0}^{J-1} \mathcal{Q}_j X(*,w), \qquad \forall w = 1,\dots n. \quad (6)$$

where $\mathcal{P}_j$ denotes the orthogonal projection on W_j. For each projection the expansion in terms of the basis functions is as follows

$$\mathcal{P}_{J_0} X(*,w) = \sum_k c_{J_0,k}(w)\, \phi_{J_0,k} \qquad \text{and}$$

$$\mathcal{Q}_j X(*,w) = \sum_k d_{j,k}(w)\, \psi_{j,k}, \qquad \forall j = J_0 \dots J-1.$$

The above coefficients are obtained by applying the Fast Wavelet Transform adapted for the MRA considered on $[0,1]$ (Sweldens & Piessens, 1993) to the coefficients given in (5).

Let Y be the random vector determined by the random coefficients from (6), that is

$$Y_J = (c_{J_0,k} : k | d_{j,k} : j = J_0, \dots, J-1; k)'.$$

Because the basis functions in (6) are orthonormal, the functional PCA of $\mathcal{P}_J X$ can be obtained from the multivariate PCA of the random vector Y_J (Aguilera *et al.*, 1995). This result establishes that the estimates in the PCA of Y_J, given by

$$\widehat{\mathrm{Cov}[Y]}\, \hat{v}_i = \widehat{\lambda}_i\, \hat{v}_i, \qquad \forall i \geq 1,$$

lead to the estimates in FPCA of $\mathcal{P}_J X$ by expressing each estimated principal factor $\widehat{f}_i$ as the linear combination of the basis functions from (6) in terms of the coefficients given by the eigenvector $\hat{v}_i$, for each $i \geq 1$.

In some real applications the sample curves are only observed at a finite set of knots in the real interval considered (Ramsay & Silverman, 1997; Aguilera *et al.*, 1995 & 1996). Therefore, the use of quadrature formulae are considered (Sweldens & Piessens, 1993) for approximating the initial inner products given in (5).

4 Application

In order to test the accuracy of the proposed wavelet approach for estimating PCA, we have considered a stochastic process whose PCA elements are known. For the Brownian motion defined on the interval $[0,1]$, it is well-known that the covariance function $R(s,t) = \min\{s,t\}$ has known principal values and factors, which are given by

$$\lambda_i = [(i-0.5)\pi]^{-2} \quad \text{and} \quad f_i(t) = \sqrt{2}\, sin\left(\lambda_i^{-1/2}\, t\right), \quad i = 1,2,\dots.$$

Thirty sample functions were simulated at the discrete knots given by $t_i = i2^{-4}$, for each $i = 0, \ldots, 2^4$. Data drawn from the Brownian motion then consist of the matrix

$$\left(X(t_i, w) : \begin{array}{c} w = 1, \ldots, 30 \\ k = 0, \ldots, 2^4 \end{array} \right).$$

To illustrate this approach, we considered the wavelet function proposed in Daubechies (1988) with 3 vanishing moments. The Daubechies' wavelets are orthogonal and compactly supported wavelet functions which are used in numerical analysis applications and whose regularity increases linearly with the number of vanishing moments (Daubechies, 1988 & 1993).

We then considered the wavelet approximation on V_4, denoted in the previous section as $\mathcal{P}_4 X$, and obtained the corresponding wavelet decomposition

$$\mathcal{P}_4 X = \mathcal{P}_3 X + \mathcal{Q}_3 X = \sum_{j=0}^{3} \mathcal{Q}_j X + \mathcal{P}_0 X$$

by using the Fast Wavelet Transform. Finally, the estimation in the FPCA for two wavelet approximations of X, $\mathcal{P}_4 X$ and $\mathcal{P}_3 X$ was obtained. Their estimated principal values appear in Table 1, and plots of the first two estimated principal factors obtained from $\mathcal{P}_4 X$ are shown in Figure 1.

No. PC	1	2	3	4	5	6
Exact	0.40528	0.04503	0.01621	0.00827	0.00500	0.00335
V_4	0.32799	0.03222	0.01696	0.00590	0.00446	0.00307
V_3	0.31964	0.03192	0.01570	0.00509	0.00389	0.00241

Table 1. Comparison among some exact variances of the Brownian motion components and their estimates by wavelet functional PCA (at resolution 4 and 3)

Acknowledgement
This research was supported by Project PB96-1436, Dirección General de Enseñanza Superior, Ministerio de Educación y Cultura, Spain.

References

Aguilera, A.M., Gutiérrez, R., Ocaña, F.A. & Valderrama, M.J. (1995). Computational approaches to estimation in the principal component analysis of a stochastic process. *Applied Stochastic Models & Data Analysis*, **11**, 279-299.

Aguilera, A.M., Gutiérrez, R. & Valderrama, M.J. (1996). Approximation of estimators in the PCA of a stochastic process using B-splines. *Commun. Statist.-Simula.*, **25**, 671-690.

Alpert, B.K. (1992). Wavelets and other bases for fast numerical linear algebra. In: *Wavelets: A tutorial in theory and applications*, (ed. C.K. Chui), 181-216. London: Academic Press.

Cambanis, S. & Masry, E. (1994). Wavelet approximation of deterministic and random signals: convergence properties and rates. *IEEE Transactions on Information Theory*, **40**, 1013-1029.

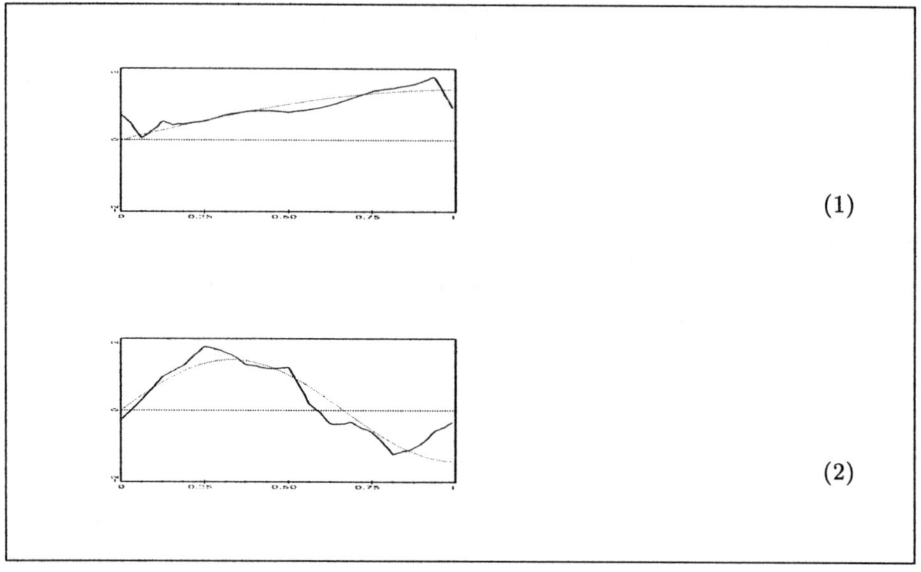

Fig. 1. Exact first and second principal factors (dashed curves) of the Brownian motion in [0, 1] and their approximations (solid curves) by wavelet analysis: (1) first principal factor; (2) second principal factor

Chui, C.K. (1990). *An Introduction to Wavelets*. San Diego: Academic Press.

Daubechies, I. (1988). Orthonormal bases of compactly supported wavelets. *Communications on Pure and Applied Mathematics*, **41**, 909-996.

Daubechies, I. (1993). Orthonormal bases of compactly supported wavelets II. Variations on a theme. *SIAM Journal of Mathematical Analysis.*, **24**, 499-519.

Dauxois, J., Pousse, A. & Romain, Y. (1982). Asymptotic theory for the principal component analysis of a vector random function: some applications to statistical inference. *J. Mult. Anal.*, **12**, 136-154.

Deville, J.C. (1974). Méthodes statistiques et numériques de l'analyse harmonique. *Annales de l'INSEE*, **15**, 3-101.

Morettin, P.A. (1996). From Fourier to wavelet analysis of time series. In: *COMPSTAT96 Proceedings in Computational Statistics* (ed. A. Prat), 111-122. Heidelberg: Physica-Verlag.

Ramsay, J.O. & Silverman, B.W. (1997). *Functional Data Analysis*. New York: Springer-Verlag.

Sweldens, W. & Piessens, R. (1993). Calculation of the wavelet decomposition using quadrature formulae. In: *Wavelets: An Elementary Treatment of Theory and Applications* (ed. T.H. Koornwinder), 139-160. Singapore: World Scientific.

Wong, E. & Hajek, B.K. (1985). *Stochastic Processes in Engineering Systems*. New York: McGraw Hill.

Quasi-Monte Carlo EM Algorithm for MLEs in Generalized Linear Mixed Models

Jian-Xin Pan and Robin Thompson

Statistics Department, IACR-Rothamsted, Harpenden, Hertfordshire, AL5 2JQ, UK

Abstract. Inferences for generalized linear mixed models are greatly hampered by the intractable integrated likelihood. In this paper numerical integration based on Quasi-Monte Carlo method is used to approximate the integral of the EM algorithm and then to fit the models. The proposed algorithm is computationally straightforward and easily implemented, and yields satisfactory estimates, compared to Monte Carlo approximation and Gauss-Hermite quadrature with the same computational cost.

Keywords. EM algorithm, generalized linear mixed models, Quasi-Monte Carlo method, random effects

1 Introduction

Let $\boldsymbol{y}$ and $\boldsymbol{b}$ denote the vectors of response and random effects with length n and q, respectively. Suppose the conditional density of $\boldsymbol{y}$ given $\boldsymbol{b}$ belongs to *a regular exponential family* $\exp\{\boldsymbol{\eta}^{\tau}\boldsymbol{y} - d(\boldsymbol{\eta}) + c(\boldsymbol{y})\}$ with a conditional expectation $E(\boldsymbol{y}|\boldsymbol{b}) = \boldsymbol{\mu}$, where the vector of random effects $\boldsymbol{b}$ is assumed to follow a normal distribution with a zero mean and an unknown variance-covariance matrix $\boldsymbol{\Sigma} > \boldsymbol{0}$, i.e., $\boldsymbol{b} \sim N_q(\boldsymbol{0}, \boldsymbol{\Sigma})$. The random effects are incorporated into the distribution via the linear predictor $\boldsymbol{\eta} = \boldsymbol{X\beta} + \boldsymbol{Zb}$, where $\boldsymbol{X}$ and $\boldsymbol{Z}$ are $n \times p$ and $n \times q$ explanatory matrices associated with the fixed effects $\boldsymbol{\beta}$ and the random effects $\boldsymbol{b}$, respectively. A *link function* $g(\bullet)$ is introduced to link the conditional expectation $\boldsymbol{\mu}$ and the linear predictor $\boldsymbol{\eta}$ through $g(\boldsymbol{\mu}) = \boldsymbol{\eta}$. This class of models is known as *generalized linear mixed models* (GLMMs).

A more general case of GLMMs is where several mutually independent random effects $\boldsymbol{b}_1, \boldsymbol{b}_2, \cdots, \boldsymbol{b}_c$ are incorporated simultaneously into the linear predictor $\boldsymbol{\eta} = \boldsymbol{X\beta} + \boldsymbol{Z}_1\boldsymbol{b}_1 + \cdots + \boldsymbol{Z}_c\boldsymbol{b}_c$, where $\boldsymbol{b}_l \sim N_{q_l}(\boldsymbol{0}, \boldsymbol{\Sigma}_l)$ $(l = 1, \cdots, c)$. The simplest but commonly used covariance structure for each $\boldsymbol{\Sigma}_l$ is of the form $\boldsymbol{\Sigma}_l = \sigma_l^2 \boldsymbol{I}_{q_l}$, where $\sigma_l^2 > 0$ is unknown and $\boldsymbol{I}_{q_l}$ is the identity matrix with size $q_l \times q_l$ $(l = 1, \cdots, c)$. This kind of GLMM is denoted as *Model I*. *Model II* is where the variance-covariance matrix $\boldsymbol{\Sigma}_l$ is unstructured $(l = 1, 2 \cdots, c)$.

Obviously, GLMMs are extensions of generalized linear models in the sense that some random effects are incorporated into the linear predictor. On the other hand, they are also naturally viewed as an extension of linear mixed models to the generalized linear arguments (see, e.g., McCullagh & Nelder, 1989). Such extensions can be used to accommodate correlated and overdispersed data in a variety of disciplines particularly in longitudinal studies and cluster samplings (e.g., Breslow & Clayton, 1993). Also, GLMMs widen extensively the application ranges of statistical models to other branches such as animal breeding experiments and biology sciences, where either binary or binomial data with within-cluster correlation are commonly concerned (e.g., Thompson, 1990). Statistical inferences for GLMMs, however, are greatly hampered by the need for numerical integrations since the integrals involved such as integrated likelihood have no analytically closed forms in general.

For GLMMs, the likelihood function of the regression coefficient $\boldsymbol{\beta}$ and the dispersion component $\boldsymbol{\Sigma}$ is of the form

$$L(\boldsymbol{\beta}, \boldsymbol{\Sigma}) = \int f(\boldsymbol{y}|\boldsymbol{b}, \boldsymbol{\beta}) f(\boldsymbol{b}|\boldsymbol{\Sigma}) d\boldsymbol{b}, \tag{1}$$

where $f(\boldsymbol{y}|\boldsymbol{b}, \boldsymbol{\beta})$ is the conditional density of $\boldsymbol{y}$ given $\boldsymbol{b}$ and $f(\boldsymbol{b}|\boldsymbol{\Sigma})$ is the density of random effects $\boldsymbol{b}$ which only depends on $\boldsymbol{\Sigma}$. In general, the maximum likelihood estimates (MLEs) of the parameters $\boldsymbol{\beta}$ and $\boldsymbol{\Sigma}$, which maximize the likelihood (1) with respect to $\boldsymbol{\beta}$ and $\boldsymbol{\Sigma}$, cannot be calculated analytically since the integral is intractable especially for high-dimensional random effects. An effective approximation to the MLEs of $\boldsymbol{\beta}$ and $\boldsymbol{\Sigma}$ is hence needed to analyze GLMMs.

The EM-algorithm is one approach to calculate the MLEs iteratively. Here each iteration consists of two steps, E-step (expectation) and M-step (maximization). For GLMMs, the E-step computes the conditional expectation of the log-likelihood

$$Q(\boldsymbol{\phi}|\boldsymbol{\phi}') = \int \log f(\boldsymbol{y}, \boldsymbol{b}|\boldsymbol{\phi}) f(\boldsymbol{b}|\boldsymbol{y}, \boldsymbol{\phi}') d\boldsymbol{b}, \tag{2}$$

where $\boldsymbol{\phi}$ consists of the parameters $\boldsymbol{\beta}$ and $\boldsymbol{\Sigma}$, $f(\boldsymbol{y}, \boldsymbol{b}|\boldsymbol{\phi})$ is the joint likelihood of $\boldsymbol{y}$ and $\boldsymbol{b}$ depending on $\boldsymbol{\phi}$, and $f(\boldsymbol{b}|\boldsymbol{y}, \boldsymbol{\phi}')$ is the posterior density of $\boldsymbol{b}$ which only depends on a previously fixed value $\boldsymbol{\phi}'$, for example, the estimate from the previous cycle in the EM algorithm. The estimate of $\boldsymbol{\phi}$ at the present cycle can be obtained by maximizing (2). For some specific GLMMs, the conditional expectation (2) can be expressed as an analytically closed form and the approximate MLEs of the parameters can be obtained straightforwardly. In the general setting, however, it remains intractably even for the normal random effects. A Laplace approximation applied to the E-step was studied by Steele (1996) and Monte Carlo approximation was implemented for (2) by McCulloch (1997). Since those approximations suffer from problems of estimate accuracy or heavy computational effort, a relatively simple but effective approximation of the MLEs for GLMMs is still needed.

2 Quasi-Monte Carlo integration

Suppose $f(\bullet)$ is an integrable function on the q-dimensional unit cube $C^q = [0,1)^q$. Consider the integral $I(f) = \int_{C^q} f(\boldsymbol{x}) d\boldsymbol{x}$. The Monte Carlo integration method draws a random sample $\boldsymbol{x}_1, \boldsymbol{x}_2, \cdots, \boldsymbol{x}_n$ from the uniform distribution on C^q and then approximates the integration $I(f)$ with the unbiased estimate $\hat{I}_n(f, \mathcal{P}_n) = n^{-1} \sum_{i=1}^n f(\boldsymbol{x}_i)$, where $\mathcal{P}_n = \{\boldsymbol{x}_i : 1 \le i \le n\}$. By the strong law of large numbers $\hat{I}_n(f, \mathcal{P}_n)$ converges to $I(f)$ with probability one as $n \to \infty$. Moreover the central limit theorem guarantees that $\hat{I}_n(f, \mathcal{P}_n)$ is approximately normally distributed with convergence rate $O(n^{-1/2})$ as the sample size n becomes large. The disadvantages of Monte Carlo integration are that the rate of convergence is only of order $O(n^{-1/2})$ and the convergence is in distribution. In other words, the Monte Carlo integration approximation performs well on average, but a particular random sample that is drawn may lead to a bad approximation. Thus, new methods with a high rate of convergence, which should be deterministic, are needed as alternatives to Monte Carlo integration.

The Quasi-Monte Carlo (QMC) approximation is one such numerical integration method. The basic idea behind QMC integration is to draw a set of integration nodes which are scattered uniformally on C^q to replace the random samples in the Monte Carlo integration. In such a way the rate of convergence for the approximation can be improved significantly and the approximation error is deterministic rather than in probability. In fact, for any set of points $\mathcal{P}_n$ on C^q we have the Koksma-Hlawka inequality

$$|I(f) - \hat{I}_n(f, \mathcal{P}_n)| \leq V(f)D(\mathcal{P}_n), \tag{3}$$

where $V(f)$ is the bounded total variation of f over C^q in the sense of Hardy & Krause, which is independent of the point set $\mathcal{P}_n$ (see, e.g., Niederreiter, 1988). The quantity $D(\mathcal{P}_n)$ is a measure of evenness of spread for the point set $\mathcal{P}_n$ in C^q, defined by

$$D(\mathcal{P}_n) = \sup_{\boldsymbol{x} \in C^q} |U_n(\boldsymbol{x}) - U(\boldsymbol{x})|, \tag{4}$$

where $U(\boldsymbol{x})$ is the uniform distribution over on C^q and $U_n(\boldsymbol{x})$ is the empirical distribution of the point set $\mathcal{P}_n$. The measure $D(\mathcal{P}_n)$ in (4) is known as discrepancy of $\mathcal{P}_n$ in QMC theory. The Koksma-Hlawka inequality (3) implies that the absolute error of approximation is bounded by a constant times the discrepancy of the set of points $\mathcal{P}_n$. Therefore the set of points with the smallest discrepancy is naturally chosen in order to obtain a highly accurate approximation of integration. It can be shown that the smallest discrepancy of a set of point is of order $O((\log n)^{q-1}/n)$, which is a superior asymptotic rate to the order $O(n^{-1/2})$ given by Monte Carlo integration. Hence the main focus of QMC integration is finding efficient ways to construct the set of points with order $O((\log n)^{q-1}/n)$. For one-dimensional integration, i.e., $q = 1$, the set of QMC nodes is of the form $\mathcal{P}_n = \{(2i-1)/2n : i = 1, 2, \cdots, n\}$ with the smallest discrepancy $D(\mathcal{P}_n) = 1/2n$. When the dimension $q \geq 2$, it is not easy to find a set of points with the smallest discrepancy, but some sets of points with discrepancies close to $O((\log n)^{q-1}/n)$ are available. Among such sets of points, the *good lattice points* (GLP) and the *scrambled Halton points* (SHP) are the most common. For their detailed definitions and properties, see Fang & Wang (1994).

For GLMMs the integration involved in either the likelihood (1) or the conditional expectation (2) is of the form

$$I(f, g) = \int_{-\infty}^{+\infty} f(\boldsymbol{x})g(\boldsymbol{x})d\boldsymbol{x},$$

where $g(\boldsymbol{x})$ can be viewed as the density function of the random effects. The Monte Carlo integration for $I(f, g)$ draws a random sample $\{\boldsymbol{x}_1, \boldsymbol{x}_2, \cdots, \boldsymbol{x}_n\}$ from the distribution $g(\boldsymbol{x})$ and then uses the same form as $\hat{I}_n(f, \mathcal{P}_n)$ to approximate $I(f, g)$. In contrast, the Quasi-Monte Carlo integration constructs integration nodes from $g(\boldsymbol{x})$ with certain structures, known as *CDF-Representative points*, to replace the random sample of Monte Carlo method, based on the rule of small discrepancy. For example, CDF-Representative points of the standard normal distribution $N_q(\boldsymbol{0}, \boldsymbol{I}_q)$ can be constructed by evaluating the inverse distribution function of $N(0, 1)$ on each coordinate of the GLP points. More details can be found in Fang & Wang (1994).

3 Quasi-Monte Carlo EM algorithm

For simplicity, we write the conditional expectation (2) in an alternative form

$$Q(\phi|\phi') = \int \log f(\boldsymbol{y}, \boldsymbol{b}|\phi) f(\boldsymbol{y}|\boldsymbol{b}, \boldsymbol{\beta}') f(\boldsymbol{b}|\boldsymbol{\Sigma}') d\boldsymbol{b} \Big/ \int f(\boldsymbol{y}|\boldsymbol{b}, \boldsymbol{\beta}') f(\boldsymbol{b}|\boldsymbol{\Sigma}') d\boldsymbol{b}, \quad (5)$$

where $f(\boldsymbol{y}|\boldsymbol{b}, \boldsymbol{\beta}')$ and $f(\boldsymbol{b}|\boldsymbol{\Sigma}')$ are the conditional density of $\boldsymbol{y}$ given $\boldsymbol{b}$ and the density of $\boldsymbol{b}$, respectively, evaluated at the previous cycle estimates $\boldsymbol{\beta}'$ and $\boldsymbol{\Sigma}'$. Suppose $\boldsymbol{b}'^{(1)}, \boldsymbol{b}'^{(2)}, \cdots, \boldsymbol{b}'^{(m)}$ are CDF-Representative points of $N_q(\boldsymbol{0}, \boldsymbol{I}_q)$. Then the CDF-Representative points of $N_q(\boldsymbol{0}, \boldsymbol{\Sigma}')$, say $\boldsymbol{b}^{(1)}, \boldsymbol{b}^{(2)}, \cdots, \boldsymbol{b}^{(m)}$, can be formed by making transformations $\boldsymbol{b}^{(k)} = \boldsymbol{\Sigma}'^{1/2} \boldsymbol{b}'^{(k)}$ $(1 \le k \le m)$, and the conditional expectation (5) can be approximated by

$$\widehat{Q}(\phi|\phi') = \sum_{k=1}^{m} \omega'^{(k)} \log f(\boldsymbol{y}, \boldsymbol{b}^{(k)}|\phi), \quad (6)$$

where the weight $\omega'^{(k)}$ is given by $\omega'^{(k)} = f(\boldsymbol{y}|\boldsymbol{b}^{(k)}, \boldsymbol{\beta}') / \sum_{k=1}^{m} f(\boldsymbol{y}|\boldsymbol{b}^{(k)}, \boldsymbol{\beta}')$, which is independent of the parameter ϕ but depends on the previous cycle estimates of parameters and the QMC nodes $\boldsymbol{b}^{(1)}, \boldsymbol{b}^{(2)}, \cdots, \boldsymbol{b}^{(m)}$. The M-step of the QMC-EM algorithm maximizes the approximate conditional expectation (6) with respect to ϕ. A direct calculation shows that the score function of $\boldsymbol{\beta}$, i.e., the first derivative of (6) with respect to $\boldsymbol{\beta}$ is of the form

$$\dot{\boldsymbol{Q}}_{\beta} = \boldsymbol{X}^{\tau} \Big(\boldsymbol{y} - \sum_{k=1}^{m} \omega'^{(k)} h(\boldsymbol{X}\boldsymbol{\beta} + \boldsymbol{Z}\boldsymbol{b}^{(k)}) \Big), \quad (7)$$

where $h(\bullet)$ is the inverse of the link function $g(\bullet)$, $h(\boldsymbol{X}\boldsymbol{\beta} + \boldsymbol{Z}\boldsymbol{b}^{(k)})$ is a vector of length n with the ith component $h(\boldsymbol{x}_i^{\tau}\boldsymbol{\beta} + \boldsymbol{z}_i^{\tau}\boldsymbol{b}^{(k)})$, and $\boldsymbol{x}_i^{\tau}$ and $\boldsymbol{z}_i^{\tau}$ are the ith rows of the matrices $\boldsymbol{X}$ and $\boldsymbol{Z}$, respectively. In general, the score equation $\dot{\boldsymbol{Q}}_{\beta} = \boldsymbol{0}$ has no analytical solution and some iterative algorithm such as the Newton-Raphson algorithm or Fisher scoring algorithm must be applied. Based on (7), the observed information matrix of $\boldsymbol{\beta}$ can be calculated by

$$-\ddot{\boldsymbol{Q}}_{\beta\beta} = \boldsymbol{X}^{\tau} \Big(\sum_{k=1}^{m} \omega'^{(k)} \dot{h}(\boldsymbol{X}\boldsymbol{\beta} + \boldsymbol{Z}\boldsymbol{b}^{(k)}) \Big) \boldsymbol{X}, \quad (8)$$

where $\dot{h}(\boldsymbol{X}\boldsymbol{\beta} + \boldsymbol{Z}\boldsymbol{b}^{(k)}) = \mathrm{diag}(\dot{h}(\boldsymbol{x}_1^{\tau}\boldsymbol{\beta} + \boldsymbol{z}_1^{\tau}\boldsymbol{b}^{(k)}), \cdots, \dot{h}(\boldsymbol{x}_n^{\tau}\boldsymbol{\beta} + \boldsymbol{z}_n^{\tau}\boldsymbol{b}^{(k)})$ is a diagonal matrix and $\dot{h}(\bullet)$ denotes the derivative of $h(\bullet)$ with respect to variable "$\bullet$". The Newton-Raphson algorithm suggests that the update solution of $\boldsymbol{\beta}$ at the $(r+1)$th iteration can be obtained by the use of the estimate at the rth cycle by $\widehat{\boldsymbol{\beta}}^{(r+1)} = \widehat{\boldsymbol{\beta}}^{(r)} + [-\ddot{\boldsymbol{Q}}_{\beta\beta}]^{-1}|_{\beta=\hat{\beta}^{(r)}} [\dot{\boldsymbol{Q}}_{\beta}]|_{\beta=\hat{\beta}^{(r)}}$, where $r = 0, 1, 2, \cdots$ and $\boldsymbol{\beta}^{(0)}$ is an initial estimate of $\boldsymbol{\beta}$. The iterations are terminated when the relative error of the update solution becomes small enough. For the variance-covariance component $\boldsymbol{\Sigma}$ in each M step, on the other hand, it is not necessary to calculate the estimate iteratively since it can be expressed analytically as $\hat{\sigma}_l^2 = q_l^{-1} \sum_{k=1}^{m} \omega'^{(k)} \boldsymbol{b}_l^{(k)\tau} \boldsymbol{b}_l^{(k)}$ for *Model I* or $\widehat{\boldsymbol{\Sigma}}_l = \sum_{k=1}^{m} \omega'^{(k)} \boldsymbol{b}_l^{(k)} \boldsymbol{b}_l^{(k)\tau}$ for

Model II ($l = 1, 2, \cdots, c$). Furthermore, if we are interested in prediction of the random effects $\boldsymbol{b}$, a reasonable solution is the approximate posterior expectation $\widehat{\boldsymbol{b}} = \widehat{E}(\boldsymbol{b}|\boldsymbol{y}, \widehat{\boldsymbol{\phi}}) = \sum_{k=1}^{m} \widehat{\omega}^{(k)} \hat{\boldsymbol{b}}^{(k)}$, where the weight $\widehat{\omega}^{(k)}$ is given by $\widehat{\omega}^{(k)} = f(\boldsymbol{y}|\hat{\boldsymbol{b}}^{(k)}, \widehat{\boldsymbol{\beta}}) / \sum_{k=1}^{m} f(\boldsymbol{y}|\hat{\boldsymbol{b}}^{(k)}, \widehat{\boldsymbol{\beta}})$, $\widehat{\boldsymbol{\phi}} = (\widehat{\boldsymbol{\beta}}, \widehat{\boldsymbol{\Sigma}})$ in which $\widehat{\boldsymbol{\beta}}$ and $\widehat{\boldsymbol{\Sigma}}$ are the QMC-EM estimates, and the integration nodes $\hat{\boldsymbol{b}}^{(1)}, \hat{\boldsymbol{b}}^{(2)}, \cdots, \hat{\boldsymbol{b}}^{(m)}$ are CDF-Representative points of $N_q(\boldsymbol{0}, \widehat{\boldsymbol{\Sigma}})$.

Furthermore, it can be shown that the variance-covariance matrix of $\widehat{\boldsymbol{\beta}}$ can be estimated by

$$\widehat{\mathrm{Cov}}(\widehat{\boldsymbol{\beta}}) \approx [-\ddot{\boldsymbol{Q}}_{\beta\beta}]^{-1} + [-\ddot{\boldsymbol{Q}}_{\beta\beta}]^{-1}([-\ddot{\boldsymbol{H}}_{\beta\beta}]^{-1} - [-\ddot{\boldsymbol{Q}}_{\beta\beta}]^{-1})^{-1}[-\ddot{\boldsymbol{Q}}_{\beta\beta}]^{-1},$$

where $\ddot{\boldsymbol{Q}}_{\beta\beta}$ is given by (8) but evaluated at the QMC-EM estimates $\widehat{\boldsymbol{\beta}}$ and $\widehat{\boldsymbol{\Sigma}}$, and the missing information matrix $-\ddot{\boldsymbol{H}}_{\beta\beta}$ is approximated by $\boldsymbol{X}^{\tau} \widehat{\boldsymbol{A}} \boldsymbol{X}$ in which

$$\begin{aligned}\widehat{\boldsymbol{A}} = & \textstyle\sum_{k=1}^{m} \widehat{\omega}^{(k)} \left[h(\boldsymbol{X}\widehat{\boldsymbol{\beta}} + \boldsymbol{Z}\hat{\boldsymbol{b}}^{(k)})\right]\left[h(\boldsymbol{X}\widehat{\boldsymbol{\beta}} + \boldsymbol{Z}\hat{\boldsymbol{b}}^{(k)})\right]^{\tau} \\ & - \left[\textstyle\sum_{k=1}^{m} \widehat{\omega}^{(k)} h(\boldsymbol{X}\widehat{\boldsymbol{\beta}} + \boldsymbol{Z}\hat{\boldsymbol{b}}^{(k)})\right]\left[\textstyle\sum_{k=1}^{m} \widehat{\omega}^{(k)} h(\boldsymbol{X}\widehat{\boldsymbol{\beta}} + \boldsymbol{Z}\hat{\boldsymbol{b}}^{(k)})\right]^{\tau}.\end{aligned}$$

For *Model I*, the variance of $\hat{\sigma}_l^2$ is estimated by $\widehat{\mathrm{var}}(\hat{\sigma}_l^2) \approx 4\hat{\sigma}_l^8/(2q_l\hat{\sigma}_l^4 - \hat{v}_l)$, where $\hat{v}_l = \sum_{k=1}^{m} \widehat{\omega}^{(k)} (\hat{\boldsymbol{b}}_l^{(k)\tau} \hat{\boldsymbol{b}}_l^{(k)})^2 - (\sum_{k=1}^{m} \widehat{\omega}^{(k)} \hat{\boldsymbol{b}}_l^{(k)\tau} \hat{\boldsymbol{b}}_l^{(k)})^2$ ($l = 1, 2, \cdots, c$). For *Model II*, let $\widehat{\boldsymbol{\theta}}_l = \mathrm{vech}(\widehat{\boldsymbol{\Sigma}}_l)$, in other words, the q_l^*-variant vector consisting of all the low-triangular entries of $\widehat{\boldsymbol{\Sigma}}_l$ with $q_l^* = q_l(q_l+1)/2$. Then the asymptotic variance-covariance matrix of $\widehat{\boldsymbol{\theta}}_l$ can be approximated by

$$\widehat{\mathrm{Cov}}(\widehat{\boldsymbol{\theta}}_l) \approx [-\ddot{\boldsymbol{Q}}_{\theta_l\theta_l}]^{-1} + [-\ddot{\boldsymbol{Q}}_{\theta_l\theta_l}]^{-1}([-\ddot{\boldsymbol{H}}_{\theta_l\theta_l}]^{-1} - [-\ddot{\boldsymbol{Q}}_{\theta_l\theta_l}]^{-1})^{-1}[-\ddot{\boldsymbol{Q}}_{\theta_l\theta_l}]^{-1},$$

where $-\ddot{\boldsymbol{Q}}_{\theta_l\theta_l} = \boldsymbol{S}_{q_l}^{\tau}(\widehat{\boldsymbol{\Sigma}}_l \otimes \widehat{\boldsymbol{\Sigma}}_l)^{-1}\boldsymbol{S}_{q_l}/2$, $\widehat{\boldsymbol{\Sigma}}_l$ is the QMC-EM estimate of $\boldsymbol{\Sigma}_l$, and $-\ddot{\boldsymbol{H}}_{\theta_l\theta_l} = \boldsymbol{S}_{q_l}^{\tau}(\widehat{\boldsymbol{\Sigma}}_l \otimes \widehat{\boldsymbol{\Sigma}}_l)^{-1}\widehat{\boldsymbol{V}}_l(\widehat{\boldsymbol{\Sigma}}_l \otimes \widehat{\boldsymbol{\Sigma}}_l)^{-1}\boldsymbol{S}_{q_l}/4$ in which

$$\begin{aligned}\widehat{\boldsymbol{V}}_l = & \textstyle\sum_{k=1}^{m} \widehat{\omega}^{(k)} \left[\mathrm{vec}(\hat{\boldsymbol{b}}_l^{(k)} \hat{\boldsymbol{b}}_l^{(k)\tau})\right]\left[\mathrm{vec}(\hat{\boldsymbol{b}}_l^{(k)} \hat{\boldsymbol{b}}_l^{(k)\tau})\right]^{\tau} \\ & - \left[\textstyle\sum_{k=1}^{m} \widehat{\omega}^{(k)} \mathrm{vec}(\hat{\boldsymbol{b}}_l^{(k)} \hat{\boldsymbol{b}}_l^{(k)\tau})\right]\left[\textstyle\sum_{k=1}^{m} \widehat{\omega}^{(k)} \mathrm{vec}(\hat{\boldsymbol{b}}_l^{(k)} \hat{\boldsymbol{b}}_l^{(k)\tau})\right]^{\tau}.\end{aligned}$$

$\boldsymbol{S}_{q_l}$ is a duplication matrix with size $q_l^2 \times q_l^*$ such that $\mathrm{vec}(\boldsymbol{\Sigma}_l) = \boldsymbol{S}_{q_l}\mathrm{vech}(\boldsymbol{\Sigma}_l)$, where the notation $\mathrm{vec}(\boldsymbol{\Sigma}_l)$ denotes the $q_l^2 \times 1$ vector formed by stacking the columns of $\boldsymbol{\Sigma}_l$ under each other ($l = 1, 2, \cdots, c$).

4 Comments

In order to assess the performance of the proposed QMC-EM algorithm and compare it with the Monte Carlo EM algorithm, a simulation study was carried and several practical data sets were analyzed. In the simulation study, we followed two specific logistic-normal mixed models with one and two-dimensional random effects presented by Zeger & Karim (1991), where the Gibbs resampling method was used to explore the posterior distributions and posterior modes of parameters of the models. Those models were also simulated using the penalized quasi-likelihood (PQL) approximation of Breslow & Clayton (1993) and the modified EM algorithm (MEM) of Steele (1996). Both the two models were applied to simulated binomial responses with four distinct denominators $s = 1, 2, 4$ and 8. For each binomial denominator, 200 and 100 data sets were generated for the first and second models, respectively, each set consisting of 100 clusters with 7 responses per cluster. Due to space limitations we present here only the main conclusions drawn from our practical analysis. First, compared with the ordinary Monte Carlo EM algorithm, the QMC-EM algorithm yielded significantly smaller biases and standard errors. More importantly, the computational efforts of the QMC-EM estimates were far less than those of the Monte Carlo EM algorithm. Therefore, *the Quasi-Monte Carlo approximation is superior to the ordinary Monte Carlo method* for GLMMs in this sense. Second, like Gauss-Hermite quadrature, the QMC-EM estimates can be improved by increasing the number of integration nodes, i.e., CDF-Representative points of the random effects. Unlike Gauss-Hermite quadrature, however, the computational efforts of the QMC-EM algorithm increase only linearly rather than exponentially with the dimension of the random effects, and it gives the same efficiency of estimates as Gauss-Hermite quadrature but with less computational loads. Hence *the QMC approach can be viewed as an alternative to Gauss-Hermite quadrature* for mixed models particularly with high-dimensional random effects, where Gauss-Hermite quadrature is less appropriate since the number of integration nodes increases exponentially with the dimension of the random effects.

Acknowledgments
This research is partially supported by a grant from Agriculture and Fisheries Department, Scottish Office. Valuable comments from Professor Roger W. Payne and an anonymous referee are appreciated.

References

Breslow, N. E. & Clayton, D. G. (1993). Approximate inference in generalized linear mixed models. *Journal of the American Statistical Association*, **88**, 9-25.

Fang, K. T. & Wang Y. (1994). *Number-Theoretic Methods in Statistics*, London: Chapman and Hall.

McCullagh, P. & Nelder, J. A. (1989). *Generalized Linear Models*, 2nd edition. London: Chapman and Hall.

McCulloch, C. E. (1997). Maximum likelihood algorithms for generalized linear mixed models. *Journal of the American Statistical Association*, **92**, 162-170.

Niederreiter, H. (1988). Quasi-Monte Carlo methods for multi-dimensional numerical integration. *International Series of Numerical Mathematics*, Birk: Springer-Verlag.

Steele, B. M. (1996). A modified EM algorithm for estimation in generalized mixed models. *Biometrics*, **52**, 1295-1310.

Thompson, R. (1990). Generalized linear models and applications to animal breeding. *Advances in Statistical Methods for Genetic Improvement of Livestock.* (ed. D. Gianola & K. Hammond) Berlin: Springer-Verlag, 312-327.

Zeger, S. L. & Karim, M. R. (1991). Generalized linear models with random effects: a Gibbs sampling approach. *Journal of the American Statistical Association*, **86**, 79-86.

Applications of Smoothed Monotone Regression Splines and Smoothed Bootstrapping in Survival Analysis

Donald E. Ramirez[1] and Philip W. Smith[2]

[1] University of Virginia, Department of Mathematics, Kerchof Hall 209, Charlottesville, VA 22903 USA
[2] WindWard Technologies, Inc., 12039 Mulholland, Meadows Place, TX 77477 USA

Abstract. An estimate of a survival curve $S(x)$ for censored data is given by the non-parametric Kaplan-Meier method, which provides an estimate of the empirical survival curve and estimators of the standard errors. We use the software package Confit, developed by WTI, which is designed to produce a smoothed approximating spline, subject to imposed constraints on the function or its derivatives over an interval. The algorithm solves a constrained least-squares problem parameterized by an appropriate spline subspace (using a B-spline representation). The constraints impose some additional constraints on these coefficients that are converted into a quadratic programming problem. We will discuss the algorithm used to solve the quadratic programming problem, and give applications to illustrate our method on several data sets.

1 Introduction

In modelling data, the statistician is searching for a functional form $\phi(x)$ that the data satisfies. In certain cases, we wish to impose additional constraints on either the functional form $\phi(x)$ or its derivative $\phi'(x)$. For example, to model the cumulative distribution function $F(x)$ or the survival $S(x) = 1 - F(x)$, monotonicity is essential. For a review of some of the important statistical applications of splines see Smith (1979).

A set of basic splines which are useful in computational statistics is the set of B-splines. For a nondecreasing knot sequence $\{t_0, t_1, \ldots, t_N\}$, $B_i^0(x)$ is the indicator function for $[t_i, t_{i+1})$ and the higher order splines ($k \geq 1$) are defined recursively by

$$B_i^k(x) = \left(\frac{x - t_i}{t_{i+k} - t_i}\right) B_i^{k-1}(x) + \left(\frac{t_{i+k+1} - x}{t_{i+k+1} - t_{i+1}}\right) B_{i+1}^{k-1}(x), \quad t_i \leq x \leq t_{i+k+1}. \tag{1}$$

The B-splines are nonnegative and are normalized by $\sum_{i=0}^{N} B_i^k(x) = 1$. (See for example, DeBoor (1978, p. 110).) We will denote by $B^{N-1}(x, \{t_0, \ldots, t_N\})$ the $N-1$ degree B-spline for $\{t_0, \ldots, t_N\}$.

A closely related set of splines is given by $M_i^k(x) = kB_i^k(x)/(t_{i+m} - t_i)$ (Curry & Schoenberg, 1966). Since the M-splines are nonnegative functions, their integrals provide a set of monotone functions. Ramsay (1988) used for his monotone regression splines sums of these integrated splines with positive coefficients. Ramsay described a nonlinear algorithm using the gradient projection constrained optimization algorithm for determining the optimal coefficients. Gaylord & Ramirez (1991), building on the work of Ramsay, introduced an adaptive linear algorithm using weighted regression to force regression splines to be monotone.

Smoothing splines, pioneered by Wahba (1990), are found by minimizing a combination of the least-squares errors $\sum_{i=0}^{N}(y_i - \phi(x_i))^2$ and the L^2 norm of the generalized curvature $\int_{t_0}^{t_N}(\phi^{(m)}(x))^2 dx$. Kelly & Rice (1990) constrained the B-spline coefficients to be monotone to enforce monotonicity of the smoothing spline. Turlach (1997) has proposed calculating smoothing splines with constraints. His approach leads to a quadratic programming problem, with the infinite number of constraints replaced by a suitable finite number, which can be solved using the algorithm of Goldfarb & Idnani (1983).

In this paper, we will briefly describe the software package Confit which finds an approximating spline where the user is able to prescribe constraints on the function and/or the derivatives. We then present some applications of monotone splines that are efficiently solved with Confit.

2 Confit

Confit is a package designed to produce a smoothed approximating spline subject to a finite number of constraints on the function or its derivatives. It fits a spline function $y = S(x)$ to the data $\{(x_i, y_i) : i = 0, \ldots, N\}$ with weights $\{w_i : i = 0, \ldots, N\}$, by minimizing the error sum of squares, subject to specified shape constraints. Specifically, it solves the problem: minimize $\sum_{i=0}^{N} w_i(S(x_i) - y_i)^2$ subject to $\alpha_j \leq S^{(p_j)}(x) \leq \beta_j$ on $a_j \leq x \leq b_j$, for $j = 1, \ldots, m$. The algorithm involves a least-squares problem which is solved for the coefficients of an appropriate B-spline. When constraints are imposed on the approximating spline, the problem is converted into a quadratic programming problem that Confit solves with amazing speed. For any given constraint on an interval, Confit checks for compliance on a grid of 500 equally spaced points. The actual computation involves building a sequence of optimization problems by adding the most violated point one at time (for each constraint) until compliance is observed. To create a finer mesh, the interval can be subdivided. Confit uses an Excel compatible spreadsheet for its data entry and output.

3 Survival curves for censored exponential data

The data used here is simulated data from an exponential distribution with mean $\mu = 24$ and sample size $N = 25$. The (five) values > 70 are considered as being right-censored values. An estimate y of the survival curve $S(x)$ is given by the Kaplan-Meier method. This method also provides an estimate of the standard error for these estimates. We choose the inverse of the square of the standard error for the weights w. This follows the procedure for a logistic response function (see Neter *et al.*, p. 363).

Table 3.1. Censored Exponent Data with $m = 24$ and $N = 25$

x	0	1	2	5	6	9	10	12	14	17
y	.96	.92	.88	.84	.72	.68	.64	.60	.56	.52
w	651	339	237	186	124	115	109	104	101	100

x	18	20	21	31	35	38	41	49	63	70
y	.48	.44	.40	.36	.32	.28	.24	.20	.16	.12
w	100	101	104	109	115	124	137	156	186	237

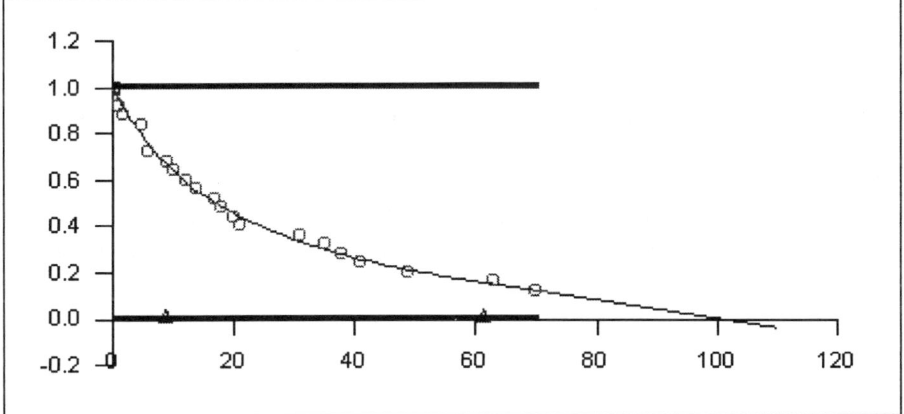

Fig. 3.1. B-Spline approximation for censored exponential data

To model this data, we introduce the four constraints: (1) $S(0) = 1$, (2) $0 \leq S(x) \leq 1$ on $[0, 70]$, (3) $S'(x) \leq 0$ on $[0, 70]$, and (4) $S''(x) = 0$ on $[70, 120]$. The first and second constraints are the cdf constraints; the third constraint is the monotonicity constraint. The fourth constraint extends the spline linearly downward to the x-axis.

Using the defaults for Confit, the B-spline approximation of the data produces a fourth degree spline approximation with eight B-splines with knot sequence $\{0,0,0,0,0,27.5,55,82.5,110,110,110,110,110\}$ and coefficients $\{1,.654,.410,.218,.137,.041,-.011,-.037\}$. Thus $S(x) = B^4(x,\{0,0,0,0,0,27.5\}) + .654B^4(x,\{0,0,0,0,27.5,55\}) + \cdots - .037B^4(x,\{82.5,110,110,110,110,110\})$.

We can estimate quantiles for the data from the smoothed spline. The estimates for the 50%, 95%, and 99% quantiles are 17.1 (16.6), 88.0 (71.9), and 97.8 (110.5), respectively, where the actual values are shown in parentheses following. The plot of the monotone spline approximation is shown in Figure 3.1.

4 Survival curves for Weibull data

The data here are $N = 25$ values from a Weibull distribution with $\alpha = 1.5$ and $\beta = 1$. The cumulative hazard function $H(x) = -\log(1 - F(x)) = (x/\beta)^\alpha$ provides another application for smoothed monotone regression splines. The hazard function $h(x) = H'(x) = (\alpha/\beta)(x/\beta)^{\alpha-1}$ is assumed to be nonnegative and monotone increasing ($\alpha > 1$). The ordered data $\{x_{(1)}, \ldots, x_{(N)}\}$ is paired with the empirical distribution by

$$F(x_{(i)}) = (i - 1/3)/(N + 1/3). \tag{2}$$

This follows the convention of Hoaglin *et al.* (1983, p. 44) for quantiles. Confit is used to fit a monotone spline to the data $\{(x_{(i)}, y_i) : i = 1, \ldots N\}$ with $y_i = -\log(1 - F(x_{(i)})) = H(x_{(i)})$. The constraints required are (1) $H(x) \geq 0$, (2) $h(x) = H'(x) \geq 0$, and (3) $h'(x) \geq 0$ for monotonicity of the hazard function. Using the defaults for Confit, the B-spline approximation of the data produces a fourth degree spline approximation with nine B-splines with 14 knots.

The survival curve estimate is recovered from $S(x) = \exp(-H(x)) = \exp(-(x/\beta)^\alpha)$. To extend the spline approximation of $S(x)$ downward to the x-axis with an exponential curve; the corresponding values for α and β used are

$$\alpha = -x_{(N)} h(x_{(N)}) / \log(S(x_{(N)})) \tag{3}$$

$$\beta = (\alpha x_{(N)}^{\alpha-1} h(x_{(N)}))^{1/\alpha} \tag{4}$$

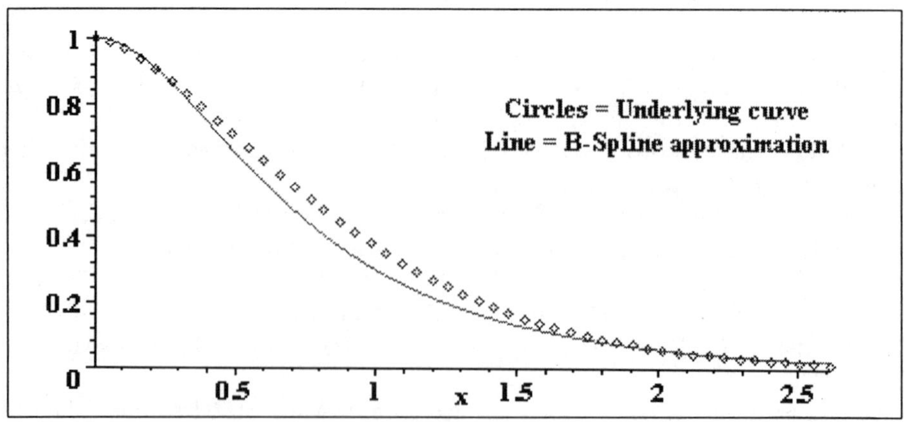

Fig. 4.1. B-Spline approximation for Weibull data with underlying curve

We can estimate quantiles for the data from $S(x)$. The estimates for the 50%, 95%, and 99% quantiles are 0.679 (0.783), 2.09 (2.08), and 3.07 (2.77), respectively, where as before the actual values are shown in parentheses following. The plot of the monotone approximation is shown in Figure 4.1 with $S(x)$ the lower curve from $[0.24, 2.04]$.

5 Density estimation

The mouse autopsy data set ($N = 40$) from Hoel (1972, Table 1: Other Causes) is a most challenging data set for density estimation. We wish to estimate the empirical distribution function $F(x)$ using a smoothed monotone spline. The pairs of values for the spline approximation are $\{(x_{(i)}, F(x_{(i)}) : i = 1, \ldots, 40\}$ using Equation 2. The x-values range from 40 to 763. Density estimation on this data set often yields negative estimates for the pdf in the interval $[70, 100]$. The constraints we used for our B-spline approximation were (1) $0 \leq F(x) \leq 1$ on $[40, 763]$, (2) $F'(x) = f(x) \geq 0$ on $[70, 100]$, (3) $F''(x) = f'(x) = 0$ on $[763, 770]$, and (4) $F''(x) = 0$ on $[25, 40]$. Constraints (1) and (2) are the cdf constraints, while constraints (3) and (4) extend $F(x)$ linearly. Using the defaults for Confit, the B-spline approximation of the cdf $F(x)$ for the mouse data produces a fourth degree spline approximation with 21 B-splines and 26 knots. The plot of the density estimation $f(x) = F'(x)$ is shown in Figure 5.1.

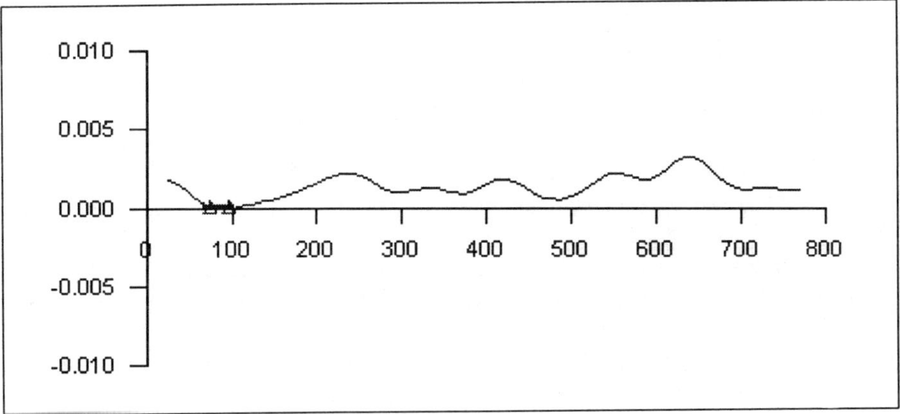

Fig. 5.1. B-Spline approximation for the density of the mouse data

6 Smoothed bootstrapping

The seminal work of Efron (1982) popularized the bootstrap that has now become a common tool for use in exploring the assessment of errors in statistical estimation problems. Consider the problem of estimating the maximum value from a set of five values from the censored exponential data above. Since the data is censored, the range is not defined when a censored value is used in the resampled data. To avoid this difficulty, one can use the monotone regression spline approximation $S(x)$ for the empirical distribution function $F(x)$, which provides a semiparametric model of the underlying distribution. For a uniformly distributed sequence $\{u_1, \ldots, u_N\}$ from $[0,1]$ the resampled data consist of $\{S^{-1}(u_1), \ldots, S^{-1}(u_N)\}$. This procedure has been used in Gaylord & Ramirez (1991). Resampling with 100 sets of five values, we found that 90% of the range values fell in the interval $[18, 92]$.

Acknowledgements
The first author was partially supported by the Center for Advanced Studies at the University of Virginia. A 90-day free trial version of Confit is available at the web site http://web.wti.net/~wti.

References

Curry, H. B. & Schoenberg, I. J. (1966). On Polya frequency functions. IV. The fundamental spline functions and their limits. *J. Analyse Math.*, **17**, 71-107.

DeBoor, C. (1978). *A Practical Guide to Splines.* New York: Springer-Verlag.

Efron, B. (1982). *The Jackknife, the Bootstrap, and Other Resampling Plans.* Philadelphia: SIAM.

Gaylord, C. & Ramirez, D. (1991). Monotone regression splines for smoothed bootstrapping. *Computational Statistics Quarterly*, **6**, 85-97.

Goldfarb, D. & Idnani, A. (1983). A numerically stable dual method for solving strictly convex quadratic programs. *Mathematical Programming*, **27**, 1-33.

Hoaglin, D. Mosteller, F. & Tukey, J. (1983). *Understanding Robust and Exploratory Data Analysis.* New York, John Wiley.

Hoel, D. (1972). A representation of mortality data by competing risks. *Biometrics,* **28**, 475-488.

Kelly, C. & Rice, J. (1990). Monotone smoothing with applications to dose-response curves and the assessment for synergism. *Biometrics*, **46**, 1071-1085.

Neter, J., Wasserman, W. & Kutner, M. (1983). *Applied Linear Regression Models.* Homewood, Illinois, Richard D. Irwin, Inc.

Ramsay, J. (1988). Monotone regression splines in action. *Statistical Science*, **4**, 425-461.

Smith, P. L. (1979). Splines as a useful and convenient statistical tool. *Amer. Statistician*, **33**, 57-62.

Turlach, B. (1997). Constrained smoothing splines revisited. *Statistics Research Report* No. SRR 008-97, Centre for Mathematics and its Applications, Australian National University.

Wahba, G. (1990). *Spline Models for Observational Data.* Philadelphia: SIAM.

Comparing the Fits of Non-Nested Non-Linear Models

Gavin J.S. Ross

Statistics Department, IACR Rothamsted, Harpenden, Herts AL5 2JQ, UK

Abstract. A simulation method is proposed for comparing the fits of non-nested non-linear models. Given the fitted parameters and residual variance for the preferred model, data sets are simulated and the preferred and rival models fitted. The residual sums of squares or deviances are transformed to approximate normality, and the mean and variance of the difference are estimated to give a test of the better fit of the preferred model. The roles can then be reversed.

Keywords. Non-linear models, non-nested models

1 Introduction

1.1 Nested and non-nested models

Statistical models are said to be *nested* if one (with p parameters) is a special case of another (with $p+q$ parameters), usually expressible as a set of constraints on the parameters of the more general model. The extra q parameters fitted normally increase the residual log-likelihood, and twice the difference in log-likelihoods yields the familiar likelihood-ratio test statistic, distributed asymptotically as χ^2 with q degrees of freedom. Familiar examples are polynomials of degree $p-1$ and $p+q-1$, and general multiple regression models. Less familiar are hierarchies of non-linear models, where the special cases are sometimes less obvious without the use of calculus: the exponential curve $E(y) = a + b.\exp(-kx)$ includes the special case of the straight line as k tends to zero.

Models of different algebraic form but the same number of parameters cannot be nested. The exponential curve and the rectangular hyperbola are of broadly similar shape: a given set of data may be fitted by either model, and sometimes one fits better than the other, even when it is not the true model. Even when one model fits very much better than the other, we have no simple statistical test that the difference in log-likelihoods is significant. Even when the number of parameters are different, if the model forms are unrelated the simpler model may fit better than the form with more parameters. For example an exponential curve may fit better than a high order polynomial when the data tend to an asymptote.

1.2 Cox's Likelihood Ratio Test

The problem of comparing the fits of non-nested models was addressed by Cox (1962). Whereas there are well-established tests for nested models, in particular the likelihood ratio test for the significance of the increase in log-likelihood due to the extra parameters fitted, no such procedure is possible if the models are from separate families. In non-linear regression, in particular the fitting of non-linear curves, it is often of interest to know which of two or more rival curves best represents the data. While there may sometimes be rival theoretical reasons for the choice of models, as in Cox's example of the one-hit and two-hit models for quantal responses, more often the choice is empirically based.

Cox defined a statistic based on the difference between the log likelihood ratios for two rival models, of which one was the preferred model. The distribution of this statistic was derived analytically for certain special cases, in particular where the models were rival frequency distributions. For non-linear regression models the analysis is likely to be much more complicated, and therefore a more direct approach is required. In the age of high-speed computing it is not very time-consuming to generate and fit enough simulated data sets to obtain an approximate idea of whether one model is generally a better fit than is a rival model.

More recent approaches to the problem are described in Kent (1986) and Seber & Wild (1989). Typically the models are embedded in a composite model with extra parameters. The recent paper of Victoria-Feser (1997) should be noted.

1.3 Sampling distribution of residual sum of squares

In the simplest case of two nonlinear regression models with i.i.d. Normal errors, namely a preferred model

$$M_1: \qquad E(\mathbf{y}) = f_1(\theta, \mathbf{X})$$

and an alternative model

$$M_2: \qquad E(\mathbf{y}) = f_2(\phi, \mathbf{X})$$

where the parameter vectors θ and ϕ have the same length, p, and the functions f and g are unrelated, and the observation vector Y=**y** is of length n, and the matrix **X** is a set of one or more independent variables, the estimation problem may be represented geometrically in R_n as that of dropping a perpendicular from the data point Y onto each of the two *Solution Loci* or *Expectation Surfaces* F_1 and F_2. The solution loci are of dimension p and may be curved or locally approximately linear. If the models are similar in effect the solution loci are close together in the neighbourhood of Y.

If T is a point on F_1 representing the 'true' expectations corresponding to a particular value of θ according to model M_1, we can simulate a data set by adding a random Normal vector **z** to produce a sample point Z, whose squared distance from T is proportional to χ^2 with n d.f. We then fit the two models to the sample data, represented by dropping perpendiculars from Z onto F_1 and F_2 respectively. The residual sums of squares are then the squared lengths of the perpendiculars, and we wish to estimate how frequently the preferred model fits better than its rival.

As a simple example in two dimensions, let (y_1, y_2) be a sample from the model M_1 with expectation (θ, θ^2) and independent $N(0,\sigma^2)$ residuals, to be compared with the alternative model M_2 with expectation $(\phi, 2\phi)$. If the true value of θ is 1.6, then we can generate sample data points centred at (1.6, 2.56) distributed within circular contours of equal probability depending on the value of σ. From each sample point we fit each model by dropping a perpendicular onto the solution loci which are F_1: $y_2 = y_1^2$, and F_2: $y_2 = 2y_1$. We assume that there is negligible probability of generating points such that there is more than one perpendicular onto F_1, a situation studied by Ross (1990, 1992) illustrating zones of potential non-uniqueness of estimates. Sample space is then divided into two regions by the locus C of points from which the two solution loci are equidistant. The probability that M_2 will be favoured over M_1 is the probability of finding sample points beyond the critical locus C. The geometry is illustrated in Figure 1 below.

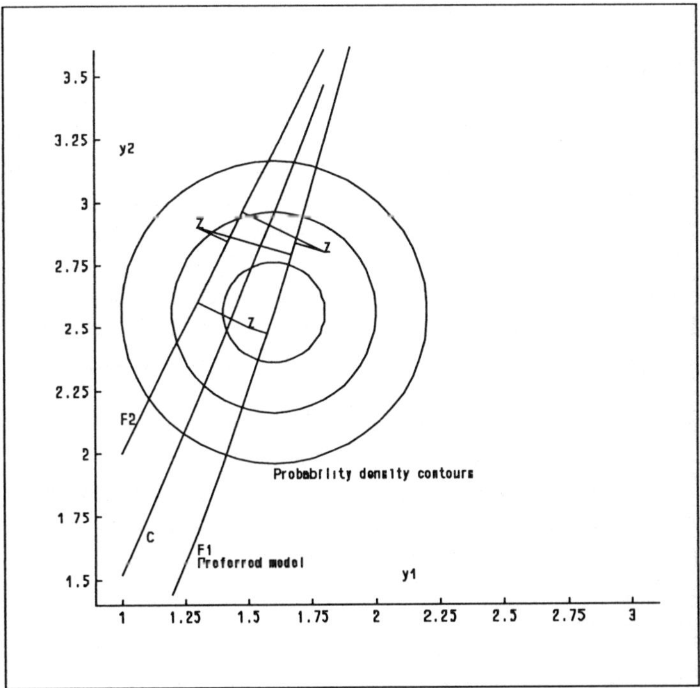

Fig.1. Representation in data space of fits to two models: from sample points Z drop perpendiculars onto loci F_1 and F_2; if points Z lie below the critical locus C, model M_1 is preferred

2 Application to nonlinear regression models

2.1 A Cube Root Deviance statistic

While it is possible to estimate by simulation the proportion of cases in which the sample data shows a better fit to the preferred model than to the rival model, it is convenient to use a comparison statistic that is approximately normally distributed, in which case we need only estimate its mean and standard deviation from a relatively small number of simulations. Since the distribution of the residual deviance or residual sum of squares after fitting the 'correct' model is known to be approximately proportional to χ^2 with $n-p$ degrees of freedom, where p parameters are fitted to n observations, we can use the Wilson-Hilferty or cube root transformation to obtain a statistic that is approximately Normally distributed. The corresponding distribution for the fits to the alternative model will be a non-central χ^2 with non-centrality due to the lack of fit of the expectations under model 1 when fitted by model 2. However the cube root transformation applied to moderately non-central χ^2 distributions still yields an approximately Normal result, and so it is reasonable to propose the following test statistic for comparing the fits of two alternative non-nested models:

$$D = (\text{Model 2 deviance})^{1/3} - (\text{Model 1 deviance})^{1/3}$$

and to simulate enough data sets to estimate the mean and standard deviation of D, at the same time reporting the proportion of cases for which D is negative.

It should be noted that when fitting non-linear models there is a finite probability of failure to find a unique solution, and we must rely on computing software to provide the best fit possible. It will always be possible to construct examples with large sampling variation or parameters close to some critical boundary for which the method does not apply. However in most practical cases where there is adequate data and a reasonable fit the method should be considered valid.

It is further recommended that when the variance σ^2 is unknown and only estimated from the data (preferably by replication of x-values), the simulations about the expectations should use the t-distribution with $n-p$ degrees of freedom, rather than the Normal distribution. Pseudo-random t-deviates may be obtained from transformations of rectangular pseudo-random numbers.

2.2 Exponential and alternative curves

As a practical example consider the Dugong data used by Ratkowsky (1983), p.101, to illustrate the problem of fitting the negative exponential curve

M_1: $E(y_i) = a + b.\exp(-kx_i)$

where x is age and y is length, in unstated units. The sample size n is 27, and the fitted parameters are: $a = 2.667$, $b = -0.9725$, $k = 0.1353$, and the residual mean square is 0.007782 on 24 d.f.

It is also possible to fit the inverse linear or rectangular hyperbola,

M_2: $E(y_i) = a + b / (1 + d\, x_i)$

for which the least squares estimates of the parameters are: $a = 2.857$, $b = -1.281$, $d = 0.1902$, and the residual mean square is 0.007307, also on 24 d.f.

Note that the rectangular hyperbola actually fits better in this case than the preferred exponential model. However we may still test for the difference between the fits, as follows: we first generate random t-deviates (with 24 d.f.) which are multiplied by 0.0882 (which is $\sqrt{0.007782}$) and added to the expectations under M_1; then we fit both models to the simulated data and compute the statistic D. We may then reverse the roles of the two models and repeat the process. The results of 1000 simulations were as follows:

Assuming M_1 to be true, the mean value of D was 0.006242 and its standard deviation was 0.012822, and in 29.1% of cases D was negative. Under the normality assumption we would expect 31.3% negative cases, but this discrepancy is not significant in a sample of 1000.

Assuming M_2 to be true, the mean value of D was 0.033526 and its standard deviation was 0.030617, and in 14.3% of cases D was negative (compared with 13.7% under the normality assumption).

We can therefore conclude that there is no evidence that either model is to be preferred on internal grounds alone. There may of course be good theoretical reasons for preferring one model to another. For example a similar test on these data comparing the exponential with the quadratic polynomial showed that the quadratic was better in 16% of cases, yet we would not consider the quadratic acceptable in view of its lack of an asymptote as age x increases.

2.3 General experience with the method

Compared with testing for significance of extra parameters in nested models, the D test is extremely weak, in the sense that where the choice of alternative model is not too inconsistent with the data, a non-significant result is very common. This implies that in order to discriminate between rival models we require a high degree of precision, or large sample size in the original data set.

Apart from the examples above, the following curves are suitable for comparison:

Three- or four-parameter sigmoid growth curves: Logistic, Gompertz, Inverse quadratic.

Asymmetric curves with a maximum: Cubic, exponential + linear, non-rectangular hyperbola.

Compartment models: sums of exponentials, Gamma model (compound exponential).

The method may be modified for weighted observations or for non-normal error distributions such as log normal data, quantal responses or counts. The modification for weighted normal data is straightforward, but the other cases require more effort to obtain the appropriate sampling schemes.

3 Implementation in nonlinear software

3.1 Maximum Likelihood Program, MLP

While it is possible to implement the method using any program with the capability of fitting both models to data, storing the fits and generating random numbers, it will be found that the problem for the user will be in organising the calculations and in storing and analysing the results. One would not wish to have to scan an output file with 2000 sets of standard output.

The method has therefore been implemented in the author's Maximum Likelihood Program (MLP) (Ross, 1987) Version 3.09 as a simple option within the FIT CURVE module. The user merely has to specify the preferred model, say CMODEL=EXP (exponential), the rival or extra model, say CEM=LOL (i.e. Linear over Linear), and a new option for Non-nested testing:

CNN=samples, seed

which specifies the number of simulations and a seed for the first random number. Provided the models have the same number of parameters and constraints (such as fixed asymptotes or origins) the program will fit both models, and then proceed to perform the simulations and fits suppressing all output, with a final report on the mean and standard deviation of D, together with the percentage of cases in which D is negative. If either model fails to fit a limit is put on the value of D.

3.2 Conclusion

The method described above is generally valid and very useful, but for ease of use it is recommended to use software for which the method has been specially programmed.

References

Cox, D.R. (1962). Further results on tests of separate families of hypotheses. *J. R. Statist. Soc.* B, **24**, 406-424.

Kent, J.T. (1986). The underlying structure of nonnested hypothesis tests. *Biometrika,* **73**, 333-343.

Ratkowsky, D.A. (1983). *Nonlinear Regression Modeling.* New York: Dekker.

Ross, G.J.S. (1987). *Maximum Likelihood Program, version 3.08.* Oxford: NAG Ltd.

Ross, G.J.S. (1990). *Nonlinear estimation.* New York: Springer Verlag.

Ross, G.J.S. (1992). Graphical methods for constructing data sets for which likelihood functions have multiple optima. *J. Stat. Plan. and Inf,* **30**, 107-133.

Seber, G.A.F. & Wild, C.J. (1989). *Nonlinear regression.* New York: Wiley.

Victoria-Feser, M.-P. (1997). A robust test for non-nested hypotheses. *J.R. Statist. Soc.* B, **59**, 715-727.

The Deepest Fit

Peter J. Rousseeuw[1] and Stefan Van Aelst[1,2]

[1] Department of Mathematics and Computer Science, Universitaire Instelling Antwerpen (UIA), Universiteitsplein 1, B-2610 Wilrijk, Belgium

[2] Research Assistant with the FWO, Belgium

1 Introduction

Recently, Rousseeuw & Hubert (1996) defined the depth of a regression fit relative to the data. This concept of *regression depth* immediately leads to a new robust regression estimator which we call *the deepest fit.* Quite simply, it is the fit with largest depth. Therefore, it can be seen as a generalization of the univariate median. We construct an algorithm to compute the deepest fit in simple regression, and illustrate it with examples. For any bivariate data set Z_n the deepest fit has depth at least $n/3$, and a breakdown value of at least $1/3$. Around the deepest fit we construct depth envelopes which generalize the quantiles around the univariate median.

2 Deepest fit

We consider a data set $Z_n = \{\mathbf{z}_i = (x_{i1}, \cdots, x_{i,p-1}, y_i); i = 1, \ldots, n\} \subset I\!R^p$, to which we want to fit a hyperplane $H_{\boldsymbol{\theta}}$ of the form $y = \theta_1 x_1 + \ldots + \theta_{p-1}x_{p-1} + \theta_p$ with $\boldsymbol{\theta} = (\theta_1, \ldots, \theta_p) \in I\!R^p$. We denote the x-part of each data point $\mathbf{z}_i$ by $\mathbf{x}_i = (x_{i,1}, \ldots, x_{i,p-1}) \in I\!R^{p-1}$. The residuals of Z_n relative to the hyperplane $H_{\boldsymbol{\theta}}$ are denoted as $r_i = r_i(\boldsymbol{\theta}) = y_i - \theta_1 x_{i1} - \cdots - \theta_{p-1}x_{i,p-1} - \theta_p$.

The regression depth of a hyperplane $H_{\boldsymbol{\theta}} \subset I\!R^p$ relative to the data set $Z_n \subset I\!R^p$ is defined as the smallest number of observations that need to be removed to make $H_{\boldsymbol{\theta}}$ a *nonfit.* Therefore, we always have $0 \leq rdepth(\boldsymbol{\theta}, Z_n) \leq n$. We call $H_{\boldsymbol{\theta}}$ a nonfit if there exists an affine hyperplane V in $\mathbf{x}$-space such that no $\mathbf{x}_i$ belongs to V, and such that $r_i > 0$ for all $\mathbf{x}_i$ in one of its open halfspaces and $r_i < 0$ for all $\mathbf{x}_i$ in the other open halfspace.

In p dimensions a fit $\boldsymbol{\theta}$ with maximal regression depth relative to a data set $Z_n \subset I\!R^p$ always passes through at least p points of Z_n (otherwise it could be made deeper by slightly tilting it until it does fit p points).

In p dimensions the **deepest fit** regression estimator $T^*_n(Z_n)$ is defined as

$$\begin{aligned} T^*_r(Z_n) &= \underset{\boldsymbol{\theta}}{argmax}\ rdepth(\boldsymbol{\theta}, Z_n) \\ &= \underset{\boldsymbol{\theta}^{(i_1,\cdots,i_p)}}{argmax}\ rdepth(\boldsymbol{\theta}^{(i_1,\cdots,i_p)}, Z_n) \end{aligned} \tag{1}$$

where $\boldsymbol{\theta}^{(i_1,\cdots,i_p)}$ is the fit passing through the observations $\mathbf{z}_{i_1}, \ldots, \mathbf{z}_{i_p}$. When there are several $\boldsymbol{\theta}^{(i_1,\cdots,i_p)}$ with the same (maximal) rdepth, then the average of those $\boldsymbol{\theta}^{(i_1,\cdots,i_p)}$ is taken because simulations have shown that this increases the finite-sample efficiency. Note that we do not make any assumptions about the type of error distribution to define the deepest fit. The deepest fit is scale equivariant, regression equivariant and affine equivariant according to the definitions in Rousseeuw & Leroy (1987).

Definition (1) yields a straightforward algorithm to compute the deepest fit T_r^*. For simple regression we first sort the x_i in $O(nlogn)$ time and for all $O(n^2)$ pairs of observations $\{\mathbf{z}_i, \mathbf{z}_j\}$ we determine the fit $\boldsymbol{\theta}^{ij}$ passing through the two observations. For each $\boldsymbol{\theta}^{ij}$ we compute $rdepth(\boldsymbol{\theta}^{ij}, Z_n)$ in $O(n)$ time with the algorithm of Rousseeuw & Hubert (1996), and keep the fit with highest rdepth. This yields T_r^* in $O(n^3)$ time and $O(n)$ storage. It is an open question whether the deepest line can be computed exactly in less than $O(n^3)$ time. In more than two dimensions, the regression depth of a fit can be computed by the algorithms of Rousseeuw & Struyf (1998). We are currently constructing an approximate algorithm to compute the deepest fit in higher dimensions with a lower time complexity.

In two dimensions, the next theorem gives upper and lower bounds for the regression depth of the deepest fit.

Theorem 1. *At any data set $Z_n \subset \mathbb{R}^2$ it holds that*

$$\left\lceil \frac{n}{3} \right\rceil \leq rdepth(T_r^*(Z_n), Z_n) \leq n. \tag{2}$$

For simple regression Hubert & Rousseeuw (1997) constructed the catline which is an estimator that always has regression depth at least $\lceil n/3 \rceil$.

Conjecture 1. *At any data set $Z_n \subset \mathbb{R}^p$ with $p \geq 3$ it holds that*

$$\left\lceil \frac{n}{p+1} \right\rceil \leq rdepth(T_r^*(Z_n), Z_n) \leq n. \tag{3}$$

We can see T_r^* as a generalization of the univariate median. (Indeed, if all $x_i = 0$ the intercept of T_r^* is exactly the median of the y_i.) However, T_r^* is quite different from L^1 regression. For instance, T_r^* has a positive breakdown value and is not attracted by a leverage point (unlike L^1).

Example 1: **Stars data.** Figure 1 contains the Herzsprung-Russell diagram of a star cluster in the direction of Cygnus (see Rousseeuw & Leroy, 1987). The logarithm of the star's light intensity is plotted versus the logarithm of its surface temperature. In this plot we see the deepest fit T_r^* which fits the main sequence of stars, and the L^1 line which is strongly attracted by the four giant stars in the upper right corner.

3 Depth envelopes

Around the deepest fit one can construct *depth envelopes* E_k for $k \geq 2$, given by (for $p = 2$):

$$E_k = \{(x, y); \min_{\boldsymbol{\theta}}(\theta_1 x + \theta_2) \leq y \leq \max_{\boldsymbol{\theta}}(\theta_1 x + \theta_2)\} \tag{4}$$

where $\boldsymbol{\theta} = (\theta_1, \theta_2)$ needs to satisfy $rdepth(\boldsymbol{\theta}, Z_n) \geq k$. It can easily be shown that it is sufficient to consider all fits $\boldsymbol{\theta}^{ij}$ through two data points for which $rdepth(\boldsymbol{\theta}^{ij}, Z_n) \geq k$. So definition (4) is equivalent with the expression

$$E_k = \{(x, y); \min_{\boldsymbol{\theta}}(\theta_1 x + \theta_2) \leq y \leq \max_{\boldsymbol{\theta}}(\theta_1 x + \theta_2)\}. \tag{5}$$

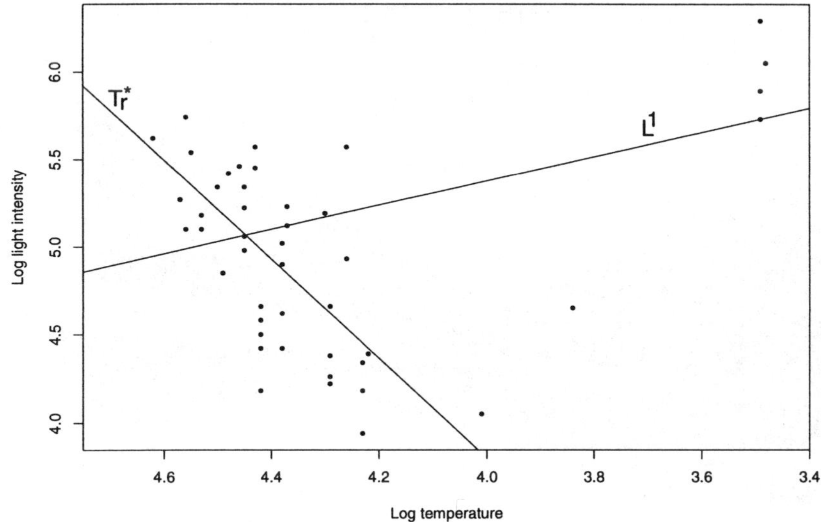

Fig. 1. Herzsprung-Russell diagram of a star cluster in the direction of Cygnus, with the deepest fit T_r^* and the L^1 fit which is attracted by the giant stars

The upper and lower boundaries of the envelope E_k thus consist of line segments. The upper boundary is convex, and the lower boundary is concave. Like the deepest fit, also the depth envelopes do not depend on assuming a particular type of error distribution.

Example 2: **Skeena River data.** In Figure 2 the number of recruits is plotted versus the number of spawners from 1940 until 1967 for the Skeena River salmon stock (Carroll & Ruppert, 1988). Figure 2 also shows the deepest fit T_r^* with depth 12 and the depth envelopes for k =4,7, and 10. Note that the set of envelopes always provides a (coarse) ordering of the observations. On the right hand side of Figure 2 we have indicated the percentage of the data lying on or below each envelope boundary.

4 Robustness properties

4.1 Breakdown value

The *breakdown value* (see Hampel *et al.*, 1986)of an estimator T is the smallest proportion of contaminated observations that can carry the estimator T beyond all bounds. The finite-sample breakdown value of any estimator T_n is defined by $\varepsilon_n^*(T_n, Z_n) = \min\{\frac{k}{n}; \sup_{Z'_n} ||T_n(Z'_n) - T_n(Z_n)|| = \infty\}$ where Z'_n ranges over all data sets obtained by replacing any k observations of Z_n by arbitrary values.

Theorem 2. *If Conjecture 1 holds, and the* $\mathbf{x_i}$ *are in general position, i.e. no more than p-1 of the* $\mathbf{x_i}$ *lie in any (p-2)-dimensional affine subspace of*

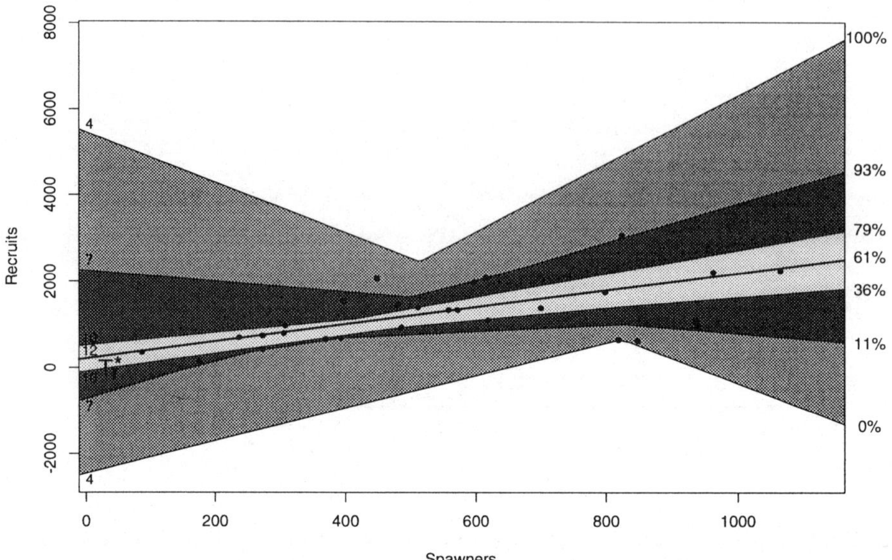

Fig. 2. The Skeena River data set ($n = 28$), its deepest fit T_r^* (with depth 12) and its depth envelopes for k =4,7 and 10. To the left of each envelope boundery its value of k is listed, and to its right the (cumulative) percentage of the data lying on or below it

$\mathbb{R}^{p-1}$, *then*

$$\varepsilon_n^*(T_r^*, Z_n) \geq \frac{1}{n}\left(\left\lceil \frac{n}{p+1} \right\rceil - p + 1\right) \approx \frac{1}{p+1}. \tag{6}$$

Note that for the case $p = 2$ the conjecture has been proven, so in simple regression the finite-sample breakdown value is approximately 1/3.

The following theorem shows that the breakdown value converges almost surely to 1/3 when some assumptions are made about the distribution that generates the samples. Conjecture 1 is not needed here.

Theorem 3. *Let* $Z_n = \{(\mathbf{x}_1, y_1), \ldots, (\mathbf{x}_n, y_n)\}$ *be a sample from a distribution* H *on* $\mathbb{R}^p$ $(p \geq 3)$ *with a strictly positive density that satisfies* $\text{med}_H(\text{y} - \mathbf{x}^t\theta | \mathbf{x} = \mathbf{x}_0) = 0$ *for all* $\mathbf{x}_0 \in \mathbb{R}^{p-1}$. *Then*

$$\varepsilon_n^*(T_r^*, Z_n) \xrightarrow[n\to\infty]{a.s.} \frac{1}{3}. \tag{7}$$

4.2 Sensitivity functions

The influence function (see Hampel *et al.*, 1986) of an estimator T at a distribution H is an asymptotic concept that measures the effect on T of adding an observation at $\mathbf{z}$. In the bivariate case we use the *averaged permutation-stylized sensitivity function* $APSF_n$ defined by (Rousseeuw *et al.*, 1995) as a finite-sample version of the influence function, as explained below.

For any estimator T the sensitivity function measures the (standardized) effect of adding an observation $\mathbf{z}$ to the sample $Z_n = \{\mathbf{z}_i; i = 1, \ldots, n\}$, i.e.

$$SF_n(\mathbf{z}, T, Z_n) = n(T_{n+1}(\mathbf{z}_1, \ldots, \mathbf{z}_n, \mathbf{z}) - T_n(\mathbf{z}_1, \ldots, \mathbf{z}_n)). \tag{8}$$

The resulting sensitivity function strongly depends on the sample Z_n, but we alleviate this effect by using a permutation-stylized data set $Z(\pi) = \{(x_i^s, x_{\pi(i)}^s); i = 1, \ldots, n\}$ where $x_i^s = \Phi^{-1}(\frac{i}{n+1})$ and where π is a random permutation on $\{1, \ldots, n\}$. Finally, the effect of the particular permutation π is tempered by averaging the sensitivity function over a collection of random permutations, leading to

$$APSF_n(\mathbf{z}) = \underset{\pi}{\text{average}}\; SF_n(\mathbf{z}, T, Z(\pi)). \tag{9}$$

Figure 3a shows the sensitivity surface of the deepest fit slope and Figure 3b that of the deepest fit intercept, both for $n = 20$, obtained by generating $m = 350$ random permutations. Note that both sensitivity functions are bounded.

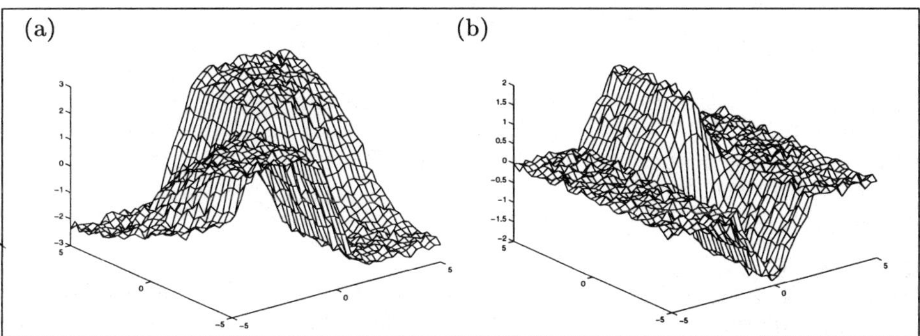

Fig. 3. (a) Averaged permutation-stylized sensitivity function $APSF_n$ of the deepest fit slope for $n = 20$; (b) $APSF_n$ of the deepest fit intercept for $n = 20$

4.3 Efficiency

To investigate the efficiency of the deepest fit in the bivariate case we have generated $m = 10,000$ samples of various sample sizes n from $N_2(0, I)$, each time computing the deepest fit $(b_{T_r^*}^{(k)}, a_{T_r^*}^{(k)})$ for $k = 1, \ldots, m$. Table 1 lists the bias which is the average of the computed slopes, and the n-fold variance given by

$$n \underset{k=1,\ldots,m}{\text{variance}}\; b_{T_r^*}^{(k)}. \tag{10}$$

Analogous results for the intercept are also given. We have also compared the efficiency of the deepest fit with that of the L^1 line. The asymptotic variance of the L^1 estimator is derived in Bassett & Koenker (1978). For the bivariate normal distribution $N_2(0, I)$ the asymptotic variances of the L^1 line become

Table 1. Bias, n-fold variance and efficiency of the deepest fit, when applied to bivariate guassian data. The simulation results are based on 10,000 samples

	slope $b_{T_r^*}$			intercept $a_{T_r^*}$		
		n-fold	relative		n-fold	relative
n	bias	variance	efficiency	bias	variance	efficiency
10	0.00581	2.614	60.01%	0.00040	1.999	78.85%
20	0.00016	1.987	79.01%	0.00481	1.876	83.37%
30	0.00157	1.916	81.19%	0.00236	1.848	84.50%
50	0.00034	1.971	79.66%	0.00130	1.817	86.41%
70	-0.00007	1.776	88.40%	0.00097	1.855	84.64%
100	0.00139	1.760	89.20%	-0.00034	1.843	85.19%
200	-0.00125	1.751	89.66%	-0.00061	1.858	84.50%
300	-0.00108	1.765	88.95%	-0.00067	1.904	82.46%
400	-0.00000	1.754	89.51%	-0.00081	1.886	83.25%
500	-0.00062	1.782	88.10%	-0.00053	1.885	83.29%

$V(b_{L^1}, H) = V(a_{L^1}, H) = 1.571$. We can now compute relative efficiencies given by

$$RE_n(b_{T_r^*}, b_{L^1}) = \frac{V(b_{L^1})}{n \operatorname*{variance}_{k=1,\dots,m} b_{T_r^*}^{(k)}}. \tag{11}$$

These are also listed in Table 1, and they are in agreement with the asymptotic results of He & Portnoy (1997). From these results we see that the efficiency of the deepest fit is close to the efficiency of the L^1 line, and as we saw before the deepest fit is more resistant against outliers and leverage points.

References

Bassett, G. & Koenker, R. (1978). Asymptotic Theory of Least Absolute Error Regression. *Journal of the American Statistical Association*, **73**, 618-621.

Carroll, R.J. & Ruppert, D. (1988). *Transformation and Weighting in Regression.* New York: Chapman and Hall.

Hampel, F.R., Ronchetti, E.M., Rousseeuw, P.J. & Stahel, W.A. (1986), *Robust Statistics: the Approach based on Influence Functions.* New York: John Wiley.

He, X. & Portnoy, S. (1997). Asymptotics of the Deepest Line. Technical report, Department of Statistics, University of Illinois at Urbana-Champaign.

Hubert, M. & Rousseeuw, P.J. (1997). The Catline for Deep Regression. Technical report, U.I.A., Belgium. Accepted for *Journal of Multivariate Analysis.*

Rousseeuw, P.J., Croux, C. & Hössjer, O. (1995). Sensitivity Functions and Numerical Analysis of the Repeated Median Slope. *Computational Statistics*, **10**, 71-90.

Rousseeuw, P.J. & Hubert, M. (1996). Regression Depth. Technical report, U.I.A., Belgium.

Rousseeuw, P.J. & Leroy, A.M. (1987). *Robust Regression and Outlier Detection.* New York: John Wiley.

Rousseeuw, P.J., & Struyf, A. (1998). Computing Location Depth and Regression Depth in Higher Dimensions. Technical report, U.I.A., Belgium. Accepted for *Statistics and Computing.*

Partially Linear Models: A New Algorithm and some Simulation Results

Michael G. Schimek

Department of Medical Informatics, Statistics and Documentation, Karl-Franzens-University of Graz, A-8010 Graz, Austria

Abstract. The problem of estimation in partially linear models is studied. We introduce an $O(n)$ smoothing spline algorithm which extends the approaches of Speckman (1988) and Green & Silverman (1994). It is known that the partial spline concept of Green & Silverman is asymptotically biased. In a Monte Carlo study we compare the small sample properties of the two approaches. The main outcome is that both concepts work well for uncorrelated predictor variables.

Keywords. Algorithms, asymptotic bias, cubic smoothing spline, generalized cross-validation, partially linear model, partial spline, semiparametric regression, simulations

1 Introduction

We consider a semiparametric regression model with a predictor function consisting of a parametric linear component and a nonparametric component involving an additional predictor variable. For this model a number of algorithms have been proposed: Green (1985), Speckman (1988), Wahba (1990), Green & Silverman (1994) and others. Here we study two cubic smoothing spline estimators corresponding to proposals of Speckman (1988) and Green & Silverman (1994). In both cases the parametric as well as the nonparametric component, and the generalized cross-validation criterion can be evaluated in linear time.

Suppose that responses $y_1, \ldots, y_n$ have been obtained at non-stochastic values $t_1, \ldots, t_n$ of a predictor variable t. The response and predictor values are connected by

$$y_i = \mathbf{u}_i^T \gamma + f(t_i) + \epsilon_i \tag{1}$$

for $i = 1, \ldots, n$, where $\mathbf{u}_1, \ldots, \mathbf{u}_n$ are known k-dimensional vectors, γ is an unknown parameter vector, $f \in C^2[0,1]$ is an unknown smooth function, and the $\epsilon_1, \ldots, \epsilon_n$ are independent, zero mean random variables with a common variance σ^2. Further it is assumed that $0 \leq t_1 < \ldots < t_n \leq 1$.

In vector-matrix form we have

$$\mathbf{y} = \mathbf{U}\gamma + \mathbf{f} + \epsilon$$

where $\mathbf{y} = (y_1, \ldots, y_n)^T$, $\mathbf{U}^T = [\mathbf{u}_1, \ldots, \mathbf{u}_n]$, $\mathbf{f} = (f(t_1), \ldots, f(t_n))^T$ and $\epsilon = (\epsilon_1, \ldots, \epsilon_n)^T$. The goal is to efficiently estimate the parameter vector γ, the function $\mathbf{f}$, and the mean vector $\mu = \mathbf{U}\gamma + \mathbf{f}$.

2 The two approaches

We take advantage of the fact that the cubic smoothing spline is a linear estimator. The fitted values for data $\mathbf{z} = (z_1, \ldots, z_n)^T$ are of the form

$$\hat{\mathbf{g}} = (g_\lambda(t_1), \ldots, g_\lambda(t_n))^T = \mathbf{S}\mathbf{z} \tag{2}$$

where g_λ is a natural cubic spline with knots at $t_1, \ldots, t_n$ for a fixed smoothing parameter $\lambda > 0$, and $\mathbf{S}$ a known symmetric (smoother) matrix that depends on λ.

For smoothing spline-based estimation in the partially linear model a solution can be obtained by minimizing the sum of squares equation

$$SS(\mathbf{b}, g) = \sum_{i=1}^{n} (y_i - \mathbf{u}_i^T \mathbf{b} - g(t_i))^2 + \lambda \int_a^b [g''(t)]^2 dt$$

over $\mathbf{b} \in \mathcal{R}^k$ and $g \in C^2[0,1]$. The resulting estimator is called a partial spline (see Wahba, 1990, Chapter 6).

For a prespecified value of λ the corresponding estimators for $\mathbf{f}$, γ and μ can be obtained by (subscript p denotes *partial*)

$$\gamma_p = (\tilde{\mathbf{U}}^T \mathbf{U})^{-1} \tilde{\mathbf{U}}^T \mathbf{y},$$
$$\mathbf{f}_p = \mathbf{S}(\mathbf{y} - \mathbf{U}\gamma_p)$$

and

$$\mu_p = \mathbf{f}_p + \mathbf{U}\gamma_p = \mathbf{H}_p \mathbf{y}$$

for

$$\mathbf{H}_p = \mathbf{S} + \tilde{\mathbf{U}}(\tilde{\mathbf{U}}^T \mathbf{U})^{-1} \tilde{\mathbf{U}}^T$$

and

$$\tilde{\mathbf{U}} = (\mathbf{I} - \mathbf{S})\mathbf{U}$$

with $\mathbf{S}$ introduced in (2). Green & Silverman (1994, Chapter 4) follow a similar concept of estimation.

Rice (1980) demonstrated that the partial spline estimator is generally biased for the optimal λ choice when the components of $\mathbf{U}$ depend on t. This asymptotic bias can be larger than the standard error.

Applying results due to Speckman (1988) the bias can be avoided if one instead uses the estimators (subscript s denotes *Speckman*)

$$\gamma_s = (\tilde{\mathbf{U}}^T \tilde{\mathbf{U}})^{-1} \tilde{\mathbf{U}}^T (\mathbf{I} - \mathbf{S})\mathbf{y},$$
$$\mathbf{f}_s = \mathbf{S}(\mathbf{y} - \mathbf{U}\gamma_s)$$

and

$$\mu_s = \mathbf{f}_s + \mathbf{U}\gamma_s = \mathbf{H}_s \mathbf{y}$$

for

$$\mathbf{H}_s = \mathbf{S} + \tilde{\mathbf{U}}(\tilde{\mathbf{U}}^T \tilde{\mathbf{U}})^{-1} \tilde{\mathbf{U}}^T (\mathbf{I} - \mathbf{S}),$$

and

$$\tilde{\mathbf{U}} = (\mathbf{I} - \mathbf{S})\mathbf{U}$$

with $\mathbf{S}$ from (2).

While the Speckman approach constructs an estimator of γ after removing the influence of t (i.e. the nonparametric predictor) from both the $\mathbf{u}_i$ and y, the partial spline approach removes t–information only from the $\mathbf{u}_i$. For the difference between the two estimation concepts, when arbitrary linear smoothers are applied, see also Schimek (1997, p.182f).

3 Smoothing parameter choice and model estimation

A data-driven choice for the smoothing parameter λ can be obtained by minimizing the generalized cross-validation criterion

$$GCV(\lambda) = \frac{n\|\mathbf{y} - \mu\|^2}{(n - trace(\mathbf{H}))^2}$$

with $\|\cdot\|$ the Euclidean norm. In the partial spline case the computation of $trace(\mathbf{H_p})$, in the case of Speckman of $trace(\mathbf{H_s})$ is required.

The goal is the efficient calculation of the estimators and the trace of $\mathbf{S}$, hence of $\mathbf{H}$. The idea is to apply de Boor's (1978) algorithm for the cubic spline fit, Hutchinson's & de Hoog's (1985) algorithm for the computation of the trace of $\mathbf{S}$ (all in $O(n)$ steps), and to make extensive use of Cholesky decomposition.

Now let us consider Speckman's estimation concept. The quantity γ_s can be calculated by first transforming $\mathbf{y}$ and $\mathbf{U}$ to $\tilde{\mathbf{y}} = (\mathbf{I} - \mathbf{S})\mathbf{y}$ and $\tilde{\mathbf{U}} = (\mathbf{I} - \mathbf{S})\mathbf{U}$. Then an ordinary least squares regression of $\tilde{\mathbf{y}}$ on $\tilde{\mathbf{U}}$ is carried out with the coefficient vector γ_s. Cholesky decomposition is used to factorize $\tilde{\mathbf{U}}^T\tilde{\mathbf{U}}$ as $\mathbf{T}^T\mathbf{T}$ for $\mathbf{T}$, an upper triangular matrix ($\mathbf{T}$ can be used for other computations as well).

Having γ_s it is easy to calculate $\mathbf{f}_s$ and μ_s. The trace of $\mathbf{H}_s$ can be obtained in a similar fashion. All quantities can be computed in linear time. Technical details and the efficient calculation of other quantities such as the variance σ^2 are described in Bissett *et al.* (1998). They also derive an algorithm to calculate $trace(\mathbf{H}_s)$ (required for GCV) in $O(n)$ operations.

Similar considerations apply to the partial spline concept of Green & Silverman, also calculated in $O(n)$ operations. An additional complication is $\tilde{\mathbf{U}}^T\mathbf{U} = \mathbf{U}^T(\mathbf{I} - \mathbf{S})\mathbf{U}$ in the matrix system. Again this expression can be factorized by means of Cholesky decomposition. Different from Speckman's estimator, γ_p and $trace$ $(\mathbf{H}_p)$ cannot be evaluated using the ordinary least squares technique.

The algorithm introduced here is implemented in a FORTRAN subroutine (compiled with the Watcom FORTRAN 77 compiler, version 10.6) which can be called by an S–Plus function.

4 The Monte Carlo simulations

A Monte Carlo experiment based on our algorithm was carried out. We compared the partial spline approach with Speckman's approach. Our interest was to study the small sample behaviour of the two estimation concepts, assuming lack of correlation between $\mathbf{U}$ and t with regard to asymptotic bias.

From Heckman (1986) we know that the parametric component in the partially linear model can be estimated with a parametric rate of convergence. Rice (1986) made clear that this is solely true when the variables in the parametric and the variables in the nonparametric components are uncorrelated. In general, the variance of $\hat{\gamma}$ decreases at a parametric rate (i.e. $\sim cn^{-1}$), whereas the bias decreases at a nonparametric rate (i.e. $\sim cn^{-\alpha}$ and $\alpha < 1$) except for special situations such as in Heckman (1986).

4.1 Outline

For the simulations we studied a simple version of the semiparametric regression model given in equation (1): $\mathbf{u}_i$ and γ in the parametric component are scalars. We assumed sample sizes $n = 100$ and 200, predictor values $t_i = (i - 0.5)/n$ on $[0,1]$, and errors $\epsilon_i \sim$ NID(0,1). For the nonparametric component we selected a function $f(t_i) = m_n f_0(t_i)$ where $f_0 = 4.26(\exp(-3.25t_i) - 4(\exp(-6.5t_i) + 3(\exp(-9.75t_1))$ with $\max|f(t)| = 9$ (i.e. $m_n = 9$) which is depicted in Figure 1 (for $n = 200$).

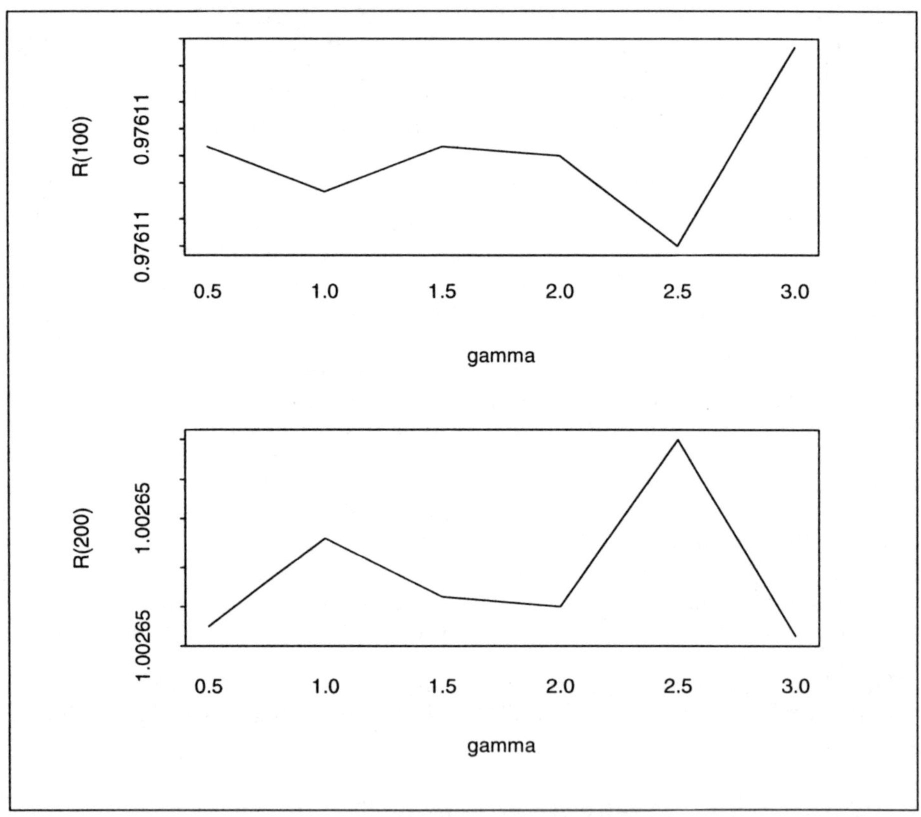

Fig. 1. The nonparametric function f

We considered a setting as in Heckman (1986) where the parametric predictor is defined as a random variable with $u_i \sim$ NID(0,1), hence no correlation between $\mathbf{U}$ and t. Further we assumed values for the regression coefficient $\gamma = \{0.5, 1, 1.5, 2, 2.5, 3\}$. A coefficient around $\gamma = 2.5$ produces an approximate balance between the parametric random contribution U and the nonparametric contribution t to the simulated data. For the two sample sizes we generated 500 replications. Each simulation run was performed for the partial spline approach and the Speckman approach using GCV as outlined earlier.

All the calculations were carried out in S–Plus for Windows on a pentium platform under Microsoft Windows 95.

4.2 Results

The results are reported in summary statistics of the estimated regression coefficient $\hat{\gamma}$ and the ratio of the squared biases

$$R = \frac{bias^2(\hat{\gamma}_p)}{bias^2(\hat{\gamma}_s)},$$

where subscript p denotes the partial spline approach and subscript s Speckman's approach, calculated over 500 replications. R is plotted as a function of the true γs.

The estimation results for the regression coefficients from both approaches are summarized in Table 1. The obtained results differ little with respect to the true regression coefficients. With Speckman's approach we are slightly better off when the sample size is small ($n = 100$). It is not possible to decide whether this is due to differences in estimation or in the asymptotic behaviour. According to Heckman (1986) we cannot expect much of a difference, at least from an asymptotic point of view.

Table 1. True regression coefficients γ and mean estimates $E(\hat{\gamma})$ over 500 replications from the partial spline approach versus Speckman's approach

γ	$E(\hat{\gamma})$			
	$n = 100$		$n = 200$	
	Partial	Speckman	Partial	Speckman
0.5	0.5013386	0.5009869	0.5053151	0.5053290
1	1.0013386	1.0009869	1.0053151	1.0053290
1.5	1.5013387	1.5009869	1.5053151	1.5053291
2	2.0013386	2.0009869	2.0053151	2.0053290
2.5	2.5013386	2.5009870	2.5053151	2.5053291
3	3.0013386	3.0009869	3.0053151	3.0053290

The squared bias ratios R obtained are plotted as a function of γ for both sample sizes in Figure 2. There is little variation of R independent of the sample size over the range of γ values. The Rs are always near one which indicates that the bias behaviour is quite the same for the partial spline approach and Speckman's approach.

Further research is required considering correlation between the parametric and the nonparametric predictor variables in different settings. Preliminary work on correlated predictor variables in small samples suggests that both approaches are less successful.

Acknowledgement

Part of this research was carried out during a sabbatical stay at the Statistics Department of the Texas A&M University at College Station (under a partnership contract with the Karl-Franzens University of Graz). Randy Eubank's hospitality and the use of his FORTRAN code for spline fitting are greatly acknowledged. Thanks go also to Thomas Yee for his advice on some messy S–Plus problems with the Watcom compiler.

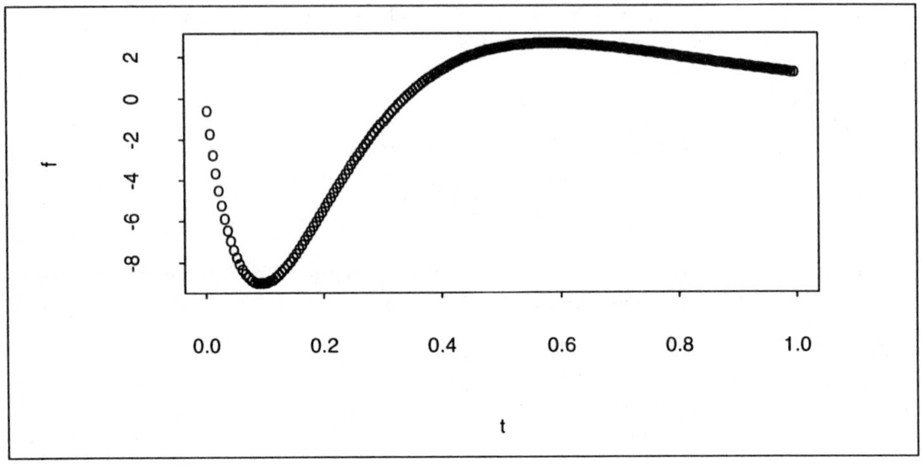

Fig. 2. Ratios R for $n = 100$ (top) and $n = 200$ (bottom) against γ

References

Bisset, K. K., Eubank, R. L., Kambour, E. L., Kim, J. T., Reese, C. S. & Schimek, M. G. (1998). Estimation in partially linear models. *CSDA*, in press.

De Boor, C. (1978). *A practical guide to splines.* New York: Springer–Verlag.

Green, P. (1985). Linear models for field trials, smoothing and cross-validation. *Biometrika*, **72**, 527-537.

Green, P. & Silverman, B. W. (1994). *Nonparametric regression and generalized linear models. A roughness penalty approach.* London: Chapman and Hall.

Heckman, N. (1986). Spline smoothing in a partly linear model. *J. R. Statist. Soc. B*, **48**, 244-248.

Hutchinson, M. F. & de Hoog, F. R. (1985). Smoothing noisy data with spline functions. *Numer. Math.*, **47**, 99-106.

Rice, J. (1986). Convergence rates for partially splined models. *Statist. and Prob. Letters*, **4**, 203-208.

Schimek, M. G. (1997). Non- and semiparametric alternatives to generalized linear models. *Computational Statistics*, **12**, 173-191.

Speckman, P. (1988). Kernel smoothing in partial linear models. *J. R. Statist. Soc. B*, **50**, 413-436.

Wahba, G. (1990). *Spine models for observational data.* Philadelphia: SIAM.

Induction of Graphical Models from Incomplete Samples

Paola Sebastiani[1] and Marco Ramoni[2]

[1] Department of Actuarial Science and Statistics, City University, UK
[2] Knowledge Media Institute, The Open University, UK

Abstract. Current methods to estimate conditional probabilities from incomplete data rely on iterative algorithms, such as the EM algorithm and Gibbs Sampling, which, although very reliable, pose convergence problems and assume that data are missing at random. This paper describes a deterministic method, called *Bound* and *Collapse* (BC), which relaxes the assumption that data are missing at random, does not pose problem of convergence rate and detection, and has a computational cost independent of the number of missing data.

1 Introduction

The estimation of conditional probabilities from a sample plays a central role in a variety of statistical applications, particularly in the quantification of association between categorical variables. This paper will focus on Directed Graphical Models (DGM): a subset of hierarchical log-linear models which have the representational advantage of being associated to a directed acyclic graph representing marginal and conditional independences between the variables. A DGM is defined by a set of *variables* $\mathcal{X} = \{X_1, \ldots, X_I\}$ with states $c_1, ..., c_I$ and a directed acyclic graph defining a model $\mathcal{M}$ of conditional dependencies among the elements of $\mathcal{X}$. From a qualitatitive viewpoint, directed links pointing from the set of variables $\Pi_i = \{X_{i1}, ..., X_{iq}\}$ to the *child* variable X_i represent the stochastic dependence of X_i on Π_i, that is called the set of *parents* of X_i. Absence of a directed link between two variables is then interpreted in terms of conditional independence, see Whittaker (1990) for further details. Let $\pi_{ij} = (x_{i1j}, ..., x_{iqj})$ denote a *configuration* of parents Π_i. The stochastic dependence of X_i on Π_i is quantified by the conditional probabilities $\theta_{ijk} = p(x_{ik}|\pi_{ij}, \theta)$, $i = 1, ..., I$, $k = 1, ..., c_i$, $\sum_k p(x_{ik}|\pi_{ij}) = 1$ for all j, $\theta = (\theta_{ijk})$. Thus $\theta_{ij1}, ..., \theta_{ijc_i}$ is the conditional distribution of $X_i|\pi_{ij}, \theta$. Suppose we are given a random sample of n cases $\mathcal{D}$ and a DGM $\mathcal{M}$, from which we wish to estimate θ. When the sample is complete, closed form solutions allow efficient estimation of θ. The marginal and conditional independences in the graph induce a factorization of the likelihood function $l(\theta) = \prod_{ijk} \theta_{ijk}^{n_{ijk}}$, where n_{ijk} is the frequency of (x_{ik}, π_{ij}) in $\mathcal{D}$. Thus, the Maximum Likelihood estimates (MLE) of θ_{ijk} are then $\hat{\theta}_{ijk} = n_{ijk}/n_{ij}$ where $n_{ij} = \sum_k n_{ijk}$ is the frequency of π_{ij}. The Bayesian approach generalizes the MLE by introducing a flattening constant $\alpha_{ijk} > 0$ for each frequency, so that the estimate is computed as

$$\hat{\theta}_{ijk} = \frac{\alpha_{ijk} + n_{ijk} - 1}{\alpha_{ij} + n_{ij} - c_i} \tag{1}$$

where $\alpha_{ij} = \sum_k \alpha_{ijk}$. When the prior distribution of $(\theta_{ij1} ..., \theta_{ijc_i})$ is Dirichlet, with *hyperparameters* $(\alpha_{ij1}, ..., \alpha_{ijc_1})$, (1) is the posterior mode of θ_{ijk}. Unfortunately, simplicity and efficiency of these closed form solutions are lost when the sample is incomplete, that is, some entries are reported as unknown. In this case, the exact estimate of θ_{ijk} is the mixture of the estimates given by (1) for each possible complete sample, and the computational cost of this operation grows exponentially in the number of missing data. The EM algorithm (Dempster *et al.*, 1997) and Gibbs Sampling (GS) (Geman & Geman, 1984) are currently regarded as the most viable solutions to the problem of missing data. However, they can be both trapped into local minima and the convergence detection can be difficult. Furthermore, they rely on the assumption that data are missing at random (MAR): within each parent configuration, the available data are a representative sample of the complete sample and the distribution of missing data can be therefore inferred from the available entries (Little & Rubin, 1987). When this assumption fails, and the missing data mechanism is *not ignorable* (NI), the accuracy of these methods can dramatically decrease. Finally, the computational cost of these methods depends heavily on the absolute number of missing data, and this can prevent their scalability to large samples.

This paper introduces a deterministic method, called *Bound* and *Collapse* (BC), to estimate conditional probabilities from an incomplete sample. The method *bounds* the set of possible estimates consistent with the available information by computing the minimum and the maximum estimate that would be obtained from all possible completions of the sample. This process returns probability intervals containing all possible estimates consistent with the available information. These bounds are then *collapsed* into a unique value via a convex combination of the extreme points with weights depending on the assumed pattern of missing data.

2 Method

The intuition behind BC is that an incomplete sample is still able to constrain the possible estimates within a set and that the assumed pattern of missing data, encoded as probability of missing data, can be used to select a point estimate within the set of possible ones.

2.1 Bound

Let X_i be a variable in $\mathcal{X}$ with parent variables Π_i. Denote by $n_{ij?}$ the frequency of cases in which only the entry on the child variable is missing, by $n_{i?k}$ the frequency of cases in which only the parent configuration is unknown but it can be completed as π_{ij}, and by $n_{i??}$ the frequency of cases in which the entries X_i, Π_i are unknown and they can be completed as (x_{ik}, π_{ij}). An example is given in Figure 1 for the DGM specified by three binary variables $\mathcal{X} = \{X_1, X_2, X_3\}$, with X_1 and X_2 marginally independent and both parents of X_3. The parent configuration in case x_2 is incomplete and it can be completed as $(2,1)$ or $(2,2)$. The whole parent-child configuration in case x_5 is incomplete and it can be completed as either $(1,1,1)$, or $(1,1,2)$, or $(1,2,1)$ or $(1,2,2)$. Define *virtual* frequencies as: $n^{\bullet}_{ijk} = n_{ij?} + n_{i?k} + n_{i??}$ and $n_{\bullet ijk} = n_{ij?} + \sum_{h \neq k=1}^{c_i} n_{i?h} + n_{i??}$. It can be shown (Ramoni & Sebastiani, 1996) that $n^{\bullet}_{ijk}$ is the maximum achievable frequency of (x_{ik}, π_{ij}) in the incomplete sample that is used to compute the maximum of $p(x_{ik}|\pi_{ij})$. The virtual frequency $n_{\bullet ijk}$ is used to compute the minimum estimate of

$p(x_{ik}|\pi_{ij})$, since it corresponds to distributing the incomplete cases to increase the frequencies of the states (x_{ih}, π_{ij}) $h \neq k$ without increasing the frequency of (x_{ik}, π_{ij}). Figure 1 gives an example for the DGM described above with $\pi_{3j} = (1,1); (1,2); (2,1); (2,2)$. It has been proved (Ramoni &

case	X_1	X_2	X_3
x_1	1	2	2
x_2	2	?	1
x_3	?	1	2
x_4	?	?	1
x_5	1	?	?

$\Rightarrow$

$$n^{\bullet}_{311} = 2 \; n^{\bullet}_{321} = 2$$
$$n^{\bullet}_{331} = 2 \; n^{\bullet}_{341} = 2$$
$$n^{\bullet}_{312} = 2 \; n^{\bullet}_{322} = 1$$
$$n^{\bullet}_{332} = 1 \; n^{\bullet}_{342} = 0$$

Fig. 1. Virtual frequencies $n^{\bullet}_{3jk}$ and $n_{\bullet 3jk}$ consistent with the incomplete database

Sebastiani, 1996) that the estimate of $p(x_{ik}|\pi_{ij})$ is bounded by:

$$p_{\bullet ijk} = \frac{\alpha_{ijk} + n_{ijk} - 1}{\alpha_{ij} + n_{ij} + n_{\bullet ijk} - c_i} \leq \hat{\theta}_{ijk} \leq p^{\bullet}_{ijk} = \frac{\alpha_{ijk} + n_{ijk} + n^{\bullet}_{ijk} - 1}{\alpha_{ij} + n_{ij} + n^{\bullet}_{ijk} - c_i}$$

This probability interval contains all and only the possible estimates consistent with $\mathcal{D}$, and therefore, it is sound and it is the *tightest* estimable interval. The width of each interval is a function of the virtual frequencies, so that it accounts for the amount of information available in $\mathcal{D}$ about the parameter to be estimated, and it represents an explicit measure of the quality of probabilistic information conveyed by the sample about a parameter.

2.2 Collapse

The second step of BC collapses the intervals estimated in the *bound* step into point estimates using a convex combination of the extreme estimates with weights depending on the assumed pattern of missing data. We assume that information about missing data is encoded as a probability distribution describing, for each variable in the sample, the probability of a completion as $p(x_{ik}|\pi_{ij}, X_i = ?) = \phi_{ijk}$, where $k = 1, ..., c_i$ and $\sum_k \phi_{ijk} = 1$. Note that this is only a part of the information required about the distribution of missing data: a full description of the pattern of missing data requires knowledge of the probabilities $p(\pi_{ij}|\Pi_i = ?)$ and $p(x_{ik}|\pi_{ij}, \Pi_i = ?)$, as well as ϕ_{ijk}. If we exclude, amongst the possible patterns of missing data, those extreme mechanisms that yield the lower bounds $p_{\bullet ijk}$, the probabilities ϕ_{ijk} is sufficient to obtain accurate estimates of θ_{ijk}. This limitation of the pattern of missing data allows us to derive new local lower bounds from the maximum probabilities as follows. Each maximum probability $p^{\bullet}_{ijk}$ is obtained when all incomplete cases that could be completed as (x_{ik}, π_{ij}) are attributed to (x_{ik}, π_{ij}), and the observed frequencies of the other states of X_i given π_{ij} are not augmented. Thus, when $p(x_{ik}|\pi_{ij}) = p^{\bullet}_{ijk}$, $p(x_{ik}|\pi_{ij}) = p_{k\bullet ijl}$ for $l \neq k$ where $p_{k\bullet ijl} = (\alpha_{ijl} + n_{ijl} - 1)/(\alpha_{ij} + n_{ij} + n^{\bullet}_{ijk} - c_i)$. The maximum probabilities induce c_i probability distributions: $\{p^{\bullet}_{ijk}, p_{k\bullet ijh}, k \neq h\}$ $k = 1, \ldots, c_i$. The distribution of missing entries in terms of ϕ_{ijk} identifies a point estimate $\hat{p}_{ijk}$ as:

$$\sum_{l \neq k} \phi_{ijl} p_{l \bullet ijk} + \phi_{ijk} p^{\bullet}_{ijk}. \qquad (2)$$

These estimates define a probability distribution since $\sum_{k=1}^{c_i} \hat{p}_{ijk} = 1$. The intuition behind (2) is that the upper bound of $p(x_{ik}, \pi_{ij})$ is obtained when all incomplete cases are completed as (x_{ik}, π_{ij}). Thus, if $p(x_{ik}|\pi_{ij}, X_i = ?) = 1$ for a particular k, then (2) will return the upper bound of the interval probability as estimate of $p(x_{ik}|\pi_{ij})$, and $p_{k \bullet ijh}$ as estimates of $p(x_{ih}|\pi_{ij})$, $h \neq k$. This case corresponds to the assumption that data are *systematically* missing about x_{ik}, i.e. entries are missing with probability 1. On the other hand, when no information on the mechanism generating missing data is available, and therefore all patterns of missing data are equally likely, then $\phi_{ijk} = 1/c_i$. As the number of missing entries decreases, $p^{\bullet}_{ijk}$ and $p_{h \bullet ijk}$ approach $(\alpha_{ijk} + n_{ijk} - 1)/(\alpha_{ij} + n_{ij} - c_i)$, so that, when the sample is complete, (2) returns the exact estimate $\hat{\theta}_{ijk}$. As the number of missing entries increases then $p_{h \bullet ijk} \to 0$, for all l, and $p^{\bullet}_{ijk} \to 1$, and (2) approaches the prior probability ϕ_{ijk}: coherently nothing is learned from a sample in which all entries on (X_i, π_{ij}) are missing. Finally, if $n^{\bullet}_{ijk} = n^{\bullet}_{ij}$ for all k, then (2) simplifies to $(\alpha_{ijk} + n_{ijk} + n^{\bullet}_{ij}\phi_{ijk} - 1)/(\alpha_{ij} + n_{ij} + n^{\bullet}_{ij} - c_i)$ which is the *expected* posterior mode. When data are MAR, incomplete samples within

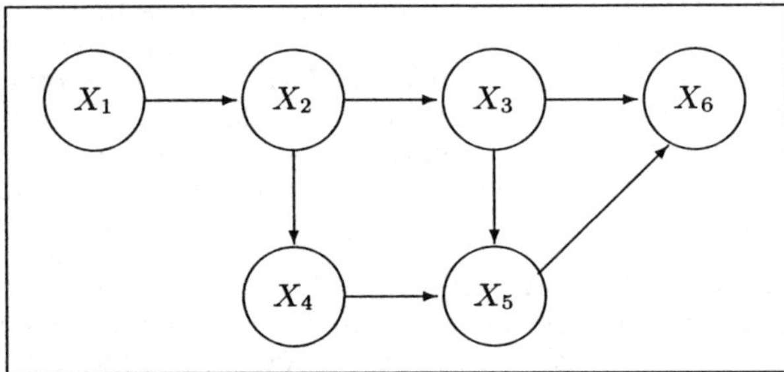

Fig. 2. The DGM used for the evaluation

parent configurations are *representative samples* of the complete but unknown ones, so that the probability ϕ_{ijk} can be estimated from the available data as $\hat{\phi}_{ijk} = \alpha_{ijk} + n_{ijk} - 1)/\alpha_{ij} + n_{ij} - c_i)$. Then $\hat{\phi}_{ijk}$ can be used to compute (2). Thus, when unreported data are MAR, BC estimates are corrections of the estimates computed from the observed data. In particular, if $n^{\bullet}_{ijk} = n^{\bullet}_{ij}$ then (2) reduces to (1), and to the standard MLE when $\alpha_{ijk} = 1$ (Little & Rubin, 1987).

3 Experimental Evaluation

Aim of these experiments is to evaluate the accuracy of BC compared to EM and GS in a sample of 1841 cases reported by Whittaker (1990, page 261). All variables are binary and they represent *Anamnesis* (X_1), *Strenuous Mental Work* (X_2), *Ratio of Beta and Alpha Lipoproteins* (X_3), *Strenuous Physical Work* (X_4) *Smoking* (X_5) and *Systolic Blood Pressure* (X_6). The sample is complete and we extracted the most probable DGM (Figure 2) using the K2 algorithm (Cooper & Herskovitz, 1992) with uniform prior probabilities. We then used the DGM to run two different learning tests in order to evaluate the accuracy of BC relative to GS and EM, using the implementation of accelerated EM in GAMES (Thiesson, 1995) and the implementation of GS in BUGS (Thomas *et al.*, 1992). The aim of the first test was to compare the estimation accuracy of the three methods when data fulfill the MAR assumption. For this purpose, four incomplete samples were created by incrementally deleting data from the complete sample. A vector ψ of 15 numbers between 0 and 1 was randomly generated, and elements of ψ were taken as the probability of deleting the occurrences of each variable X_i, independently of its value, given the parent configuration, in the the 10%, 20%, 30% and 40% of the sample. This

	Missing at Random						Non Ignorable					
Estimation Errors	9%			37%			3%			9%		
	GS	EM	BC	GS	EM	BC	GS	EM	BC	GS	EM	BC
Min	0	0	0	2	1	1	0	0	0	3	3	3
1st Qu.	1	1	1	8	8	9	3	3	3	17	17	16
Median	4	4	3	18	19	21	8	8	8	57	56	57
3rd Qu.	7	7	10	31	30	41	21	20	20	116	116	116
Max	35	34	37	119	117	229	112	111	112	917	907	918
X Ent.	12	11	18	336	319	473	78	78	78	3583	3501	3587
Prediction Errors												
Min	0	1	1	66	49	18	1	0	0	0	0	0
1st Qu.	21	28	13	88	97	175	272	275	273	1186	1131	1186
Median	46	51	21	220	224	199	314	315	314	1436	1440	1438
3rd Qu.	59	62	31	404	396	414	355	359	356	1696	1756	1697
Max	87	94	89	719	699	858	669	678	672	4786	4750	4789

Table 1. Summary statistics of the results obtained in the two tests

process generated four samples with 9% (1004), 18% (2035), 28% (3041) and 37% (4092) missing entries. The rationale of the second test was to compare the robustness of these methods as the missing data mechanism is NI. Four samples were generated from the complete sample by deleting respectively 25%, 50%, 75% and 100% of the entries ($X_5 = 2, X_6 = 1$) with probability 0.9, and ($X_5 = 2, X_6 = 2$) with probability 0.1. This process generated 4 samples with 3% (278), 5% (532), 7% (790) and 9% (1030) missing entries. The estimation accuracy was then measured by comparing the exact joint probability distribution of $(X_1, ..., X_6)$ to those learned from the incomplete data using GS, EM and BC, under the MAR assumption. The threshold for the EM was 10^{-4}. Results of GS are based on a burn-in of 5,000 iterations and a successive sample of 5,000 cases. The predictive accuracy was evaluated by comparing the 43 predictive probabilities of X_6 obtained by the three methods to those calculated by the DGM extracted from the complete sample. Some results are in Table 1, where the first five rows report summary statistics of the absolute difference (multiplied by 10^4) between the 64 exact joint probabilities and those obtained from the DGMS learned with GS, EM and BC. The sixth row gives the cross entropy between joint probability distributions. Summary statistics of the absolute errors in prediction are given in the second half of the table. The three methods give overall equivalent results under both the MAR and NI assumptions and they all suffer a decrease in accuracy when missing data are NI. However, BC provides bounds on the predicted values which reflect their reliability and can be taken into account during the reasoning process. It must be also remarked that BC does not rely *per se* on the MAR assumption and that the available information about the missing data could have been exploited by BC to achieve a better performance. The

Missing entries	GS	EM	BC
278	00:15:22	00:00:30	00:00:14
532	00:23:00	00:01:10	00:00:13
790	00:29:42	00:01:50	00:00:14
1004	00:37:47	00:01:50	00:00:13
1030	00:29:12	00:02:00	00:00:13
2035	01:09:47	00:03:12	00:00:13
3041	01:56:02	00:04:18	00:00:14
4056	04:26:07	00:07:26	00:00:13

Table 2. Execution time in *hours:minutes:seconds* for all eight samples

main difference among the three methods highlighted by the experiments is the execution time and, most of all, the shape of its growth curve (Table 2).

References

Cooper, G.F. & Herskovitz, E. (1992). A Bayesian method for the induction of probabilistic networks from data. *Machine Learning*, **9**, 309–347.

Dempster, A., Laird, D. & and Rubin, D. (1977). Maximum likelihood from incomplete data via the EM algorithm. *Journal of the Royal Statistical Society, Series B*, **39**, 1–38.

Geman, S. & Geman, D. (1984). Stochastic relaxation, Gibbs distributions and the Bayesian restoration of images. *IEEE Transactions on Pattern Analysis and Machine Intelligence*, **6**, 721–741.

Little, R.J.A. & Rubin, D.B. (1987). *Statistical Analysis with Missing Data.* New York: Wiley

Ramoni, M. & Sebastiani, P. (1996). Robust learning with missing data. Technical Report KMi-TR-28, Knowledge Media Institute, The Open University.

Thiesson, B. (1995). Accelerated quantification of Bayesian networks with incomplete data. In: *Proceedings of first international conference on knowledge discovery and data mining*, 306–11. San Mateo: Morgan Kaufman.

Thomas, A., Spiegelhalter, D.J. , & Gilks, W.D. (1992). Bugs: A program to perform Bayesian inference using Gibbs Sampling. In: *Bayesian Statistics 4*, 837–42. Oxford: Clarendon Press.

Whittaker, J. (1990). *Graphical Models in Applied Multivariate Statistics.* New York: Wiley.

Locally and Bayesian Optimal Designs for Binary Dose–Response Models with Various Link Functions

D.M.Smith[1] and M.S.Ridout[2]

[1] Centre for Statistics in Medicine, Institute of Health Sciences, Old Road, Headington, Oxford, OX3 7LF, UK.

[2] Horticultural Research International, East Malling, West Malling, Kent, ME19 6BJ, UK.

Abstract. A FORTRAN77 program for finding optimal designs for binary dose–response experiments is described. It is an enhancement of the program of Chaloner & Larntz (1988). The program finds locally and Bayesian optimal designs for models with a wide range of link functions. For Bayesian designs the parameters may have uniform, beta or bivariate normal prior distributions. The design criteria include D–optimality, and minimizing the variance of the slope parameter or of a percentile.

Keywords. Optimal design, generalized linear models, dose-response models

1 Introduction

Chaloner & Larntz (1988) describe a FORTRAN77 program for finding locally and Bayesian optimal designs for logistic regression with a single explanatory variable. In this paper a much enhanced version of their program (DESIGNV1) is described that provides a wider range of link functions, prior distributions and criterion functions. The program is useful for designing binary dose–response experiments where the probability of response π has the form

$$\pi(x;\boldsymbol{\theta}) = F(\alpha + \beta x) = F(z) \tag{1}$$

where $\boldsymbol{\theta}^T = (\alpha, \beta)$ is a vector of unknown parameters, $F()$ is a cumulative distribution function, and $z = \alpha + \beta x$. The inverse function $F^{-1}()$ is termed the *link function.* DESIGNV1 includes the various link functions considered by Ford, Torsney & Wu (1992). An alternative parametrization is $z = \beta(x - \mu)$, $\boldsymbol{\theta}^T = (\mu, \beta)$, where $\mu = -\alpha/\beta$. For several of the link functions, μ is the ED50, *i.e.* the value of x for which the probability of response is 0.5. The parametrization being used will be clear from the context.

Optimal designs minimize some function of the expected Fisher information matrix, termed the *criterion function.* However, for nonlinear models, such as (1), the information matrix depends on the unknown parameters $\boldsymbol{\theta}$. As a result the criterion function cannot be optimized directly. Two approaches have been widely used to get round this problem. *Locally optimal designs* arise when the unknown parameters in the criterion function are replaced by the experimenter's "best guess" of the true values. *Bayesian optimal designs* instead require the uncertainty about the parameters to be expressed as a prior distribution, with the optimal design chosen to minimize the expectation of the criterion function over the prior distribution. Chaloner & Verdinelli (1995) provide an extensive review of Bayesian design.

In this paper the components of DESIGNV1 are described. The designs that can be generated using it are illustrated. In the Discussion a brief comparison is made with the recent program of Spears, Brown & Atkinson (1997).

2 Link functions

The following table shows the link functions that are available in DESIGNV1.

Link function	$F(z)$	Comment
Logit	$1/(1 + exp(-z))$	
Probit	$\Phi(z)$	Φ normal c.d.f.
Double exponential	$\frac{(1 + s)}{2} - \frac{s}{2}exp(-\|z\|)$	$s = \text{sign}(z)$
Double reciprocal	$\frac{(1 + s)}{2} - \frac{s}{2}\left(\frac{1}{1 + \|z\|}\right)$	$s = \text{sign}(z)$
Complementary log–log	$1 - exp[\, -exp(z)\,]$	
Skewed logit	$1/(1 + exp(-z))^m$	$m = 1/3, 2/3, 3/2, 3$

Ford, *et al.*, (1992), and Sitter & Wu (1993), consider locally D–optimal designs for these link functions. The designs have two support points, except for the double exponential and double reciprocal link functions which have three support points. Guadard, Karson, Linder & Tse (1993) consider the robustness of locally D–optimal designs for the skewed logit model to misspecifications of the parameter values.

3 Prior distributions

A choice of prior distributions is available for the two parameters (α, β) or (μ, β). First, the parameters may have fixed values, leading to locally optimal designs. Second, the two parameters may have independent uniform distributions or, more generally, independent beta distributions. Finally, the parameters may have a bivariate normal distribution. This is useful for two–stage experiments, where the asymptotic normal distribution of the parameter estimates at the first stage can be used as the prior distribution at the second stage (Spears, *et al.*, 1997).

4 Criterion functions

Four criterion functions are available. All are functions of the expected information matrix. Let $\xi = \left\{ \begin{matrix} x_1, x_2, \cdots, x_k \\ w_1, w_2, \cdots, w_k \end{matrix} \right\}$ denote a design with distinct support points $x_1, \cdots, x_k$ and weight w_i at x_i, where $0 \leq w_i \leq 1$ and $\sum w_i = 1$. For $\boldsymbol{\theta}^T = (\mu, \beta)$ the elements of the information matrix $\boldsymbol{I}(\xi; \boldsymbol{\theta})$ are

$$i_{\mu\mu} = \beta^2 \sum \frac{w_i f_i^2}{\pi_i(1 - \pi_i)} \quad i_{\mu\beta} = -\beta \sum \frac{w_i f_i^2 (x_i - \mu)}{\pi_i(1 - \pi_i)} \quad i_{\beta\beta} = \sum \frac{w_i f_i^2 (x_i - \mu)^2}{\pi_i(1 - -\pi_i)}$$

where $f_i = F'(\beta(x_i - \mu))$ and the summations are for $i = 1, \ldots, k$. For $\boldsymbol{\theta}^T = (\alpha, \beta)$ the elements of the matrix $\boldsymbol{I}(\xi;\boldsymbol{\theta})$ are

$$i_{\alpha\alpha} = \sum \frac{w_i f_i^2}{\pi_i(1-\pi_i)} \qquad i_{\alpha\beta} = \sum \frac{w_i f_i^2 x_i}{\pi_i(1-\pi_i)} \qquad i_{\beta\beta} = \sum \frac{w_i f_i^2 x_i^2}{\pi_i(1-\pi_i)}$$

where $f_i = F'(\alpha + \beta x_i)$.

The various criterion functions, $\phi(\xi)$ are now given, together with a directional derivative function $h(x)$ which occurs in the appropriate form of the General Equivalence Theorem (see, *e.g.* Atkinson & Donev, 1992, for this theorem). The function $h(x)$ is used in checking the optimality of a design. If the design is optimal the function $h(x)$ has a maximum value of zero, attained at, and only at, the support points of the design. The functions $\phi(\xi)$ and $h(x)$ are given for locally optimal designs. For Bayesian designs they are replaced by their expectation over the prior distribution of the parameters.

4.1 Log(determinant) of the information matrix (D–optimality)

The criterion to be minimised is $\phi(\xi) = -log\,(det\boldsymbol{I}(\xi;\boldsymbol{\theta}))$ and

$$h(x) = tr(\boldsymbol{I}(x;\boldsymbol{\theta})\ \boldsymbol{I}(\xi;\boldsymbol{\theta})^{-1}) - 2\ . \tag{2}$$

The matrix $\boldsymbol{I}(x;\boldsymbol{\theta})$ represents the information matrix with $\xi^T = \{x \quad 1\}$.

4.2 Asymptotic variance of a percentile (D_A–optimality)

Let γ denote the $100\mathrm{p}^{th}$ percentile of $F(), i.e. F(\gamma) = \mathrm{p}$. Percentiles are often referred to as *effective* or *lethal* doses *e.g.* ED50, LD50). Chaloner & Larntz (1989) considered the ED50 and ED95. The criterion to be minimised is $\phi(\xi) = tr\left(\boldsymbol{B}(\boldsymbol{\theta})\boldsymbol{I}(\xi;\boldsymbol{\theta})^{-1}\right)$.

$$\text{For } \boldsymbol{\theta} = (\mu, \beta)\ ,\quad \boldsymbol{B}(\theta) = \begin{pmatrix} 1 & -\gamma/\beta^2 \\ --\gamma/\beta^2 & \gamma^2/\beta^4 \end{pmatrix}\ , \tag{3}$$

$$\text{and for } \boldsymbol{\theta} = (\alpha, \beta)\ ,\quad \boldsymbol{B}(\theta) = \begin{pmatrix} 1/\beta^2 & (\gamma-\alpha)/\beta^3 \\ (\gamma-\alpha)/\beta^3 & (\gamma-\alpha)^2/\beta^4 \end{pmatrix} \tag{4}$$

where γ is $F(\text{percentile})^{-1}$ so for the 50th percentile (LD50) $\gamma = F^{-1}(0.50)$ ($= 0$ for those link functions symmetric about the LD50). The function $h(x)$ is given by

$$h(x) = tr(\boldsymbol{B}(\theta)\ \boldsymbol{I}(\xi;\theta)^{-1}\ \boldsymbol{I}(x;\theta)\ \boldsymbol{I}(\xi;\theta)^{-1}) + \phi(\xi) \tag{5}$$

where $\phi(\xi)$ is the criterion (variance of LD50) value at the optimal design.

Locally optimal designs for this criterion sometimes have only a single support point (Wu, 1988). Flournoy (1993) describes a practical example in which the experiment was designed to estimate the 10th percentile in a logistic model.

4.3 Average asymptotic variance over a range (V–optimality)

It is assumed that there is equal interest in a range of percentile values, the limits of which are specified by the user (Chaloner & Larntz, 1989, considered the range 27% to 73%). The functions $\phi(\xi)$ and $h(x)$ are as in Section 4.2, but in the matrix $\boldsymbol{B}$, terms involving γ are replaced by their expectation, assuming γ to be uniformly distributed over the range of interest.

4.4 Asymptotic variance of the slope β (D_S–optimality)

The functions $\phi(\xi)$ and $h(x)$ are as in Section 4.2, but $\boldsymbol{B} = \begin{pmatrix} 0 & 0 \\ 0 & 1 \end{pmatrix}$.

5 Computational details

DESIGNV1 retains the structure and terminology of the Chaloner & Larntz (1988) program, though much code has been changed and added. The Nelder–Mead simplex algorithm is used for optimization. Various subroutines from Statlib (statlib@lib.stat.cmu.edu) have been used; a version of the simplex algorithm (written by D.E.Shaw, R.W.M.Wedderburn, A.J.Miller); the logarithm of the gamma function of AS245 (Macleod, 1989); the normal distribution NPROB (written by A.J.Miller); the normal deviate PPND16 of AS241 (Wichura, 1988); the symmetrix matrix inversion SYMINV of AS7(Healy, 1968b); and Cholesky decomposition CHOL of AS6 (Healy, 1968a).

The Bayesian design criteria calculate expectations over prior distributions. A 16–point Gauss–Legendre rule was used for the independent uniform and beta prior distributions, and a 16–point Gauss–Hermite rule was used for the bivariate normal prior.

Some locally optimal designs are known to have a certain nature, *e.g.* the doses may be symmetrical about the ED50. In these instances the program searches for the optimal design only within this class of design.

6 Example

Figure 1 gives the nature of the locally optimal designs for the double reciprocal link, the criterion function being the average variance over a percentile range.

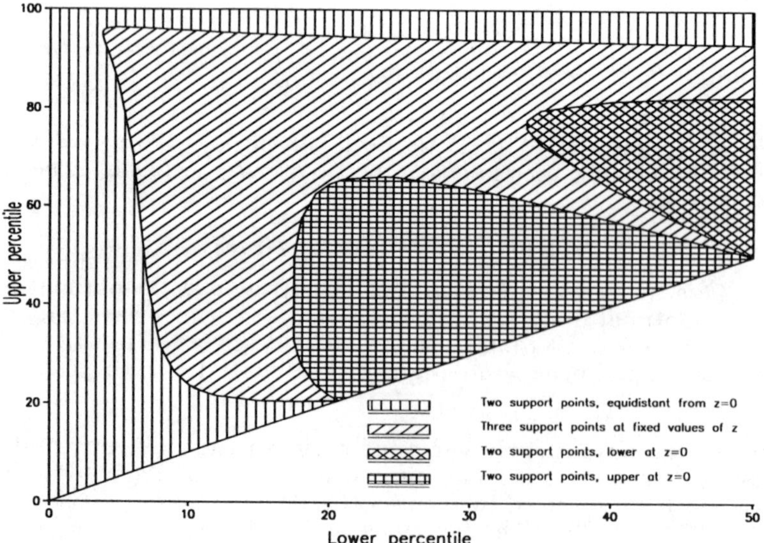

Fig. 1. Average variance over a percentile range; double reciprocal link

It shows how the nature of the design varies with the percentile interval, for intervals whose lower limit is less than 50%. Results for intervals whose lower limit exceeds 50% may be found by symmetry. The support points for the three point design are those that give the probability of response as 0.207, 0.5 and 0.793. A similar figure is available for the double exponential link function.

7 Discussion

The practical usefulness of locally optimal designs is limited by their dependence on the unknown underlying parameter values. Moreover, they often involve too few support points to allow the fit of the model to be assessed. However, they do provide a benchmark against which other, more practical, designs can be assessed (Ford, Titterington & Kitsos, 1989). The use of prior probability distributions to represent uncertainty about parameter values is a natural way to proceed. As an alternative, Sitter (1992) proposed a minimax procedure to obtain designs that are robust to poor guesses about the true parameter values.

The Bayesian design framework arises naturally in multi–stage experiments where the prior information at later stages is based on estimates of the parameters from earlier stages. The bivariate normal prior was introduced with this in mind. However, DESIGNV1 does not address the question of how the overall resources should be allocated amongst the different stages. Ridout (1995) addresses this issue for a specific problem involving the complementary log–log function link function.

As observed by Chaloner & Larntz (1989), as the spread of the prior distribution increases, so too does the number of support points of the optimal design. Typically optimization of $\phi(\xi)$ is difficult, with numerous local optima existing, when the design has more than seven or eight support points. In addition, $h(x)$ becomes very flat over a wide range, indicating the difficulty of finding the optimal design. From a practical point of view this may be unimportant as the design found, even if not optimal, will often be almost as efficient as the optimal design.

It has been assumed that the user is able to specify a prior distribution for the parameters $\boldsymbol{\theta}$. Bedrick, Christensen & Johnson (1996) argued that often it will be easier to specify prior distributions for the probabilities of response to particular doses, deriving from these the implied prior distribution for $\boldsymbol{\theta}$. DESIGNV1 could be extended to offer this feature.

The program SINGLE of Spears, *et al.*, (1997), is available from Statlib. This program offers the logistic and log–log link functions, and two design criteria *i.e.* minimization of the variance of a percentile or of the slope parameter β. The way in which the criterion functions are defined differs slightly from DESIGNV1 and the Chaloner & Larntz (1988) program (see Discussion of Spears *et al.*, 1997). Both SINGLE and DESIGNV1 offer the choice of (α, β) or (μ, β) parametrizations and provide similar sets of prior distributions. Different strategies apply for finding optimal designs. SINGLE has an automated procedure for determining the number of support points. In DESIGNV1 this must be done manually. However, in DESIGNV1 it is possible to tabulate $h(x)$ to check on the optimality of a design. When the design is not optimal, $h(x)$ can provide information about its efficiency (Atwood, 1969).

References

Atkinson, A.C. & Donev, A.N. (1992). *Optimum Experimental Designs.* Oxford: Oxford University Press.

Atwood, C.L. (1969). Optimal and efficient designs of experiments. *Annals of Mathematical Statistics*, **40**, 1570–1602.

Bedrick, E.J., Christensen, R. & Johnson, W. (1996). A new perspective on priors for generalized linear models. *Journal of the American Statistical Association*, **91**, 1450–1460.

Chaloner, K. & Larntz, K. (1988). Software for logistic regression experiment design. In: *Optimal Design and Analysis of Experiments* (ed. Y. Dodge, V.V. Fedorov & H.P.Wynn), 207–211. Elsevier Science Publishers B.V. (North–Holland).

Chaloner, K. & Larntz, K. (1989). Optimal Bayesian design applied to logistic regression experiments. *Journal of Statistical Planning and Inference*, **21**, 191–208.

Chaloner, K. & Verdinelli, I. (1995). Bayesian Experimental Design: A Review. *Statistical Science*, **10**, 273–304.

Flournoy, N. (1993). A clinical experiment in bone marrow transplantation: estimating a percentage point of a quantal response curve. In: *Case Studies in Bayesian Statistics.* (ed. C. Gatsonis, J. Hodges, R. Kass & N. Singpurwalla), 324–326. New York: Springer.

Ford, I., Titterington, D.M. & Kitsos, C.P. (1989). Recent Advances in Nonlinear Experimental Design. *Technometrics*, **31**, 49–60.

Ford, I., Torsney, B. & Wu, C.F.J. (1992). The use of a canonical form in the construction of locally optimal designs for non–linear problems. *Journal of the Royal Statistical Society* B, **54**, 569–583.

Gaudard, M.A., Karson, M.J., Linder, E. & Tse, S.K. (1993). Efficient designs for estimation in the power logistic quantal response model. *Statistica Sinica*, **3**, 233–243.

Healy, M.J.R. (1968a). Triangular decomposition of a symmetric matrix. Algorithm AS6. *Applied Statistics*, **17**, 195–197.

Healy, M.J.R. (1968b). Inversion of a positive semi–definite symmetric matrix. Algorithm AS7. *Applied Statistics*, **17**, 198–199.

Macleod, A.J. (1989). A Robust and Reliable Algorithm for the Logarithm of the Gamma Function. Algorithm AS245. *Applied Statistics*, **38**, 397–402.

Ridout, M.S. (1995). Three–stage designs for seed–testing experiments. *Applied Statistics*, **44**, 153-162.

Sitter, R.R. (1992). Robust designs for binary data. *Biometrics*, **48**, 1145–1155.

Sitter, R.R. & Wu, C.F.J. (1993). Optimal Designs for Binary Response Experiments: Fieller, D, and A Criteria. *Scandinavian Journal of Statistics*, **20**, 329–341.

Spears, F.M., Brown, B.W. & Atkinson, E.N. (1997). The effect of incomplete knowledge of parameter values on single– and multiple–stage designs for logistic regression. *Biometrics*, **53**, 1–10.

Wichura, M.J. (1988). The Percentage Points of the Normal Distribution. Algorithm AS241. *Applied Statistics*, **37**, 477–484.

Wu, C.F.J. (1988). Optimal design for percentile estimation of a quantal response curve. In: *Optimal Design and Analysis of Experiments* (ed. Y. Dodge, V.V. Fedorov & H.P. Wynn), 213–223. Elsevier Science Publishers B.V. (North–Holland).

Statistics Training and the Internet

Mike Talbot

Biomathematics and Statistics Scotland
The King's Buildings, Edinburgh, EH9 3JZ, UK
email: mike@bioss.sari.ac.uk http://www.bioss.sari.ac.uk/~mike/

Abstract. This paper examines the role of the Internet in teaching and learning statistics. It describes how the World-Wide Web is being used to support and enhance statistics coursework. Some opinions are offered on ways in which teaching practices might develop to exploit the technology.

Keywords. statistics, education, Internet, WWW

1 Introduction

Statistical science and practice are evolving rapidly under the impact of computing developments. The major influence until now has come from exploiting increased computational power, e.g. iterative minimisation methods, exploratory data analysis, stochastic simulation and Bayesian methods, image processing, neural networks. However the major influence on statistics may come in future, not from faster computers, but from improved communications via the World-Wide Web (WWW) and the Internet.

Although the WWW was initially seen as a distributed multimedia system, the facility for using scripting languages and Java applets means that interaction can be built into WWW applications, and much of the current interest in the WWW in academia is based on its potential for interactive teaching.

The WWW has much to promise in the teaching of statistics since it provides learning environments that are consistent with many of the recommendations of recent pedagogical research. For example, the statistical education reforms recommended in the US by the ASA/MAA Statistics Focus Group (Cobb, 1992) highlight some key principles:

- emphasize statistical thinking, e.g. the strategy of exploratory data analysis, the design of comparative experiments, logical inference from analyses;
- present more data and concepts but less theory and fewer recipes, i.e. provide conceptual rather than mathematical explanations.

These principles, which good teachers will recognise and practise, are readily implemented in a computer-based learning environment such as that provided by the WWW

This paper provides a personal overview of a very rapidly expanding field. In the following pages, Internet sites are identified by a number in square parentheses while the full address can be found in the References section.

2 Information and computing resources

The WWW is the largest and most diverse information resource in the world and it is growing daily as individuals and organisations provide access to their sources. Its usefulness in statistical education can be categorised under three headings:

Data Sources: The handling of real data must be an integral part of learning statistics. In the past it has been difficult for teachers to provide the quantity and complexity of examples which give students 'hands-on' experience of analysing substantial bodies of relevant and up-to-date data. Increasingly, national statistical offices and other information providers are allowing access via the Internet to their archives from which selections can be retrieved on to local computers with the minimum of effort. For example, it is possible to obtain from the US Census Bureau [1] statistics on crime in US counties and, separately, information on demographic structures in those counties. Information from the Bureau is not confined to the US: for example, it also has data on population changes in Russia by region. Other agencies which provide data that are both interesting and useful as classroom material include Eurostat, FAO and WHO [2].

Many of the classical data sets referenced in the statistical literature are also available on-line, a good access point being Stat-Lib, the WWW server operated by Carnegie Mellon University [3]. Some of these datasets are presented as case studies, and include the background to the problem and the circumstances in which the data were collected, e.g. DASL (Data and Story Library) from Cornell University [4].

Electronic Journals: Recent years have seen the development of electronic journals with the submission, reviewing and distribution, of accepted papers, all being done via the Internet. The economics of publishing a wide range of journals, and of maintaining libraries in the face of pressure on budgets, allied with the convenience to users of calling up papers rapidly from their desks, together represent strong forces which are bound to promote the electronic journal movement.

Statistical journals already on-line include Journal of Statistics Education [5], InterStat [6], Journal of Statistical Software [7]. Some other journals have already established on-line archives where algorithms, datasets and other information can be accessed.

Software: It is not generally appreciated how great an impact the WWW may have on the way that statistical software is used and accessed. Already statistical calculators can be called up freely from WWW pages, e.g. from the University of Amsterdam [9]. This allows a user to perform many standard parametric and nonparametric multiple-group comparison tests. By means of copy-and-paste operations it is possible to transfer data to the HTML page, submit the data for analysis, and view the results. Other WWW-based software is available for: performing power calculations e.g. from UCLA [9]; producing experimental design randomisations e.g. from the John Innes Centre, UK [10]; performing multivariate analysis e.g. NetMul from University of Lyon [11].

These programs, and until recently most WWW-based statistical programs, have adopted what is known as a forms-based interface using CGI programs to call the statistical algorithms written in C, Fortran or similar languages. While such programs work well they have a major drawback for statistics in allowing only limited interaction between the user and software. The Java language developed by Sun, overcomes this problem by providing the basis for secure, dynamic WWW applications which can run in most standard computing environments.

Several substantial Java statistical developments are under way. A good example is WebStat [12], developed at the University of South Carolina by Webster West and others, which provides simple graphical and data exploratory tools in an integrated environment. Other innovative statistical developments include XploRe from the Humboldt University of Berlin [13] which permits Java applets to access the XploRe graphical and computational tools for statistical analysis. SAS is developing facilities (JConnect) to allow users to call software located on SAS servers from within Java applets on a local computer. This facility might be useful in situations where an individual or organisation requires occasional use of software tools but cannot justify the purchase of the standard license. Other major statistical developers also have Java developments planned or in progress.

3 Developing new learning methods

Can the WWW add value to learning in more radical ways than merely through the support of existing practices? After all, it may be argued that the WWW is just television linked to computers and television has had only a limited role in formal education. There are at least two areas where the WWW may radically change the teaching of statistics.

Interactive Courseware on-line: Training in quantitative methods is particularly suited to delivery by computer-aided-learning (CAL) techniques and it is not surprising that there have been a number of initiatives on this topic e.g. STEPS (Redfern & Bradford, 1994), QUERCUS [14].

Modern authoring tools make it possible to produce interesting and dynamic software that allows students to work at their own pace and which can place statistics in the context of the student's subject area. Initially these CAL tools have used CD ROM as the delivery technology but now the WWW is the favoured medium and there are many courses in statistics on the Internet, available free of charge: at the WWW address listed in the title of this paper you can find links to more than 100 courses. Some of these are little more than lecture notes on-line but there is some innovative material which is bound to expand in time. Lecturers starting out on a new course might well benefit from viewing what is available.

There are some deficiencies in current WWW technology as far as statistics and CAL is concerned. At present the HTML code which controls the format of WWW pages cannot represent mathematical expressions. Proposals for extending HTML to include mathematical symbols are under consideration, but in the meantime devices such as image files can be employed.

Problem-solving Discussion Groups: Recent research has demonstrated the value of cooperative problem-solving in the acquisition of statistical skills, e.g. Garfield (1993), Keeler & Steinhorst (1995). Though requiring a much higher commitment from the lecturer, experience suggests that cooperative problem-solving produces higher grades for students and engenders a more positive attitude towards the subject.

The potential for virtual classrooms and virtual apprenticeships has yet to be explored in statistical education. These might work as follows: students establish a dialogue with an employer who is looking for an apprentice; the students are provided with access to the statistical data and to background information; in collaboration with the employer investigations are planned, executed, data analysed and reports presented while using the Internet as the main communications medium.

A virtual apprenticeship of this kind has several advantages:

- it is possible for many students to share in an applied investigation;
- it teaches students a collaborative approach to problem solving;
- students learn from each other; it helps develop analytical thought.

4 Continuing professional development

There is increasing awareness of the need to update the skills of those at work. This applies particularly to professional specialists, e.g. scientists, engineers, who often rely for their skills in quantitative methods on what they were taught at graduate level some years earlier. These skills quickly become obsolescent in a rapidly-changing computer-based field. Such specialists are often located in small dispersed groups in universities, colleges, research institutes or industrial enterprises. Traditional courses in specialist or more advanced techniques are difficult to organise, or justify, for such small groups.

Recognition of the problems outlined above has inspired a collaborative WWW-based training initiative under the title "Statistics and Mathematics as Advanced Research Tools (SMART)" [15]. SMART aims to provide a cost-effective way for experts in a topic to present that technique to those in their own or other disciplines in a convincing manner through the exploitation of WWW tools. The system is organised as an encyclopaedia with topic modules cross-referenced by application discipline and methodological classification. The aims of the modules are to help users:

- understand quickly the essence of a technique;

- evaluate its usefulness in their work;
- apply the method using standard application software, e.g. SAS, S-Plus, Genstat;
- find references for further study, both on-line and off-line;
- make contact with specialists in the technique;
- feed back their experience of the module so as to improve the training material.

The SMART system is modular in structure: links can be made to more than one application program doing the same task; the text and audio presentations can be readily adapted to handle new languages; the examples used for illustration can be substituted to suit a new application area.

At present there are few modules available for exploration within SMART. However, over time it is hoped to build up a critical mass of modules which would make the system a worthwhile reference resource.

5 Discussion

The WWW offers exceptional opportunities, and challenges, to the statistical community. The opportunities lie in two main directions.

Deepening Understanding: Statistics is generally seen to be a particularly difficult subject to learn, yet the concepts underpinning many of the techniques are simple. The difficulties often stem from an overly mathematical and non-problem-oriented approach to the teaching of the subject. The WWW with its strong multimedia elements demands a more graphical and interactive approach that is well suited to conveying concepts and demonstrating problems. The challenge for teachers is to think in these terms and to start to collect the audio-visual material necessary for WWW presentations in their area of expertise.

Quality Improvement: The openness of the WWW presents organisational dilemmas. Institutions may be reluctant to freely share the fruits of their investment in preparing teaching material. But at the same time they wish others to know if they are producing good course material. The solution may come in two ways. Firstly, educational funding agencies may encourage teaching institutions to cooperate in the preparation and use of training material. Secondly, the institutions with good material may be prepared to share the fruits of their work but only after a delay of an academic year or more, in order to maintain an academic advantage. The principal outcome from such cooperation and striving for competitive advantage is likely to be the development of high-quality training material which is used at many campuses.

6 Conclusions

The WWW will never replace human teachers with their ability to interact with individuals, to identify and overcome difficulties in understanding, and to inspire students. Nevertheless, the strengths of the WWW lie in providing a supporting role in classwork, freeing teachers from tasks that computers can do better. The WWW may be an indispensable tool when dealing with large classes or individuals at dispersed locations. One way or the other the WWW is likely to transform the teaching of statistics in the decade to come.

Acknowledgements
BioSS receives grant-aid support from the Scottish Office Agriculture, Environment, and Fisheries Department.

References

Cobb, G. (1992). *"Teaching Statistics" in Heeding the Call for Change: Suggestions for Curricular Action*, ed L.A. Steen, Washington DC. Mathematical Association of America.

Garfield, J. (1993). *Teaching Statistics Using Small-Group Cooperative Learning*. Journal of Statistics Education, **1**. [4].

Keeler, C.M. & Steinhorst, R.K. (1995). *Using Small Groups to Promote Active Learning in the Introductory Statistics Course: A Report from the Field.* Journal of Statistics Education, **3**. [4].

Redfern, E.J. & Bedford, S.E. (1994). *Teaching and Learning through Technology - The Development of Software for Teaching Statistics to Non-Specialist Students*. Dutter, R. and Grossman, W. (eds). Compstat-Proceedings in Computational Statistics. Physica-Verlag, 408-414.

Velleman, P.F. & Moore, D.S. (1996). *Multimedia for Teaching Statistics: Promises and Pitfalls*. The American Statistician, **50**(3), 217-225.

Internet Addresses

1 http://www.census.gov:80/
2 http://rcade.essex.ac.uk/other_data_sources.htm
3 http://lib.stat.cmu.edu/
4 http://lib.stat.cmu.edu/DASL
5 http://www.stat.ncsu.edu/info/jse/
6 http://interstat.stat.vt.edu/InterStat
7 http://www.stat.ucla.edu/journals/jss
8 http://fonsg3.let.uva.nl:8001/Service/Statistics.html
9 http://www.stat.ucla.edu/~jbond/HTMLPOWER/index.html
10 http://www.uea.ac.uk/nrp/jic/edgar/
11 http://pbil.univ-lyon1.fr/ADE-4/NetMul.html
12 http://www.stat.sc.edu/~west/webstat
13 http://wotan.wiwi.hu-berlin.de/xplore/index.html
14 http://www.stams.strath.ac.uk/external/CAL/quercus/index.html
15 http://www.bioss.sari.ac.uk/smart/unix/moutline.htm

Jointly Modelling Longitudinal and Survival Data

Jeremy M. G. Taylor and Yan Wang

Department of Biostatistics, UCLA, Los Angeles, CA, 90095, USA

Abstract. In many clinical and epidemiologic studies, disease markers are measured periodically and used to monitor progression to the onset of disease. Examples of this are CD4 counts and viral load measures in AIDS and PSA values in prostate cancer. We develop a joint model for analysis of both longitudinal and survival data. We use a longitudinal model for continuous data which incorporates a mean structure dependent on covariates, a random intercept, a correlated stochastic process and measurement error. The model is based on an integrated Ornstein-Uhlenbeck (IOU) stochastic process, which is an underlying AR(1) process for the derivatives of the observations. This stochastic process represents a family of covariance structures with a random effects model as one special case and Brownian motion as another. The regression model for the event time data is a time-dependent proportional hazards model, in which the longitudinal marker is a time-dependent variable and includes other covariates as well. An algorithm using Gibbs sampling and Metropolis-Hastings steps is developed for fitting the model. The algorithm requires drawing a value for the IOU stochastic process at every time point for each individual. Judicious choice of parametrisation and prior distributions is needed for an efficient algorithm. The approach is tested in a simulation study and applied to AIDS data.

Keywords. AIDS data, longitudinal models, Markov chain Monte Carlo, survival models

1 Introduction

There has been a considerable amount of research on statistical methods for the analysis of longitudinal data and censored survival data considered separately, but much less on considering them jointly. There are a number of reasons why it is important to model these two aspects jointly (Jewell & Kalbfleisch, 1992). If there is an underlying disease state which affects both the longitudinal marker process and the endpoint then joint modelling of the two processes will be more efficient than considering them separately. Not modelling these jointly can lead to biased estimates in both the longitudinal and the survival models, and also to incorrect variance estimates. The joint modelling procedures allow one to more accurately predict future failure times (Berzuini & Larizza, 1996) by incorporating information from the longitudinal data (Malani, 1995), and to correct the bias induced by informative censoring. The joint modelling also allows an assessment of whether the disease marker might be used as a surrogate endpoint (Prentice, 1989) or as an auxiliary variable (Fleming *et al.*, 1994) in a clinical trial.

Prior work on the joint modelling of longitudinal marker data and failure time endpoints (DeGruttola & Tu, 1994; Pawitan & Self, 1993; Wulfsohn

& Tsiatis, 1997; Faucett & Thomas, 1996; Berzuini & Larizza, 1996) has generally used the standard random effects model and simple parametric models for the survival distribution.

The joint model we develop incorporates (i) a flexible stochastic process determining the covariance structure of the observations, (ii) the ability to handle measurement error in the marker process, (iii) a time dependent proportional hazards model for the endpoint with non-parametrically specified baseline hazard function, (iv) additional covariates that can influence both the progression of the marker and the hazard of the event and (v) the ability to handle unequally spaced and unbalanced marker observations.

Recent work (Taylor *et al.*, 1994) has shown that the standard random intercept plus random slope model for longitudinal data can be much improved upon for CD4 counts in AIDS data, by a model which allows individual trajectories to vary. This is achieved by using a model in which the covariance structure is described by an integrated Ornstein-Uhlenbeck (IOU) stochastic process. This model gives a plausible and interpretable description of the pattern of CD4 decline, better prediction of future CD4 values and is more parsimonious.

2 Notation and Model

For subject i ($i = 1, ..., n$) the observed marker data, or some transformation of these data, consists of $Y_i(t_{i1}), ..., Y_i(t_{iJ_i})$ measured at times $t_{i1}, ..., t_{iJ_i}$, covariates are denoted by X_i and P_i, the failure time data consists of the pair (s_i, δ_i) where s_i denotes the failure time and $\delta_i = 1$ for observed events and $\delta_i = 0$ for censored failure times.

For the longitudinal model we will assume $Y_i(t_{ij}) = Z_i(t_{ij}) + e_{ij}$, where $e_{ij} \sim N(0, \sigma_e^2)$. The quantity $Z_i(t_{ij})$ can be thought of as the "true" value of the marker. We assume

$$Z_i(t_{ij}) = a_i + bt_{ij} + \beta X_i + W_i(t_{ij})$$

where $a_i \sim N(\mu_a, \sigma_a^2)$. This model can be easily expanded to include interactions between X_i and t_{ij}. The term $W_i(t_{ij})$ is an IOU process, with covariance function between observations at times s and t given by

$$\frac{\sigma^2}{2\alpha^3}[2\alpha min(s,t) + \exp(-\alpha t) + \exp(-\alpha s) - 1 - \exp(-\alpha|t-s|)] \quad (1)$$

The term $W_i(t_{ij})$ implies that each person's observed path is viewed as a separate realisation of a stochastic process. The parameters α and σ^2 control the amount of smoothness of a person's path, without imposing specific shapes on the path. The inverse parameter $1/\alpha$ has an interpretation as the time scale over which slopes of the marker value are associated, i.e. the correlation between the slopes of CD4 at times t and s is $\exp(-\alpha|t-s|)$. A feature of this model is that for α tending to zero the model corresponds to a random effects model, and at large values of α the model corresponds to Brownian motion. For numerical reasons we reparametrised (α, σ^2) as (α, θ), where $\theta = \sigma^2/\alpha^2$. In order to facilitate estimation of the IOU process, we approximate the continuous function $W_i(t)$ by the values W_{ij} at a set of N_W grid points.

The model for the failure time process is a proportional hazards model in which the hazard is given by

$$\lambda(t) = \lambda(t|Z_i(t), P_i) = \lambda_0(t)exp(\gamma Z_i(t) + \omega P_i)$$

where $\lambda_0(t)$ is the baseline hazard, i.e. the hazard depends on the "true" value of the marker. The function $\lambda_0(t)$ is parametrized as piecewise constant over N_λ intervals.

3 Likelihood

The full likelihood is a product of the likelihood from the longitudinal portion of the model and the likelihood from the failure time part of the model. The longitudinal likelihood is a product of conditional densities in this hierarchical model. The parameters are $\mu_a, \sigma_a^2, b, \beta, \sigma_e^2, \alpha, \theta, a_i (i = 1, ..., n)$ and $W_{ij}(i = 1, ..., n; j = 1, ..., N_W)$. Using the notation [.|.] for conditional probabilities, the likelihood multiplied by the priors for the longitudinal model is

$$\left(\prod_i \left(\prod_{j=1}^{J_i}[Y_{ij}|a_i, b, W_{ij}, \beta, \sigma_e^2]\right)[a_i|\mu_a, \sigma_a^2][W_i|\alpha, \theta]\right)[b][\mu_a][\sigma_a^2][\beta][\alpha][\theta][\sigma_e^2].$$

The IOU process manifests itself as the prior $[W_i|\alpha, \theta]$, which is multivariate normal with a parametric covariance structure given by equation (1). We use standard prior distributions for the regression coefficients and variances. Specifically we use uninformative flat priors for μ_a, b, β, $log(\sigma_e^2)$, $log(\sigma_a^2)$ and $log(\theta)$.

For the failure time part of the model the likelihood multiplied by the prior is

$$\left(\prod_i [s_i, \delta_i|\lambda_0(t), \gamma, \omega]\right)[\lambda_0(t)][\gamma][\omega]$$

where

$$[s_i, \delta_i|\lambda_0(t), \gamma, \omega] = (\lambda(s_i))^{\delta_i} \exp(-\int_0^{s_i} \lambda(t)dt).$$

Note that we use the full likelihood rather than the partial likelihood, because we are interested in both the regression coefficients and the baseline hazard. The integration in the above expression is a sum over grid points because we have discretized time. We use flat priors for γ and ω. We use independent log uniform priors for $\lambda_{01}, \lambda_{02}, ..., \lambda_{0N_\lambda}$.

4 Conditional Distributions

The Markov chain Monte Carlo methodology (Geman & Geman, 1984; Gelfand & Smith, 1990) proceeds from this likelihood by iterating through all of the parameters and sampling from each in turn from its conditional distribution given all of the other parameters and observations. The total number of parameters is very large because a value of W_{ij} is sampled for each grid point for each person. For many of the parameters sampling from the required conditional distribution is not easy because expressions involving the parameter are in both likelihoods.

For example, for the parameter a_i, the relevant terms in the likelihood from the longitudinal part of the model are

$$\exp(-\sum_j (Y_{ij} - (a_i + bt_{ij} + \beta X + W_i(t_{ij})))^2/(2\sigma_e^2)) \exp(-(a_i - \mu_a)^2/(2\sigma_a^2)).$$

The relevant terms in the failure time likelihood are

$$(\lambda(s_i))^{\delta_i} \exp(-\int_0^{s_i} \lambda(t)dt),$$

where $\lambda(t) = \lambda_0(t)\exp(\gamma(a_i + bt + \beta X_i + W_i(t)) + \omega P_i)$ and the integral is replaced by a sum in our discrete formulation.

The product of these likelihoods is the product of a normal and a survival function which is a non standard form. We draw from the required conditional distribution using a Metropolis-Hastings step (Besag *et al.*, 1995). We first draw a proposal for a_i using the conditional distribution for a_i defined by the longitudinal part of the likelihood, then decide whether to accept or reject this by including the failure time part of the likelihood. The other location type parameters (b, β, W_{ij}) are sampled in a similar way. One slight difference for W_{ij} is that for each i the parameters are sampled in a block

The parameter μ_a has a normal conditional distribution, θ, σ_a^2 and σ_e^2 have inverse Gamma distributions and the conditional distribution for λ_{0j} is a Gamma distribution. The parameter α and the log relative risk parameters of the failure time model, γ and ω are sampled using a Metropolis-Hastings step. For γ and ω the proposed density is a normal with the previous value as the mean, and the variance chosen to ensure good mixing of the Markov chain. For α we experimented with using adaptive rejection Metropolis sampling, but preferred the simple Metropolis-Hastings step.

5 Numerical Issues

The algorithm was tested on both a simulated and an AIDS data set. The simulated data sets consisted of 110 subjects with an average of 7 longitudinal measurements per person and between 20% and 50% censoring rate. The MCMC sequence was typically run for between 5,000 and 20,000 iterations. Convergence of the sequence was very fast and not dependent on the prior distribution for some parameters $(\mu_\alpha, \beta, b, \omega, \sigma_e^2, \sigma_a^2, \theta, \lambda_0)$. Convergence of the MCMC was much slower or mixed less well for α and γ. The prior for α was found to play a strong role in determining its posterior distribution, but did not influence the posterior distribution of the other parameters, except for a slight influence on σ_e^2, σ_a^2 and θ. Three classes of priors were considered for α, a Gamma prior, one with $log(\alpha)$ as uniform and one with $\alpha/(1+\alpha) \sim$ beta where the parameters of the beta were chosen to be close to a uniform.

Centring on the covariates was also found to lead to marker improvement in the convergence properties of the algorithm. In particular we use $X_i - \bar{X}$ instead of X_i in equation 1, and use $Z_i(t) - \bar{Y}$ and $P_i - \bar{P}$ in equation 2. The number of steps used for λ_0 was at most 8. The efficiency of the estimate of λ_0 decreased as the number of steps increased, as did the efficiency of γ and ω to a lesser degree.

6 Results

Simulations indicate that the joint modelling approach leads to good coverage rates of confidence intervals and reduced bias and greater efficiency compared to separate longitudinal and proportional hazards modelling, particularly for the parameters of the failure time model. Whether there was any benefit from joint modelling and how much the benefit was depended on the design and parameter values.

When applied to CD4 count and AIDS event time data from seroconverters in the Multicenter AIDS Cohort Study (Taylor *et al.*, 1994) the method shows the importance of the CD4 level prior to HIV infection as a covariate influencing disease progression. This covariate significantly influenced μ_a but not ω or an $X_i t_{ij}$ interaction term, i.e. it influenced the post HIV infection CD4 value but did not significantly influence the rate at which CD4 declined or the hazard of AIDS after adjusting for the current CD4 value.

References

Berzuini, C. & Larizza, C. (1996). A unified approach for modelling longitudinal and failure time data, with application to medical monotoring. *IEEE Transactions on Pattern Analysis and Machine Intelligence*, **18:2**, 109-123.

Besag, J., Green, P., Higdon, D. & Mengersen, K. (1995). Bayesian computation and stochastic systems. *Statistical Science*, **10**, 3-66.

deGruttola, V & Tu, X.M. (1994). Modelling progression of CD4 lymphocyte count and its relationship to survival time. *Biometrics*, **50**, 1003-1014.

Faucett, C.L. & Thomas, D.C.(1996). Simultaneously modeling censored survival data and repeatedly measured covariates: a Gibbs sampling approach. *Statistics in Medicine*, **15**, 1663-1685.

Fleming, T.R., Prentice, R.L., Pepe, M.S. & Glidden, D. (1994). Surrogate and auxiliary endpoints in clinical trials, with potential applications in cancer and AIDS research. *Statistics in Medicine*, **13**, 955-968.

Gelfand, A. E. & Smith, A. F. M. (1990). Sampling-based approaches to calculating marginal densities. *Journal of the American Statistical Association*, **85**, 398-409.

Geman, S. & Geman, D. (1984). Stochastic relaxation, Gibbs distributions, and the Bayesian restoration of images. *IEEE Transactions on Pattern Analysis and Machine Intelligence*, **6**, 721-741

Jewell, N.P. & Kalbfleisch, J.D. (1992). Marker Models in Survival Analysis and Applications to Issues Associated with AIDS. In:*AIDS Epidemiology: Methodological Issues* (ed. N.P. Jewell, K. Dietz & V.T. Farewell), 211-230. Boston: Birkhauser.

Malani H.M. (1995). Modification of the re-distribution to the right algorithm using disease markers. *Biometrika*, **82**, 515-526.

Pawitan, Y. & Self, S.(1993). Modeling disease marker processes in AIDS. *Journal of the American Statistical Association*, **88**, 719-726.

Prentice, R.L. (1989). Surrogate endpoints in clinical trials: definition and operational criteria. *Statistics in Medicine*, **8**, 431-440.

Taylor, J.M.G., Cumberland, W.G. & Sy, J.P. (1994). A stochastic model for analysis of longitudinal AIDS data. *Journal of the American Statistical Association*, **89**, 727-736.

Wulfsohn, M.S. & Tsiatis, A.A. (1997). A joint model for survival and longitudinal data measured with error. *Biometrics*, **53**, 330-339.

Representing Solar Active Regions with Triangulations

Michael J. Turmon and Saleem Mukhtar

Machine Learning Systems Group, Jet Propulsion Laboratory, Pasadena, CA 91109, USA

Abstract. The solar chromosphere consists of three classes which contribute differently to ultraviolet radiation reaching the earth. We describe a data set of solar images, means of segmenting the images into the constituent classes, and a novel high-level representation for compact objects based on a triangulated spatial 'membership function'. Such representations are fitted in a variable-dimension Markov chain Monte Carlo scheme.

Keywords. Triangulation, chromosphere, Markov chain Monte Carlo, image segmentation

1 Introduction

The solar atmosphere is comprised of several features having various characteristics, and distinct physical origin. The most apparent are active regions, which are associated with sunspots in the photosphere and plages in the chromosphere. These plages can be quite large in extent, and show themselves by their strong magnetic field and altered light emissions. To a lesser extent, so does the chromospheric network, which is an evenly-distributed cell-patterned web of enhanced activity. The remainder of the surface shows only ordinary fluctuations and is termed quiet sun. See Figure 1 and Zirin (1988) for more on chromospheric features.

The three classes contribute differently to the ultraviolet radiation reaching Earth's upper atmosphere, with the plages and magnetic network giving the largest contribution. This radiation cannot be sensed directly from the ground but the features giving rise to it can be; they are used as proxy inputs to models of solar irradiance. These models are crucial to understanding phenomena such as global warming and photochemical decomposition processes in the upper atmosphere; see Withbroe (1994).

Further, much can be learned about solar irradiance by relating irradiance changes observed via satellite to region evolution identified in spatially-resolved images. Current understanding of these effects, and of plage evolution in general, is of a qualitative sort and a more refined description of anticipated plage shapes and the evolution of plage regions would be of value.

The features studied here, as well as related photospheric phenomena, are observed by many instruments on Earth and a few in space. The primary source of data for this study is the set of CaII K full-disk spectroheliograms that has been collected daily at Sacramento Peak National Solar Observatory from the mid-sixties onward. The images are recorded on photographic film, an interval of which (from the mid-eighties forward) has been digitized to 2K$\times$2K pixels.

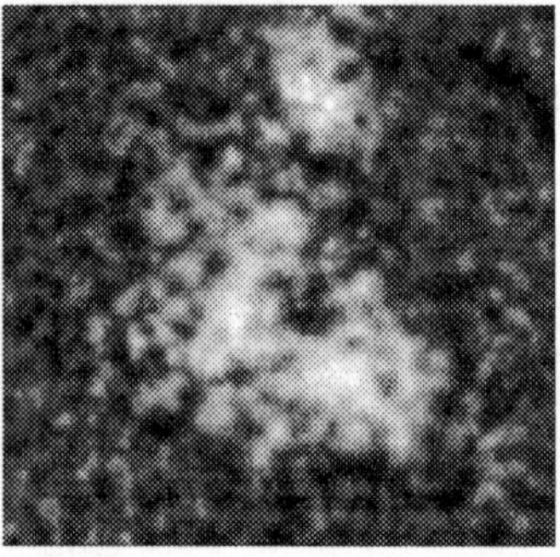

Fig. 1. A chromospheric image from 15 July 1992 shows both a decayed plage pair in the lower-right quadrant of the sun and a younger, more concentrated plage at upper left; a detail image of the latter is in the second panel

Currently, scientists typically either apply a threshold across the flattened image to determine plage areas, or manually surround the plages with polygons. The first method, while simple and objective, ignores all spatial information that is available. The second method clearly uses a large amount of side information possessed by the scientists, but is also highly subjective, difficult to even describe, and hard to repeat. We will describe a more objective and automatic procedure based on a hierarchical model of image features and formation.

We introduce our paper with an overview of the model and method we have used. While the Bayesian framework is not universally appropriate for inference problems, in the situation at hand the prior information is so apparent that approximating it is preferred to neglecting it. So, following Grenander (1991, e.g.), we establish a Bayesian formalism for a hierarchical representation of plages in three levels.

With each pixel of the observed image $\mathbf{y}$ we associate a small-integer label determining its class; these labels are $\mathbf{x}$. The labelling in $\mathbf{x}$ captures the information needed to, for example, determine how much of the chromosphere is plage. The plages themselves are large-scale phenomena which are not well-captured by pixel-level rules, so their representation should bind nearby plage sites into a cluster of heightened activity. Furthermore, even experts have uncertainty in precisely delineating plage regions, so the plage description should express this equivocation. Accordingly, the plage is represented by a membership function h across the image space, with large values indicating increased confidence that a site is plage. To combine these quantities, let there be a Markov relationship between the three levels of the stochastic model so that

$$P(h, \mathbf{x}, \mathbf{y}) = P(h)P(\mathbf{x} \mid h)P(\mathbf{y} \mid \mathbf{x}) \tag{1}$$

The interpretation is that an underlying, large-scale activity pattern h occurs, giving rise to a fine-scale pattern $\mathbf{x}$. The latter is then responsible for the observed image $\mathbf{y}$. In the next section we detail the model; then we describe the scheme for inference and provide some representative results.

2 Image representation and modelling

Denote a generic spatial position by $s = [s_1\, s_2] \in \bar{N} = [0,1]^2$. Observations are made at a lattice of sites $N \subset \bar{N}$. The class labels $\mathbf{x} = \{x_s\}_{s \in N}$ take values in the set $\{\mathtt{P}, \mathtt{N}, \mathtt{B}\}$, while entries in the corresponding observation $\mathbf{y}$ are each real-valued. We work from the data backward in defining the factors of (1).

Conditioned on the labelling, the likelihood factors as

$$P(\mathbf{y} \mid \mathbf{x}) = \prod P(y_s \mid x_s) \quad . \tag{2}$$

Labelled images supplied by scientists suggest the three densities $P(y \mid x)$ are lognormal, so, up to an additive universal constant,

$$-\log P(\mathbf{y} \mid \mathbf{x}) = \sum_{s \in N} \left(\frac{(\log y_s - \mu_{x_s})^2}{2\sigma_{x_s}^2} + \log \sigma_{x_s} + \log y_s \right) \quad . \tag{3}$$

For $P(\mathbf{x} \mid h)$, we use the "Potts model", an ordinary Markov random field smoothness prior (Besag, 1974), modified so that the membership function $h(s) \in [0,1]$ favours the event $\{x_s = \mathtt{P}\}$:

$$-\log P(\mathbf{x} \mid h) = K_h + \beta \sum_{s \sim s'} 1(x_s \neq x_{s'}) + \alpha \sum_{s \in N} |1(x_s = \mathtt{P}) - h(s)| \quad . \tag{4}$$

The relation $s \sim s'$ is true for "neighbouring" sites in N. On our rectangular lattice, sites are neighbours if they adjoin vertically, horizontally, or diagonally. Here $\alpha \geq 0$ indicates the influence of h, $\beta \geq 0$ favours agreement among labels, and K_h is an appropriate normalizing constant.

To represent a plage, or a cluster of related plages, we propose a tent-like structure defined by a triangulated planar graph

$$\begin{aligned} &G = (V, E, h) && \\ &V \subset \bar{N} && \text{a vertex set} \\ &E \subset \bar{N}^2 && \text{an edge relation} \\ &h \colon V \to [0,1] && \text{a height function} \end{aligned} \tag{5}$$

The height function extends to all of $\bar{N}$ by linear interpolation across the faces of the pyramids. This structure models the "degree of membership" of a given pixel in the plage class and allows the binding of nearby plage regions into one coherent object. We note that, if the height function is thresholded at a given level, the resulting shape is a cluster of regions bounded by polygons — the same way scientists currently delimit plage regions manually. See the diagram below.

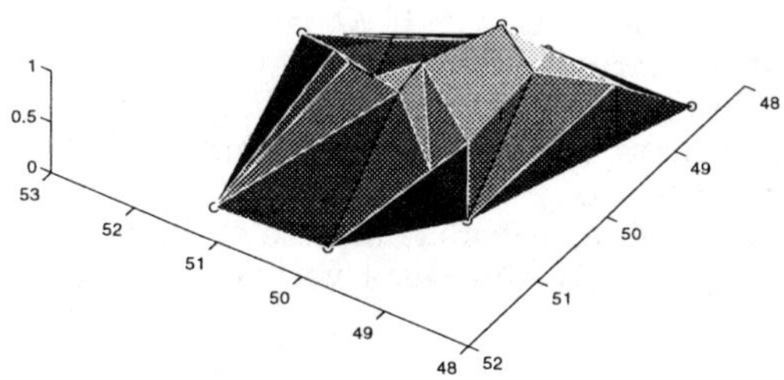

To define a probability distribution on membership functions, we generate each as the interpolated version of the Delaunay triangulation of iid points, uniform in $\bar{N}$. These points comprise V, and E is generated mechanically as the triangulation of V. Heights in $[0,1]$ are then assigned independently to the members of V to form tie-points. The probability density of such a membership function is induced by the one on V:

$$P(h) = Z^{-1} e^{-\gamma \operatorname{card}(V_h)} \quad . \tag{6}$$

A computational advantage of this scheme is that additions, deletions, and adjustments of one vertex have a local effect on the triangulation. Also, the penalty in log-probability paid by joining two separated graphs is the sum of component penalties, so that separated plages co-exist independently.

3 Inference

This describes the "synthesis problem"; the complementary "analysis problem" focuses on the posterior

$$P(h, \mathbf{x} \,|\, \mathbf{y}) = P(h, \mathbf{x}, \mathbf{y}) / P(y) \propto P(h, \mathbf{x}, \mathbf{y}) \quad .$$

Sampling from this distribution is a sufficient basis for any other inference scheme; we have in mind principally MAP. Adopting the well-known Markov chain Monte Carlo outlook (Besag *et al.* , 1995, for example), we sample alternately from $\mathbf{x}$ and h. The former is easily accomplished via the well-known Gibbs sampler, so we concentrate on updates to h with $\mathbf{x}$ held fixed.

One technical difficulty is the normalizing constant K_h which figures in the posterior. Existing Monte Carlo techniques for estimating K_h (Potamianos & Goutsias, 1997) simply involve sampling from $P(\mathbf{x} \,|\, h)$, but the computation involved for this is too large to justify the effort. At present we have assumed that the variation of K_h with respect to h is negligible compared to the designed variation in $P(h, \mathbf{x}, \mathbf{y})$, leading to an approximate posterior $\pi(h, \mathbf{x})$ with negative log-probability (excluding constant terms)

$$\beta \sum_{s \sim s'} 1(x_s \neq x_{s'}) + \alpha \sum_{s \in N} |1(x_s = \mathrm{P}) - h(s)| + \\ \sum_{s \in N} \left(\frac{(\log y_s - \mu_{x_s})^2}{2\sigma_{x_s}^2} + \log \sigma_{x_s} \right) + \gamma \operatorname{card}(V_h) \tag{7}$$

having a minimum at $(\hat{h}, \hat{\mathbf{x}})$. MAP inference proceeds, as noted, by alternately varying $\mathbf{x}$ and h while decreasing a temperature parameter.

Updates of h correspond to altering the vertex list, and are done with simple Metropolis-Hastings steps. Such a step proposes a new state h', computes $\rho(h, h') := \pi(h', \mathbf{x}) / \pi(h, \mathbf{x})$, and probabilistically accepts or rejects h' largely on this basis; this results in a Markov transition kernel $Q(v, dv')$ on the composite vertex-list set $\mathcal{V} = \cup_k \mathcal{V}_k$. If Q is designed properly, it has the posterior π as its stationary distribution. Beyond the obvious restrictions that Q be aperiodic and irreducible, it is sufficient that Q maintains detailed balance: under π, the mass moving directly from $A \subseteq \mathcal{V}$ to B equals that moving in the reverse direction.

First we describe a set of operators complete enough to ensure irreducibility. A *vertex move* operator M chooses a vertex at random and displaces it

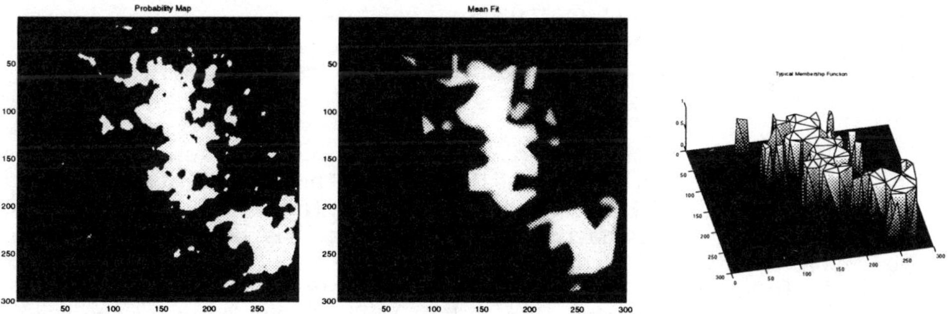

Fig. 2. Plage probability; mean inferred membership function; sample membership function

randomly. A *vertex raise* operator R raises or lowers a vertex at random. To allow movement between the constituent spaces of $\mathcal{V}$, we have *add* operators A_k, and corresponding *kill* operators A'_k, which move back and forth between $\mathcal{V}_k$ and $\mathcal{V}_{k+1}$.

Next, we define a transition kernel Q on the basis of these operators; this kernel is a "hybrid sampler" composed of each of the three move-types (M, R, A/A'). In each epoch in the simulation, one such move-type is chosen at random. Ensuring detailed balance within each move-type yields detailed balance in the superposition. Obtaining detailed balance in types M and R is trivial provided the distribution of the additive displacement is symmetric. (Modular addition will eliminate edge conditions.) Operators M and R are accepted with probability $\min(1, \rho(h, h'))$.

Obtaining detailed balance of A_k, A'_k is more complex because the flow between two different Euclidean spaces must be equalized. Following recent work of P. Green (1995), we find the chance of accepting a proposed deletion of v^* via A'_k should be the lesser of unity and

$$\rho(h, h') \times \frac{P(\text{select } A_k)}{P(\text{select } A'_k)} \times \frac{p_v(v^*)}{1/(k+1)} \quad . \tag{8}$$

(Here p_v is a density used to choose a new point for an add operation; in practice it is used to focus attention on interesting parts of the image.) The intuition is simple: the more likely it is to attempt deletion, the less likely we must be to accept it. The more likely it is to add v^* back in, the more willing we are to delete it. The factor of $k+1$ comes from the random choice of which vertex to delete: when v^* is added via A_k, there is one chance in $k+1$ that a subsequent application of A'_k will consider v^* for deletion.

4 Computational aspects and results

Initialization is important since a small feature may become hidden in a large triangle so that that π is not increased by any single vertex addition. The initialization procedure should therefore ensure locality of the effects of changes. A procedure that has proven effective is to initially replace the term of π enforcing agreement between h and the plage probability with one penalizing per-triangle inhomogeneity:

$$\sum_T |T|\, q_T(1-q_T) \quad \text{, with} \quad q_T := |T|^{-1} \sum_{s\in T} 1(x_s = \mathtt{P})$$

and $|T|$ the number of pixels in triangle T. The modified criterion subdivides the image during an initial phase of 1000 epochs; then it is gradually replaced by the final criterion in a secondary stage twice this length. By the end of the second stage, a satisfactory basin of $\pi(h, \mathbf{x})$ has been found and the Metropolis iteration proceeds as described above.

Finally, to speed the sampling process the indicator $1(x_s = \mathtt{P})$ above is replaced with its expectation $P(x_s = \mathtt{P} \mid y_s)$. This is analogous to the use of conditional expectation in the ICE algorithm of A. Owen (1986) and allows the sampler to directly access the uncertainty in the label, instead of reacting to its probabilistic fluctuations as Gibbs iterations proceed.

Sample results for fitting a rather complex plage pair are shown in Figure 2. Fits with $\gamma = 2$, $\alpha = 0.4$ were obtained from a total of 30 000 Metropolis proposals taking 170 seconds of computation time on a Sun Ultrasparc. Roughly 175 proposals/sec are made by exploiting the significant cancellation in the quotient $\rho(h, h')$: only the changed triangles need be reconsidered. As desired, the membership function has suppressed the small-scale features and identified the two main objects and their principal outliers.

Acknowledgments
This work was carried out by the Jet Propulsion Laboratory, California Institute of Technology, under contract with the National Aeronautics and Space Administration. Thanks to Eric Mjolsness of JPL for helpful suggestions on the image representations used here, and to Judit Pap of the UCLA Department of Physics and Astronomy for help in understanding the scientific context of this problem. Delaunay triangulations were computed with the Triangle package authored by J. R. Shewchuck.

References

Besag, J. (1974). Spatial interaction and the statistical analysis of lattice systems. *J. R. Statist. Soc. Ser. B*, **36**, 192–236.

Besag, J., Green, P., Higdon, D., & Mengersen, K. (1995). Bayesian Computation and Stochastic Systems. *Statistical Science*, **10**, 3–66.

Green, P. J. (1995). Reversible jump Markov chain Monte Carlo computation and Bayesian model determination. *Biometrika*, **82**(4), 711–732.

Grenander, U., Chow, Y., & Keenan, D. (1991). *Hands: A Pattern-Theoretic Study of Biological Shapes.* New York: Springer.

Owen, A. (1986). Discussion of Ripley, "Statistics, images, and pattern recognition". *Canad. Jour. Statist.*, **14**, 106–110. Article covers pp. 83–111.

Potamianos, G., & Goutsias, J. (1997). Stochastic Approximation Algorithms for Partition Function Estimation of Gibbs Random Fields. *IEEE Trans. Inform. Theory*, **43**(6), 1948–1966.

Withbroe, G. L., & Kalkofen, W. (1994). Solar Variability and its Terrestrial Effects. *Pages 11–19 of: The Sun as a Variable Star: Solar and Stellar Irradiance Variations.* Cambridge: Cambridge Univ.

Zirin, H. (1988). *Astrophysics of the Sun.* Cambridge: Cambridge Univ.

A General Form for Specification of Correlated Error Models, with Allowance for Heterogeneity

S J Welham[1], R Thompson[1] and A R Gilmour[2]

[1] Statistics Department, IACR-Rothamsted, Harpenden AL5 2JQ, UK
[2] Orange Agricultural Institute, NSW Agriculture, Forest Road, Orange 2800, Australia

1 Introduction

This paper considers various correlated error structures in the linear mixed effects model

$$y = X\alpha + \Sigma Z_i u_i + e$$

with data y, fixed effects α with design matrix X, several sets of random effects $\{u_i, i=1...c\}$ with design matrices Z_i, and random error e. The sets of random effects u_i correspond to individual model terms, and u_i and e are mutually independent with $\text{cov}(u_i,u_j)=\sigma^2\gamma_{ij}G_{ij}$, $\text{var}(e)=\sigma^2 R$ and $\text{cov}(u_i,e)=0$, where G_{ij} and R may contain unknown parameters. This generates the variance matrix for the data $V(y)$ as

$$V = \sigma^2 (\Sigma_{ij} \gamma_{ij} Z_i G_{ij} Z_j' + R)$$

In the traditional linear mixed model $\gamma_{ij}=0$ for $i\neq j$, and both the G_{ii} and R are identity matrices, which assumes that sets of random effects are independent with common variance within sets. In this case, the terms $Z_i Z_i'$ can generate a limited range of covariance structures between units of y, for example, equal correlation between units within groups where columns of Z_i are indicator variables for levels of a factor. Or where columns of Z_i are true covariate values, $Z_i Z_i'$ can generate a highly structured known correlation pattern depending on the values of Z_i. However, a wide range of models exist where more general structured or unstructured correlations are required: repeated measurements, spatial analysis (in >1 dimension), multivariate analysis, random coefficient regression and animal models, and combinations of these.

Our approach generates the (scaled) covariance structures G_i and R using direct products, where the components of the direct product correspond to the factors defining the individual model terms $Z_i u_i$ and the residual e. Cullis & Gleeson (1991) used a direct product structure to model independent spatial correlation patterns in two directions within field experiments, and this construction fits naturally within the structure of the models listed above. This general specification also allows for combination of correlation structures, for example, repeated measurements in a spatial layout.

2 Examples of direct product construction

2.1 Repeated measurements data

For example, consider an analysis of repeated measurements where data have been taken weekly over 5 weeks from a set of 14 subjects. It is likely that data taken from the same subject will be correlated, with correlation decreasing over time, but that subjects will be independent. Data units (and hence e) are completely indexed by the model term `Subject.Week`. The residual e can correspondingly be written in terms of sub-vectors e_i for subject i at times 1...5, with some common covariance structure C imposed on the sub-vectors e_i to model correlation over time. Independence between subjects, i.e. between the e_i, is retained giving $\mathrm{var}(e_i)=C$ and $\mathrm{cov}(e_i,e_j)=0$. The resulting variance matrix on e can be written as a direct product of an identity matrix and the covariance matrix C:

$$e = \begin{pmatrix} e_1 \\ e_2 \\ \cdot \\ \cdot \\ \cdot \\ e_{14} \end{pmatrix}, \qquad \mathrm{var}(e) = R = \begin{pmatrix} C & 0 & 0 & \ldots & 0 \\ 0 & C & 0 & \ldots & 0 \\ 0 & 0 & C & \ldots & 0 \\ \cdot & \cdot & \cdot & \cdot & \cdot \\ 0 & 0 & 0 & \ldots & C \end{pmatrix} = I_{14} \otimes C$$

So the variance model for the residual can be constructed by considering the components of the term: independence between subjects (I_{14}) crossed with common correlation within subjects (C). In this case, no other random terms are required to describe the structure of the variance model.

2.2 Spatial analysis of field experiments

The repeated measurements example above naturally generates a block diagonal variance matrix V, but it is easy to find examples where more complex structures arise by combining variance models. For example, consider the analysis of a field experiment laid out as 10 rows of 15 columns, where the object is to model spatial variation in both directions across the experiment to obtain more accurate standard errors. In this case this residual term `Row.Column` takes the form $C_R \otimes C_C$.

2.3 Random coefficient regression

In some longitudinal data sets, individual profiles appear to increase linearly over time, but with obvious variation in slope between subjects within treatment groups. In this case, a natural model for the data consists of a common linear trend over time for treatment groups plus random variation about the intercept and slope for subjects. In this case, the random model can be written `Subject+Subject.Time`. The random intercept (a) and slope (b) effects are assumed to have variances var(a)=$\sigma_a^2 I_n$ and var(b)=$\sigma_b^2 I_n$. To make the fitted variance model invariant to the scale of time measurement, correlation is imposed between the intercept and slope for each subject, i.e. $G_{ab} = I_n$, giving

$$\text{cov}\begin{pmatrix} a_1 \\ \cdot \\ a_n \\ b_1 \\ \cdot \\ b_n \end{pmatrix} = \begin{pmatrix} \sigma_a^2 G_{aa} & \sigma_{ab} G_{ab} \\ \sigma_{ab} G_{ba} & \sigma_b^2 G_{bb} \end{pmatrix} = \begin{pmatrix} \sigma_a^2 I_n & \sigma_{ab} I_n \\ \sigma_{ab} I_n & \sigma_b^2 I_n \end{pmatrix} = \begin{pmatrix} \sigma_a^2 & \sigma_{ab} \\ \sigma_{ab} & \sigma_b^2 \end{pmatrix} \otimes I_n = U \otimes I_n$$

and thus can also be specified via the direct product construction. Here the direct product is formed by linking two model terms with equal numbers of effects and assuming the same form for correlations within and across the two terms (intercept and slope).

2.4 Multivariate analysis

Multivariate datasets often consist of several traits measured on a set of individuals. In this case, the multivariate random model is the univariate random model crossed with an unstructured covariance matrix representing correlation across traits. This generates direct product forms of both R and the G_i.

2.5 Animal model

Where measurements have been taken on a group of related animals (or plants) and the pedigree is known, it may be possible to attribute variation to genetic relationships. For example, terms **`animal+dam`** may be fitted to represent variation due to the genetic effect of the animal and mother, respectively. In each term, the additive genetic relationships between individuals can be accounted for by use of a known covariance matrix A which refers to all animals in the pedigree. Correlation between animal and dam effects are also required, leading to model

$$\text{cov(animal,dam)} = U \otimes A$$

where U is a 2×2 unstructured matrix. This uses the same construction as random coefficient regression to link two separate model terms of the same length, but with $G_{ij}=A$ for all i,j.

3 Extensions to models: multivariate and multi-site

An extension to standard model formulae is required to make specification easy for the user. In a multivariate example, with t traits measured on n individuals, it is generally more efficient to organise the data as t vectors of length n, i.e. with the measurements on each trait held in parallel with the factors/covariates defining the model for each individual. For example, data in the repeated measurement example 2.1 can be specified either as a vector of length 70, or as five parallel vectors of length 14. In the latter case, a **`Subject`** factor can be used to define the 14 individuals, but since data from different weeks are held in parallel, no **`Week`** factor can be defined. In this case a virtual factor must be defined to specify model terms across time. Using the generic term **`'Trait'`** for the virtual factor, this gives error term **`'Trait'.Subject`** with form $C \otimes I$ as before. This method can also be used to specify differential effects of covariate **`X`** across traits using term **`'Trait'.X`**, or to specify effects of covariates specific to one trait, eg. **`'Trait$3'.X`**.

The structure of multi-site data, i.e. data from similar (but not identical) experiments at different sites, is often more complex. Experiments at different sites may have different numbers of units and so cannot be recorded in parallel vectors, explanatory variables may differ between sites, and the structure of the data (and hence the random model) may differ between sites. In this case, it cannot be assumed that a common covariance structure holds at all sites, and the residual error term becomes a direct sum rather than a direct product, although within sites the direct product structure still holds. In this case, if sites are held as separate data sets, then it is not possible to directly identify factors held in common, and there is no advantage in joint as opposed to separate analysis of the experiments. If sites are appended as a single data set, with missing values where appropriate and an indexing factor defining experiments, then terms can be estimated jointly across or separately within experiments as required.

4 Limitations of the direct product approach

There are several structures not easily modelled using the direct product construction. For example, consider an unbalanced repeated measurements data set, where each subject has had a different number of measurements at their own unique set of times. First note that if a random coefficient regression model is used, then the model can be fitted as before. However, if the approach of Section 2.1 is used, the natural specification of the error term is now

$$\boldsymbol{e} = \begin{pmatrix} \boldsymbol{e}_1 \\ \boldsymbol{e}_2 \\ \cdot \\ \cdot \\ \cdot \\ \boldsymbol{e}_{14} \end{pmatrix}, \qquad \mathrm{var}(\boldsymbol{e}) = \boldsymbol{R} = \begin{pmatrix} C_1 & 0 & 0 & \dots & 0 \\ 0 & C_2 & 0 & \dots & 0 \\ 0 & 0 & C_3 & \dots & 0 \\ \cdot & \cdot & \cdot & \cdot & \cdot \\ 0 & 0 & 0 & \dots & C_{14} \end{pmatrix}$$

where although the C_i may be based on the same models, they are different sizes and calculated at different time points. It is easily seen that this is a simple case of the multi-site analysis described in Section 3, with subjects in the place of sites, and with common error models and parameters.

In spatial analysis, the direct product construction assumes that data has been produced on a regular grid. If this is not the case, then the error structure cannot take direct product form and in these cases, covariance structures generated from distances between points can be used.

5 Variance heterogeneity

In spatial data, it is often satisfactory to use $G_i = \gamma_i B_i$ where B_i is a correlation matrix, since constant variance over the site is often a reasonable assumption. For repeated measurements data, it may be expected that variance will change (often increase) over time. We consider two possible approaches to extending standard correlation models, using the example of an auto-regressive model of order 1, to incorporate changing variance over time.

In some circumstances, it is reasonable to assume that the covariance structure can be represented as a scaled correlation matrix $D^{0.5}BD^{0.5}$, for some diagonal matrix D, and correlation matrix B (here representing an AR(1) process). However, if it is reasonable to suppose that the errors do follow an AR(1) process, there is no natural interpretation for this scaling mechanism. A more natural form can be found by considering the structure of the process. The AR(1) process can be written as

$$y_{t+1} = \phi y_t + e_{t+1}$$

or, in vector terms,

$$Ly = e$$

where

$$L = \begin{pmatrix} 1 & 0 & 0 & 0 & 0 & \ldots \\ -\phi & 1 & 0 & 0 & 0 & \ldots \\ 0 & -\phi & 1 & 0 & 0 & \ldots \\ 0 & 0 & -\phi & 1 & 0 & \ldots \\ . & . & . & . & . & . \end{pmatrix}$$

If V=var(y) and D=var(e), this gives

$$LVL' = D$$

or

$$V^{-1} = L'D^{-1}L$$

If D is an identity matrix, this is still the AR(1) process which is generalized by allowing the process variance to change over time, so that D is some positive diagonal matrix. If restrictions on the matrix L are also relaxed, so that the coefficient ϕ is allowed to change in time, this becomes an antedependence model of order 1. In this sense, allowing heterogeneity of the form $V^{-1} = L'D^{-1}L$ results in a model midway between the auto-regressive and antedependence models, with a simple interpretation in terms of the process variance.

6 Use of random cubic spline terms to model correlation

Use of random cubic spline terms gives an alternative to imposing a given form of covariance model to each random term (Verbyla *et al.*, 1998). The smoothing parameter of the cubic spline is estimated using REML, and this gives a flexible covariance model, determined by the data. This also gives a method of investigating nonlinearity in the data, since pattern not accounted for in the fixed model will be incorporated into the random spline term. The nonlinear trend for a term can then be represented by the BLUPs (Best Linear Unbiased Predictors) for the cubic spline term.

7 Implementation

The programs ASREML (Gilmour *et al.*, 1997) and Genstat (Payne *et al.*, 1998) use a common core algorithm to estimate parameters in models of the form described above. The core algorithm performs REML estimation of variance parameters via an efficient algorithm using sparse matrix methods and average information (AI) optimisation (Gilmour *et al.*, 1995). The program interfaces to the core differ: ASREML offers a very basic but flexible interface, whereas the Genstat interface is more user-friendly. Both interfaces provide spline models and outside heterogeneity ($D^{0.5}GD^{0.5}$), but the core algorithm requires extension to provide inside heterogeneity (LDL'). The Windows implementation of Genstat will also allow provision of menus for standard cases, e.g. balanced repeated measurements.

References

Cullis, B.R. & Gleeson A.C. (1991). Spatial analysis of field experiments – an extension to two dimensions. *Biometrics*, **47**, 1449-1460.

Gilmour, A.R., Thompson, R. & Cullis B.R. (1995). Average Information REML, an efficient algorithm for variance parameter estimation in linear mixed models. *Biometrics*, **51**, 1440-1450.

Gilmour, A.R., Thompson, R., Cullis, B.R. & Welham S.J. (1997) *ASREML.* Biometric Bulletin. NSW Agriculture.

Payne, R.W., Lane, P.W., Baird, D.B., Gilmour, A.R., Harding, S.A., Morgan, G.W., Murray, D.A., Thompson, R., Todd, A.D., Tunnicliffe-Wilson, G., Webster, R. & Welham, S.J. (1998). *Genstat 5 Release 4.1 Reference Manual Supplement.* Numerical Algorithms Group, Oxford.

Verbyla, A.P., Cullis, B.R., Kenward, M.G. & Welham, S.J. (1998). Smoothing splines in the analysis of designed experiments and longitudinal data. *Applied Statistics*, in press.

Exploratory Data Analysis with Linked Dotplots and Histograms

Adalbert F.X. Wilhelm

Institut für Mathematik, Universität Augsburg, 86135 Augsburg, Germany

Keywords. Brushing, linked views, histogram, dotplot, cluster, outlier

1 Linked low-dimensional views

Data sets are currently increasing in their number of variables as well as their number of observations. Standard exploratory tools for multivariate data analysis, like the scatterplot matrix, have problems when dealing with such data. The screen space available gives only enough resolution for a scatterplot matrix of four variables. Overplotting of tied or close observations is present in scatterplots even for medium sized data sets. As an alternative the use of linked low-dimensional views has been suggested in the literature. In this paper we look at some particular data analytic problems and investigate the usefulness of linked dotplots and histograms for these problems. We restrict ourselves mainly to two-dimensional phenomena so that we can easily check our conclusions by scatterplots.

Histograms and dotplots have been proven to be useful instruments in visualizing one-dimensional data, but both methods also have some drawbacks. Ties or overlapping points can destroy the usefulness of dot plots. The choice of number and placement of the bins in the histogram is in no way unique and, depending on the data, might lead to rather different visual impressions. However, these drawbacks are in some sense complementary: one strength of the histogram is in looking at highly overlapping data, the strength of the dotplot is that it is rather robust against changes of scale.

A key feature of high interaction graphics is the availability of linked views, see Eick & Wills (1995). Linked plots have become rather popular in the form of scatterplot brushing (Becker & Cleveland, 1987), but for other plot types they are available in only a few packages, like DATA DESK or MANET. While the construction of linking is straightforward in scatterplots, where each observation is visualized by an individual symbol, the linking of aggregated displays is done by overlaying a plot of the same type for the selected points over the plot for all points.

The intuitive and standard way to use such linked views is to look for interesting subsets in one plot and then to see how these subsets look in the context of other views. If we use this strategy with univariate views we will almost surely only detect one-dimensional structure. But the capabilities of linked univariate views are claimed to reach beyond that point. By systematically subsetting one variable one can compare, for some other variables, the conditional distributions of these subsets with the distribution of the entire sample. Instead of comparing the selected subset with the entire sample one can also compare it to the complementary subset. In dynamic painting the focus is on the differences of two subsequent graphs, comparing the current image with a mental copy of the previous one.

Stuetzle (1991) described painting in an abstract way for the linking of scatter plots. We extend his description to plots that do not show each observation individually. Let us assume we have a sample of i.i.d. observations of a pair of one-dimensional random variables (X, Y) taken at individuals in the set Ω. Suppose that the distribution of each variable is portrayed in a histogram and that the histogram of X is the active plot. The standard selection techniques available in interactive graphics offer only the possibility to select subsets of the population Ω whose observed values for X fall in a union of classes determined by the bins. Let $\mathcal{A} = (A_i)_{i \in I}$, I a finite index set, denote the partition of the set Ω that is induced by the choice of the bin size for the histogram of X. Assume further that we have currently selected a set of bins in the active plot; that means we have selected a subset $\mathcal{X}_A$ of the image $\mathcal{X}$ of the random variable X, or equivalently, a subset $A = \cup_{j=1}^{a} A_i, \{1, \ldots, a\} \in I$ of the underlying sample population Ω. We then superimpose a histogram for the selected subset A in all connected plots, that is we draw a histogram for $\mathcal{Y}_A = \{y(\omega) : \omega \in A\}$ on top of the histogram for Y. If X and Y are independent, the conditional distribution $P^{Y|X \in A}$ is identical to the unconditional distribution of Y. So for any measurable sets $A \subset \Omega$ and $B \subset \mathcal{Y}$ we have the following independence properties:

$$P(Y \in B \mid X \in \mathcal{X}_A) = P(Y \in B)$$

In practice, however, we are not able to visualize all measurable sets; the only sets we can see are based on the partitions of Ω induced by the various plot scales. Let $\mathcal{B} = (B_j)_{j \in J}$ denote the partition induced by the histogram of Y. What we are actually judging are the distributions

$$P_Y(B_j \mid A_i) \text{ versus } P_Y(B_j) \quad \forall i \in I, j \in J.$$

In static painting, we fix the index i of the conditioning event, and our eyes run over the linked plots, mentally passing through the index set J, searching for non-uniformities in the relation between the original histogram and the overlaid part. In dynamic painting, while we move the brush over the active plot, we pass through the index set I and try to detect changes in the distribution of the highlighted objects. So, in this case, we look simultaneously at sets of the partitions $\mathcal{A}$ and $\mathcal{B}$ and focus on the differences between two subsequent graphs, comparing the current image with our mental copy of the previous image. Actually, in many cases the judgment is more qualitative than quantitative and will be mainly based on the magnitude of changes. The results of our graphical exploration by conditioning depends heavily on the partitions that are accessible in the conditioning plot. We can only condition on unions of the basic partition elements defined by the graph type and the scale. Without overplotting or tied observations, dotplots offer the set of all possible partitions, because each observation is represented by an individual point in the graph. The accessible partitions offered by histograms vary with the bin width and the anchor point. So, interactively changing these parameters also changes the accessible partitions and hence the amount of structure that we can find.

2 Geometric structure

Visualization of point clouds in high-dimensions aims particularly at identifying the existence of lower-dimensional structure. Carr & Nicholson (1988) translated dimensionality into terms of constraints on the display variables. They investigated three petals generated by

$$x = \cos(u) * \cos(3u) + \text{noise}$$
$$y = \sin(u) * \sin(3u) + \text{noise}.$$

In a scatterplot the structure can be seen immediately. Do we have a chance to reconstruct the two-dimensional picture by linking one-dimensional marginal plots? To produce the petals we have used 1000 equidistant points in the interval $[0, \pi]$. In one-dimensional scatterplots much of the structure remains obscure due to the heavy overplotting of points. We therefore prefer linked marginal histograms. While we brush over the x-values in the histogram (see Figure 1 top row) the constraints in the y-variable (see Figure 1 second row) become apparent. The holes in the petals become visible as well as the

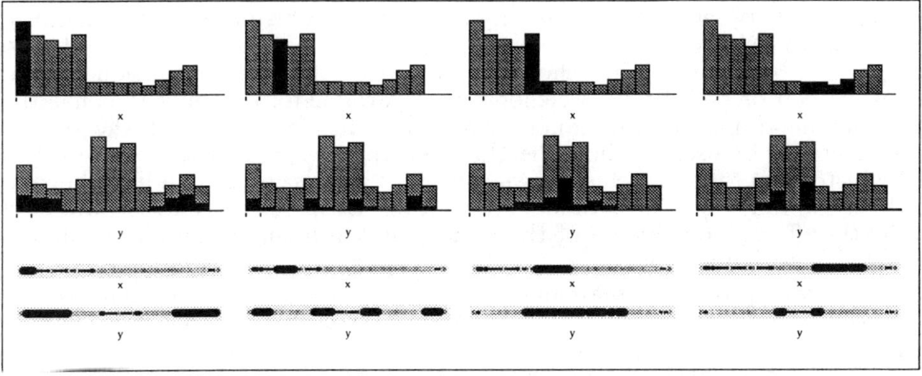

Fig. 1. Linked marginal histograms to depict two-dimensional structure

intersection point of the three petals in the centre. It is easy to recognize four branches for low x-values in contrast to two branches for high x-values. In the dotplots (see Figure 1 bottom rows) the gaps are seen as well, but overplotting and highlighting cover much of the information. With some effort we are able to reconstruct the entire structure of the petals from one-dimensional marginal views. In higher dimensions the procedure works in the same way, with the only difference that there is no perfect two-dimensional summary plot to compare with - as the scatterplot provides for two-variate data.

3 Detecting outliers and clusters

Single univariate views can only show clusters that can be discriminated by a line perpendicular to the plot axis. The standard example for this deficiency are clusters that split along the diagonal in two dimensions. Those can easily be depicted in a scatterplot but they can not be seen in any of the standard projections. Linking also enables us to see such clusters in the marginal views. While brushing over one of the variables the highlighted points in the other variable will split up and show the clusters. Dotplots as well as histograms can show this effect (see Figure 2). But histograms smooth the data and

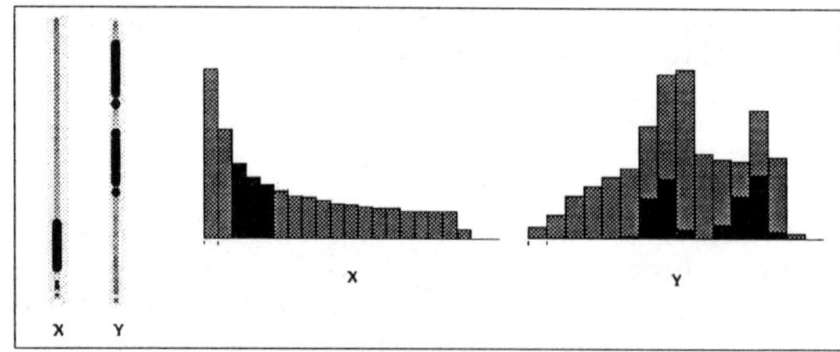

Fig. 2. Linking shows two-dimensional clusters that can not be seen in the projections

do not show small gaps between the clusters as clearly as dotplots do. A separation between clusters can only be shown in a histogram if the distance between the clusters in this variable is greater than the bin width. It also depends which variable we choose for brushing. For two-dimensional data, clusters can be seen best if we choose that variable for brushing that has the smaller one-dimensional distance between the clusters. Since then the greater distance can be used to show the clusters. Also the smaller the bin width in the histogram the more clusters we can see. There is certainly a limit beyond which too many cluster artifacts are created. Dotplots, in contrast, are only slightly affected by change of the scale, but the highlighting unfortunately covers the original plot.

How do we proceed in searching for three-dimensional clusters that can not be seen in any two-dimensional projection? We then have to combine selections and to condition on two variables searching for a subset that produces a sufficient gap within the highlighted points of the third variable.

An outlier can be seen as a second cluster that only consists of one point. Therefore, the above mentioned clustering procedure is also valid for outlier detection, see Figure 3.

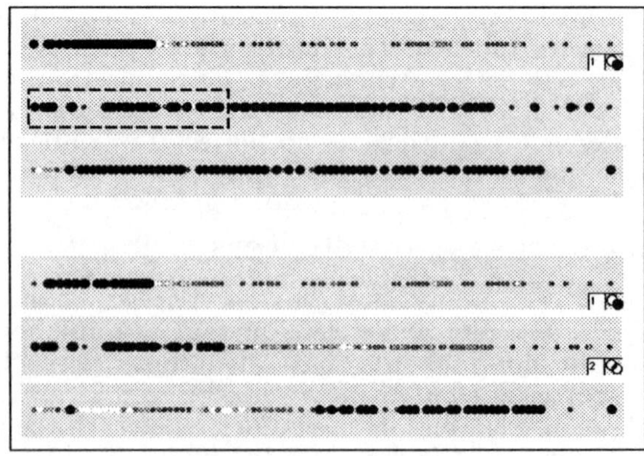

Fig. 3. Searching for outliers in three dimensions that can not be seen in lower dimensions

4 Dependencies and group comparisons

The classic attempt to portray functional dependencies between variables is to use a scatter plot. For two variables this works excellently, and for medium sized data sets there is not much more to ask for. Dynamic graphics extended the scatter plot to three variables, and Carr *et al.* (1987) augmented the scatter plot with ray glyphs and colour to show four dimensions. However, these efforts did not gain widespread acceptance and therefore most of the visualization is still limited to two or three dimensions.

Linking seems to be a quite natural choice to portray the relationship between explanatory and response variables in a regression model and also to overcome some of the dimension restrictions. A 4-dimensional data set with three explanatory and one response variable can be displayed with a 3-D rotating plot linked to a one-dimensional dot plot. How much can we see if we just use one-dimensional views? We demonstrate the univariate response case only, but two or three-dimensional responses can be dealt with analogously. Brushing the response variable and checking the corresponding highlighting in the plots for the explanatory variables corresponds to inverse regression; brushing an explanatory variable and linking to the response variable falls in line with partial response plots. A monotone relationship will immediately strike in the eye by the simultaneous change in the location of the highlighting, see Figure 4, but we can not distinguish between linear and nonlinear relationships. It is also impossible to detect the conditional linearity in the explanatory variables which can be seen in scatterplots.

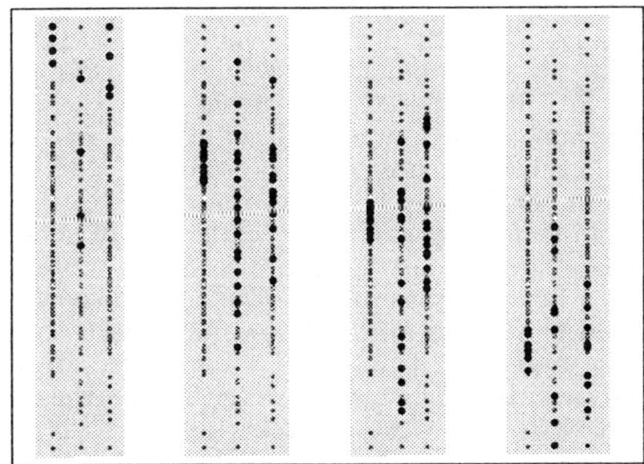

Fig. 4. Searching for functional relationship between one response and two explanatory variables

Also comparing groups is an easy task. Two different groups are selected in Figure 5 (left and right group of nine dotplots) . The resulting points are highlighted in the dot plots and they show that the groups cluster in all the variables. Some variables can be used as single discriminator between the two groups. Other variables need to be combined to give a sufficient discrimination between the groups. For such a large number of variables drawing histograms or scatterplots brings us soon to the limits of the screen space. Dotplots here offer a more efficient way. In addition, structure found in multiple dotplots is easier to interpret than structure found in pairwise scatterplots.

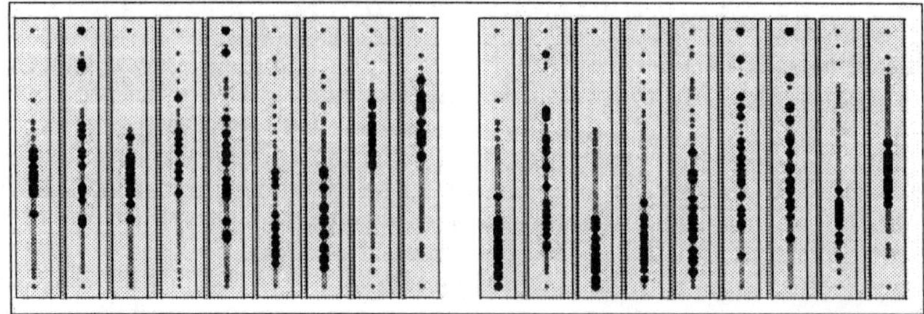

Fig. 5. Comparing groups in multiple dotplots

5 Conclusion

Linked low-dimensional views can be used to detect two- or higher-dimensional structures. Clusters and outliers can be found as easily as in scatterplots or rotating plots, but in addition interpretation is much easier. Especially in comparison to high-dimensional graphical techniques like the Grand Tour, linked low-dimensional views give an immediate explanation for the structure that has been found. On the other hand, the analyst has to be careful when using linked views, since the results will heavily depend on the conditioning sets. To detect higher-dimensional structure the software has to allow the combination of selections made in various plots. Obviously, it is impossible for the human being to reconstruct all of the higher-dimensional structure by just looking at low-dimensional projections, but the main features can be easily derived. Another advantage of low-dimensional views is that they are familiar to many scientists not only statisticians, and that the structure that has been detected can easily be interpreted and communicated to other people. The use of linked views is similar to the use of conditional plots but linked views are more flexible and can more easily be modified. For data sets with many variables but small numbers of observations linked dotplots are preferable. With data sets containing large numbers of observations, when high overplotting is present, dotplots and scatterplots are no longer able to give a consistent view. Linked histograms can handle these data pretty well and should be used with large sample sizes.

References

Becker, R.A. & Cleveland, W.S. (1987). Brushing Scatterplots. *Technometrics*, **29**, 127-142.

Carr, D.B. & Nicholson, W. (1988). Explor4: A program for exploring four-dimensional data using stereo-ray glyphs, dimensional constraints, rotation and masking. In: *Dynamic Graphics for Statistics*, 309-329. Belmont, CA: Wadsworth Inc.

Carr, D.B., Littlefield, R., Nicholson, W. & Littlefield, J. (1987). Scatterplot matrix techniques for large N. *JASA*, **82**, 424-436.

Eick, S.G. & Wills, G.J. (1995). High interaction graphics. *European Journal of Operational Research*, 445-459.

Stuetzle, W. (1991). Odds Plots: A graphical aid for finding associations between views of a data set. In: *Computing and Graphics in Statistics*, 207-217. New York: Springer.

An Object-Oriented Approach to Local Computation in Bayes Linear Belief Networks

Darren J. Wilkinson

Department of Statistics, University of Newcastle, Newcastle upon Tyne, NE1 7RU, England.

Email: d.j.wilkinson@newcastle.ac.uk

Abstract. A generalised conditional independence property based on zero partial correlations can be used to define Bayes linear graphical models, which may then be used as the basis for local computation. As for the full Bayesian case, local computation actually takes place on the clique-tree of the triangulated moral graph. An object-oriented framework for local computation is described which can be implemented via pure message-passing between objects representing adjacent clique-tree nodes. Messages consist of small matrices and vectors containing information about the current observation, making updating and propagation very fast. A sequential implementation, BAYES-LIN, is described.

Keywords. Bayes linear methods, clique-trees, conditional independence, continuous variable belief networks, Gaussian models, graphical models, local computation, parallelism

1 Introduction

Bayes linear methods are an alternative to conventional Bayesian statistics which acknowledge the difficulties associated with the full modelling, specification, and conditioning required by distributional Bayesian statistics. They instead try to make best possible use of partial specifications, based on means, variances and covariances. Unsurprisingly, much of the theory is formally identical to inference in multivariate Gaussian Bayesian networks, but the interpretation of results is generally different.

In this paper, an object-oriented framework for local computation in Bayes linear networks is described. The algorithms are ripe for implementation on massively parallel hardware, as not only can computations proceed in parallel in different branches of the tree, but information may be introduced into different parts of the tree simultaneously, allowing incorporation of different pieces of information to proceed in parallel.

2 Bayes linear graphical models

2.1 Bayes linear methods

The Bayes linear approach to inference is motivated by both practical and theoretical considerations. The practical considerations arise from the need to develop methods of Bayesian analysis based on partial prior specifications for problems which are sufficiently complex that we are unable either to make full prior specification or to carry out the full posterior analysis. The theoretical considerations follow from the value in creating an essentially geometric approach to belief specification and inference, which treats general random objects in a formally similar way to simple random quantities. For an overview of Bayes linear analysis, see Goldstein (1998) and for a discussion of Bayes linear computing, see Goldstein & Wooff (1995). The foundations of the theory are discussed in Goldstein (1997) and references therein. For a complete discussion of a purely geometric approach to Bayes linear local computation and diagnostic graphical displays, see Goldstein & Wilkinson (1997).

2.2 Bayes linear adjustment

Suppose we have a collection $C = \{C_1, C_2, \ldots\}$ of random quantities. For each quantity, we specify a prior mean and variance, and between each pair of quantities we specify a prior covariance. These specifications are made directly, treating expectations, rather than probability, as the primitive quantity; see, for example, the development in de Finetti (1974). The collection of all expectation, variance and covariance specifications relating to C is known as the *belief structure* for C, as it contains all aspects of belief needed for a Bayes linear analysis. Note that coherence requires the specified variance matrix to be non-negative definite. The *adjusted expectation* of a random quantity X, given a collection $D = \{D_1, \ldots, D_k\}$, written $\mathrm{E}_D(X)$, is the linear combination $\mathrm{E}_D(X) = \sum_{i=0}^k h_i D_i$ which minimises $\mathrm{E}[(X - \sum_{i=0}^k h_i D_i)^2]$, over all collections $h = (h_0, h_1, \ldots, h_k)$, where D_0 is the unit constant, *i.e.* $D_0 = 1$.

If B and D are finite vectors of random quantities, then the *adjusted expectation vector* $\mathrm{E}_D(B)$, the vector of adjusted expectations for the elements of B by D, is

$$\mathrm{E}_D(B) = \mathrm{E}(B) + \mathrm{Cov}(B, D)[\mathrm{Var}(D)]^{-1}[D - \mathrm{E}(D)], \tag{1}$$

as this is the linear combination of the data which minimises the expected quadratic loss. Similarly, the *adjusted variance matrix*, $\mathrm{Var}_D(B) = \mathrm{Var}[B - \mathrm{E}_D(B)]$ is given by

$$\mathrm{Var}_D(B) = \mathrm{Var}(B) - \mathrm{Cov}(B, D)[\mathrm{Var}(D)]^{-1}\mathrm{Cov}(D, B). \tag{2}$$

By an obvious notational extension, we may now define the *adjusted covariance matrix* to be the obvious sub-matrix of the adjusted variance matrix, and hence for vectors B, C and D is given by

$$\mathrm{Cov}_D(B, C) = \mathrm{Cov}(B, C) - \mathrm{Cov}(B, D)[\mathrm{Var}(D)]^{-1}\mathrm{Cov}(D, C). \tag{3}$$

From a Bayesian viewpoint, adjusted expectation is a simple and tractable approximation to a full Bayes posterior expectation, based on limited prior specification, which is exact in certain important special cases. In particular, Bayes linear analysis is exact for Gaussian models, and one feature of interest in the Bayes linear approach is to observe how much of the simplicity of the Gaussian analysis may be preserved without making the Gaussian assumptions.

2.3 Bayes linear separation

Full Bayesian graphical models are defined in terms of the conditional independencies present in the model. From a Bayes linear viewpoint, we do not wish to make such strong statements about the joint probability density function, but instead use a tractable equivalent, based on the specifications made. Consequently we define the following separation property:

$$A \perp\!\!\!\perp C | B \iff \mathrm{Cov}_B(A, C) = 0. \tag{4}$$

If $A \perp\!\!\!\perp C | B$ we say that B *separates* A and C. Note that disjointness of A, B and C is not required. Goldstein (1990) shows that this separation property is a *generalised conditional independence* property. Smith (1990) shows that such generalised conditional independence properties behave computationally as a full conditional independence property, and in particular, that graphs defined using it may be manipulated in the same way as full conditional independence graphs.

2.4 Graphs

Graphs have been shown to be a particularly effective way of highlighting and exploiting conditional independencies over a collection of random quantities.

A directed acyclic graph is a Bayes linear graphical model if each node represents a vector of random quantities, and any two nodes are separated by the union of their parents.

A Bayes linear moral graph can be obtained from a Bayes linear graphical model by marrying parents and dropping arrows. Such a graph has the property that a group of nodes, B, separates node groups A and C if every path from a node in A to a node in C passes through a node in B.

A Bayes linear clique-tree may be formed from a Bayes linear moral graph by triangulating the moral graph (ensuring decomposability) and forming the clique-tree. The clique-tree is the tree of cliques of the triangulated moral graph, and is also known as the junction tree. These constructions and their properties are discussed further in Goldstein & Wilkinson (1997).

3 Local computation

3.1 Algorithm

In principle, equations (1), (2) and (3) allow us to carry out Bayes linear adjustments for models of any size. In practice, the storage and computation issues associated with very large problems (say, adjusting by more than 1,000 observations) make these equations impractical in such a situation.

Pearl (1988) gives an overview of the key ideas behind the use of graphical models for local computation. As for full Bayesian local computation (Lauritzen, 1992), it turns out that the computations can be localised if they are carried out on the clique-tree of the triangulated moral graph. The reason is that the clique tree has very important separation properties, described shortly.

At each node N of the clique-tree, $\mathrm{E}(N)$ and $\mathrm{Var}(N)$ are stored. Suppose that a subset, D, of one of the clique tree nodes is to be observed. We wish to know the effect of observing D on each of the clique-tree nodes. In particular, we will want to know $\mathrm{E}_D(N)$ and $\mathrm{Var}_D(N)$ at each node N of the tree. We can see from equations (1) and (2) that this will require knowledge of $\mathrm{Cov}(N, D)$ at each node N, and so we need a way to propagate this to each node of the tree.

Suppose that the clique-tree is connected, and that F and G are adjacent nodes on the clique-tree. Then removal of the arc between them will form exactly two trees. Let E be the union of all nodes in the tree containing F, and H be the union of all nodes in the tree containing G. The adjacent nodes F and G will have variables in common. Let $S = F \cap G$, the variables in common. We then have

$$E \perp\!\!\!\perp H | S \tag{5}$$

and S is known as the *separator* of F and G. This is the key property required for efficient local computation on the clique tree.

Returning to the problem of calculating $\mathrm{Cov}(N, D)$ at each node N of the tree, we can use this property of separators in order to propagate this covariance matrix around the tree. The algorithm is as follows.

1. Start at any node M containing D. The matrix $\text{Cov}(M, D)$ is just a sub-matrix of $\text{Var}(M)$ and hence is "known" by node M.
2. Suppose that $\text{Cov}(M, D)$ is known at node M, but not known at some adjacent node N. Node M should pass the message consisting of M (the list of variables represented by the node M) and $\text{Cov}(M, D)$ to node N.
3. When node N receives a message from some node M, it can compute the separator between M and N as $S = M \cap N$. By the property of the separator, we have $N \perp\!\!\!\perp D|S$, and so
$$\text{Cov}(N, D) = \text{Cov}(N, S)[\text{Var}(S)]^{-1}\text{Cov}(S, D). \tag{6}$$
This can be computed by N, since $\text{Cov}(N, S)$ and $\text{Var}(S)$ are sub-matrices of $\text{Var}(N)$, and $\text{Cov}(S, D)$ is a sub-matrix of $\text{Cov}(M, D)$.
4. Re-label N to be M, and return to step 2.

In this way, the covariance with the data can be propagated throughout the tree via message-passing between adjacent clique-tree nodes. Once this has been done, calculation of adjusted expectation and variance, and associated interpretive and diagnostic summaries is relatively straight forward. Full details of this algorithm and its purely geometric formulation may be found in Goldstein & Wilkinson (1997).

Note that a key feature of a Bayes linear analysis, which distinguishes it from (say) a more conventional Gaussian analysis, is the range of summary and diagnostic measures associated with the adjustment process. Therefore, local computation of these measures is as important as the computation of adjusted expectation and variance. Fortunately, once $\text{Cov}(N, D)$ has been propagated, all of these measures may be computed using only information local to adjacent nodes.

3.2 Parallelism

Note that there is no problem associated with propagating messages in different branches of the tree in parallel — since the clique-tree is a tree, the messages will not interfere with one another. More importantly, however, if data are to be introduced into many nodes, this introduction may take place simultaneously, and propagation of all messages may proceed in parallel. This is because adjustments may be made sequentially, and the order of adjustment does not matter, so the order messages arrive at a particular node is unimportant. In this latter case, however, some node-locking or scheduling of adjustments will be required in order to ensure that two messages are not processed by a node concurrently.

4 Object-oriented implementation

4.1 Object classes and hierarchy

The message-passing local computation algorithm described in the previous section is ideal for implementation in an object-oriented programming language. Objects should be defined representing clique-tree nodes. The objects should have slots for the various attributes of the node, such as the list of variables represented by the node, and the associated prior and "current" expectation vectors and variance matrices. It should also have a list of adjacent nodes, so that it knows which objects to pass messages to, and should also have slots for information relating to the currently propagating data, such as the data itself, the variance matrix for the data, and the covariance matrix between the node and the data. These node objects will also have associated methods for receiving, processing and passing propagating messages, and for calculating adjustments and associated interpretive and diagnostic summaries.

Ideally there will also be objects representing the nodes of the corresponding moral graph, as well as the associated directed graph from which the moral graph, and hence clique-tree are derived. Since these objects will share many data types and methods, these objects should all be descendants of a common object hierarchy for Bayes linear structures. Figure 1 gives an overview of a possible hierarchy for such object classes.

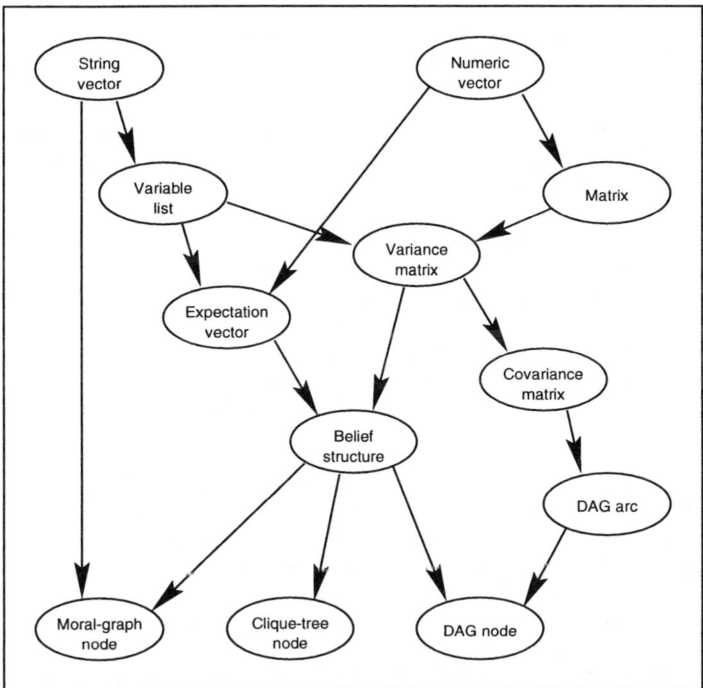

Fig. 1. Object hierarchy for Bayes linear local computation

By binding expectation vectors and variance/covariance matrices with the variables they represent, intelligent behaviour can be built into the matrix computations. For example, consider the following pre-multiplication of an expectation vector by a variance matrix:

$$\begin{pmatrix} A & 10 & 2 & 1 \\ B & 2 & 10 & 2 \\ C & 1 & 2 & 10 \end{pmatrix} \times \begin{pmatrix} B & 5 \\ C & 10 \\ D & 15 \\ E & 20 \end{pmatrix} = \begin{pmatrix} B & 70 \\ C & 110 \end{pmatrix}.$$

This intelligent behaviour of the objects leads to exceptionally neat forms for the expressions required by the local computation algorithm and associated interpretive and diagnostic calculations. Note also that the message-passing algorithm for local computation outlined in Section 3.1 requires only the passing of a simple covariance matrix object between nodes. Again, binding variable labels to the matrix simplifies the algorithm. Since a belief structure is just the collection of all partial specifications relating to a collection of variables, it corresponds to the expectation vector and variance matrix for that collection. For example, a belief structure for the variables A, B and C might take the form

$$\begin{pmatrix} A & 0 & 10 & 2 & 1 \\ B & 5 & 2 & 10 & 2 \\ C & 10 & 1 & 2 & 10 \end{pmatrix}$$

and will, of course, have methods associated with it which return its component expectation vector and variance matrix. Similarly, the objects representing clique-tree nodes will inherit all of the methods associated with the belief structure for the represented variables, but will contain additional slots to hold information relating to adjacent nodes, *etc.*, and additional methods which carry out the local computation algorithms.

4.2 BAYES-LIN

The author is currently developing a suite of software tools for sequential Bayes linear local computation. The implementation is known as BAYES-LIN, and is implemented in LISP-STAT as a set of modules. LISP-STAT is an object-oriented environment for statistical computing, described in Tierney (1990). BAYES-LIN provides a collection of object prototypes appropriate for modelling, locally computing and diagnosing adjustment in Bayes linear belief networks. Interpretive and diagnostic graphics are produced to aid understanding of the adjustment process. These are displayed on the nodes of the moral graph (rather than the clique-tree), as this is where they live most naturally. BAYES-LIN is documented in Wilkinson (1997) and is freely available over the Internet from `http://www.ncl.ac.uk/~ndjw1/bayeslin/`

References

de Finetti, B. (1974). *Theory of Probability, Vol. 1.* Chichester: Wiley.

Goldstein, M. (1990). Influence and belief adjustment. In: *Influence Diagrams, Belief Nets and Decision Analysis.* Chichester: Wiley.

Goldstein, M. (1997). Prior inferences for posterior judgements. In: *Structures and Norms in Science.* Pordrecht: Kluwer.

Goldstein, M. (1998). Bayes linear analysis. In: *Encyclopedia of Statistical Sciences.* Update volume 3. Chichester: Wiley.

Goldstein, M. & Wilkinson, D.J. (1997). Bayes linear analysis for graphical models: The geometric approach to local computation and interpretive graphics. *University of Newcastle, Department of Statistics Preprint*, STA97,21.

Goldstein, M. & Wooff, D.A. (1995) Bayes linear computation: concepts, implementation and programming environment. *Statistics and Computing*, **5**, 327–341.

Lauritzen, S.L. (1992). Propagation of probabilities, means, and variances in mixed graphical association models. *Journal of the American Statistical Association*, **87**(420), 1098–1108.

Pearl, J. (1988). *Probabilistic Reasoning in Intelligent Systems.* Morgan Kaufmann.

Smith, J.Q. (1990). Statistical principles on graphs. In: *Influence Diagrams, Belief Nets and Decision Analysis.* Chichester: Wiley.

Tierney, L. (1990). *LISP-STAT: An Object-Oriented Environment for Statistical Computing and Dynamic Graphics.* Chichester: Wiley.

Wilkinson, D.J. (1997). BAYES-LIN: An object-oriented environment for Bayes linear local computation. *University of Newcastle, Department of Statistics Preprint*, STA97,20.

Index of Authors

Index of Keywords

References are to the first pages of papers that have a given keyword in their lists of keywords, or in their titles. Some additional references have been added.